HYDROGEOPHYSICS

Water Science and Technology Library

VOLUME 50

The titles published in this series are listed at the end of this volume.

Hydrogeophysics

Edited by

YORAM RUBIN
University of California at Berkeley,
CA, U.S.A.

and

SUSAN S. HUBBARD
Lawrence Berkeley National Laboratory,
Berkeley, CA, U.S.A.

Springer

A C.I.P. Catalogue record for this book is available from the Library of Congress.

ISBN-10 94-017-8330-6 (HB) Springer Dordrecht, Berlin, Heidelberg, New York
ISBN-10 1-4020-3102-5 (e-book) Springer Dordrecht, Berlin, Heidelberg, New York
ISBN-13 978-94-017-8330-9 (HB) Springer Dordrecht, Berlin, Heidelberg, New York
ISBN-13 978-1-4020-3102-1 (e-book) Springer Dordrecht, Berlin, Heidelberg, New York

Published by Springer,
P.O. Box 17, 3300 AA Dordrecht, The Netherlands.

Cover design by Yoram Rubin and Susan Hubbard. Graphics by Zhangshuan Hou.

Printed on acid-free paper

Yoram Rubin dedicates this book to Atalya, with love

Susan S. Hubbard dedicates this book to Cahit Çoruh, for his geophysical tutelage and lifelong influence

Acknowledgments

Hydrogeophysics was conceived as a vehicle for disseminating some of the background and current research presented at a NATO funded Hydrogeophysics Advanced Study Institute, which was held in the Czech Republic in 2002. We thank Philippe Baveye for initiating the development of this Institute, NATO for funding the Institute, and Miroslav Kobr for serving as a wonderful host for our meeting.

Thirty five authors from nine countries, who are leaders in their respective areas, contributed to *Hydrogeophysics*. We are grateful to this outstanding author group for their efforts in developing their chapters and for reviewing other chapters. We also thank the external reviewer group, including: Alex Becker, Niels Christianson, Tom Daley, Sander Huisman, Rosemary Knight, Zbigniew Kabala, Niklas Linde, Seiji Nakagawa, Peter Styles, and Vitaly Zlotnik. We particularly thank Lee Slater for reviewing several chapters within demanding timeframes.

Production of this book could not have been possible without the superb editing provided by Daniel Hawkes. We thank Zhangshuan Hou for helping with the book cover graphics, and Bryce Peterson for assisting with the book index. This work was supported in part by United States' NSF grant EAR-0087802 made by the Hydrologic Science Division and by DOE DE-AC03-76SF00098.

Yoram Rubin and Susan Hubbard
Editors

TABLE OF CONTENTS

Hydrogeophysical Case Studies

Hydrogeophysical Frontiers

Background
and
Hydrogeology

1 INTRODUCTION TO *HYDROGEOPHYSICS*

SUSAN S. HUBBARD[1] and YORAM RUBIN[2]

[1]*Lawrence Berkeley National Laboratory, Earth Sciences Division, Berkeley, CA 94720 USA.* sshubbard@lbl.gov
[2]*Department of Civil and Environmental Engineering, UC Berkeley, Berkeley, CA 94720, USA*

In this chapter, we discuss the need for improved hydrogeological characterization and monitoring approaches, and how that need has provided an impetus for the development of an area of research called hydrogeophysics. We briefly describe how this research area has evolved in recent years in response to the need to better understand and manage hydrological systems, provide discussions and tables designed to facilitate navigation through this book, and discuss the current state of the emerging discipline of hydrogeophysics.

1.1 Evolution of Hydrogeophysics

The shallow subsurface of the earth is an extremely important geological zone, one that yields much of our water resources, supports our agriculture and ecosystems, and influences our climate. This zone also serves as the repository for most of our municipal, industrial, and governmental wastes and contaminants, intentional or otherwise. Safe and effective management of our natural resources is a major societal challenge. Contaminants associated with industrial, agricultural, and defense activities in developed countries, the increased use of chemical pollutants resulting from the technological development of countries with evolving market economies, the increased need to develop sustainable water resources for growing populations, and the threat of climate change and anthropogenic effects on ecosystem all contribute to the urgency associated with improving our understanding of the shallow subsurface.

Many agencies and councils have recently described the pressing need to more fully develop tools and approaches that can be used to characterize, monitor, and investigate hydrogeological parameters and processes in the shallow subsurface at relevant spatial scales and in a minimally invasive manner (e.g., the National Research Council, 2000; U.S. Department of Energy, 2000; U.S. Global Change Research Program, 2001). Recognizing this need, NATO funded a Hydrogeophysics Advanced Study Institute, which was held in the Czech Republic in 2002. At that workshop, an international group of researchers gathered to review advances and obstacles associated with using geophysical methods to improve our understanding of subsurface hydrogeological parameters and processes. This book, *Hydrogeophysics*, was conceived at that meeting as a vehicle for disseminating some of the background and current research associated with hydrogeophysics.

Y. Rubin and S.S. Hubbard (eds.), Hydrogeophysics, 3–21.

It is widely recognized that natural heterogeneity and the large spatial variability of hydraulic parameters control infiltration within the vadose zone, groundwater storage, and the spread of contaminants. It is also well recognized that natural heterogeneity is typically great and can have multiple spatial scales of variability, and that those different hydrogeological objectives require different levels of characterization. In some cases, reconnaissance characterization approaches that capture the major features of the study site are sufficient, while other investigations require much more detailed information about geological unit boundaries and properties. For example, in applications such as water resource management or regional mass-balance analysis, characterization of an aquifer through its geometry and effective properties may be sufficient. However, studies of contaminant transport at the local scale generally require a detailed understanding of the spatial distribution of hydraulic conductivity.

Conventional sampling techniques for characterizing or monitoring the shallow subsurface typically involve collecting soil samples or drilling a borehole and acquiring hydrological measurements (Chapter 2 of this book) or borehole geophysical logs (Chapter 10). When the size of the study site is large relative to the scale of the hydrological heterogeneity, or when the hydrogeology is complex, data obtained at point locations or within a wellbore may not capture key information about field-scale heterogeneity; especially in the horizontal direction. Similar to how biomedical imaging procedures have reduced the need for exploratory surgery, integrating more spatially extensive geophysical methods with direct borehole measurements holds promise for improved and minimally invasive characterization and monitoring of the subsurface. Integration of geophysical data with direct hydrogeological measurements can provide a minimally invasive approach to characterize the subsurface at a variety of resolutions and over many spatial scales.

The field of hydrogeophysics has developed in recent years to investigate the potential that geophysical methods hold for providing quantitative information about subsurface hydrogeological parameters or processes. Hydrogeophysical investigations strive to move beyond the mapping of geophysical “anomalies” to provide information that can be used, for example, as input to flow and transport models. Hydrogeophysics builds on previous experience and developments associated with the mining and petroleum industries, which in the past 50 years have relied heavily on geophysical data for exploration of ore and hydrocarbons, respectively. Because geophysics has been used as a tool in these industries for so long, there is a relatively good understanding of optimal data acquisition approaches and petrophysical relationships associated with the consolidated, high pressure, and high temperature subsurface environments common to those industries. However, these subsurface conditions are quite different from the shallow, low temperature, low pressure, and less consolidated environments that typify most hydrogeological investigation sites. Indeed, the contrasts in shallow subsurface properties are often quite subtle compared to those of the deeper subsurface. Because of the different subsurface conditions, the relationships that link geophysical attributes to subsurface parameters, as well as the geophysical attribute ranges, can vary dramatically between the different types of environments.

In recent years, geophysical methods have been used to assist with many hydrogeological investigations. Some of the first books devoted to the use of geophysical methods for environmental problems were published more than a decade ago (e.g., Keys, 1989; Ward, 1990), and there have since been many journal papers and conference proceedings published on the topic. We define hydrogeophysics as the use of geophysical measurements for mapping subsurface features, estimating properties, and monitoring processes that are important to hydrological studies, such as those associated with water resources, contaminant transport, and ecological and climate investigations. Hydrogeophysical research has been also performed under other rubrics, such as near-surface or shallow geophysics. Although the geophysical methods and estimation approaches are sometimes similar, hydrogeophysical studies focus solely on hydrological objectives, rather than the use of geophysical data for other near-surface investigations, such as those associated with transportation, hazard assessment, engineering, and archaeology.

Characterization data can be collected from many different platforms, such as from satellites and aircraft, at the ground surface of the earth, and within and between wellbores. Discussions in this book are limited to characterization investigations performed at the laboratory (or 'point') scales ($\sim 10^{-4}$ to 1 m), local scales ($\sim 10^{-1}$ to 10^{2} m) and regional scales ($\sim 10^{1}$ to 10^{5} m); we exclude investigations that are performed at the (smaller) molecular and the (larger) global scales. The scale category definitions are approximate and are intended to convey the typical spatial extent associated with *investigations* rather than to imply specific notions of measurement support scales for individual instruments, averaging or dimensionality. Figure 1.1 illustrates relative length scales and relative resolutions associated with different geophysical (G) and hydrological (H) data acquisition catagories discussed in this book. The relative resolution categories describe the approximate length scale associated with measurements collected using the different acquisition approaches, such as the diameter of a zone of influence associated with wellbore logs, the width of pixels used when inverting crosshole measurements, or the Fresnel zone of wave-based surface geophysical approaches. Figure 1.1 also illustrates that typically a trade-off exists between the resolution of measurements collected using a method having a certain acquisition geometry and the typical scale of investigation for that method. For example, core measurements provide information about a very small portion of typical field investigations at a very high resolution, while surface geophysical techniques often provide lower resolution information but over larger spatial scales. It is important to realize that the scale and resolution of a particular characterization technique may vary, depending on the acquisition parameters and geometry employed. For example, electrical resistivity measurements are commonly collected from core samples, electrical logs, between wellbores, and from the ground surface; electrical measurements can thus be useful for point, local, and regional scale investigations, and can have a range of resolutions. Additionally, varying the acquisition parameters of a particular method can change the resolution, even if the acquisition geometry stays the same. For example, collection of crosshole seismic travel-time measurements between wellbores that have a great separation distance and that have large source-receiver station spacing within the wellbores will result in a lower possible spatial resolution than seismic crosshole data collected using close wellbore spacing and source-receiver station spacing.

The choice of which characterization or technique or acquisition approach to use for a particular investigation is made by considering many factors, such as: the objective of the investigation relative to the sensitivity of different geophysical methods to that objective; the desired level of resolution; conditions at the site; time, funds, and computational resources available for the investigation; experience of the investigator; and availability of other data. Data are often collected in a sequential fashion, where lower-resolution acquisition approaches are used for reconnaissance investigation, followed by higher-resolution approaches to provide more detailed information as needed. Because there may be several plausible subsurface scenarios that could produce a given geophysical response, and because different geophysical techniques are sensitive to different properties, a combination of different geophysical techniques is often used to characterize the subsurface. Combinations of hydrogeological and geophysical data can be used to yield a better understanding of shallow subsurface parameters and processes over various scales and resolutions. The use of combined data sets to improve subsurface characterization is more fully described in the case study chapters (see Section 1.2 below), as well as in Chapter 17.

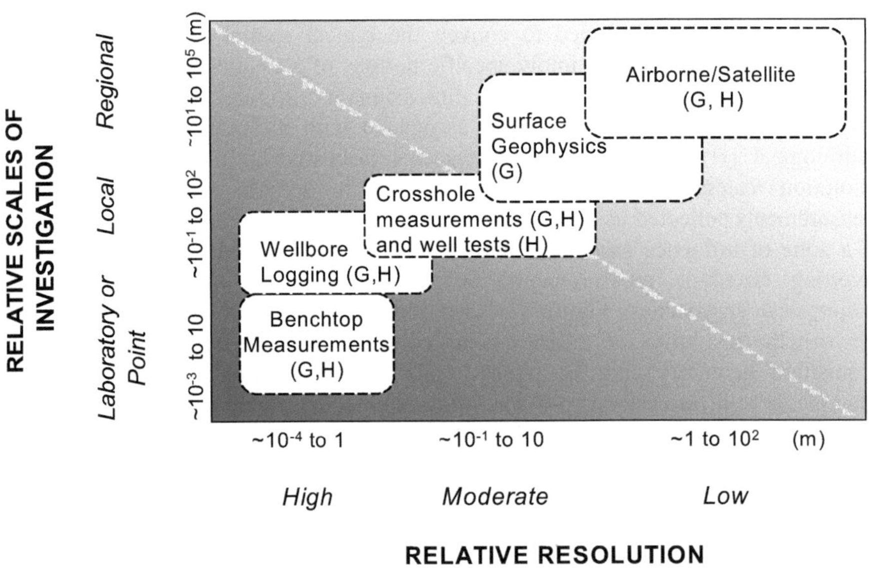

Figure 1.1 Chart showing trade-off between the relative resolution of the information obtained using different yet typical geophysical (G) and hydrological (H) measurement acquisition approaches and the relative scale of the investigations for which those acquisition geometries are typically used. Benchtop measurements refer to those such as core, column, and tank measurements collected within a laboratory.

Table 1.1 lists several common geophysical and hydrological characterization methods, which are classified according to their acquisition category (and thus by their typical investigation scales and resolution). The attribute that is typically obtained from each method is given, along with some examples of hydrogeological objectives for which each method is particularly well suited. Additionally, this table lists chapters in this book where these methods are discussed; more information on what is covered in subsequent chapters will be given below. Although some of the techniques listed under hydrological methods actually rely on geophysical measurements (e.g., time-domain reflectometers), because they have traditionally been used by hydrologists rather than geophysicists, they are commonly considered to be hydrological measurement techniques.

Table 1.1. Common geophysical and hydrological characterization methods that are used to assist in hydrogeological investigations, and associated chapter references

Acquisition Approaches	Characterization Methods	Attributes Typically Obtained	Examples of Hydrogeological Objectives	Reference Chapter(s)
Geophysical Methods				
Airborne	Remote sensing	Many attributes, including electrical resistivity, gamma radiation, magnetic and gravitational field, thermal radiation, electromagnetic reflectivity	Mapping of bedrock, fresh-salt water interfaces and faults, assessment of regional water quality	11
Surface	Sesimic refraction	P-wave velocity	Mapping of top of bedrock, water table, and faults	8
	Seismic reflection	P-wave reflectivity and velocity	Mapping of stratigraphy, top of bedrock and delineation of faults/fracture zones.	8, 12, 13
	Electrical resistivity	Electrical resistivity	Mapping of aquifer zonation, water table, top of bedrock, fresh-salt water interfaces and plume boundaries, estimation of hydraulic anisotropy, and estimation/monitoring of water content and quality.	5, 13, 14
	Electromagnetic	Electrical resistivity	Mapping of aquifer zonation, water table and fresh-salt water interface, estimation/monitoring of water content and quality.	6, 12, 14
	GPR	Dielectric constant values and dielectric contrasts	Mapping of stratigraphy and water table, estimation and monitoring of water content	7, 13, 14

Acquisition Approaches	Characterization Methods	Attributes Typically Obtained	Examples of Hydrogeological Objectives	Reference Chapter(s)
Crosshole	Electrical resistivity	Electrical resistivity	Mapping aquifer zonation, estimation/monitoring of water content and quality	5, 13, 14
	Radar	Dielectric constant	Estimation/monitoring of water content and quality, mapping aquifer zonation	7, 13, 14, 17
	Seismic	P-wave velocity	Estimation of lithology, fracture zone detection	9, 13, 17
Wellbore	Geophysical well logs	Many attributes, including electrical resistivity, seismic velocity and gamma activity	Lithology, water content, water quality, fracture imaging	2, 10, 12, 13, 16, 17
Laboratory/ Point	Electrical, seismic, dielectric, and x-ray methods	Many attributes, including electrical resistivity, seismic velocity and attenuation, dielectric constant and x-ray attenuation	Development of petrophysical relationships, model validation, investigation of processes and instrumentation sensitivity	4, 9, 11, 15, 17
Hydrological Methods				
Well tests	Pump, slug tests, and single wellbore hydraulic tests	Change in water level	Estimation of hydraulic conductivity, specific storage	2
Crosshole	Hydraulic tracer tests	Tracer concentration	Hydraulic conductivity, dispersivity, fast flow paths	2, 13
Crosshole	Hydraulic tomography	Drawdown measurements	Hydraulic conductivity	2
Wellbore	Flowmeter tests	Water flow	Hydraulic conductivity	2, 13, 17
Wellbore	Neutron probes	Back-scattered neutron counts	Water content	14
Laboratory	Permeameters	Hydraulic head	Hydraulic conductivity	2
Laboratory	Sieves	Grain size distribution	Estimation of hydraulic conductivity	2
Laboratory/ Point	Time domain reflectometer	Dielectric constant	Water content	2, 14, 15

1.2 Organization of *Hydrogeophysics*

This book covers the fundamentals of the discipline from both the hydrogeological and geophysical perspectives, provides guidelines for application of different approaches and methods, and includes extensive examples of hydrogeophysical applications through case studies. We have divided the book into four parts, each of which contains multiple chapters:

I. The Fundamentals of Hydrological Characterization Methods (Chapters 2 and 3)
The concepts of subsurface hydrology that are important to a subsurface scientist are covered in the subsequent two chapters of the first part of this book. These chapters focus on hydrological data acquisition and measurement analysis, as well as geostatistical approaches. We assume that the reader has already been introduced to the key concepts in hydrogeology, and as such, they are not reviewed in this book. There are several excellent references that are available to those readers requiring hydrological background information, such as Freeze and Cheery (1979), DeMarsily (1986), Dagan (1989), Jury et al. (1991), Delleur (1999), Schwartz and Zhang (2003), and Rubin (2003).

II. The Fundamentals of Geophysical Characterization Methods (Chapters 4–11)
Part II focuses on describing the various geophysical techniques often used for hydrogeological characterization and monitoring, including electrical, controlled source electromagnetic, ground penetrating radar (GPR), seismic, well logging, and remote sensing methods. Although the most widely used geophysical characterization methods are covered, there are other useful but less commonly used methods that are not discussed in this part (such as spontaneous potential, gravity, and radiomagnetotelluric methods). Two of the chapters in this part focus on the relationships between geophysical attributes (seismic and electrical) and hydrogeological properties. Unlike most geophysical textbooks, the geophysical methods and petrophysical discussions presented in this part of the book emphasize the theory, assumptions, approaches, and interpretations that are particularly important for hydrogeological applications.

III. Hydrogeophysical Case Studies (Chapters 12–15)
Part III incorporates the fundamentals of hydrogeology and geophysics, which are presented in the first two parts of the book, into a series of hydrogeophysical case studies. The case-study chapters are organized by decreasing scale of investigation. Case studies are presented that describe the use of geophysical approaches for characterization at the regional scale, the local scale, and the laboratory scale, where the notions of scale follow those presented in Figure 1.1. For example, the local-scale case-study chapters utilize wellbore, crosshole, tracer, and surface geophysical data. The local-scale chapters have been further divided into two hydrological regimes: Chapter 13 describes investigations that have been carried out in the saturated subsurface, while Chapter 14 describes hydrogeophysical investigations performed in the vadose zone. Finally, the laboratory-scale investigations primarily focus on the use of core measurements for characterizing and monitoring hydrogeological properties, and also include discussions on the use of laboratory measurements for developing petrophysical relationships and assessing spatial sensitivities of measurements. The case studies are intended to highlight the potential and limitations of different characterization methods for various hydrogeological applications over a range of investigation scales, and to give many references to other such hydrogeophysical investigations.

IV. Hydrogeophysical Frontiers (Chapters 16 and 17)
The final two chapters in the last part of the book focus on two frontier areas in hydrogeophysics. Chapter 16 describes emerging technologies in hydrogeophysics (such as surface nuclear-magnetic-resonance techniques). Chapter 17 describes the use of stochastic approaches to integrate geophysical and hydrological data. As will be discussed in Section 1.3, data integration is perhaps the largest obstacle in hydrogeophysics. Stochastic methods provide a systematic framework for assessing or handling some of the complexities that arise in fusing disparate data sets, such as those associated with spatial variability, measurement error, model discrimination, and conceptual model uncertainty. Although stochastic approaches have been used extensively within the hydrological community, such experience within the hydrogeophysical community is still limited, and thus stochastic integration of hydrological and geophysical data is considered to be a frontier area in hydrogeophysics.

Examination of Table 1.1 suggests that geophysical data are used to assist with a wide range of hydrological objectives. These objectives can be broadly categorized into three key areas:

1. Hydrogeological mapping
2. Hydrological parameter estimation
3. Monitoring of hydrological processes

These three categories are briefly described below, each followed by a single example as well as a table that describes where other examples are given in this book.

Hydrogeological Mapping
Subsurface hydrogeological mapping, or inferring information about the geometry of subsurface units or interfaces, is often performed using surface and airborne geophysical data. It has great utility in regional as well as local hydrological investigations (for example, those associated with water resources and contaminant transport studies). Several recent publications highlight the importance of characterizing aquifer geometry for hydrogeological investigations (e.g., Bristow and Jol, 2003; Bridge and Hyndman, 2004). One common mapping approach is to integrate sparse borehole data, which provide direct and detailed information about subsurface hydrogeology as a function of depth at a single location, with less invasive and more laterally continuous geophysical data (such as are available using surface or crosshole acquisition geometries) to extrapolate information away from the boreholes. In some cases, this procedure is routine, whereas in other cases it can be quite challenging. Figure 1.2 shows an example of hydrogeological mapping, where 100 MHz GPR data were first calibrated using wellbore lithology information, and then used to delineate the subsurface stratigraphy in an Atlantic Coastal Plain aquifer of Virginia (modified from Hubbard et al., 2001). As shown in Figure 1.2, within the top 15 m of this sedimentary subsurface, the GPR profile reveals the water table, key unconformities, several stratigraphic packages, and the top of a subregional aquitard.

Table 1.2 lists several different hydrogeological mapping applications discussed in subsequent chapters of this book. Although other geophysical methods could have been used to investigate different hydrogeological mapping objectives (Table 1.1), this and all other subsequent tables list only those objectives and methods that are further discussed in this book. Depending on the method, "mapping" may pertain to the mapping of interfaces (as is typically the case with surface seismic and GPR methods), or the mapping of layers or units (as is the case with surface electrical or electromagnetic methods). We emphasize that the spatial extent of the investigation is controlled by the acquisition geometry and parameters. For example, it might be of interest to focus on a smaller area identified on the regional-scale GPR line shown in Figure 1.2. To do this, we could collect a grid of GPR data over only a small portion of the study area imaged in Figure 1.2, using closer line and trace spacing and possibly higher frequency antennae. Hydrogeological boundaries that were mapped using such a detailed GPR grid at the same site where the regional GPR line (Figure 1.2) was collected were used to construct a layered local-scale aquifer model, which was subsequently used to constrain flow and transport simulations (Scheibe and Chien, 2003).

Table 1.2. Geophysical methods and acquisition approaches that are used for mapping various hydrogeological features and that are further described in this book

Mapping Objective	Acquisition Approach	Method Used (and chapter where example is discussed)
Geometry of subsurface interfaces or units	Airborne	Gamma-Ray Spectroscopy (11)
	Surface	Electrical Resistivity (12, 16)
		Electromagnetic (6, 12)
		GPR (7, 13, 14)
		Seismic Reflection (12)
	Crosshole	Electrical Resistivity (14)
		GPR (13)
		Seismic (13)
	Wellbore	Wellbore logs (10, 12, 14)
Water Table	Surface	Electrical Resistivity (14)
		Electromagnetic (14)
		Seismic Reflection (13)
		GPR (7, 13, 14)
Top of Bedrock	Airborne	Electromagnetic (11)
		Gamma-Ray Spectroscopy (11)
		Magnetics (11)
	Surface	Electromagnetic (6)
		Seismic (8)
Fresh-Salt Water Interface	Airborne	Electromagnetic (11)
	Surface	Electromagnetic (12)
	Wellbore	EM Induction Logs (10)
Plume Boundaries	Airborne	Electromagnetic (11)
	Surface	GPR (7, 13)
Faults/Fracture Zones	Airborne	Magnetics (11)
	Surface	Seismic (12)
		Electromagnetic (6)
		GPR (13)
	Wellbore	Wellbore methods (2, 10)

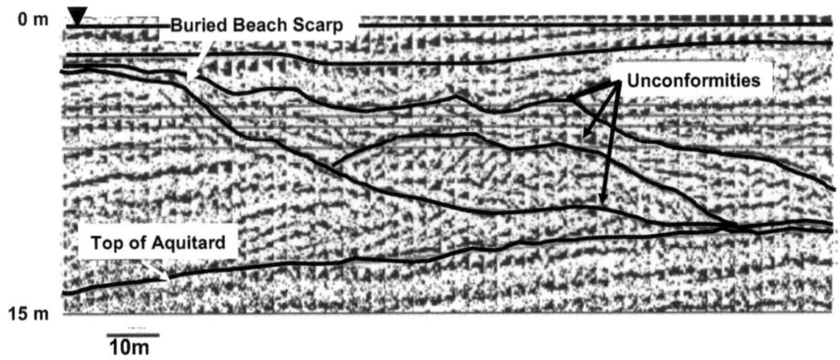

Figure 1.2. Example of the use of surface 100 MHz GPR data for mapping shallow subsurface features (modified from Hubbard et al., 2001)

Hydrogeological Parameter Estimation

Numerical modeling of fluid flow and transport is often used to test hypotheses or to assist subsurface management decisions. For example, mathematical models are used to simulate contaminant plume transport over time and to design effective remediation plans. They are also used to simulate aquifer depletion versus recharge over time and to design sustainable resource extraction programs. Effective hydrological hypothesis testing using mathematical modeling requires accurate and sufficient hydrological parameter input. Given the difficulty of obtaining sufficient characterization data for model parameterization, it is generally felt that computational flow and transport abilities have surpassed the ability to adequately parameterize the synthetic aquifer domains. As a result, the predictive capability of these models is limited, and a significant disconnect has developed between the state-of-the-art associated with modeling and the use of these models to manage practical problems on a regular basis.

Calibrated geophysical attributes have been used to provide dense and high-resolution estimates of hydrogeological parameters that can be used as input to numerical flow and transport models or to obtain a better understanding of hydrological processes and systems. The ability to estimate subsurface hydrogeological properties using a particular geophysical approach is a function of geophysical technique, the relation between the property under consideration and geophysical attributes, the magnitude of the property variations and contrasts, and the spatial scale of the changes relative to the scale of the objective target. The success of a particular geophysical approach for estimating hydrogeological parameters also depends on the geophysical data acquisition parameters and geometry employed, the data reduction or inversion procedures, and the availability of petrophysical relationships for linking the hydrogeological and geophysical attributes. As will be discussed below in Section 1.3, estimation of hydrogeological parameters using geophysical data is a current topic of research, and there are many obstacles associated with these procedures. Nonetheless, several case studies have illustrated the potential that geophysical data hold for these applications. An example of parameter estimation is shown in Figure 1.3 (modified from Chen et al., 2004), which illustrates estimates of lithology and mean values of sediment Fe(II) obtained using 100 MHz tomographic radar amplitude data within a Monte Carlo Markov Chain estimation approach. Estimates such as these can be very useful in parameterizing reactive flow and transport models used for predicting contaminant or bacterial transport, or for assessing redox conditions. As shown in Table 1.3, many other examples of hydrogeological parameter estimation using geophysical data will be discussed in subsequent chapters of the book.

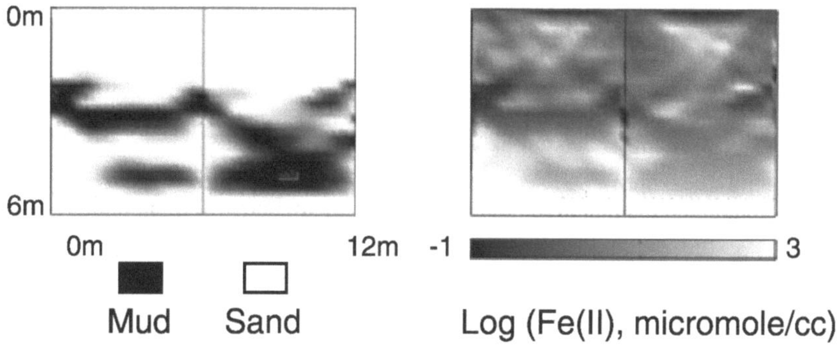

Figure 1.3. Estimates of lithology (left) and mean sediment Fe(II) distribution (right) obtained using tomographic GPR amplitude data within a Monte Carlo Markov Chain approach (from Chen et al, 2004)

Table 1.3. Geophysical methods and acquisition approaches that are used for estimating various hydrogeological parameters and that are further described in this book

Hydrogeological Parameter	Acquisition Approach	Method Used (and chapter where example is discussed)
Water Content	Airborne	Remote Sensing (11)
	Surface	NMR (16)
		Electrical Resistivity (14, 16)
		Electromagnetics (14)
		GPR (14)
	Crosshole	Electrical Resistivity (14)
		GPR (14)
	Benchtop	TDR (16)
Water Quality	Airborne	Remote Sensing (15)
	Surface	Electrical Resistivity (13, 14)
		Electromagnetic (13, 14)
		GPR (7)
	Crosshole	Electrical Resistivity (5, 13)
		GPR (7)
		Well Logs (10)
	Benchtop	X-ray attenuation (15)
Permeability or Hydraulic Conductivity	Well Tests	Hydraulic Well Tests (2)
	Crosshole	Tracer tests (2, 13)
		Hydraulic Tomography (2)
		GPR Tomography (13, 17)
		Seismic Tomography (13)
	Wellbore	Logs and Well Tests (2, 13, 17)
	Benchtop	Core measurements (2)
Hydrogeological Spatial Correlation Parameters	Surface	GPR (13)
	Crosshole	Seismic (13)
		GPR (13)

Hydrological Process Monitoring

In addition to geophysical data collected at one point in time to characterize systems in a static sense, geophysical data can be collected at the same location as a function of time. Observing the data as "time difference" sections (measurements collected at an earlier time subtracted from measurements collected at a later time), or "time-lapse" sections, enhances the imaging of subtle geophysical attribute changes caused by natural or forced system perturbations. Time-lapse imaging also decreases the dependence of the geophysical measurements on both the static geological heterogeneities and on data inversion procedures and associated artifacts. Difference images are most commonly obtained from crosshole data sets, but have also been obtained using measurements acquired with surface, airborne, and single borehole acquisition modes. These data sets have been used to elucidate dynamic transformations and thus have the potential to assist in understanding hydrological processes. An example is given in Figure 1.4, which illustrates the use of time-lapse, crosshole 100 MHz GPR velocity data collected over a period of a year at an important aquifer in the United Kingdom (from Binley et al., 2002a). These estimates show how high rainfall, which occurred in March 1999, increased the subsurface moisture content. This figure also shows the emergence of a drying front in the fall and winter of 1999, and the influence of an impeding layer (that exists between 6 and 7 m depth) on the infiltrating water front. Monitoring using such approaches is extremely useful for understanding complex hydrological processes. We expect that the use of noninvasive or minimally invasive geophysical measurements, collected over a prolonged time period, will become a standard monitoring approach for improving our understanding of subsurface processes and our ability to manage natural resources. Examples of other process-monitoring investigations discussed in subsequent chapters are listed in Table 1.4.

Table 1.4. Geophysical methods and acquisition approaches that are used for monitoring hydrological parameters and that are further described in this book

Monitoring Objective	Acquisition Approach	Method Used (and chapter where example is discussed)
Water Content	Surface	GPR (13, 14)
		Electromagnetic (14)
		Electrical Resistivity (14)
	Crosshole	Electrical Resistivity (14)
		GPR (14)
Water Quality	Surface	GPR (14)
	Crosshole	Electrical Resistivity (5, 13, 14)
		GPR (13)
	Benchtop	Seismic Amplitude (15)
		Electrical Resistivity (15)

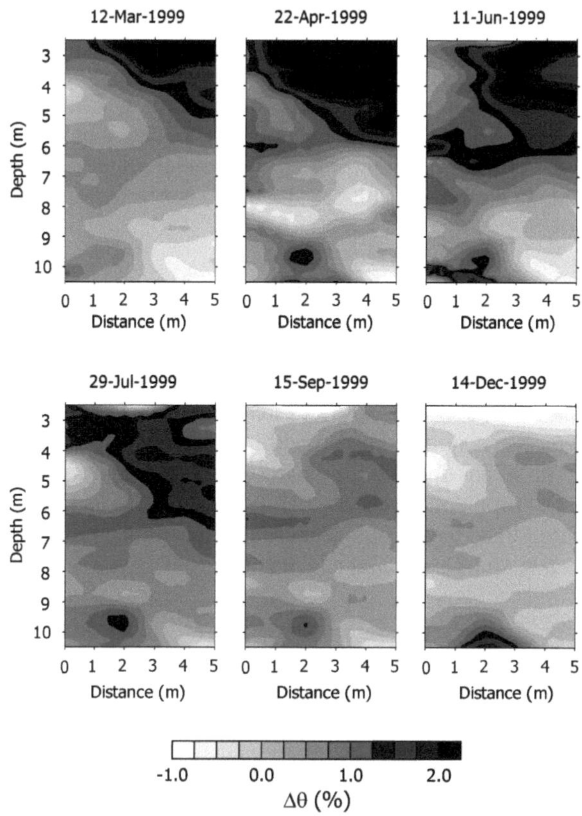

Figure 1.4. Percent change in volumetric water content obtained over time using time-lapse GPR velocity information associated with natural precipitation and drying. Changes in moisture content are shown relative to those estimates obtained using GPR data collected on February 1, 1999 (from Binley et al., 2002a).

1.3 Hydrogeophysical Challenges

In the last decade, many advances associated with near-surface geophysics have been made that facilitate the use of geophysical data for hydrogeological characterization. These advancements have improved digital technology for acquisition, improved many geophysical methods for near-surface data acquisition and processing, and improved computational speed and capabilities associated with processing, inversion, and modeling and visualization of geophysical data. Nonetheless, there are many obstacles that still hinder the *routine use* of geophysics for hydrogeological characterization. A short discussion of some of the current key gaps in hydrogeophysics is given below.

1.3.1 INTEGRATION OF GEOPHYSICAL AND HYDROGEOLOGICAL MEASUREMENTS

Integration of geophysical and hydrogeological measurements poses perhaps the largest obstacle for hydrogeophysics. To perform integration, we must confront issues of scale, non-uniqueness, and uncertainty. We must also have the available petrophysical relationships by which to integrate these measurements (refer to Section 1.3.2 below) and have a working knowledge of different estimation approaches. Many different integration and estimation approaches have been used to fuse hydrogeological and geophysical data. The choice of which approach to use typically depends on data density, project objectives, and interpreter background. Integration can be based on data, models, and/or concepts, and can be approached using deterministic and/or stochastic methods.

Probably the most familiar approach to geophysicists is one that capitalizes on expert skills and intuition to merge hydrogeological and geophysical data sets. This methodology allows for incorporating information that is often very difficult to quantify, such as knowledge associated with geophysical signatures and depositional processes. Using a team of experts to best interpret subsurface structures, stratigraphy, and hydraulic properties is quite common within the petroleum industry. However, with this approach, it is often difficult to quantify the uncertainty associated with the components of the problem, such as the conceptual model, the hydrogeological parameter estimate, and the geophysical data inversion procedure. Because this approach requires a team of experts, it can also be prohibitive for single or small groups of investigators.

Stochastic methods provide a systematic framework for assessing or handling some of the complexities that arise when integrating geophysical data. For example, Bayesian methods permit incorporation of a-priori information into the estimation approach, and provide a framework for incorporating additional data as they become available. The development and application of stochastic methods for integration of hydro- and geophysical data is considered to be a frontier area, and is the focus of Chapter 17. Geostatistics is probably the most widely used stochastic approach for incorporating hydrological and geophysical data. As will be discussed in Chapter 3, these techniques rely on spatial correlation information to interpolate between measurements in a least-squares sense. Because natural geologic materials often exhibit strong spatial organization, these methods have been widely used within the hydrological sciences.

Although many advances have been made in recent years in integrating geophysical and hydrogeological data for improved subsurface characterization, there is still significant room for estimation improvement in many of these areas, as briefly described below:

- Although geophysical methods can be used to investigate the subsurface over various scales (Figure 1.1), because quantifying the measurement support scale is not straightforward, integration of multiple types of measurements that have different support scales is challenging.
- Although it is intuitive and obvious that inversion of geophysical data can benefit from hydrogeological constraints, true joint inversion of both data sets is rarely performed in practice. What is typically implied by joint inversion is often in fact a sequential process, whereby the geophysical data are interpreted and converted to a form in which they can act as a constraint on the hydrogeological inversion. Recent examples of joint inversion using

geophysical-hydrological data include those studies given by Chen et al. (2003), Kowalsky et al. (2003), and Linde et al. (2004).

- Most hydrogeophysical studies use either a single type of geophysical data set or two data sets independently. Because geophysical measurements often respond to more than one parameter (such as the sensitivity of electrical measurements to pore fluid and sediment texture, as will be described in Chapter 4), fusion of multiple sources of information has the potential to be much more beneficial for characterizing and monitoring (e.g., Binley et al., 2002b).
- As will be discussed in Chapter 17, errors in geophysical data stem from geophysical data acquisition and inversion procedures (e.g., Peterson, 2001) as well as from petrophysical models (as will be discussed in Section 1.3.2) and integration approaches. Incorporation of spatially variable geophysical error into an inversion procedure, and assessment of the impact of these errors on the hydrogeological parameter estimation results is a current topic of research.
- The majority of the progress associated with hydrogeological parameter estimation using geophysical data has been achieved using high-resolution crosshole (GPR, ERT, seismic) methods, which provide high-resolution information but typically sample over a small area of investigation. To be useful for large-scale contaminant remediation or watershed hydrological investigations, hydrogeological parameter estimates must be obtained from more spatially exhaustive surface geophysical data.

1.3.2 PETROPHYSICAL RELATIONSHIPS

Many of the relationships that link hydrogeological parameters (such as mean grain size, porosity, type and amount of pore fluid) within typical near-surface environments to geophysical attributes (such as seismic velocity, electrical resistivity, and dielectric constant) are not well understood. Although a significant amount of research has been performed to investigate the relationships between geophysical and lithology or pore fluid within consolidated formations common to petroleum reservoirs and mining sites, very little petrophysical research has been devoted to the less consolidated, low-pressure and low-temperature environments common to hydrogeological investigations. As will be discussed in Chapters 4 and 9 of this book, in some circumstances we can anticipate the responses of some geophysical methods based on physical models and theory. For some parameters, such as seismic velocity and permeability, it is clear that no universal relationship can exist for all types of geologic materials (refer to Chapter 9 of this book). However, in cases where the geophysical attribute predominantly responds to a single hydrogeological parameter, some guidelines can be developed. As will be described in Chapter 4, relationships such as Topp's law (which relates dielectric constant values to moisture content; Topp et al., 1980) and Archie's law (which relates bulk electrical conductivity to the electrical conductivity of water-saturated geologic materials; Archie, 1942) are widely used in hydrogeophysics.

Different approaches have been used to develop site-specific petrophysical relationships that can be used in conjunction with geophysical data for hydrogeological investigations. One common approach is to develop site-specific empirical relationships between the geophysical measurements and the parameters of interest, using co-located field data (such as at a wellbore) or by developing relationships using site material within the laboratory. However, developing petrophysical relationships using co-located field data is often problematic, because of the different sampling scales, measurement directions, and errors associated with wellbore hydrological measurements and geophysical "measurements" located at or near a wellbore (e.g., Day-Lewis and Lane, 2004). As will be described in Chapter 15, investigations of the relationships between geophysical and hydrogeological parameters are often performed at the laboratory scale. At that scale, a better understanding of the sensitivities and spatial and temporal resolutions of the geophysical method can be quantified, as well as the manner in which properties are averaged within the sample volume. However, upscaling between the laboratory scales and field scales (where the developed relationships must be employed) remains a challenge to hydrogeophysics (e.g., Moysey et al., 2003).

At both the laboratory and field scale, several investigations have illustrated how single geophysical attribute measurements can be associated with more than one hydrogeological condition (Marion, 1993; Copty et al., 1993; Knoll, 1995; Hubbard et al., 1997; Prasad, 2002). These studies suggest that it is not unusual for non-uniqueness to exist (even under idealized conditions of error-free measurements in natural systems), and that the ambiguity is only exacerbated by measurement errors. As will be discussed in Chapter 17, stochastic approaches have been developed to deal with such non-unique conditions. However, there have been few field-scale tests of such approaches.

Because of an ever-increasing recognition of the complex and dynamic nature of natural systems, recent studies have begun to investigate the potential of geophysical methods for estimating biogeochemical properties, as well as their influence on hydrological properties (e.g., Atekwana et al., 2004; Williams et al., 2004). For example, products such as gases, precipitates, and biofilms can be formed during remediation approaches such as *in situ* oxidation or bioremediation, and these products can dramatically alter the hydraulic conductivity of the system. Although these studies suggest that geophysical approaches hold potential for such investigations, much more research is necessary to assess the relationships between geophysical attributes and biogeochemical properties.

1.3.3 OTHER CHALLENGES

In addition to the key scientific challenges associated with data integration and petrophysical relationships, there are many other outstanding issues associated with the use of geophysical data for improved subsurface characterization and monitoring. Many of the outstanding issues associated with individual geophysical methods or approaches are described in the following chapters. These challenges include the need to improve our understanding of less conventional geophysical approaches or attributes for hydrogeological investigations (such as GPR amplitude data, induced polarization, spontaneous polarization, seismic surface waves, surface nuclear magnetic resonance, and electroseismic approaches), and a better assessment of the accuracy of geophysical data inversion approaches in the presence of noise and in real three-dimensional, noisy systems. We expect that the need for improved understanding of the shallow subsurface will continue to drive research associated with improving our understanding of different geophysical attributes and the development of geophysical instrumentation for hydrogeological investigations.

1.4 Looking Forward

A new discipline of hydrogeophysics has evolved, aimed at improved simultaneous use of geophysical and hydrogeological measurements for hydrogeological investigations. By combining measurements made from hydrologic instruments with geophysical methods, accurate subsurface characterization and monitoring can be achieved, with high temporal and spatial resolution over a range of spatial scales. Discussions in Section 1.3 above focused on some of the many scientific issues within hydrogeophysics that are not well understood and that will require many years of research to develop sufficiently. Nonetheless, as will be described in the following chapters, there have been several successful examples of using geophysical and hydrogeological measurements for improved understanding of the near subsurface.

Hydrogeophysics, being ultimately an applied science, must strike a balance between improving our understanding of basic principles while implementing pragmatic solutions to subsurface problems. Thus, there is a need within the hydrogeophysics community for researchers who will advance our understanding of the fundamental principles, as well as for researchers who can apply such principles toward subsurface characterization and develop pragmatic approaches when the fundamental principles are subject to uncertainty. For geophysical methods to be used more routinely for hydrogeological characterization and monitoring, we will need to broaden the group of subsurface scientists that routinely relies on geophysical methods. This seems like a daunting obstacle, given that there are many nuances associated with the quantitative use of geophysical data for subsurface investigations, and that improper use of geophysical methods may reduce the overall credibility of such approaches, as discussed by Greenhouse (1991). The hydrogeophysics community, government agencies, and professional societies should strive to collaborate with practitioners, so that advanced characterization techniques can be more routinely incorporated into hydrogeological studies. As suggested by the National Research Council (2000), a dedicated and long-term effort is needed to investigate and document the utility of noninvasive characterization methods for a wide variety of applications, with an emphasis on research performed by multidisciplinary teams; increased accurate automation of data acquisition, data processing, and decision-making for real-time field interpretation; and improved methods for transferring advanced tools and approaches to field practitioners.

Increasing population and pollution, decreasing natural resources, and changing climatic conditions all contribute to the urgency associated with improving our understanding of hydrogeological parameters and processes. Although there are many obstacles associated with the implementation of hydrogeophysical approaches, scientific and otherwise, there is already a substantial body of evidence that illustrates the promise of hydrogeophysical approaches for improved subsurface characterization and monitoring. We expect that advancements in hydrogeophysics will continue through increased research, cross-disciplinary education and collaboration, and publications. *Hydrogeophysics* is intended to provide an introduction to new researchers to the field, to be used as a resource for researchers already active in the field, and to serve as a springboard for advances in the field that are needed to improve our understanding and management of the earth's shallow subsurface.

Acknowledgments

The authors gratefully acknowledge Andrew Binley, Tom Daley, and Niklas Linde for reviewing this chapter and support from DOE-AC03-76SF00098 and NSF EAR-0087802.

References

Archie, G.E., The electrical resistivity log as an aid in determining some reservoir characteristics, *Trans. Amer. Inst. Mining Metallurgical and Petroleum Engineers*, *146*, 54–62, 1942.

Atekwana, E.A., D.D. Werkema, Jr., J.W. Duris, S. Rossbach, E.A. Atekwana, W.A. Sauck, D.P. Cassidy, J. Means, and F.D. Legall, In-situ apparent conductivity measurements and microbial population distribution at a hydrocarbon-contaminated site, *Geophysics, 69*(1), 56–63, 2004.

Binley, A., P. Winship, L.J. West, M. Pokar and R. Middleton, Seasonal variation of moisture content in unsaturated sandstone inferred from borehole radar and resistivity profiles, *J. Hydrology, 267*(3–4), 160–172, 2002a.

Binley, A., G. Cassiani, R. Middleton, and P. Winship, Vadose zone model parameterisation using cross-borehole radar and resistivity imaging, *J. Hydrology. 267*(3–4), 147–159, 2002b.

Bridge, H., and D. Hyndman (eds.), *Aquifer Characterization*, Society of Economic and Petroleum Geologists (in press), 2004.

Bristow, C.S., and H.M. Jol, (editors), *Ground Penetrating Radar in Sediments*, Geological Society, London, Special Publications, 211, 2003.

Chen, J., S. Hubbard, M. Fienen, D. Watson, and T.L. Mehlhorn, Estimating hydrogeological zonation at a NABIR field research center study site using high-resolution geophysical data and Markov Chain Monte Carlo methods, Eos. Trans. AGU, *84*(46), Fall Meet. Suppl., Abstract H21F-04, 2003

Chen, J., S. Hubbard, Y. Rubin, C. Murray, E. Roden, and E. Majer, Geochemical characterization using geophysical data: A case study at the South Oyster Bacterial Transport Site in Virginia, Water Resources Research, V40, WI2412, doi: 1029/2003WR002883, 2004.

Copty, N., Y. Rubin, and G. Mavko, Geophysical-hydrogeological identification of field permeabilities through Bayesian updating, *Water Resour. Res., 29*(8), 2813–2825, 1993.

Dagan, G., *Flow and Transport in Porous Formations*, Springer-Verlag, N.Y., 1989.

Day-Lewis, F.D. and J.W. Lane Jr., Assessing the resolution-dependent utility of tomograms for geostatistics, *Geophysical Research Letters, 31*(7), L07503, doi:10.1029/2004GL019617.

Delleur, J.W., *The Handbook of Groundwater Engineering*, CRC Press, Washington D.C., 1998.

De Marsily, G., *Quantitative Hydrogeology*, Academic Press, N.Y., 1986.

Freeze, R.A., and J.A. Cherry, *Groundwater*, Prentice-Hall, 1979.

Greenhouse, J.P., Environmental geophysics: It's about time, *The Leading Edge of Exploration, 10*(1), 32–34, 1991.

Hubbard, S., J. Chen, J. Peterson, E. Majer, K. Williams, D. Swift, B. Mailliox, and Y. Rubin, Hydrogeological characterization of the D.O.E. Bacterial Transport Site in Oyster, Virginia, using geophysical data, *Water Resour. Res.*, *37*(10), 2431–2456, 2001

Hubbard, S. S., and Y. Rubin, Ground penetrating radar assisted saturation and permeability estimation in bimodal systems, *Water Resour. Res.*, *33*(5), 971–990, 1997.

Jury, W.A., W.R. Gardner, and W.H. Gardner, *Soil Physics*, John Wiley & Sons, Inc. 1991.

Keys, W.S., *Borehole Geophysics Applied to Ground-water Investigations*, National Water Well Association, 1989.

Kowalsky, M.B., Y. Rubin, J. Peterson, and S. Finsterle, Estimation of flow parameters using crosshole GPR travel times and hydrological data collected during transient flow Experiments, *Eos Trans. AGU, 84*(46), Fall Meet. Suppl., Abstract H21F-07, 2003.

Linde, N., S. Finsterle, and S. Hubbard, Inversion of hydrological tracer data constrained by tomographic data, *Eos Trans. AGU, 85*(17), Jt. Assem. Suppl., Abstract NS13A-04 , 2004.

Marion, D., A. Nur, H. Yin, and D. Han, Compressional velocity and porosity in sand-clay mixtures, *Geophysics,* 57, 554–563, 1992.

Moysey, S., and R. Knight, Modeling the field-scale relationship between dielectric constant and water content in heterogeneous media, *Water Resour. Res., 40*, W03510, doi: 10.1029/2003WRR002589, 2004.

National Research Council, *Seeing into the Earth*, National Academy Press, Washington D.C., 2000.

Peterson, J.E., Jr., Pre-inversion corrections and analysis of radar tomographic data, *Journal of Env. and Eng. Geophysics*, *6*, 1–8, 2001.

Prasad, M., Velocity-permeability relations within hydraulic units, *Geophysics*, *68*(1), 108–117, 2003.

Rubin, Y., *Applied Stochastic Hydrogeology*, Oxford, N.Y., 2003.

Scheibe, T., and Y. Chien, An evaluation of conditioning data for solute transport prediction *Ground Water*, *41*(2), 128–141, 2003

Schwartz, F.W., and H. Zhang, *Fundamentals of Ground Water*, Wiley & Sons, New York, 2003.

Topp, G.C., J.L. Davis, and A.P. Annan, Electromagnetic determination of soil water content: Measurements in coaxial transmission lines, *Water Resour. Res.*, *16*, 574–582, 1980.

U.S. Department of Energy's Environmental Management Science Program, National Research Council, *Research Needs in Subsurface Science*, National Academy Press, Washington D.C., 2000.

U.S. Global Change Research Program, USGCRP Water Cycle Study Group, *A Plan for a New Science Initiative on the Global Water Cycle*, 2001.

Ward, S.H., *Geotechnical and Environmental Geophysics*, SEG Investigations in Geophysics No. 5, S.H. Ward, Editor, 1990.

Williams, K.H., D. Ntarlagiannis, L.D. Slater, S. Hubbard, and J.F. Banfield, Monitoring microbe-induced sulfide precipitation under dynamic flow conditions using multiple geophysical techniques, in submission to *Nature*, 2004.

2 HYDROGEOLOGICAL METHODS FOR ESTIMATION OF SPATIAL VARIATIONS IN HYDRAULIC CONDUCTIVITY

JAMES J. BUTLER, JR.

Kansas Geological Survey, 1930 Constant Ave., Campus West, University of Kansas, Lawrence, KS 66047, U.S.A.

2.1 Introduction

2.1.1 CHAPTER OVERVIEW

Virtually every hydrogeologic investigation requires an estimate of hydraulic conductivity (K), the parameter used to characterize the ease with which water flows in the subsurface. For water-supply investigations, a single estimate of K averaged over a relatively large volume of an aquifer will usually suffice. However, for water-quality investigations, such an estimate is often of limited value. A large body of work has demonstrated that spatial variations in K play an important role in controlling solute movement in saturated flow systems (e.g., Sudicky and Huyakorn, 1991; Zheng and Gorelick, 2003). Numerous studies have shown that information about such variations is required to obtain reliable predictions of contaminant transport and to design effective remediation systems. Varieties of methods have been used in efforts to acquire this information. The primary purpose of this chapter is to describe these methods and assess the quality of the information that they can provide. Later chapters will discuss how geophysics can augment the information obtained with these approaches.

In this chapter, three classes of methods will be discussed. The first class, designated here as "traditional approaches," consists of methods that have been used for a number of decades to acquire information about hydraulic conductivity for water-supply investigations. The second class, designated as "current generation," consists of approaches that have been developed in the last decade or two for the specific purpose of acquiring information about spatial variations in K. The third class, designated as "next generation," consists of methods that are currently in various stages of development. For each method, the underlying principles of the approach will be described, followed by a brief discussion of its major advantages and limitations. Each method will be illustrated with data collected at an extensively studied field site. The use of data from the same site facilitates the discussion of the relative advantages of the various approaches, as well as the type of information each can provide. Although the focus of this chapter will be on methods for investigations in shallow unconsolidated aquifers, many of these methods are also of value for investigations in consolidated materials and in units that would not be classified as aquifers. Techniques primarily used in other hydrogeologic settings are briefly discussed at the end of this chapter.

Y. Rubin and S.S. Hubbard (eds.), Hydrogeophysics, 23–58.

2.1.2 FIELD SITE

The methods described in this chapter were evaluated at a research site of the Kansas Geological Survey. This site, the Geohydrologic Experimental and Monitoring Site (GEMS), is located in the floodplain of the Kansas River just north of Lawrence, Kansas, in the central portion of the United States (Figure 2.1). GEMS has been the site of extensive research on flow and transport in heterogeneous formations (e.g., McElwee et al., 1991; Butler et al., 1998a,b, 1999a,b; 2002; Bohling, 1999; Schulmeister et al., 2003).

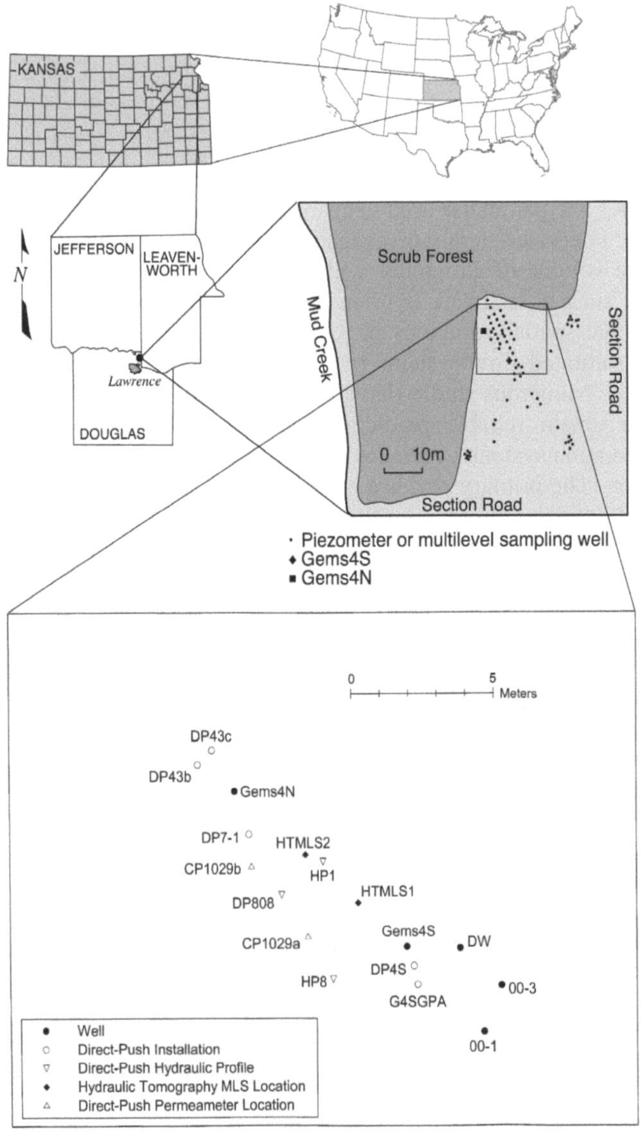

Figure 2.1. Site location map for GEMS(only locations referred to in text are shown in inset)

This work enables the techniques discussed here to be evaluated in a relatively controlled field setting. The shallow subsurface at GEMS consists of 22 meters of unconsolidated Holocene sediments of the Kansas River alluvium that overlie and are adjacent to materials of Pennsylvanian and late Pleistocene age, respectively. Figure 2.2 displays a cross-sectional view of the shallow subsurface with electrical conductivity logging data obtained using a direct-push unit (Butler et al., 1999b), and a geologic interpretation from core and logging data. As shown in that figure, the heterogeneous alluvial facies deposits at GEMS consists of 11.5 m of primarily clay and silt overlying 10.7 m of sand and gravel. The methods described in this chapter were applied in the sand and gravel interval, which is hydraulically confined by the overlying clay and silt.

The subarea of GEMS used in this work is depicted in the inset of Figure 2.1. This inset displays the locations of conventional wells (Gems4N, Gems4S, DW, 00-1, and 00-3), multilevel sampling wells (HTMLS1 and HTMLS2), and various direct-push activities.

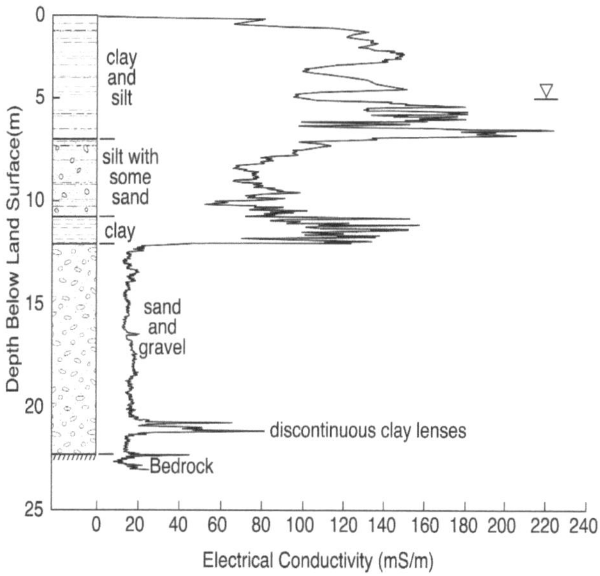

Figure 2.2. Generalized GEMS stratigraphy with electrical conductivity log from G4SGPA (after Butler et al., 1999b)

The conventional wells were all constructed out of PVC and installed with hollow-stem augers, with 0.168 m outer diameter (OD) flights for Wells 00-1 and 00-3 (both wells—0.051 m inner diameter [ID]) and 0.286 m OD flights for other wells (Gems4N and Gems4S–0.102 m ID, DW–0.127 m ID). A natural filter pack was used in all cases, and all wells were grouted across the overlying silt and clay interval. The multilevel sampling wells (MLS) were constructed out of extruded PVC (seven-channel) pipe and installed with a direct-push unit (OD of direct-push pipe–0.083 m; OD of MLS–0.041 m). Various configurations were used for the direct-push activities, as will be discussed in Sections 2.3 and 2.4.

2.2 Traditional Approaches

A variety of methods have been used to obtain information about hydraulic conductivity for water-supply investigations. Many of these methods have also been utilized to assess spatial variations in K. Those methods are described in this section.

2.2.1 PUMPING TESTS

The constant-rate pumping test is the most commonly used method to obtain information about the transmissive nature of an aquifer for water-supply investigations. In this approach, a central well is pumped at a near-constant rate, while changes in water level are measured at that and nearby wells. The changes in water level, termed drawdown, are analyzed using various models of the well-formation configuration (Streltsova, 1988; Kruseman and de Ridder, 1990; Batu, 1998). Several investigations (Butler and Liu, 1993; Meier et al., 1998; Sánchez-Vila et al., 1999) have shown that a pumping test will yield a hydraulic conductivity estimate that represents an average of K over a relatively large volume of an aquifer. Thus, little information can be gleaned about variations in K on the scales of interest for solute-transport investigations.

A series of short-term pumping tests performed at GEMS can be used to illustrate the limitations of this approach for assessment of spatial variations in K. Figure 2.3a depicts the drawdown at direct-push installation DP4S produced by a constant rate of pumping at Well DW. The Cooper-Jacob method (Cooper and Jacob, 1946) is used to fit a straight line to the latter portions of the drawdown record. The resulting hydraulic conductivity estimate of 116 m/d is a reasonable value for the average K of a sand and gravel sequence (Freeze and Cherry, 1979). Figure 2.3b depicts the drawdown at direct-push installation DP7-1 produced by a constant rate of pumping at Well Gems4N. Although a backpressure adjustment between 100 and 200 seconds and the commencement of pumping at a nearby well at approximately 600 seconds complicate the analysis, the Cooper-Jacob method can still be used to fit a straight line to an extensive portion of the drawdown record. The resulting K estimate is again 116 m/d. The similarity in the K estimates from the two tests, which is in keeping with the findings of the previously cited studies, demonstrates that little information about variations in hydraulic conductivity can be obtained using K estimates from analyses of conventional pumping tests performed in nearby wells. Although K estimates will vary little, estimates of the specific storage parameter (S_s) can vary considerably. For example, the S_s estimates, obtained from analyses of drawdown at DP4S and DP7-1, differ by a factor of 36 (3.81×10^{-5} m^{-1} and 1.06×10^{-6} m^{-1}, respectively). This variation in S_s is produced by a number of factors, including variations in the K of the material between the pumping well and the observation point (Butler, 1990; Schad and Teutsch, 1994; Sánchez-Vila et al., 1999). However, given the uncertainty about the factors contributing to the variation in S_s, it is extremely difficult to extract information about variations in K from it.

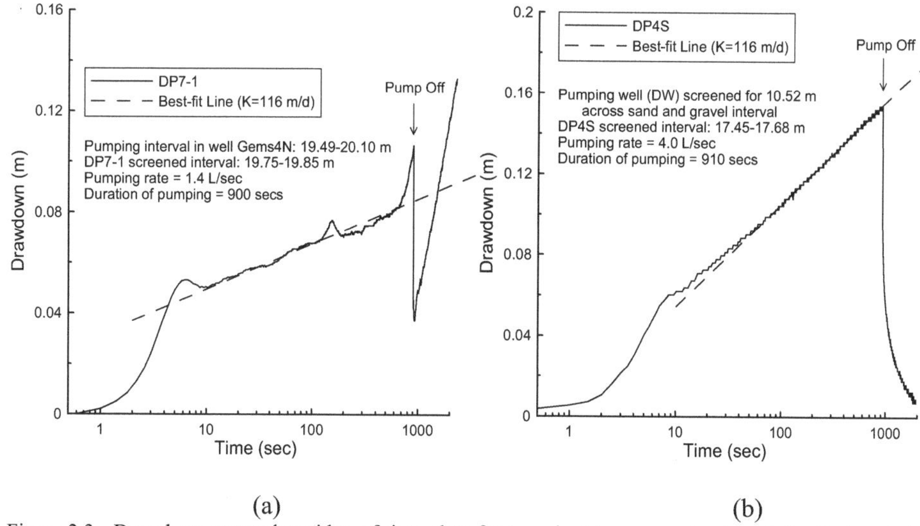

Figure 2.3. Drawdown versus logarithm of time plots for two short-term pumping tests (after Butler et al., 2002): (a) 3/16/99 test at Well DW; (b) 8/13/99 test at Well Gems4N (all depths in this and following figures are with respect to the top of casing at Well Gems4S)

2.2.2 SLUG TESTS

The slug test is a commonly used method for acquiring information about K for both water-supply and water-quality investigations. This approach, which is quite simple in practice, consists of measuring the recovery of head in a well after a near-instantaneous change in head at that well. Head data are analyzed using various models of the well-formation configuration (see Butler [1998] for a detailed description of field and analysis procedures). The slug test provides a K estimate that is heavily weighted towards the properties of the material in the immediate vicinity of the screened interval (Beckie and Harvey, 2002). Thus, at a site with an extensive network of wells, the slug test can be a valuable tool for describing spatial variations in K (Yeh et al., 1995). However, considerable care must be taken in all stages of the test process to obtain reasonable K estimates (Butler et al., 1996; Butler, 1998). In particular, the quality of the K estimates is heavily dependent on the effectiveness of well-development activities (Butler and Healey, 1998; Butler et al., 2002). Failure to give appropriate attention to well-development procedures will often result in K estimates that bear little resemblance to reality.

The potential and pitfalls of this approach can be illustrated using a slug test performed at GEMS. Figure 2.4 is a normalized head (deviation from static head normalized by the initial head change) versus time plot from a test performed in Well 00-1. The data were analyzed using a high-K extension of the Hvorslev (1951) model that incorporates the inertial mechanisms that give rise to the oscillatory head data (Butler, 1998; Butler et al., 2003). The resulting K estimate of 224 m/d is close to twice the large volumetric average obtained from the pumping tests, demonstrating that slug tests can be used to assess the K of discrete zones that may act as preferential pathways for or barriers to solute movement. However, as stated above, appropriate attention must be given to all phases of the test procedure to obtain reliable K estimates. In highly permeable intervals, such as the test interval at Well 00-1, a number of issues must be considered.

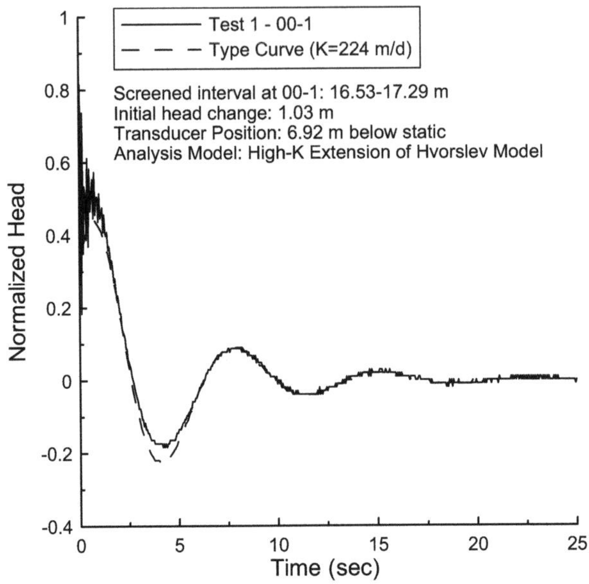

Figure 2.4. Normalized head versus time plot for 3/18/94 slug test #1 at Well 00-1 (data from Butler, 1998)

For example, the pressure transducer at Well 00-1 was located 6.92 m below static, so, as shown by McElwee (2001) and Zurbuchen et al. (2002), water-column acceleration effects on transducer readings must be considered. Furthermore, as shown by Butler et al. (2003), analysis using a model that neglects inertial mechanisms, such as the conventional forms of the Hvorslev (1951) and Bouwer and Rice (1976) models, can result in a significant overestimation of K. Butler et al. (2003) discuss these and related issues, and provide a series of guidelines for the performance and analysis of slug tests in highly permeable intervals. More general guidelines are provided in Butler et al. (1996) and Butler (1998). Adherence to these guidelines is necessary to obtain reliable K estimates. Note that the deviation between the test data and the best-fit type curve in the vicinity of the trough at four seconds on Figure 2.4 is likely a product of the nonlinear responses discussed by Butler (1998) and McElwee and Zenner (1998).

2.2.3 LABORATORY ANALYSES OF CORE SAMPLES

Laboratory analyses of samples collected during drilling is a common method for acquiring information about the properties of a formation. Estimates of the saturated hydraulic conductivity are obtained using various permeameter methods or relationships based on particle-size analyses. Permeameter methods involve running water through a core under either a constant or variable hydraulic gradient. The constant-gradient (constant-head) permeameter is primarily used for materials of moderate or high K, while the variable-gradient (falling-head) approach is primarily used for materials of low K. In either case, considerable care must be taken in all phases of the experimental procedures. Entrapped air, mobilization and redeposition of fine fractions, non-Darcian flow, and use of a permeant fluid at different temperature, pressure, and biochemical conditions than the natural *in situ* fluid will often lead to artificially low estimates of K (Klute and Dirksen, 1986). In most cases, water is run parallel to the vertical axis of the core, so the K estimate is for the vertical component of hydraulic conductivity. Permeameter experiments can be performed on either the original sample, if relatively undisturbed, or on a repacked sample. Hydraulic conductivity estimates obtained from the original samples tend to be lower than those obtained from the repacked cores, a difference that is commonly attributed to the greater structural control in the original samples.

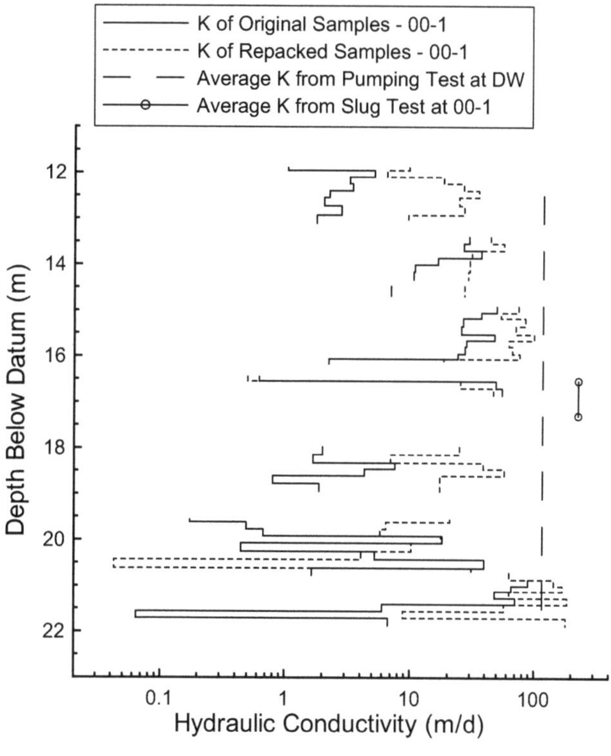

Figure 2.5. Hydraulic conductivity versus depth plots at Well 00-1 (core data from Butler and McElwee, 1996)

Results of analyses of core samples from Well 00-1 can be used to illustrate the limitations of permeameter-based approaches. Figure 2.5 is a plot of K estimates obtained from permeameter analyses of the original and repacked core samples. In addition, the average K obtained from the pumping test at Well DW and the K from the slug test at Well 00-1 are shown for reference. The averages for the permeameter analyses of the original and repacked cores are 16% and 39%, respectively, of the K determined from the pumping test. Every K estimate obtained from the original cores is below the average value from the pumping test, a situation that has often been reported by others (e.g., Table 1 of Rovey, 1998). The increase in K by a factor of 2.4 between the original and repacked cores is undoubtedly caused by the repacking process removing thin layers of low-K material that exert a strong influence on the original core estimates. The underprediction of K in the repacked samples with respect to the pumping-test estimate is likely caused by a combination of imperfect laboratory procedures, heterogeneity, and incomplete sample recovery in zones of high K. This latter possibility is reflected in the incomplete recovery in the zone opposite the screened interval of the well, which the slug test results indicate is of a relatively high K. Recovery of relatively intact core samples is a difficult task in highly permeable intervals. Although specialized devices have been developed for this purpose (e.g., Zapico et al., 1987; McElwee et al., 1991; Stienstra and van Deen, 1994), complete recovery is rarely possible. For example, the recovery at Well 00-1 was 72%.

The second approach commonly used for estimating K from core samples is based on relationships between hydraulic conductivity and various physical properties of the samples. A large number of empirical and theoretical relationships have been developed for estimation of K from particle-size statistics, which are assumed to be a reflection of the pore-size distribution. The most common empirical relationships are based on some measure of effective grain size. For example, the relationship developed by Hazen (Freeze and Cherry, 1979) uses d_{10}, the particle diameter at which 10% of the grains by weight are less than this diameter and 90% are greater ($K = Cd_{10}^2$ where C is a coefficient depending on grain size and sorting). Theoretical relationships have been developed from the Navier-Stokes equations and models of the porous medium. For example, the relationship developed by Kozeny-Carman (Bear, 1972) uses d_m, the geometric mean grain size calculated from $\sqrt{d_{84}d_{16}}$, and porosity (n) to estimate hydraulic conductivity ($K = [n^3/(1-n)^2][d_m^2/180]$ for K in m/d and d_m in mm). Although procedures for calculating particle-size distributions and porosity are well developed (Gee and Bauder, 1986; Danielson and Sutherland, 1986), determining the appropriateness of a given relationship for a particular site can be difficult.

Results of the laboratory analyses of the core samples from Well 00-1 help illustrate the uncertainty inherent in these empirical and theoretical relationships. Figure 2.6a is a plot of the permeameter results for the repacked samples, the K estimates determined from the Hazen equation (C = 864 for a well-sorted sand with K in m/d and d in mm), and the K estimates from the pumping and slug tests. The average K from the Hazen analysis is 147 m/d, 27% larger than that from the pumping test, a difference that could be attributed to heterogeneity, the vastly different scales of the two approaches, and uncertainty regarding the appropriate value for the coefficient C. The Hazen K estimates approach the slug-test K for the screened interval at Well 00-1, but incomplete recovery prevents a full comparison. Although the Hazen K values appear reasonable, use of additional relationships casts doubt on these values. Figure 2.6b compares results obtained using the Hazen and Kozeny-Carman equations. The average K value from the Kozeny-Carman equation is 110 m/d, which is within 6% of that obtained from the pumping test. Use of other relationships (not shown here) reveals a continuing lack of consistency between methods. For example, the empirical relationship of Bedinger (1961) produces an average K of 111 m/d, but in certain intervals, values differ significantly from those obtained using the Kozeny-Carman equation.

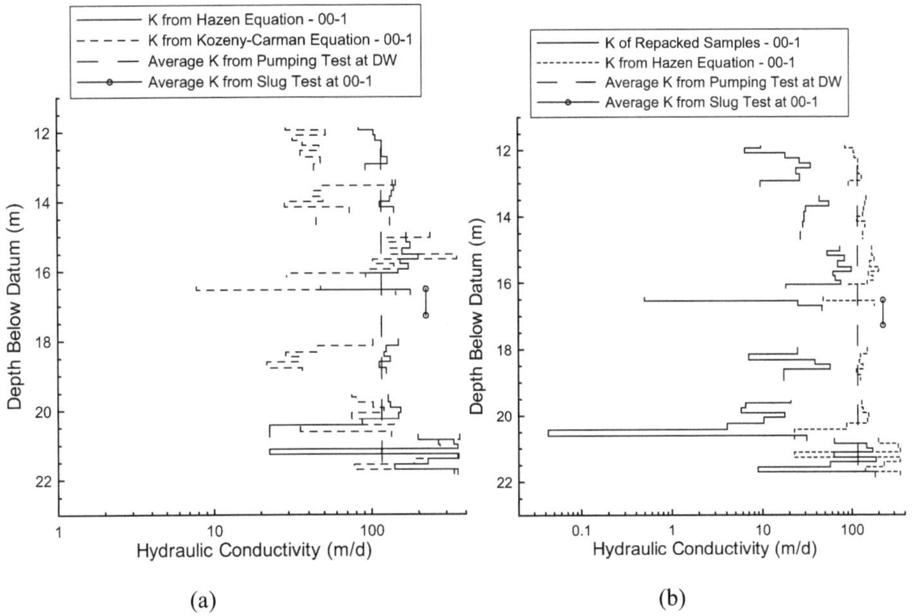

Figure 2.6. Plots of hydraulic conductivity versus depth (Butler and McElwee, 1996; previously unpublished KGS data): (a) Estimates from permeameter and particle-size analyses; (b) Estimates from particle-size analyses

2.2.4 GEOPHYSICAL LOGGING

Geophysical logging is a common means of acquiring information about relative variations in hydraulic conductivity (Keys, 1990; also, see Chapter 10 of this volume). In unconsolidated formations, variations in hydraulic conductivity are often produced by variations in clay content, which can be detected using a variety of logging tools. However, the relatively large averaging volume of wellbore logging tools used in unconsolidated formations limits their utility. Thus, although large variations in clay content between a clay and sand unit are readily detectable, characterization of smaller variations within a single unit is rarely possible. Recently, geophysical logging tools have been incorporated into direct-push equipment (Christy et al., 1994). The averaging volume of these direct-push logging tools can be significantly smaller than conventional wellbore logging tools, enabling valuable information to be obtained about small-scale variations in clay content.

The two most common types of geophysical logs used to characterize variations in clay content in unconsolidated formations are natural gamma and electrical conductivity. Natural-gamma logs provide a record of natural-gamma radiation versus depth. This radiation is quantified by counting the gamma particles passing through a scintillation crystal in a certain time interval. A high natural-gamma reading is generally associated with clay-rich intervals, while a low reading is generally associated with sands and gravels (Keys, 1990, 1997). The stochastic nature of the radiation process, coupled with the speed at which logs are normally run, introduces a great deal of noise into the natural-gamma data. Although various filtering approaches (e.g., Savitzky-Golay algorithm (Press et al., 1992)) can reduce this noise, they often do so at the cost of a loss in vertical resolution. The spatial averaging produced by filtering can largely be avoided by running the logs at very low speeds (<1 m/min). That approach, however, is rarely implemented in practice.

Electrical-conductivity or resistivity logs are commonly used to detect variations in clay content, fluid-filled porosity, and water chemistry. When variations in groundwater chemistry and porosity are small, changes in electrical conductivity are primarily a function of changes in the clay-sized fraction. Although conventional logging tools have relatively large averaging volumes, electrical-conductivity sensors incorporated in direct-push equipment can detect layers as small as 0.025 m in thickness (Schulmeister et al., 2003). Unlike natural-gamma logs, electrical-conductivity logs are not affected by logging speed.

A series of wellbore and direct-push geophysical logs have been performed at GEMS to assess variations in clay content in the unconsolidated sequence. Figure 2.7 (top) is a plot of natural-gamma radiation versus depth at Well 00-3. The high-frequency fluctuations observed on Figure 2.7 (top) are most likely noise, as discussed above, and not related to variations in clay content. Although the log in Figure 2.7 (top) was run at a standard logging speed, natural-gamma logs at GEMS have been run at speeds as low as 0.15 m/min. The high-frequency fluctuations are dramatically reduced at the very low speeds, indicating that the fluctuations are primarily artifacts of the radiation process and logging speed. Figure 2.7 (bottom) supports this interpretation, because the natural-gamma fluctuations at Well 00-3 are not strongly correlated with variations in the fine fraction at Well 00-1 (2.3 m from Well 00-3). In addition, the high-resolution electrical-conductivity log at G4SGPA (2.9 m from 00-3) shows little variation except near the top and bottom of the sand and gravel interval. Comparison with Figure 2.6 indicates that K variations in the sand and gravel are, for the most part, not a product of variations in the fine fraction. In such cases, natural-gamma and electrical-conductivity logs may be of little use for characterizing relative variations in K.

Hydraulic conductivity variations can also be the product of variations in porosity, which may be assessed using a variety of conventional geophysical logs. However, the interpretation of those logs in terms of K is difficult when the particle-size distribution and clay fraction vary as well (see Chapter 10 of this volume).

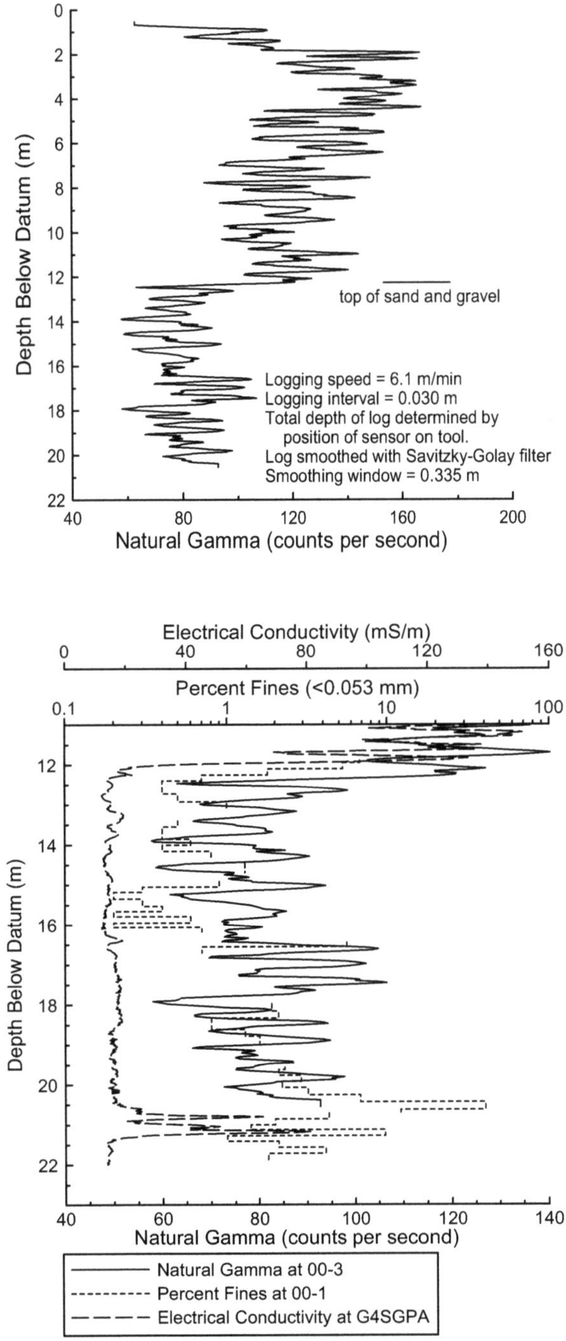

Figure 2.7. (top) Plot of natural gamma vs depth, Well 00-3; (bottom) Expanded view of the sand and gravel interval.

2.2.5 SUMMARY OF TRADITIONAL APPROACHES

The traditional approaches appear to have difficulty providing K estimates at the level of reliability and detail needed to characterize spatial variations in K for water-quality investigations. Pumping tests do produce reliable K estimates, but the estimates are large volumetric averages. Laboratory analyses can provide information at a very fine scale, but there are many questions about the reliability of the K estimates obtained with those analyses. Geophysical logs can provide valuable information about relative K variations produced by changes in clay content and, to a lesser extent, porosity, but are of limited value otherwise. Although the slug test has the most potential of the traditional approaches for detailed characterization of K variations, most sites do not have the extensive well network required for effective application of this approach. The use of slug tests for characterizing K variations is more fully explored in the following sections where modifications of the traditional approach are described.

2.3 The Current Generation

Casual scrutiny of sedimentary strata in outcrop reveals a pronounced anisotropy in the degree of continuity of most observable media characteristics. In general, strata tend to have significant continuity in the lateral direction (parallel to bedding), but very little in the vertical (perpendicular to bedding). Experimental studies in unconsolidated sedimentary sequences have found that hydraulic conductivity varies in a similar manner (e.g., Smith, 1981; Sudicky, 1986; Hess et al., 1992). Given this apparent large anisotropy in the continuity or correlation structure of hydraulic conductivity, much valuable information about K variations at a site can be gained through a limited number of profiles of hydraulic conductivity versus depth. In an effort to obtain such profiles, a variety of techniques have been developed to characterize the vertical variations in K along the screened interval of a well. These techniques, which have been increasingly applied in recent years, are reviewed in this section.

2.3.1 DIPOLE-FLOW TEST

The dipole-flow test (DFT), first described by Kabala (1993), is a single-well hydraulic test involving use of the three-packer tool shown in Figure 2.8 (left). A pump in the central pipe of the middle packer transfers water from the upper chamber to the lower, setting up a recirculatory pattern in the formation. Pressure transducers placed in the upper and lower chambers (transducers labeled UT and LT in Figure 2.8 (left), respectively) measure the head change in each chamber. A control transducer (CT in Figure 2.8 [left]) above the tool is used to detect short-circuiting through fittings in the upper packer or along a near-well disturbed zone. Zlotnik and Ledder (1996) and Zlotnik and Zurbuchen (1998) developed the theory, equipment, and field methodology for the steady-state form of the DFT. They found that the radial component of hydraulic conductivity (Kr) can be estimated from the total head change in the two chambers at steady state (Δh; Figure 2.8 [right]) using an approximate equation:

$$K_r = \frac{Q}{2\pi(\Delta h)\Delta} \ln\left[\frac{4a\Phi(\lambda)\Delta}{e r_w}\right] \tag{2.1}$$

where a = anisotropy ratio, $(K_r/K_z)^{0.5}$; K_z = vertical component of hydraulic conductivity; and $\Phi(\lambda)$ = dipole shape function, ranging from 0.5 to 1.0 depending on λ (L/Δ ratio). Zlotnik and Ledder (1996) show that Equation (2.1) is a reasonable approximation under conditions met in most field applications. Using this formula, the vertical variation in K_r can be estimated through a series of DFTs between which the tool is moved a short distance in the well. An estimate of the anisotropy ratio, which is rarely known, is required to calculate K_r. In most cases, however, an anisotropy ratio of one is assumed for lack of better information. Butler et al. (1998a) discuss a number of additional practical issues that must be considered for successful application of the DFT. Transient forms of the DFT have been proposed (Kabala, 1993), but have not proven successful in field applications (Hvilshøj et al., 2000). Such approaches have the most potential in units of moderate to low K, where the time to steady state is greater than the few to tens of seconds found in sand and gravel intervals.

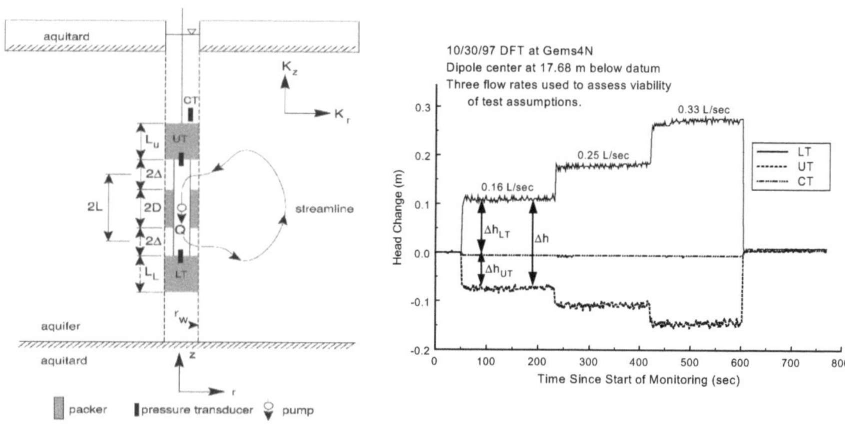

Figure 2.8. (left) Schematic of dipole-flow test; (right) Plot of head change versus time for a dipole-flow test at Gems4N (after Butler et al., 1998a)

A series of DFTs performed at Wells Gems4N and Gems4S can be used to demonstrate the advantages and limitations of this approach. Figure 2.9a is a plot of the hydraulic conductivity profiles obtained at these two wells using the DP_2 configuration of Zlotnik and Zurbuchen (1998). The significant difference between the Gems4N and Gems4S profiles at the top of the aquifer is consistent with differences in natural-gamma logs from the two wells (Butler et al., 1998a). The average K obtained from the two DFT profiles is considerably less than that obtained from the pumping tests (dashed vertical line). For example, the average K from the DFT at Gems4S is 73 m/d, which is 63% of that from the pumping tests. This difference could be a result of a number of factors, including insufficient testing near the bottom of the interval where the K appears highest (Figures 2.5 and 2.6), heterogeneity, and incomplete well development. In this case, the first two factors appear to be the most probable contributors to the difference. Insufficient testing, which is undoubtedly the most important contributor, occurs because of the length of the DFT tool (3.19 m) and the termination of the screened interval above the bottom of the sand and gravel. As will be shown later (Section 2.3.2), complete testing across this interval could have produced an average K_r estimate much closer to that obtained from the pumping test.

The steady-state DFT analysis is based on the assumption of a homogeneous formation. However, the profiles displayed in Figure 2.9a indicate that this assumption may not be appropriate at Gems4N and Gems4S. Thus, an analysis approach more appropriate for heterogeneous formations is needed. Zlotnik and Ledder (1996) have demonstrated that head gradients are largest near the two chambers, as would be expected in a convergent flow system. The hydraulic conductivity of the portions of the formation in the vicinity of a chamber should therefore have the greatest influence on the head changes in that chamber. Thus, a more appropriate analysis method for the DFT in heterogeneous formations would be to estimate K_r using the head changes in a single chamber. Zlotnik and Ledder (1996) have derived a single-chamber form of the DFT formula:

$$K_r = \frac{Q}{4\pi(\Delta h_{sc})\Delta} \ln\left[\frac{4a\,\Phi(\lambda)\Delta}{er_w}\right] \tag{2.2}$$

where Δh_{sc} is the head change in a single chamber (Δh_{LT} or Δh_{UT}, Figure 2.8b). Zlotnik et al. (2001) and Zurbuchen et al. (2002) have demonstrated the increased level of detail that can be obtained with the single-chamber formula. This increased level of detail was also observed at Gems4S (Figure 2.9a). Note that the upper chamber should be used for this analysis to avoid an underestimation in K_r resulting from the inadvertent injection of fine materials into the lower chamber (i.e., formation of a low-K skin during the DFT).

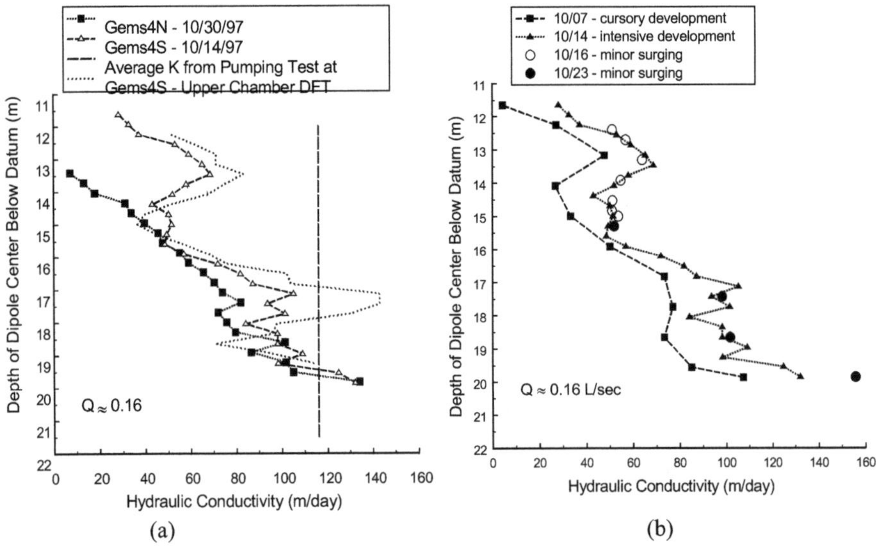

Figure 2.9. (a) Hydraulic conductivity profiles from DFTs at Gems4N and Gems4S; (b) Impact of well-development activities on DFT profiles at Gems4S (after Butler et al., 1998a)

Zlotnik and Zurbuchen (1998) emphasize several important advantages of the DFT for field applications. These include (1) no water is added or removed from the well during a test program, (2) the scale of the region influenced by the test can be readily defined and controlled, (3) K_r estimates can be obtained using a simple steady-state formula, and (4) relatively low flow rates can be used in high K media, so that the well losses associated with other methods are avoided. In addition, the dipole tool can be configured so that the influence of a high-conductivity zone created by the filter pack is minimized (Peursem et al., 1999). However, as with any single-well hydraulic test, the results are highly dependent on the effectiveness of well-development activities. During the drilling process, a considerable amount of fine debris will be concentrated in the near-well portions of the formation. One of the primary goals of well-development activities is to remove this drilling-generated material, so that K_r values representative of the formation can be obtained. Figure 2.9b displays the results of DFT surveys performed at Gems4S after varying degrees of well development. The cursory development consisted of pumping at a constant rate (1.3 L/s) until an approximately clear stream of water was obtained (20 min). The intensive development consisted of stressing discrete intervals via pumping and surging. The final two phases of development consisted of a small amount of surging followed by pumping to remove the debris brought into the well. At Gems4S, the intensive development produced an increase in the magnitude of the K_r estimates (profile average increased by 33%) but minimal change in the profile shape. The minor surging produced little change except at the bottom of the well, where fine material that had accumulated during previous development activity was removed. Note that the change in the average K_r was much greater at Gems4N (increase by a factor of 6.7) as a result of clay being smeared across the sand and gravel interval during drilling. Clearly, proper attention must be given to the development of wells at which the DFT, or any other single-well hydraulic test, is to be applied. The stability of K_r profiles before and after a period of development is the most convincing demonstration of the sufficiency of the development activities.

2.3.2 MULTILEVEL SLUG TEST

The multilevel slug test (MLST) is an extension of the traditional slug test specifically developed for characterizing vertical variations in hydraulic conductivity along the screened interval of a well (Melville et al., 1991; Butler et al., 1994). This approach involves the performance of slug tests in a portion of the screen isolated with a two-packer tool (Figure 2.10 [left]). A number of tests are performed in each isolated interval (Figure 2.10 [right]) to assess the viability of test assumptions (Butler et al., 1996; Butler, 1998). Test data are analyzed as discussed in Section 2.2.2. A K profile is obtained by repeating this process as the tool is moved in steps through the screened interval. The tool depicted in Figure 2.10 (left) was specifically developed for slug tests in highly permeable aquifers, so the pneumatic method is used for test initiation to minimize the noise introduced by noninstantaneous initiation. Other initiation methods (solid slug, etc.) can be used in less permeable intervals. Note that packer circumvention can affect test results in certain conditions. Butler et al. (1994) demonstrate those conditions and recommend measures for diminishing the potential for circumvention (use of additional or longer packers).

A series of MLSTs performed at Gems4N and Gems4S can be used to demonstrate the potential of this approach. Figure 2.11 is a plot of the K profiles obtained at the two wells using the tool depicted in Figure 2.10. The average K from the multilevel slug tests at Gems4N and Gems4S is 77% and 81%, respectively, of the average K from the pumping tests. This difference from the pumping-test K is likely a result of the incomplete testing of the bottom portion of the aquifer. As shown on Figure 2.11, the bottom 1.5 m of the aquifer is not tested at either well because both wells terminate above the bottom of the aquifer. If the K value of the deepest interval tested at each well is assumed to represent the untested interval for that well, the average K value from the MLST is 90% (Gems4N) and 93% (Gems4S) of the pumping-test K. Heterogeneity and anisotropy could easily account for the remaining difference. As discussed by Butler (1998), the anisotropy ratio cannot be estimated from a slug test in the absence of observation wells, so a value for it must be assumed in the analysis. Isotropy was assumed here for lack of better information.

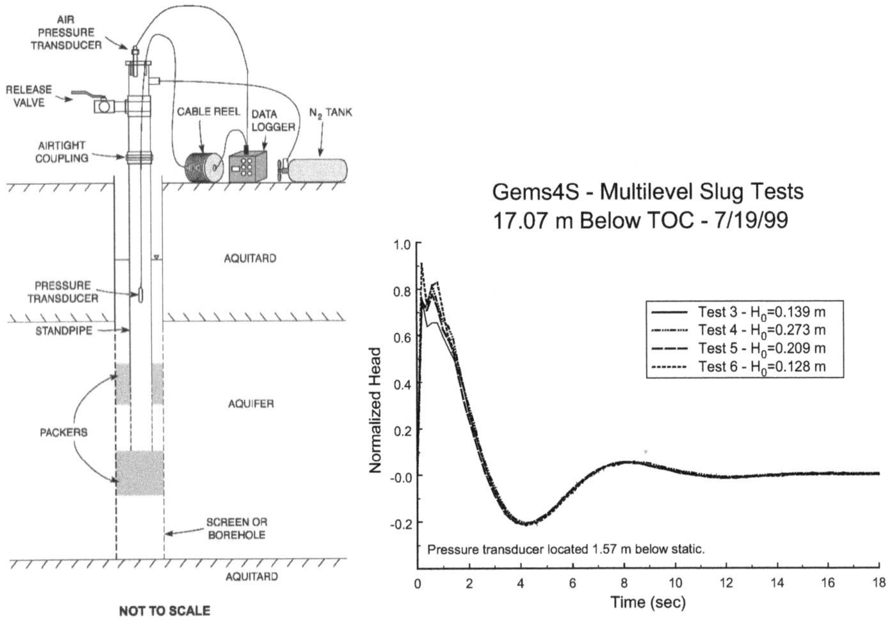

Figure 2.10. (left) Schematic of multilevel slug test configuration (after Butler, 1998); (right) normalized head versus time plot from one test interval

The upper-chamber DFT profile from Gems4S is also plotted on Figure 2.11 to demonstrate the similarity between profiles obtained with two different techniques. Note that the upper two DFT K values plotted on Figure 2.11 are questionable because of concerns about leakage through the top DFT packer during those tests. The excellent agreement between the MLST and upper-chamber DFT profiles is strong evidence of the viability of these two approaches (Zlotnik and Zurbuchen, 2003). In addition, the similarity of the average K from each profile to the average K from the pumping test, when the untested region is considered, is further evidence for the viability of these methods.

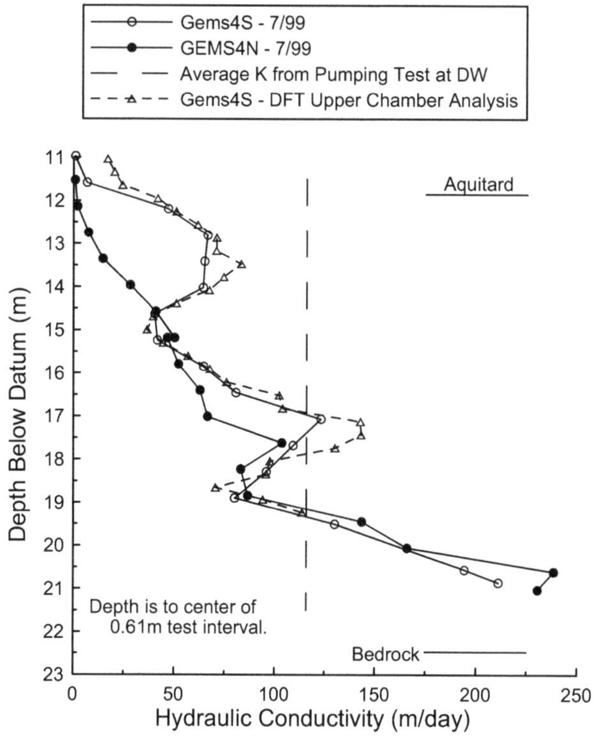

Figure 2.11. Hydraulic conductivity profiles for MLSTs at Gems4S and Gems4N and DFTs at Gems4S (Butler et al., 1998b; previously unpublished KGS data)

2.3.3 BOREHOLE-FLOWMETER TEST

The borehole-flowmeter test (BFT) is the most efficient, in terms of both field and analysis procedures, among the current generation of techniques for estimation of spatial variations in K. This approach involves pumping a well at a constant rate while measuring the vertical flow within the well using a downhole flowmeter (Figure 2.12). The flowmeter is initially positioned at the bottom of the screen and then systematically moved up the well while pumping continues. The flowmeter is kept at each depth until a stable reading is obtained (usually a few minutes). A test is completed when the flowmeter reaches the water table or the top of the screen. The record of cumulative vertical flow versus depth (Figure 2.12 and Figure 2.13a) is then processed to obtain a profile of hydraulic conductivity versus depth (Molz et al., 1989; Molz and Young, 1993). This processing has two primary steps. First, the flow between successive flowmeter positions (ΔQ) is calculated by subtracting the lower flowmeter reading from the upper one and taking any ambient flow into account. Second, hydraulic conductivity is calculated from the ΔQ record using several approaches. The simplest and most defensible of the analysis approaches is based on a study of pumping-induced flow in perfectly layered aquifers by Javandel and Witherspoon (1969). In that study, it was shown that, when the pumping well is fully screened across the aquifer, the ΔQ of an interval (ΔQ_i) will be proportional to the hydraulic conductivity (K_i) and thickness

(Δb_i) of that interval. The ratio of K_i over the average K for the aquifer ($\overline{K}$) can therefore be calculated as:

$$\frac{K_i}{\overline{K}} = \frac{\Delta Q_i / QP}{\Delta b_i / B} \tag{2.3}$$

where QP is the total pumping rate and B is the aquifer thickness. A record of K_i versus depth can then be obtained by multiplying this ratio by the average hydraulic conductivity of the aquifer near the well. Although the average K from a pumping test at the well is often used for $\overline{K}$, this procedure can introduce error into the K_i estimates, because the average K in the immediate vicinity of the well may differ from that determined from a pumping test in a laterally heterogeneous aquifer. Use of the K from a slug test at the same well should be a better approach. Young and Pearson (1995) describe different types of flowmeters and conclude that the electromagnetic flowmeter is the best for use in aquifers. Boman et al. (1997) describe field applications of the electromagnetic borehole flowmeter and discuss important practical issues. Chapter 10 of this volume discusses flowmeter use in fractured or multi-aquifer systems.

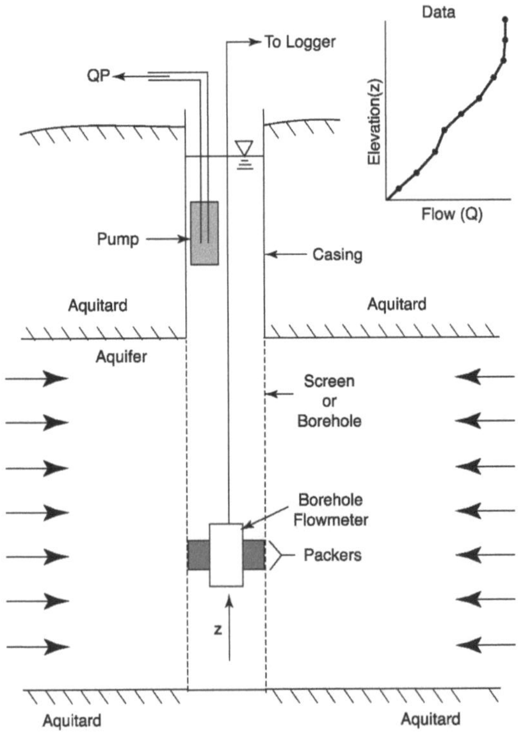

Figure 2.12. Schematic of borehole- flowmeter test (after Molz et al., 1989)

A series of BFTs performed at Wells Gems4N and Gems4S demonstrate both the potential and pitfalls of this approach. Figure 2.13a is a plot of cumulative discharge versus depth obtained at Gems4S using the 0.025 m ID electromagnetic flowmeter described in Young and Pearson (1995). Mechanical packers were attached to the device to direct flow through the meter as shown in Figure 2.12. The flow rate was measured every 0.305 m, and ΔQ was calculated from two successive readings as discussed earlier. Figure 2.13b, a plot of ΔQ versus depth at Gems4S, reveals some of the problems that can arise in highly permeable aquifers. Note the large increase in flow at the top of the screen as well as the intervals of negative ΔQ (those to left of dashed vertical line at zero). Both of these features are a result of flow bypassing the flowmeter (Figure 2.14a), a phenomenon that often occurs in highly permeable formations because the head loss through the flowmeter is comparable or greater than that produced by bypass flow in the formation (Dinwiddie et al., 1999). The increase in flow at the top of the screen is a result of the flowmeter moving into the casing where bypass flow no longer is possible. An attempt was made to minimize head losses through the flowmeter by using a small flow rate as recommended by Arnold and Molz (2000), but bypass flow is difficult to avoid in highly permeable units. An unweighted 1.52 m moving average was employed to remove the zones of negative flow on Figure 2.13b. Processing then proceeded using the smoothed data. The processed data were multiplied by the average value of the multilevel slug tests for the tested interval at Gems4S and Gems4N to produce a record of K versus depth. Figure 2.14b compares the BFT K profiles from Gems4S and Gems4N with the MLST K profile from Gems4S. Given the problems introduced by bypass flow, the agreement between the BFT and MLST profiles is quite good. Although these BFT experiments were performed by pumping water from the well, the test could also be performed in an injection mode. However, considerable care must be used to avoid injecting entrained air and sediments that can artificially lower the K in the immediate vicinity of the screen (Crisman et al., 2001).

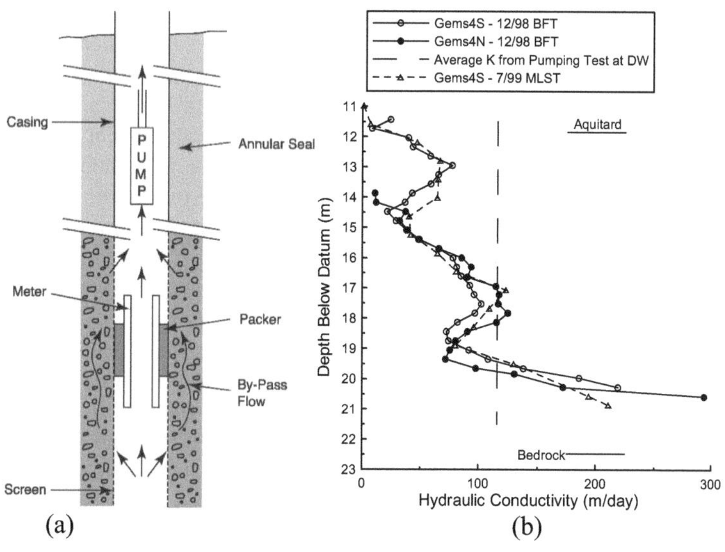

Figure 2.13. (a) Plot of cumulative flow versus depth at Well Gems4S; (b) Plot of ΔQ versus depth at Gems4S (after Butler et al., 1998b)

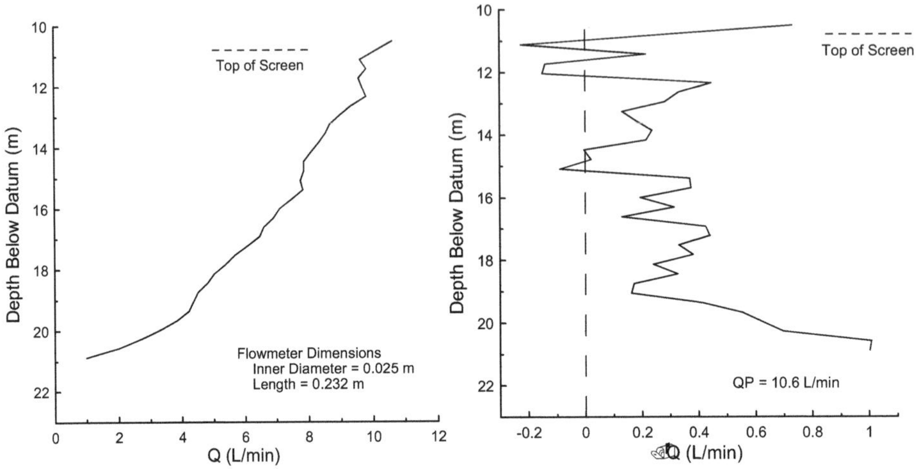

(a) (b)

Figure 2.14. (a) Schematic of bypass flow (after Boman et al., 1997); (b) K profiles from borehole-flowmeter tests at Gems4S and Gems4N and multilevel slug tests at Gems4S (Butler et al., 1998b; previously unpublished KGS data)

2.3.4 SUMMARY OF THE CURRENT GENERATION

When appropriate attention is given to the details of the field and analysis procedures, the current generation of methods for estimation of vertical variations in hydraulic conductivity are capable of providing reliable estimates of K variations along the screened interval of a well. Given the differences in the theoretical bases of the three techniques that were the primary focus of this section, the agreement between the K profiles obtained with the various methods (Figure 2.15) is very strong evidence of the viability of these approaches (Zlotnik and Zurbuchen, 2003). The similarity between the profile averages and the average K from the pumping tests further demonstrates the quality of the information that can be obtained with these approaches. The decision regarding which approach to use in a particular investigation may not be straightforward. The borehole-flowmeter test is the fastest in terms of both field and analysis procedures, but in-well hydraulics (bypass flow) can introduce considerable uncertainty in highly permeable aquifers, and the removal (pumping mode) or addition (injection mode) of water may conflict with regulatory restrictions at some sites. An important limitation of this approach is the requirement that the well be screened across the entire aquifer. The dipole-flow test is well suited for use in highly permeable aquifers, has a well-defined scale, and does not require the addition or removal of water. However, the complexity and size of the multi-packer tool can limit its utility. In addition, a relatively long-screened well is required to obtain information about K variations. The multilevel slug test is the most flexible of the three techniques in terms of the K range over which it can be applied, the size of the test interval, and the length of the well screen—but the analysis procedure is considerably more involved.

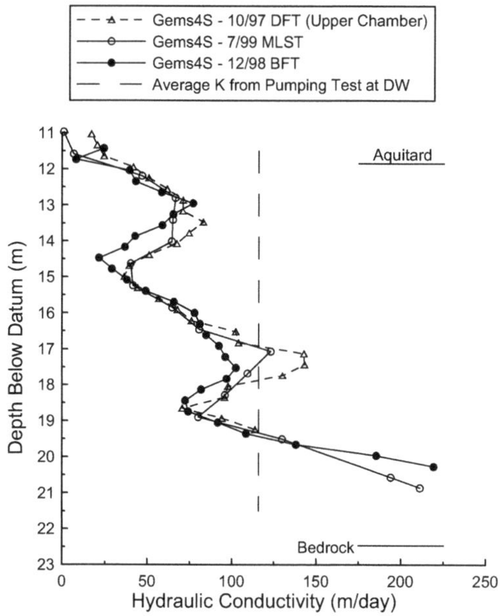

Figure 2.15. Hydraulic conductivity profiles obtained at Gems4S using the dipole-flow test, the multilevel slug test, and the borehole-flowmeter test.

2.4 The Next Generation

There are two major limitations to the current generation of methods for estimation of spatial variations in hydraulic conductivity. First, these techniques can only be used in wells, which often must be screened across a relatively large portion of the aquifer. Second, these techniques only provide information about conditions in the immediate vicinity of the well in which they are used. New techniques are being developed that provide information about K variations outside of the immediate vicinity of existing wells. These methods are described in this section.

2.4.1 DIRECT-PUSH METHODS

Most of the methods discussed so far in this chapter have involved procedures performed in wells. The restriction of these techniques to existing wells limits their utility for characterizing spatial variations in K. Over the last two decades, direct-push technology has become a widely used alternative to conventional well-based methods for site investigations in unconsolidated formations. A variety of methods have been developed that exploit the unprecedented access to unconsolidated formations provided by this technology, to obtain information about spatial variations in K without the need for permanent wells. For example, empirical relationships have been developed for estimating K from sediment classification information produced by cone penetrometer (CPT) surveys (Farrar, 1996). The resulting K values, however, are, at best, order-of-magnitude estimates for the hydraulic conductivity of the formation. Pore-pressure dissipation tests accompanying CPT surveys have been used to estimate K from a relationship between hydraulic conductivity and the consolidation properties of the formation (Baligh and Levadoux, 1980). Lunne et al. (1997), however, caution that this relationship results in relatively poor approximations of K. Pitkin and Rossi (2000) describe a method for estimation of relative variations in K by monitoring the rate and pressure of water injected during the advancement of an unshielded well point. Dietrich et al. (2003) have demonstrated that this approach can yield semi-quantitative K estimates by utilizing regressions with existing K data. A more promising direction for quantifying actual K variations is the performance of various types of hydraulic tests in direct-push equipment. Two classes of these methods will be discussed here.

2.4.1.1 *Direct-Push Slug Tests*

Slug tests can be performed in direct-push pipe using small-diameter adaptations of conventional methods (Butler et al., 2002). Systems have been developed that allow tests to be performed at one or multiple levels in a single probehole. The earliest approach was that of Hinsby et al. (1992), in which slug tests are performed in small-diameter pipe attached to an unshielded well point that is driven from the surface. This approach may be reasonable in sand and gravel sequences with little clay, but it is of questionable effectiveness in more heterogeneous settings where the buildup of fine material on the well screen can affect test responses. Butler and coworkers (Butler et al., 2002; Butler, 2002) describe a method for performing slug tests in direct-push equipment in which a screen is driven within protective steel casing to the target depth, and then exposed to the formation for the tests. If the screen is exposed in a series of steps, the vertical variations in K over the screened interval can be assessed (Figure 2.16a). The screen cannot be reshielded downhole, so the equipment has to be brought to the surface before another level can be tested. McCall et al. (2002) have developed a shielded-screen approach that enables multiple levels to be tested in a single probehole. Sellwood et al. (in review) increased the efficiency of that approach and added an electrical-conductivity probe for lithologic determination. In intervals of high hydraulic conductivity, frictional losses in the small-diameter direct-push pipe can introduce error into K estimates. Butler (2002), however, presents a procedure for accounting for those losses. Direct-push extensions of conventional pumping tests can also be utilized for estimation of vertical variations in K. Cho et al. (2000), for example, describe a constant-head pumping test method for direct-push equipment. Injection-based variants of their approach are the most promising, because there is no suction-depth restriction, and higher flows can be obtained. However, the time to steady state and other logistical issues may limit the utility of these methods in many systems. Regardless of which approach is used, the quality of K estimates is critically dependent on the efficacy of development procedures. Henebry and Robbins (2000) and Butler et al. (2002) describe development procedures designed for use in small-diameter direct-push installations.

A series of direct-push slug tests (DPSTs) performed at GEMS can be used to demonstrate the potential of this approach. Figure 16a presents a comparison between direct-push slug tests performed at DP43b and DP43c (0.84 m apart), and multilevel slug tests performed at Gems4N (DP43b and DP43c are 1.85 m and 2.16 m, respectively, from Gems4N). At both direct-push installations, three sets of slug tests were performed as the screen length was progressively increased from 0.15 m to 0.61

m. As explained by Butler (2002), the test data can be analyzed to obtain a detailed view of the K variations over the 0.61 m interval tested at the two locations. The agreement between the DPST and the MLST estimates over the same interval is quite good (within 6% at the bottom of the interval and within 12% at the top), demonstrating that direct-push slug tests can provide reliable K estimates.

Figure 2.16a shows that the DPST can be used to acquire detailed information about K variations over a single interval. However, as discussed above, the approach can also be implemented through a profiling procedure in which multiple levels are tested in a single probehole (McCall et al., 2002; Sellwood et al., in review). Figure 2.16b presents three series of direct-push slug tests performed in profiling mode between Gems4S and Gems4S (DPST test interval of 0.31 m in all cases). The MLST profile from Gems4S is also presented for comparison purposes. As shown in the figure, the DPST and MLST profiles are not in good agreement in the upper portion of the aquifer. The trough and peak in the MLST record is shallower than that in the DPST records at points A and B. This difference cannot be explained by the dissimilar length of the test intervals (0.31 m DPST versus 0.61 m MLST). Heterogeneity is undoubtedly the cause of this difference, as it is for the difference between the MLST K profiles at Gems4N and Gems4S (Figure 2.11). The DPST and MLST profiles are in much better agreement in the lower portion of the aquifer, as the trough and peak in the vicinity of C and D, respectively, are at similar depths. In this case, the difference between profiles may primarily be a function of the length of the test interval. Note that the average K from each of the DPST profiles is within 4% of that obtained from the pumping test at DW.

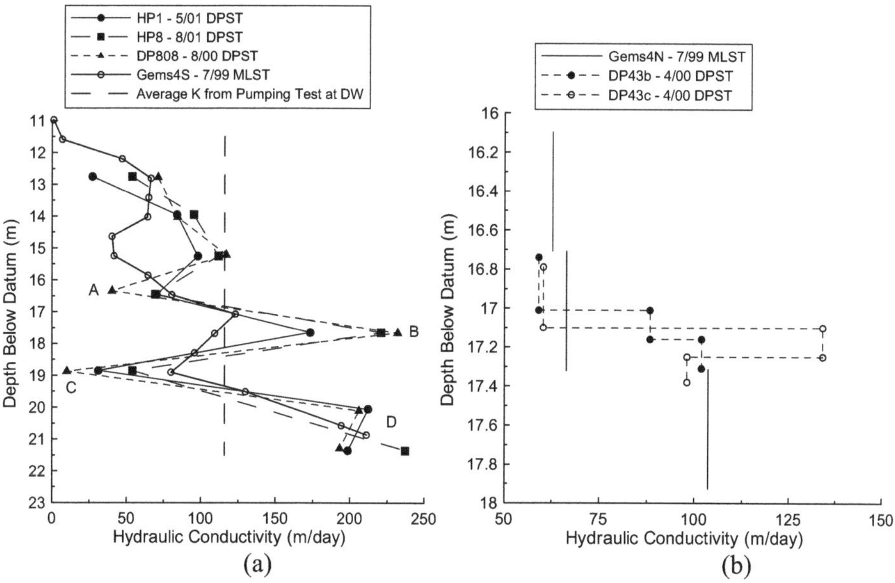

Figure 2.16. (left) Comparison of MLST K estimates from Gems4N with DPST K estimates from nearby direct-push installations (after Butler, 2002); (right) K profiles from DPST at three locations and from MLST at Gems4S (McCall et al., 2002; Sellwood et al., in review)

2.4.1.2 *Direct-Push Permeameter*

A major limitation of the direct-push slug test, as with the methods discussed in Section 2.3, is that the quality of the K estimate is highly dependent on the effectiveness of well-development activities. Unfortunately, the uncertainty regarding the effectiveness of development activities can never be completely eliminated. To address this limitation, a method has been developed that is relatively insensitive to the zone of disturbance/compaction created by the advancement of a direct-push tool (Stienstra and van Deen, 1994; Lowry et al., 1999). This method, which is referred to here as the direct-push permeameter (DPP), involves an unshielded-screen tool with pressure transducers inset into the pipe above or below the screen (Figure 2.17a). The tool is pushed into the ground while water is injected at a low rate to keep the screen clear. Upon reaching a depth at which a K estimate is desired, pushing ceases and a pumping test is performed by injecting water through the screen while monitoring pressure changes at the transducer locations. K estimates are obtained using the spherical form of Darcy's law at steady-state (constant hydraulic gradient) conditions:

$$K = \left(Q(\frac{1}{r_2} - \frac{1}{r_1}) \right) / (4\pi(p_2 - p_1)) \tag{2.4}$$

where r_i and p_i are the distance to and pressure at transducer i. Although Equation (2.4) assumes an isotropic aquifer, an anisotropic form of this equation can readily be developed for use when the anisotropy ratio is known from other information. Lowry et al. (1999) have shown that a compacted zone of lower permeability along the direct-push pipe will not influence steady-state pressure gradients and thus the K estimate. An important advantage of this approach is that the scale of the test is readily defined and controlled, because the K estimate represents the average over the interval between the transducers.

A field assessment of the direct-push permeameter at GEMS demonstrates the potential of this approach, as well as the need for further refinement (Butler et al., 2004). Figure 2.17b presents DPP K profiles obtained at CP1029a and CP1029b, and a DPST K profile obtained at DP808 (DP808 located 2.24 m and 1.72 m from CP1029a and CP1029b, respectively). As shown in the figure, the profiles are in good agreement in the lower permeability zones at the top of the aquifer and at about 19 m below datum. The agreement is quite poor in the higher K intervals. The poor agreement in the higher-K intervals is most likely a result of the lower signal-to-noise ratio of the pressure measurements in those intervals and of preferential flow along the direct-push pipe. A recent follow-up field assessment in Germany found that the DPP can provide reasonable estimates for K between 0.01 and 100 m/day (Dietrich et al., 2003), in keeping with the range reported by Stienstra and van Deen (1994). Modification of the equipment to allow use at higher K ranges is proceeding. These initial field assessments indicate that this approach should soon be capable of producing reliable estimates over the range of K values expected in most aquifers. The efficiency of the approach is particularly noteworthy. A K profile can be obtained with the direct-push permeameter much faster than with slug-test profiling methods (the DPP profiles of Figure 2.17b were each obtained in two hours, while the DPST profile was obtained in two days). The direct-push permeameter appears to be as efficient as a flowmeter survey, but without many of its limitations. Clearly, this approach has potential for characterizing spatial variations in hydraulic conductivity at a level of detail and efficiency that has not previously been possible. Note that Sørensen et al. (2002) have recently proposed a related approach using hollow-stem augers that may allow testing at greater depths than possible with direct-push technology.

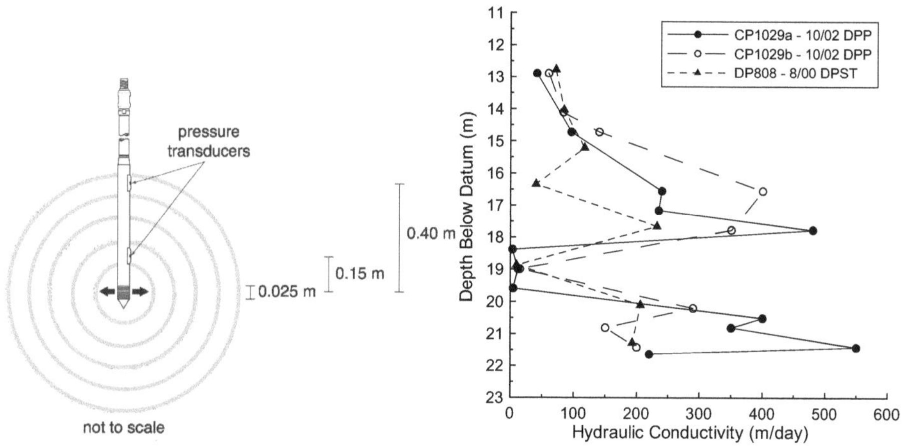

(a) (b)

Figure 2.17. (a) Schematic of direct-push permeameter; (b) K profiles from DPP at CP1029a and CP1029b and DPST at DP808 (McCall et al., 2002; Butler et al., 2004)

2.4.2 HYDRAULIC TOMOGRAPHY

Except for the pumping tests described in Section 2.2.1, all of the methods that have been discussed here provide information about conditions in the immediate vicinity of a well or probehole. Solute transport, however, depends critically on the connectivity of regions of low or high hydraulic conductivity, which may be difficult to determine from a suite of essentially point values. Although multiwell tracer tests can provide information about hydraulic conductivity variations between wells (Freyberg, 1986; Hess et al., 1992), such tests are too expensive in terms of time, money, and effort to be used on a routine basis. Thus, new, more efficient approaches are needed to provide information about spatial variations in K between wells.

Over the last decade and a half, several research groups have begun work on a new approach, hydraulic tomography, that has the potential to yield information on the spatial variations in K between wells at a level of detail that has not previously been possible (Neuman, 1987; Tosaka et al., 1993; Gottlieb and Dietrich, 1995; Butler et al., 1999a; Yeh and Liu, 2000; Liu et al., 2002; Bohling et al., 2002). This method consists of performing a series of short-term pumping tests in which the position of the stressed interval in the pumping well is varied between tests (Figure 2.18a). The sequence of tests produces a pattern of crossing streamlines in the region between the pumping and observation wells, similar to the pattern of crossed ray paths used in seismic or radar tomography (see Chapters 7 and 9 of this volume). Although a large number of drawdown measurements is required to delineate the numerous streamlines produced by the test sequence, new methods for drawdown measurement have been developed that can provide the requisite density of data in a practically feasible manner (Butler et al., 1999a; Davis et al., 2002). Bohling et al. (2002) describe a variety of methods for the analysis of the drawdown data collected during a test sequence. All of these methods involve the numerical inversion of test data, assuming a model of the aquifer structure based on existing site information. Hydraulic tomography extracts information about variations in K from vertical differences in drawdown. These differences, however, diminish as the distance between the pumping and observation wells increases. At distances greater than $2B(K_r/K_z)^{1/2}$, where B is aquifer thickness and K_r and K_z are the average radial and vertical component of K, respectively (Kruseman and de Ridder, 1990), there will rarely be a measurable difference in drawdown in the vertical direction (Bohling et al., 2002).

A series of hydraulic tomography experiments performed at GEMS illustrate the type of information that can be obtained with this approach. In these experiments, Gems4S served as the pumping well, while drawdown measurements were made at multilevel sampling wells HTMLS1 and HTMLS2, which are 2.74 m and 5.74 m, respectively, from Gems4S (Davis et al., 2002). Pumping tests (Q=1.3 L/s) were performed at 15 different intervals approximately equally spaced across the screened interval at Gems4S, each test lasting 900 seconds. Drawdown was measured at three depths in each multilevel well. The depths at which drawdown was measured were offset between the wells to allow more complete coverage across the aquifer. Drawdown records were similar in form to those depicted in Figure 2.3. The data from HTMLS1 and HTMLS2 were numerically inverted to estimate conditions between those two wells, using a seven-layer model based on a crosshole radar survey between wells Gems4N and Gems4S. As shown in Figure 2.18b, the comparison between the numerical inversion of the tomography data and the DPST profile at HP1 is quite reasonable. Differences between the HP1 profile and the inversion results are most likely caused by differences between the actual aquifer structure and the model based on the radar data, and the termination of the screened interval in well Gems4S above the bottom of the aquifer. Note that the average K from the inversion of the hydraulic tomography data is 78% of that obtained from the pumping test at DW, most likely again a result of the undersampling of the permeable lower portion of the aquifer.

Variations of the hydraulic tomography procedure described above are currently under development. For example, ongoing work is exploring the use of additional geophysical methods for better definition of the model of the aquifer structure used for inversion of test data. In addition, direct-push methods are being explored as a flexible means of installing temporary observation wells. (Figure 2.3b is an example of an initial test of that concept.) The combination of geophysical methods for structure definition and direct-push methods for data acquisition has great potential.

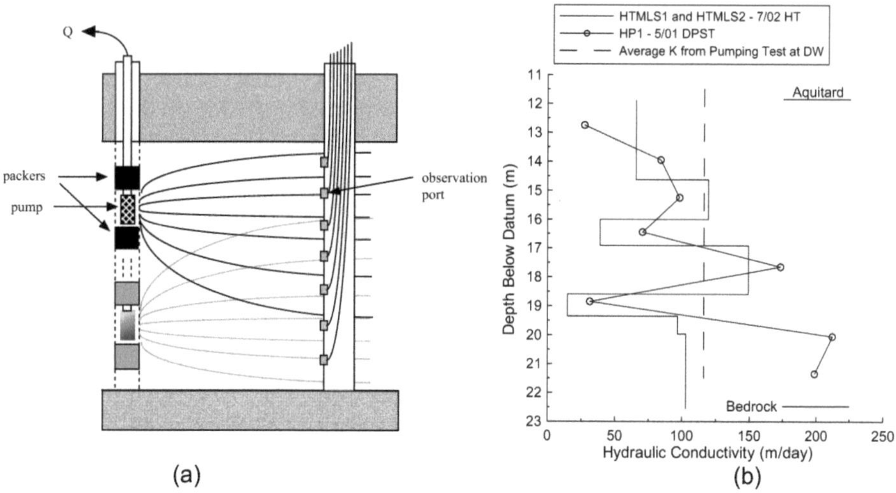

Figure 2.18. (a) Schematic of hydraulic tomography procedure (after Butler et al., 1999a); (b) K profile from hydraulic tomography experiments using HTMLS1 and HTMLS2 and K profile from DPST at HP1 (after Bohling et al., 2003; Sellwood et al., in review)

2.5 Additional Methods

The major methods for obtaining information about spatial variations in hydraulic conductivity in shallow unconsolidated aquifers have been described in this chapter. These, however, are not the only methods that can be used to estimate spatial variations in K. Several additional methods have been employed to estimate spatial variations in K in shallow unconsolidated aquifers. Moreover, a number of techniques have been specifically developed for use in other hydrogeologic settings. In this section, these additional methods are briefly described.

2.5.1 METHODS FOR SHALLOW UNCONSOLIDATED AQUIFERS

The most commonly used method for investigations in shallow unconsolidated aquifers not covered in the preceding sections is the tracer test. Chemical and physical tracers have long been used to obtain information about the permeable nature of subsurface formations. Single-well borehole-dilution and injection-withdrawal tests (Freeze and Cherry, 1979; Leap and Kaplan, 1988; see Chapter 10 of this volume) performed with and without packers have been utilized to determine the relative variation in groundwater velocity along the screened interval of a well. These variations can be related to K variations under certain conditions (e.g., constant hydraulic gradient along the screen). Various other single-well tracer tests have been proposed in the last two decades (Taylor et al., 1990; Sutton et al., 2000), but have seen relatively limited use because of their involved procedures. Multiwell tracer tests performed under natural or induced hydraulic gradients can provide information about K variations between wells (Freyberg, 1986; Hess et al., 1992; Gelhar et al., 1992). Although significant advances in the analysis of multiwell tracer tests have been achieved (e.g., Datta-Gupta et al., 2002), logistical and regulatory constraints still limit their use.

An induced-gradient tracer test, GMSTRAC1, performed at GEMS in the fall of 1994, illustrates the effort required for, the logistical problems associated with, and the potential information that can be obtained from multiwell tracer tests. A bromide tracer was introduced into a steady flow field created by pumping well DW at a constant rate (4.4 L/s). The tracer moved 14.2 m laterally from the point of injection to Well DW through a network of 24 multilevel sampling wells (17 sampling ports/well, ports distributed evenly across the sand and gravel aquifer) located to the immediate northeast of the wells shown in the inset of Figure 2.1. The test lasted 32 days, and over 6,000 samples of bromide were collected and analyzed during that period. An evaluation of the tracer data produced a profile of fluxes that was then transformed into a K profile, using the average hydraulic conductivity from the pumping tests and the assumption of horizontal flow (Bohling, 1999). Figure 2.19 is a comparison of the tracer-test K profile with those obtained from direct-push slug tests performed at locations adjacent to the sampling well network. Although the patterns of the tracer-test and slug-test profiles are in reasonable agreement in the lower half of the aquifer, the K values are not. Bohling (1999) details the problems that likely influenced the K profile obtained from the tracer test. These included the inability to introduce the tracer uniformly over the aquifer (and thus define the tracer distribution immediately after introduction), more rapid-than-expected tracer movement in the lower portion of the aquifer, and the influence of nearby pumping wells. Although the GMSTRAC1 test was impacted by a number of design and logistical problems, the resulting K profile does illustrate the potential of multiwell tracer tests for assessing spatial variations in K. Thus, despite the expense and effort required to perform multiwell tracer tests, they often can provide valuable information about interwell variations in K. However, in many cases, the approaches described in the preceding sections can provide similar information in a significantly more efficient (time, cost, and effort) fashion. To get the most information from tracer tests, attention must be given to all phases of test design and performance, and, whenever possible, tests should be performed in conjunction with geophysical surveys (see Chapter 13 of this volume).

Nuclear magnetic resonance (NMR) methods also have potential for providing information about spatial variations in hydraulic conductivity in shallow unconsolidated settings. NMR logging is commonly used in the petroleum industry to obtain information about vertical variations in K. This approach involves measuring the response of hydrogen protons to a series of imposed magnetic fields using a downhole logging tool (Coates et al., 2001). The response is a function of, among other things, the size and distribution of pores in the immediate vicinity of the borehole. Hydraulic conductivity is estimated from the pore-size distribution information with techniques similar to those described in Section 2.2.3 (White, 2000). Calibration using permeameter analyses of cores from the logged borehole is done to reduce the uncertainty associated with the K estimates. NMR logging could potentially be a useful means for rapid acquisition of information regarding relative variations in K, but logistics (e.g., size of the logging tool) and costs have greatly limited its use for shallow hydrogeologic investigations. In addition, there are concerns regarding the interpretation of NMR data in the near-surface environment owing to the sensitivity of the data to redox conditions (Bryar and Knight, 2002). Yaramanci et al. (2002) have recently demonstrated the potential of NMR surface surveys for obtaining vertical profiles of K. Further discussion of NMR methods is provided in Chapter 16 of this volume.

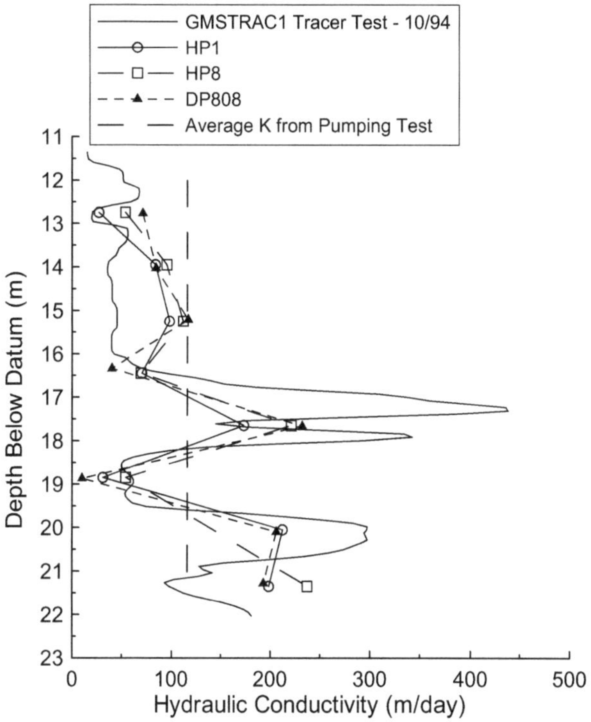

Figure 2.19. Hydraulic conductivity profile for GMSTRAC1 tracer test and DPST K profiles from nearby locations (Bohling, 1999; McCall et al., 2002; Sellwood et al., in review).

2.5.2 METHODS FOR CHARACTERIZING OTHER HYDROGEOLOGIC SETTINGS

Fractures in media of relatively low permeability often play an important role in groundwater flow and solute transport. Thus, efforts to acquire information about spatial variations in K in fractured settings often focus on determining which fractures serve as conduits for flow. Constant-head injection tests (packer tests) performed with a straddle-packer system similar to that depicted in Figure 2.18a are commonly used for this purpose (Doe et al., 1989). Constant-head tests, however, are susceptible to error introduced by low-permeability well skins and the assumption of a steady flow rate (Bliss and Rushton, 1984; Braester and Thunvik, 1984). In addition, such single-well tests cannot provide information about how the identified flow conduits are interconnected in space. Multiwell slug or pulse tests (Novakowski, 1989), and conventional pumping and tracer tests (Karasaki et al., 2000) are commonly used to assess fracture interconnectivity. Traditional interpretations of hydraulic test data, however, may not be viable because of the geometric complexity of the fracture network (Barker, 1988; Doughty and Karasaki, 2002). Multidisciplinary approaches, such as those described by Shapiro et al. (1999) and Karasaki et al. (2000), are often required to improve understanding of K variations in fractured settings. Chapter 10 of this volume describes additional approaches for identifying flow conduits in fractured media.

Many of the methods discussed in this chapter have been developed for use in aquifers. Often, however, there is a need to get information about the transmissive nature of units of relatively low K. Neuzil (1986) describes the many challenges faced by those working to obtain estimates of hydraulic conductivity in low-permeability formations. Van der Kamp (2001) summarizes the major methods for obtaining K estimates in shallow units of relatively low permeability (aquitards) and identifies the slug test as the primary tool for K estimation in that setting. Butler (1998) discusses the variants of the slug test that have been developed for low-K media. Constant-head tests are also commonly used in low-K settings (Tavenas et al., 1990). Van der Kamp (2001), however, questions the utility of constant-head tests, because they are considerably more difficult to perform than slug tests and provide little, if any, additional information.

Estimates of hydraulic conductivity are also needed for the vadose zone, where water is at pressures less than atmospheric and the pore space is usually not completely filled (saturated) with water. Stephens (1996) discusses the additional complexities confronting investigators in unsaturated media and the primary methods used for obtaining information about K in that setting. Reynolds et al. (2002) summarize the major advantages and disadvantages of these methods, while many authors have discussed the factors that can introduce error into the resulting K estimates (Flühler et al., 1976; Campbell and Fritton, 1994). Holt et al. (2002) discuss how such errors in a commonly used technique, the tension infiltrometer, can lead to a bias in the description of the spatial variations of K. The recently proposed tension permeametry approach (Shani and Or, 1995; Or et al., 2000) appears to have much potential for characterizing spatial variations in K. Further work, however, is needed to fully assess that potential.

Given the difficulties associated with efforts to estimate hydraulic conductivity in unsaturated media, many investigators have preferred to use pneumatic-based methods in the vadose zone. In this case, estimates of the air permeability of the media can be obtained using steady-state or transient extensions of conventional hydraulic tests (Baehr and Hult, 1991; Illman and Neuman, 2000, 2001). Although the degree of water saturation is commonly ignored, it should be considered in the calculation of intrinsic permeability from air-permeability estimates when the degree of saturation is above the specific retention (Weeks, 1978). The power of pneumatic approaches has recently been demonstrated by Vesselinov et al. (2001a,b), who used a suite of injection tests to perform pneumatic tomography in an unsaturated fractured tuff. Pneumatic minipermeameters have also been widely used to study permeability variations in outcrops and rock samples (Hurst and Goggin, 1995; Tidwell and Wilson, 1997; Dinwiddie et al., 2003).

2.6 Summary and Conclusions

Varieties of hydrogeologic methods are available for the estimation of spatial variations in hydraulic conductivity in shallow unconsolidated aquifers. A number of those methods were reviewed in this chapter. The potential and limitations of each method were evaluated in a highly controlled field setting. This evaluation demonstrated that the current generation of methods can provide reliable estimates of K variations along the screened interval of a well. The choice of the most appropriate technique for an investigation depends on the particulars (e.g., existing wells, the scale at which estimates are desired, available time) of that investigation. For example, the borehole-flowmeter test is the most efficient of existing techniques, while the dipole-flow test has the best-defined scale of investigation. None of the current generation of methods, however, can provide information away from the vicinity of existing wells, a severe limitation at sites with a sparse well network.

Methods are currently being developed that provide information about K variations away from the vicinity of existing wells. Direct-push methods have the potential to provide information about lateral and vertical variations in K at a level of detail that has not previously been possible. The direct-push permeameter is noteworthy because of its efficiency and its low sensitivity to conditions in the disturbed zone created by the advancement of the direct-push pipe. However, these direct-push methods only provide information about hydraulic conductivity in the immediate vicinity of the probehole. An approach for obtaining information away from existing wells or probeholes is currently under development. This approach, hydraulic tomography, can provide a detailed view of conditions between wells or probeholes in many situations. The integration of direct-push methods, crosshole geophysics, and hydraulic tomography is a particularly promising direction for future research.

Existing or in-development methods can only provide information about K variations in the immediate vicinity of a well/probehole or between relatively closely spaced wells. Additional methods are needed to provide information about the connectivity of regions of high or low hydraulic conductivity under more general conditions. Surface and crosshole geophysical methods have considerable potential in this respect. That potential is explored thoroughly in later chapters of this volume.

Acknowledgments

This work was supported in part by grants from the U.S. National Science Foundation, the U.S. Geological Survey, the U.S. Air Force Office of Scientific Research, and the Kansas Water Resources Research Institute. Any opinions, findings, and conclusions or recommendations expressed in this material are those of the author and do not necessarily reflect the views of the funding agencies. Geoprobe Systems provided significant support for the work described in Section 2.4.1. The author gratefully acknowledges Vitaly Zlotnik for providing the equipment used to perform the tests described in Section 2.3, and Gerard Kluitenberg for advice on vadose zone methods. This chapter benefited from reviews provided by Vitaly Zlotnik, Zbigniew Kabala, Geoff Bohling, Xiaoyong Zhan, and the editors of this volume.

References

Arnold, K.B., and F.J. Molz, In-well hydraulics of the electromagnetic borehole flowmeter: Further studies, *Ground Water Monit. and Remed., 20*(1), 52–55, 2000.

Baehr, A.L., and M.F. Hult, Evaluation of unsaturated zone air permeability through pneumatic tests, *Water Resour. Res., 27*(10), 2605–2617, 1991.

Baligh, M.M., and J.N. Levadoux, Pore pressure dissipation after cone penetration, Report R80-11, Dept. of Civil Engineering, Massachusetts Institute of Technology, Cambridge, Massachusetts, 1980.

Barker, J.A., A generalized radial flow model for hydraulic tests in fractured rock, *Water Resour. Res., 24*(10), 1796–1804, 1988.

Batu, V., *Aquifer Hydraulics*, Wiley, 1998,

Bear, J., *Dynamics of Fluids in Porous Media*, Dover, 1972.

Beckie, R., and C.F. Harvey, What does a slug test measure: An investigation of instrument response and the effects of heterogeneity, *Water Resour. Res., 38*(12), 1290, doi:10.1029/2001WR001072, 2002.

Bedinger, M.S., Relation between median grain size and permeability in the Arkansas River Valley, Arkansas, *USGS Prof. Paper 292*, Art. 147, p. C-31, 1961.

Bliss, J.C., and K.R. Rushton, The reliability of packer tests for estimating the hydraulic conductivity of aquifers, *Q. J. Eng. Geol., 17*, 81–91, 1984.

Bohling, G.C., Evaluation of an Induced Gradient Tracer Test, Ph.D. dissertation, Univ. of Kansas, Lawrence, Kansas, 1999,

Bohling, G.C., X. Zhan, J.J. Butler, Jr., and L. Zheng, Steady-shape analysis of tomographic pumping tests for characterization of aquifer heterogeneities, *Water Resour. Res., 38*(12), 1324, doi:10.1029/2001 WR001176, 2002.

Bohling, G.C., Zhan, X., Knoll, M.D., and J.J. Butler, Jr., Hydraulic tomography and the impact of a priori information: An alluvial aquifer field example (abstract), *Eos, 84*(46), p. F632 (also *Ks. Geol. Survey Open-File Rept. 2003-71*), 2003.

Boman, G.K., F.J. Molz, and K.D. Boone, Borehole flowmeter application in fluvial sediments: Methodology, results, and assessment, *Ground Water, 35*(3), 443–450, 1997.

Bouwer, H., and R.C. Rice, A slug test for determining hydraulic conductivity of unconfined aquifers with completely or partially penetrating wells, *Water Resour. Res., 12*(3), 423–428, 1976.

Braester, C., and R. Thunvik, Determination of formation permeability by double-packer tests, *Jour. Hydrology, 72*, 375–389, 1984.

Bryar, T.R. and R.J. Knight, Sensitivity of nuclear magnetic resonance relaxation measurements to changing soil redox conditions, *Geophys. Res. Let., 29*(24), 2197–2200, 2002.

Butler, J.J., Jr., The role of pumping tests in site characterization: Some theoretical considerations, *Ground Water, 28*(3), 394–402, 1990.

Butler, J.J., Jr., *The Design, Performance, and Analysis of Slug Tests*, Lewis Pub., 1998.

Butler, J.J., Jr., A simple correction for slug tests in small-diameter wells, *Ground Water, 40*(3), 303–307, 2002.

Butler, J.J., Jr., and J.M. Healey, Relationship between pumping-test and slug-test parameters: Scale effect or artifact? *Ground Water, 36*(2), 305–313, 1998.

Butler, J.J., Jr., and W.Z. Liu, Pumping tests in nonuniform aquifers: The radially asymmetric case, *Water Resour. Res., 29*(2), 259–269, 1993.

Butler, J.J., Jr., and C.D. McElwee, Well-testing methodologies for characterizing heterogeneities in alluvial-aquifer systems: Final technical report, *Project Report to USGS Water Resources Research Program*, U.S. Dept. of Interior, (also *Ks. Geol. Survey Open-File Rept. 95-75*), 1996.

Butler, J.J., Jr., E.J. Garnett, and J.M. Healey, Analysis of slug tests in formations of high hydraulic conductivity, *Ground Water, 41*(5), 620–630, 2003.

Butler, J. J., Jr., C.D. McElwee, and G. C. Bohling, Pumping tests in networks of multilevel sampling wells: Motivation and methodology, *Water Resour. Res., 35*(11), 3553–3560, 1999a.

Butler, J.J., Jr., C.D. McElwee, and W. Liu, Improving the quality of parameter estimates obtained from slug tests, *Ground Water, 34*(2), 480–490, 1996.
Butler, J.J., Jr., G.C. Bohling, Z. Hyder, and C.D. McElwee, The use of slug tests to describe vertical variations in hydraulic conductivity, *Jour. Hydrology, 156*, 137–162, 1994.
Butler, J.J., Jr., P. Dietrich, T.M. Christy, and V. Wittig, The direct-push permeameter for characterization of spatial variations in hydraulic conductivity: Description and field assessment (abstract), In: *Proc. of the 2004 North American Environmental Field Conf. and Exposition*, 2004.
Butler, J.J., Jr., J.M. Healey, V.A. Zlotnik, and B.R. Zurbuchen, The dipole flow test for site characterization: Some practical considerations (abstract), *Eos, 79*(17), p. S153 [Also: Ks. Geol. Survey Open-File Rept. 98-20, 1998a], www.kgs.ukans.edu/Hydro/publication /OFR98_20/index.html),
Butler, J.J., Jr., V.A. Zlotnik, B.R. Zurbuchen, and J.M. Healey, Single-borehole hydraulic tests for characterization of vertical variations in hydraulic conductivity: A field and theoretical assessment (abstract), In: *Proc. of Technical Program for the NGWA 50th National Convention*, pp. 94-95, 1998b.
Butler, J.J., Jr., J.M. Healey, G.W. McCall, E.J. Garnett, and S.P. Loheide, II, Hydraulic tests with direct-push equipment, *Ground Water, 40*(1), 25–36, 2002.
Butler, J.J., Jr., J.M. Healey, L. Zheng, W. McCall, and M.K. Schulmeister, Hydrostratigraphic characterization of unconsolidated alluvium with direct-push sensor technology, *Kansas Geological Survey Open-File Rept. 99-40* (www.kgs.ukans.edu/Hydro/Publications/OFR99_40/ index.html), 1999b.
Campbell, C.M., and D.D. Fritton, Factors affecting field-saturated hydraulic conductivity measured using the borehole permeameter technique, *Soil Sci. Soc. Am. J., 58*, 1354–1357, 1994.
Cho, J.S., J.T. Wilson, and F.P. Beck, Jr., Measuring vertical profiles of hydraulic conductivity with in-situ direct push methods, *Jour. Environ. Eng., 126*(8), 775–777, 2000.
Christy, C.D., T.M. Christy, and V. Wittig, A percussion probing tool for the direct sensing of soil conductivity, In: *Proc. of the 8th National Outdoor Action Conf.*, NGWA, 381–394, 1994.
Coates, G.R., L. Xiao, and M.G. Prammer, *NMR Logging—Principles and Applications*, Gulf Publishing, 2001.
Cooper, H.H., Jr., and C.E. Jacob, A generalized graphical method for evaluating formation constants and summarizing well field history, *Eos Trans. AGU, 27*, 526–534, 1946.
Crisman, S.A., F.J. Molz, D.L. Dunn, and F.C. Sappington, Application procedures for the electromagnetic borehole flowmeter in shallow unconfined aquifers, *Ground Water Monit. and Remed., 21*(4), 96–100, 2001.
Danielson, R.E., and P.L. Sutherland, Porosity, Methods of soil analysis, Part 1., In: *Physical and Mineralogical Methods*, ed. by A. Klute, *Agronomy Monograph 9*, American Soc. of Agronomy, 443–461, 1986.
Datta-Gupta, A., S. Yoon, D.W. Vasco, and G.A. Pope, Inverse modeling of partitioning interwell tracer tests: A streamline approach, *Water Resour. Res., 38*(6), 1079, doi:10.1029/2001WR000597, 2002.
Davis, G.A., S.F. Cain, J.J. Butler, Jr., X. Zhan, J.M. Healey, and G.C. Bohling, A field assessment of hydraulic tomography: A new approach for characterizing spatial variations in hydraulic conductivity (abstract), In: *GSA 2002 Annual Meeting Abstracts with Program, 34*(6), p. 23, 2002.
Dietrich, P., J.J. Butler, Jr., U. Yaramanci, V. Wittig, T. Tiggelmann, and S. Schoofs, 2003, Field comparison of direct-push approaches for determination of K-profiles (abstract), *Eos, 84*(46), p. F661, 2003.
Dinwiddie, C.L., N.A. Foley, and F.J. Molz, In-well hydraulics of the electromagnetic borehole flowmeter, *Ground Water, 37*(2), 305–315, 1999.
Dinwiddie, C.L., F.J. Molz, III, and J.W. Castle, A new small drill hole minipermeameter probe for in situ permeability measurement: Fluid mechanics and geometrical factors, *Water Resour. Res., 39*(7), 1178, doi:10.1029/2001WR001179, 2003.
Doe, T., J. Osnes, M. Kenrick, J. Geier, and S. Warner, Design of well-testing programs for waste disposal in crystalline rock, In: *Proc. Sixth Congress Intl. Soc. Rock Mech.*, ed. by G. Herget and S. Vongpaisal, *6*(3), 1377–1398, 1989.
Doughty, C., and K. Karasaki, Flow and transport in hierarchically fractured rock, *Jour. Hydrology, 263*, 1–22, 2002.
Farrar, J.A., 1996, Research and standardization needs for direct push technology applied to environmental site characterization, In: *Sampling Environmental Media*, ed. by J.H. Morgan, *ASTM Special Technical Publication 1282*, American Society for Testing and Materials, Philadelphia, 93–107, 1996.
Flühler, H., M.S. Ardakani, and L.H. Stolzy, Error propagation in determining hydraulic conductivities from successive water content and pressure head profiles, *Soil Sci. Soc. Am. J., 40*, 830–836, 1976.
Freeze, R.A., and J.A. Cherry, *Groundwater*, Prentice Hall, 1979.

Freyberg, D.L., A natural gradient experiment on solute transport in a sand aquifer, 2. Spatial moments and the advection and dispersion of nonreactive tracers, *Water Resour. Res., 22*(13), 2031–2046, 1986.

Gee, G.W., and J.W. Bauder, Particle-size analysis: Methods of soil analysis, Part 1., In: *Physical and Mineralogical Methods*, ed. by A. Klute, Agronomy Monograph 9, American Soc. of Agronomy, 383–411, 1986.

Gelhar, L.W., C. Welty, and K.R. Rehfeldt, A critical review of data on field-scale dispersion in aquifers, *Water Resour. Res., 28*(7), 1955–1974, 1992.

Gottlieb, J., and P. Dietrich, Identification of the permeability distribution in soil by hydraulic tomography, *Inverse Problems, 11*, 353–360, 1995.

Henebry, B.J., and G.A. Robbins, Reducing the influence of skin effects on hydraulic conductivity determinations in multilevel samplers installed with direct push methods, *Ground Water, 38*(6), 882–886, 2000.

Hess, K.M., S.H. Wolf, and M.A. Celia, Large-scale natural gradient tracer test in sand and gravel, Cape Cod, Massachusetts, 3. Hydraulic conductivity variability and calculated macrodispersivities, *Water Resour. Res., 28*(8), 2011–2027, 1992.

Hinsby, K., P.L. Bjerg, L.J. Andersen, B. Skov, and E.V. Clausen, A mini slug test method for determination of a local hydraulic conductivity of an unconfined sandy aquifer, *Jour. Hydrology, 136*, 87–106, 1992.

Holt, R.M., J.L.Wilson, and R.J. Glass, Spatial bias in field-estimated unsaturated hydraulic properties, *Water Resour. Res., 38*(12), 1311, doi:10.1029/
2002WR001336, 2002.

Hurst, A., and D. Goggin, Probe permeametry: An overview and bibliography, *AAPG Bull., 79*(3), 463–473, 1995.

Hvilshøj, S., K.H. Jensen, K.H., and B. Madsen, Single-well dipole flow tests: Parameter estimation and field testing, *Ground Water, 38*(1), 53–62, 2000.

Hvorslev, M.J., Time lag and soil permeability in ground-water observations, *U.S. Army Corps of Engrs. Waterways Exper. Sta. Bull no. 36*, 1951.

Illman, W.A., and S.P. Neuman, Type-curve interpretation of multirate single-hole pneumatic injection tests in unsaturated fractured rock, *Ground Water, 38*(6), 899–911, 2000.

Illman, W.A., and S.P. Neuman, Type curve interpretation of a cross-hole pneumatic injection test in unsaturated fractured tuff, *Water Resour. Res., 37*(3), 583–603, 2001.

Javandel, I., and P.A. Witherspoon, A method of analyzing transient fluid flow in multilayered aquifers, *Water Resour. Res., 5*(4), 856–869, 1969.

Kabala, Z.J., The dipole flow test: A new single-borehole test for aquifer characterization, *Water Resour. Res., 29*(1), 99–107, 1993.

Karasaki, K., B. Freifeld, A. Cohen, K. Grossenbacher, P. Cook, and D. Vasco, A multidisciplinary fractured rock characterization study at Raymond field site, Raymond, CA, *Jour. Hydrology, 236*, 17–34, 2000.

Keys, W.S., Borehole geophysics applied to ground-water investigations, *USGS Techniques of Water-Resources Investigations, Book 2*, Chapter E2, 1990.

Keys, W.S., *A Practical Guide to Borehole Geophysics in Environmental Investigations*, CRC Press, 1997.

Klute, A., and C. Dirksen, Hydraulic conductivity and diffusivity: Laboratory methods, Methods of soil analysis, Part 1., In: *Physical and Mineralogical Methods*, ed. by A. Klute, *Agronomy Monograph 9*, American Soc. of Agronomy, 687–734, 1986.

Kruseman, G.P., and N.A. de Ridder, Analysis and Evaluation of Pumping Test Data—*ILRI Pub. 47*, The Netherlands, Int. Inst. for Land Reclamation and Improvement, 1990.

Leap, D.I., and P.G. Kaplan, A single-well tracing method for estimating regional advective velocity in a confined aquifer: Theory and preliminary laboratory verification, *Water Resour. Res., 24*(7), 993–998, 1988.

Liu, S., T.-C. J. Yeh, and R. Gardiner, Effectiveness of hydraulic tomography: Sandbox experiments, *Water Resour. Res., 38*(4), doi:10.1029/2001WR000338, 2002.

Lowry, W., N. Mason, V. Chipman, K. Kisiel, and J. Stockton, In-situ permeability measurements with direct push techniques: Phase II Topical Report, SEASF-TR-98-207 Rept. to DOE Federal Energy Tech. Center, 1999.

Lunne, T., P.K. Robertson, and J.J.M. Powell, *Cone Penetration Testing in Geotechnical Practice*, London, Blackie Academic and Professional, 1997.

McCall, W., J.J. Butler, Jr., J.M. Healey, A.A. Lanier, S.M. Sellwood, and E.J. Garnett, A dual-tube direct-push method for vertical profiling of hydraulic conductivity in unconsolidated formations, *Environ. & Eng. Geoscience, 8*(2), 75–84, 2002.

McElwee, C.D., Application of a nonlinear slug test model, *Ground Water, 39*(5), 737–744, 2001.

McElwee, C.D., and M.A. Zenner, A nonlinear model for analysis of slug-test data, *Water Resour. Res., 34*(1), 55–66, 1998.

McElwee, C.D., J.J. Butler, Jr., and J.M. Healey, A new sampling system for obtaining relatively undisturbed cores of unconsolidated coarse sand and gravel, *Ground Water Monit. Rev., 11*(3), 182–191, 1991.

Meier, P.M., J. Carrera, and X. Sánchez-Vila, An evaluation of Jacob's method for the interpretation of pumping tests in heterogeneous formations, *Water Resour. Res., 34*(5), 1011–1025, 1998.

Melville, J.G., F.J. Molz, O. Guven, and M.A. Widdowson, Multilevel slug tests with comparisons to tracer data, *Ground Water, 29*(6), 897–907, 1991.

Molz, F.J., and S.C. Young, Development and application of borehole flowmeters for environmental assessment, *The Log Analyst, 34*(1), 13–23, 1993.

Molz, F.J., R.H. Morin, A.E. Hess, J.G. Melville, and O. Guven, The impeller meter for measuring aquifer permeability variations: Evaluation and comparison with other tests, *Water Resour. Res., 25*(7), 1677–1683, 1989.

Neuman, S.P., Stochastic continuum representation of fractured rock permeability as an alternative to the REV and fracture network concepts, In: *Proc. U.S. Symp. Rock Mech., 28th*, pp. 533–561, 1987.

Neuzil, C.E., Groundwater flow in low permeability environments, *Water Resour. Res., 22*(8), 1163–1195, 1986.

Novakowski, K.S., Analysis of pulse interference tests, *Water Resour. Res., 25*(11), 2377–2387, 1989.

Or, D., U. Shani, and A.W. Warrick, Subsurface tension permeametry, *Water Resour. Res., 36*(8), 2043–2053, 2000.

Peursem, D.V., V.A. Zlotnik, and G. Ledder, Groundwater flow near vertical recirculatory wells: Effect of skin on flow geometry and travel times with implications for aquifer remediation, *Jour. Hydrology, 222*, 109–122, 1999.

Pitkin, S.E., and M.D. Rossi, A real time indicator of hydraulic conductivity distribution used to select groundwater sampling depths (abstract), *Eos, 81*(19), S239, 2000.

Press, W. H., S.A. Teukolsky, W.T. Vetterling, and B. P. Flannery, 1992, *Numerical Recipes in FORTRAN: The Art of Scientific Computing, 2nd ed.*, Cambridge University Press, 1992.

Reynolds, W.D., D.E. Elrick, E.G. Youngs, A. Amoozegar, H.W.G. Booltink, and J. Bouma, Saturated and field-saturated water flow parameters, In: *Methods of Soil Analysis, Part 4, Physical Methods*, ed. by J.H. Dane and G.C. Topp, Soil Science Society of America Book Series No. 5, Soil Sci. Soc. of Am., Inc., 797–878, 2002.

Rovey, C.W., II, Discussion of "Relationship between pumping-test and slug-test parameters: Scale effect or artifact?" *Ground Water, 36*(6), 866–867, 1998.

Sánchez-Vila, X., P.M. Meier, and J. Carrera, Pumping tests in heterogeneous aquifers: An analytical study of what can be obtained from their interpretation using Jacob's method, *Water Resour. Res., 35*(4), 943–952., 1999.

Schad, H., and G. Teutsch, Effects of the investigation scale on pumping test results in heterogeneous porous aquifers, *Jour. Hydrology, 159*, 61–77, 1994.

Schulmeister, M.K., J.J. Butler, Jr., J.M. Healey, L. Zheng, D.A. Wysocki, and G.W. McCall, Direct-push electrical conductivity logging for high-resolution hydrostratigraphic characterization, *Ground Water Monit. and Remed., 23*(5), 52–62, 2003.

Sellwood, S.M., J.M. Healey, S.R. Birk, and J.J. Butler, Jr., Direct-push hydrostratigraphic profiling, *Ground Water, 2004* (in review).

Shani, U., and D. Or, In situ method for estimating subsurface unsaturated hydraulic conductivity, *Water Resour. Res., 31*(8), 1863–1870, 1995.

Shapiro, A.M., P.A. Hsieh, and F.P. Haeni, Integrating multidisciplinary investigations in the characterization of fractured rock, *U.S. Geological Survey Water-Resources Investigations Report 99-4018c*, pp. 669–680, 1999.

Smith, J.L., Spatial variability of flow parameters in a stratified sand, *Math. Geology, 13*(1), 1–21, 1981.

Sørensen, K.I., F. Effersø, E. Auken, and L. Pellerin, A method to estimate hydraulic conductivity while drilling, *Jour. of Hydrology, 260*, 15–29, 2002.

Stephens, D.B., *Vadose Zone Hydrology*, Lewis Pub., 1996.

Stienstra, P., and J.K. van Deen, Field data collection techniques—Unconventional sounding and sampling methods, In: *Engineering Geology of Quaternary Sediments*, ed. by N. Rengers, Balkema, pp. 41–55, 1994.

Streltsova, T.D., *Well Testing in Heterogeneous Formations*, Wiley, 1988.

Sudicky, E.A., A natural gradient experiment on solute transport in a sand aquifer: Spatial variability of hydraulic conductivity and its role in the dispersion process, *Water Resour. Res., 22*(13), 2069–2082, 1986.

Sudicky, E.A., and P.S. Huyakorn, Contaminant migration in imperfectly known heterogeneous groundwater systems, *U.S. Natl. Rept. Int. Union Geol. Geophys. 1987-1990, Rev. Geophys., 29*, 240–253, 1991.

Sutton, D.J., Z.J. Kabala, D.E. Schaad, and N.C. Ruud, The dipole-flow test with a tracer -- a new single-borehole tracer test for aquifer characterization, *Jour. Contaminant Hydrology, 44*(1), 71–101, 2000.

Tavenas, F., M. Diene, and S. Leroueil, Analysis of the in situ constant-head permeability test in clays, *Can. Geotech. Jour. 27*, 305–314, 1990.

Taylor, K., S. Wheatcraft, J. Hess, J. Hayworth, and F.J. Molz, Evaluation of methods for determining the vertical distribution of hydraulic conductivity, *Ground Water, 28*(1), 88–98, 1990.

Tidwell, V.C., and J.L. Wilson, Laboratory method for investigating permeability upscaling, *Water Resour. Res., 33*(7), 1607–1616, 1997.

Tosaka, H., K. Masumoto, and K. Kojima, Hydropulse tomography for identifying 3-D permeability distribution, In: *High Level Radioactive Waste Management: Proc. of the Fourth Annual International Conference of the ASCE*, Reston, VA, 1993.

van der Kamp, G., Methods for determining the in situ hydraulic conductivity of shallow aquitards—An overview, *Hydrogeology Jour., 9*(1), 5–16, 2001.

Vesselinov, V.V., S.P. Neuman, and W.A. Illman, Three-dimensional numerical inversion of pneumatic cross-hole tests in unsaturated fractured tuff: 1. Methodology and borehole effects, *Water Resour. Res., 37*(12), 3001–3017, 2001a.

Vesselinov, V.V., S.P. Neuman, and W.A. Illman, Three-dimensional numerical inversion of pneumatic cross-hole tests in unsaturated fractured tuff: 2. Equivalent parameters, high-resolution stochastic imaging and scale effects, *Water Resour. Res., 37*(12), 3019–3041, 2001b.

Weeks, E.P., Field determination of vertical permeability to air in the unsaturated zone, U.S. Geological Survey Prof. Paper No. 1051, 1978.

White, J., Guidelines for estimating permeability from NMR measurements, *DiaLog, 8*(1), 2000, www.lps.org.uk/dialogweb/archive/nmr_measurements_white/white.htm.

Yaramanci, U., G. Lange, and M. Hertrich, Aquifer characterisation using Surface NMR jointly with other geophysical techniques at the Nauen/Berlin test site, *Jour. Applied Geophysics, 50*, 47–65, 2002.

Yeh, T.-C. J., and S. Liu, Hydraulic tomography: Development of a new aquifer test method, *Water Resour. Res., 36*(8), 2095–2105, 2000.

Yeh, T.-C. J., J. Mas-Pla, T.M. Williams, and J.F. McCarthy, Observation and three-dimensional simulation of chloride plumes in a sandy aquifer under forced-gradient conditions, *Water Resour. Res., 31*(9), 2141–2157, 1995.

Young, S.C., and H.S. Pearson, The electromagnetic borehole flowmeter: Description and application, *Ground Water Monit. and Remed., 15*(2), 138–146, 1995.

Zapico, M., S. Vales, and J. Cherry, A wireline piston core barrel for sampling cohesionless sand and gravel below the water table, *Ground Water Monit. Rev., 7*(3), 74–82, 1987.

Zheng, C., and S.M. Gorelick, Analysis of solute transport in flow fields influenced by preferential flowpaths at the decimeter scale, *Ground Water, 41*(2), 142–155, 2003.

Zlotnik, V.A., and G. Ledder, Theory of dipole flow in uniform anisotropic aquifers, *Water Resour. Res., 32*(4), 1119–1128, 1996.

Zlotnik, V.A., and B.R. Zurbuchen, Dipole probe: Design and field applications of a single-borehole device for measurements of small-scale variations of hydraulic conductivity, *Ground Water, 36*(6), 884–893, 1998.

Zlotnik, V.A., and B.R. Zurbuchen, Field study of hydraulic conductivity in a heterogeneous aquifer: Comparison of single-borehole measurements using different instruments, *Water Resour. Res., 39*(4), 1101, doi:10.1029/2002WR001415, 2003.

Zlotnik, V.A., B.R. Zurbuchen, and T. Ptak, The steady-state dipole-flow test for characterization of hydraulic conductivity statistics in a highly permeable aquifer: Horkheimer Insel site, Germany, *Ground Water, 39*(4), 504–516, 2001.

Zurbuchen, B.R., V.A. Zlotnik, and J.J. Butler, Jr., Dynamic interpretation of slug tests in highly permeable aquifers, *Water Resour. Res., 38*(3), 1025, doi:10.1029/ 2001WR000354, 2002.

3 GEOSTATISTICS

J. JAIME GÓMEZ-HERNÁNDEZ

Institute of Water and Environmental Engineering
Universidad Politécnica de Valencia
46071 Valencia, Spain (jgomez@upv.es)

3.1 Introduction

Geostatistics started as an *ad hoc* solution for solving biased estimates of ore reserves in South African gold mines. In the 1950s, Daniel Krige (1950), a South African engineer aware of the incorrect reserve estimations obtained by the simple averaging of the sample data, developed a technique to account for the spatial correlation in the data, thus solving the biasing problem produced by the clustering of samples in the high ore areas. This technique was later studied by Matheron (1962) in France, who gave it a mathematical framework, and coined the term "kriging," as well as the term "geostatistics."

The problem solved by Krige and formalized by Matheron arises when dealing with phenomena displaying a spatial variability that can be described by an erratic component overlaying a deterministic trend. This behavior is typical of mining resources, but also of many other earth-science-related parameters and variables, such as permeability, porosity, infiltration capacity, or resistivity. Although some authors have attempted to explain the spatial variability of these parameters using genetic models (Tetzlaff and Harbaugh, 1989), in which the underlying physical processes leading to a specific spatial distribution of the parameter are explicitly modeled, their success has been limited. For this reason, the statistical approach has been favored. The unique (but known only at a few sample locations) spatial distribution of a parameter is studied as a component of a much larger set of alternative realizations. These realizations are characterized through a random function model that incorporates the spatial correlation as a statistical parameter of the model.

The extension of geostatistics to fields outside the earth sciences is due to the successful results obtained from its application first in mining, then in earth sciences. Geostatistics builds on the foundation that most parameters related to natural processes cannot be modeled by a deterministic, smooth function; they require models that incorporate a random component to capture the erratic spatial variability that cannot be deterministically described. Geostatistics first started with the aim of parameter estimation at unsampled locations, with excellent results in the estimation of ore reserves in mining deposits, or total rainfall in watersheds. These first applications relied on the existence of a relatively large data set. However, the conceptualization underlying the geostatistical model has proven to be strong enough for the extension of geostatistics to applications where data are scarce, yet where the deterministic plus

Y. Rubin and S.S. Hubbard (eds.), Hydrogeophysics, 59–83.

random model is pertinent (such as for characterizing subsurface aquifers). In these cases, some of the parameters underlying the geostatistical model cannot be data-based, but instead have to be specified on the basis of prior experience or expert knowledge. The results obtained in this data-meager scenario were also satisfactory, but introduced the need to quantify the estimation uncertainty—something that had always been a by-product of geostatistics but whose value was not fully recognized. Furthermore, it was recognized that the types of smooth fields obtained by geostatistical estimation were not directly applicable to estimating complex transfer functions, for the cases in which these transfer functions are dominated by extreme parameter values (as is the case in groundwater mass transport controlled by the presence of flow channels and flow barriers). Thus, geostatistical applications expanded from estimation to stochastic simulation, where the aim is to generate parameter fields that display spatial variability patterns resembling natural patterns.

There are many monographs devoted to geostatistics for readers interested in deepening their knowledge of this field. A very good introductory book is the one by Isaaks and Srivastava (1990), a quick (and dense) overview of geostatistics is given by Journel (1989), a more thorough presentation of geostatistics is made by Goovaerts (1997), and library or public-domain Fortran sources for geostatistical applications are found in Deutsch and Journel (1998). Other recent references include Olea (1999) and Chiles and Delfiner (1999). An excellent source of information and discussion is the web site maintained by Gregoire Dubois since 1995: http://www.ai-geostats.org.

3.2 Spatial Variability

There are natural phenomena that can be precisely described with mechanistic laws, such as the movement of planets or the occurrence of the next solar eclipse. However, the spatial variability of many earth science parameters, such as hydraulic conductivity, cannot be predicted from limited knowledge of space or time. Thus, we cannot make an estimation of hydraulic conductivity at an unsampled location, given a few measurements, with the same precision that we can predict the occurrence of the next planetary alignment. At the same time, earth science parameters do not vary at random—there is a certain structure to their variability that allows making inferences about the likely values of the parameter at an unknown location, given the knowledge at nearby locations. That is, earth science parameters display a spatial variability that is neither deterministic nor totally random. Figure 3.1 shows a typical profile of observed hydraulic conductivity variability with those characteristics.

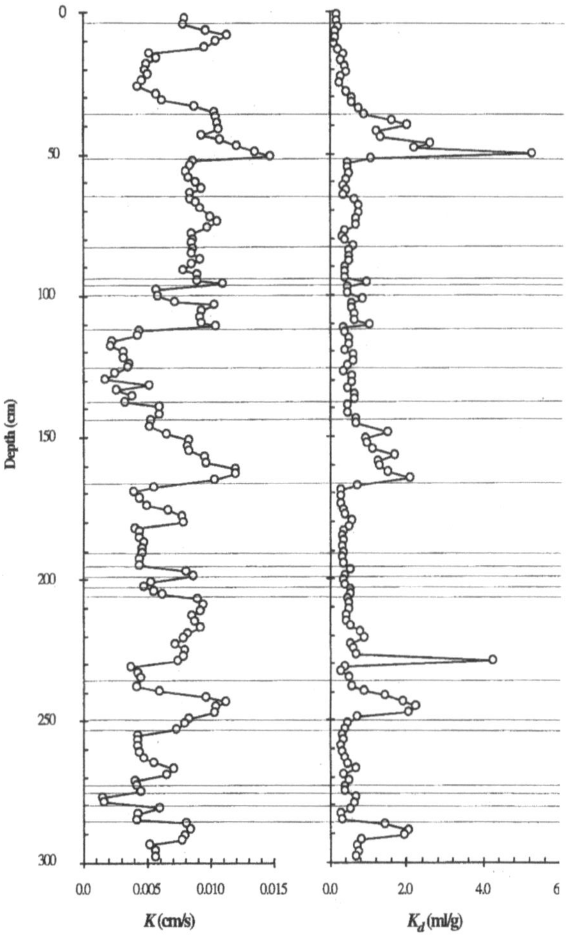

Figure 3.1. Vertical profiles of hydraulic conductivity and partition coefficient at the Borden aquifer. It can be appreciated that neither parameter is totally random or fully structured with depth (after Allen-King et al., 1998, ©American Geophysical Union).

3.3 Space Random Function

The theory of regionalized variables is especially suited for the purpose of analyzing parameters that are sampled in space and that display a random spatial variability with a certain degree of spatial correlation. The variable under study is described by a regionalized variable $z(\mathbf{u})$, which is considered a realization of a space random function $Z(\mathbf{u})$, with $\mathbf{u}$ being the space coordinate vector. The space random function is made of the set of random variables that can be defined at each location $\mathbf{u}$ with the domain of interest. Characterizing the space random function amounts to characterizing the joint behavior of all random variables, which is equivalent to saying that each random variable is characterized by its univariate distribution function, each pair of variables (at locations $\mathbf{u}_1$ and $\mathbf{u}_2$) is characterized by their bivariate distribution function, each triplet of random variables (at locations $\mathbf{u}_1$, $\mathbf{u}_2$ and $\mathbf{u}_3$) is characterized by a trivariate distribution function, and so on. The space random function is defined when all distributions for the combination of any number of random variables are defined. Since describing all these distribution functions is impossible in the most general case, some simplifications are necessary, so that the space random function can be characterized by an acceptable number of parameters. The three most important simplifications are the following:

Stationarity of the mean
The expected value of all random variables is constant:

$$E\{Z(\mathbf{u})\} = m, \ \forall \mathbf{u} \in D \tag{3.1}$$

where $E\{\cdot\}$ is the expected value operator, m is the constant mean value for all random variables, and D is the domain of interest.

Stationarity of the covariance
The covariance between two random variables depends only on the separation vector, not on their absolute position:

$$C\left(Z(\mathbf{u}_1),Z(\mathbf{u}_2)\right) = E\{Z(\mathbf{u}_1)Z(\mathbf{u}_2)\} - E\{Z(\mathbf{u}_1)\}E\{Z(\mathbf{u}_2)\} = C(\mathbf{h}), \forall \mathbf{u}_1, \mathbf{u}_2 \in D \tag{3.2}$$

where $C(\cdot)$ is the covariance operator, and $\mathbf{h} = \mathbf{u}_1 - \mathbf{u}_2$, is the separation vector. Stationarity of the mean implies stationarity of the variance; that is, the variance is constant for all random variables:

$$\mathrm{Var}(Z(\mathbf{u})) = C(Z(\mathbf{u}),Z(\mathbf{u})) = \sigma^2, \ \forall \mathbf{u} \in D \tag{3.3}$$

where σ^2 is the constant variance.

Ergodicity
To justify the inference of the values m, σ^2, and $C(\mathbf{h})$ from a set of data collected from a regionalized variable, it is necessary to assume that the space random function is not only stationary, but also ergodic in their mean, variance, and covariance. Under this assumption, expected values can be approximated by the limiting value of a spatial integral over a domain, the extent of which tends to infinity:

$$E\{Z(\mathbf{u})\} = \lim_{V \to \infty} \frac{1}{V} \int_V z(\mathbf{u}) d\mathbf{u} = m \tag{3.4}$$

$$\mathrm{Var}(Z(\mathbf{u})) = \lim_{V \to \infty} \frac{1}{V} \int_V (z(\mathbf{u}) - m)^2 d\mathbf{u} = \sigma^2 \tag{3.5}$$

$$E\left\{Z(\mathbf{u})Z(\mathbf{u}+\mathbf{h}) - m^2\right\} = \lim_{V \to \infty} \int_V \left(z(\mathbf{u})z(\mathbf{u}+\mathbf{h}) - m^2\right) d\mathbf{u} = C(\mathbf{h}) \tag{3.6}$$

These three restrictions are enough to develop a consistent statistical framework that permits the construction of spatial representations of the attribute at hand—particularly when the space random function is assumed to be multi-Gaussian, since a multi-Gaussian model is fully determined by mean and covariance.

For some parameters, it is not appropriate to use a space-random-function model with stationary variance, particularly when the overall degree of variability increases with the area of study, as in the case of fractal variables. In these cases, the stationarity in the covariance can be relaxed to stationarity of the variance of the increments:

$$\mathrm{Var}(Z(\mathbf{u}_1)-Z(\mathbf{u}_2))=E\left\{(Z(\mathbf{u}_1)-Z(\mathbf{u}_2))^2\right\}=2\gamma(\mathbf{h}),\forall\mathbf{u}_1,\mathbf{u}_2\in D \tag{3.7}$$

where $\gamma(\mathbf{h})$ is known as the variogram (sometimes also referred to as the semivariogram) function. The space random functions that satisfy this property are referred to as intrinsic.

If an intrinsic space random function is also ergodic, the variogram function can be inferred from a spatial integral computed over a single sample $z(\mathbf{u})$ by

$$E\left\{(Z(\mathbf{u})-Z(\mathbf{u}+\mathbf{h}))^2\right\}=\lim_{V\to\infty}\int_V(z(\mathbf{u})-z(\mathbf{u}+\mathbf{h}))^2 d\mathbf{u}=2\gamma(\mathbf{h}) \tag{3.8}$$

For space random functions that are stationary in the mean and covariance, the simple relationship between variogram and covariance is given by

$$C(\mathbf{h})=\sigma^2-\gamma(\mathbf{h}) \tag{3.9}$$

3.4 Stationarity and Ergodicity

Note that using a space random function to analyze the spatial distribution of a parameter that is unique but unknown does not imply anything random about the parameter itself. The use of random variables allows us to make predictions of the likely values that the parameter may have, based on the patterns of variability observed elsewhere. This probable knowledge does not predict reality, but gets us closer to it.

Space random functions are tools to gain knowledge to help in the decision-making process. In the application of geostatistics, it is important to distinguish between the space random function, which is the model, and reality, which is a unique spatial distribution of the attribute. In this sense, the stationarity and ergodicity conditions are properties of the model, never of reality. Further, in the same way that we decide to use a space random function model, we must decide whether to choose a space random function that is stationary or ergodic. It is impossible to make this decision based on the data used to infer the statistical model parameters; rather, such a decision must be based on the modeler's experience and common sense. For instance, stationarity on the mean signifies that each random variable $Z(\mathbf{u})$ at every location $\mathbf{u}$ has the same expected value. This value, however, can never be checked because only one value of each random variable exists where the random function has been sampled, from which it is impossible to infer the means of the random variables. Common sense should, for instance, rule out a stationary model when the spatial variability of the parameter shows a distinct trend, even though the specific pattern of variability could be made consistent with a stationary model.

In all cases, evaluating which model is best should be carried out outside of their statistical description. For instance, if the model is to be used to predict the spatial distribution of hydraulic conductivities, and these hydraulic conductivities are used to predict groundwater flow and mass transport, the best model is the one that produces the best flow and transport predictions.

Deciding whether or not to choose a stationary model is nontrivial. In Figure 3.1, if the only available samples of hydraulic conductivity had been taken between 110 cm and 160 cm, it would appear as if the sample values had been drawn from random variables with a nonconstant increasing mean, as a function of depth and a small variance. But when looking at the whole sample profile, it appears that such a trend in the mean may not be appropriate for the entire profile. It appears that for the entire profile, a stationary space random function with a constant mean would be more reasonable. Looking back at the segment between 110 cm and 160 cm, and considering that we had to decide on a model just for this segment, is it better to use a stationary model or a nonstationary one? The answer must always be found outside the data in the sample: Are there geological reasons to assume that hydraulic conductivity must be trending with depth for this particular formation? Is this trend also evident in samples outside the area under study? If no answers to these questions are readily available, it is best to use the simpler stationary model; and when in doubt, to use both models and compare the performance of each model against validating information that had not been used.

3.5 Univariate Statistics

Each random variable $Z(\mathbf{u}_0)$, with $\mathbf{u}_0$ being any location within the study domain, is fully characterized by its cumulative distribution function (cdf) $F_Z(z)$, which gives the probability that the variable is less than or equal to z. The cdf is the integral of the probability density function (pdf) $f_Z(z)$ for $z \in D$, $F_Z(z)=\int_D f_Z(z)dz$. These functions are generally estimated after calling for univariate stationarity of the space random function, from a single outcome of the random variables for which there are samples. The cdf's and pdf's can be described by a set of parameters, as is the case for the Gaussian distribution (defined by its mean and variance) or the uniform distribution (defined by its minimum and maximum), or nonparametrically by specifying the values of the function for a set of possible outcomes of the random variable.

3.6 Multivariate Statistics

The n-tuple of random variables $\{Z(\mathbf{u}_1), Z(\mathbf{u}_2), \ldots, Z(\mathbf{u}_n)\}$ is fully characterized by their multivariate cumulative distribution function $F_{Z1,Z2,\ldots,Zn}(z_1, z_2, \ldots, z_n)$, which gives the probability that $\{Z(\mathbf{u}_1) \leq z_1, Z(\mathbf{u}_2) \leq z_2, \ldots, Z(\mathbf{u}_n) \leq z_n\}$. This function is the integral of the multivariate probability density function $f_{Z1,Z2,\ldots,Zn}(z_1, z_2, \ldots, z_n)$. Even under the assumption that the space random function is stationary for all possible n-tuples, and ergodic, inferring a multivariate distribution function from a sparse sample of a regionalized variable $z(\mathbf{u})$ is difficult. The latest developments in geostatistics (Strebelle, 2002) point to the use of multivariate statistics for characterizing complex spatial patterns of variability, while circumventing the inference problem by limiting the value of n and by using a training image as a densely sampled alternative to the experimental sparse sample.

Of all multivariate distribution functions, the multi-Gaussian one is particularly attractive: for stationary space random functions, it is fully characterized by only a mean and a covariance function. This simplicity in its characterization is counterweighted by some characteristics, such as the lack of spatial correlation for extreme values, which have to be analyzed before deciding for the convenience of the multi-Gaussian function (Gómez-Hernández and Wen, 1998; Wen and Gómez-Hernández, 1998). When the need exists to introduce spatial continuity of extreme values, alternative multivariate models can be used, such as nonparametric models based on indicator variables (Journel, 1983; Dreiss and Johnson, 1989; Gómez-Hernández and Srivastava, 1990; Rubin, 1995), multivariate models derived from training images (Strebelle, 2002; Caers, 2002), object models (Haldorsen and Damsleth, 1990; Jussel et al., 1994; Deutsch and Wang, 1996), or a combination of these models (i.e., Bierkens, 1996)

3.7 Variogram

3.7.1 DEFINITION

The variogram is the basic tool in geostatistics for characterizing spatial continuity. It is intimately related to the covariance function, but to its advantage, it does not require prior knowledge of the mean for its computation (a fact that was highly valued in the early 1960s, when numerical calculations had to be done by hand or with very rudimentary calculators). Its formal definition is given by Equation (3.7) and could be derived from a single realization $z(\mathbf{u})$ applying Equation (3.8). The variogram is a vectorial function measuring some type of "structural" distance. The larger the variogram value, the smaller the correlation between samples, for a given direction and distance. It is, therefore, inherently anisotropic, since we should expect earth science attributes to be "structurally" more similar in directions and distances along the strata rather than orthogonal to the strata. Evaluating Equation (3.8) from a sparse sample requires using the following discrete approximation:

$$\gamma(\mathbf{h}) \approx \frac{1}{2N(\mathbf{h})} \sum_{\mathbf{u}_1-\mathbf{u}_2=\mathbf{h}\pm\Delta\mathbf{h}} \left((z(\mathbf{u}_1) - z(\mathbf{u}_2) \right)^2 \qquad (3.10)$$

This approximation assumes that the data are not overlaid on a regular grid, and thus it is difficult to find many pairs of data separated exactly by vector **h**. For this reason, the value of the variogram for a given vector **h** is approximated by the semiaverage of the squared difference of all pairs separated by a vector that falls within $\mathbf{h}\pm\Delta\mathbf{h}$, in which $\Delta\mathbf{h}$ is a tolerance vector large enough to ensure that the number of pairs $N(\mathbf{h})$ is acceptable. In practice, it is customary to specify the number of directions in which to compute the variogram, and then decide on an angular tolerance plus a lag tolerance to estimate the variogram value from the data. Figure 3.2 shows an example of an experimental variogram computed from a two-dimensional sparse data set for two given directions.

With triangles, the variogram was computed using all pairs separated by vectors oriented at 0°±45°; with squares, the variogram was computed with pairs oriented at 90°±45°. In this particular data set, the spatial correlation is stronger in the 90° directions, as indicated by the smaller variogram values for the shorter distances. The practice of experimental variography is difficult—it is well discussed at book length in Isaaks and Srivastava (1990). The public domain code VARIOWIN (Pannatier, 1996) for variogram analysis is available from http://www-sst.unil.ch/research/variowin/index.html.

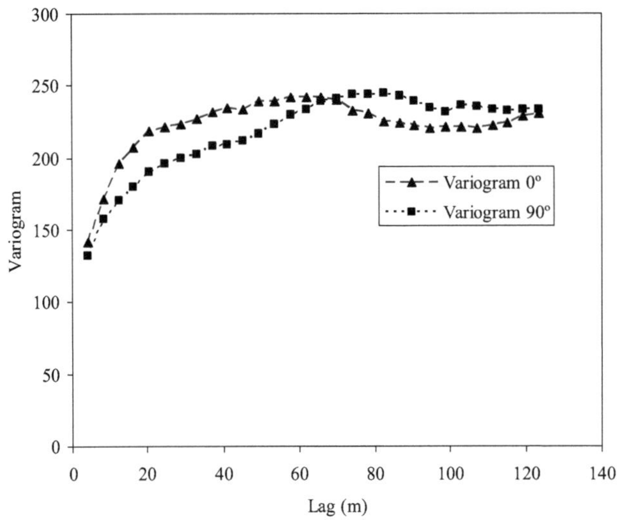

Figure 3.2. Experimental variograms computed on a 2-D data set for the 0° and 90° directions, with angular tolerance of ±45° and distance tolerance of ±5 m

The general shape of the variograms is that of an increasing function with lag, which levels out after a certain distance. This shape is an implicit representation of the two components of each regionalized variable. Some structure is reflected in the shorter lag distances. At greater lags, the variogram stabilizes, signifying that the variable is no longer structurally correlated at those distances, and that the observed differences between pairs at those lengths are the result of the overall variability of the variable.

3.7.2 MODELS

The experimental variogram as depicted in Figure 3.2 is of little use beyond establishing the directions with the larger and smaller continuities, and determining the distances at which the spatial correlation disappears. It does not provide the value of the variogram for distances or directions different from the ones for which it has been computed. It is not enough to make a simple linear interpolation between the symbols in Figure 3.2 and then extrapolate this interpolation in between directions, because the variogram, as a vectorial function, must satisfy the statistical restriction of conditional non-negativeness. This restriction is necessary to ensure that the kriging systems that will be discussed later have a solution, and that the kriging variances obtained by the expressions that will be presented below are positive (Chistakos, 1992, pp. 65–66, 255–256; Goovaerts, 1997, p. 87). Only a few functions satisfy this restriction; the most common one-dimensional functions are displayed in Figure 3.3.

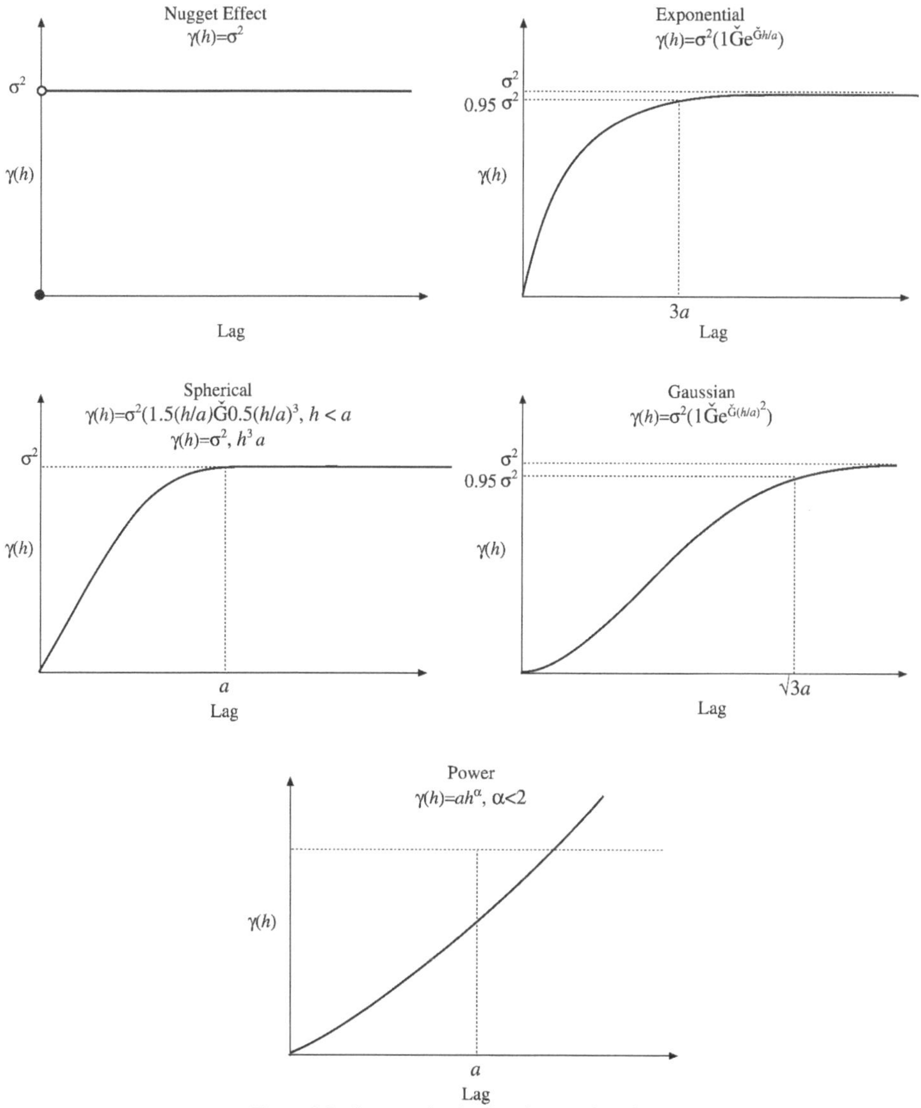

Figure 3.3. Some authorized variogram functions

In addition to the functions displayed in Figure 3.3, any linear combination having positive coefficients of valid variogram functions is also a valid variogram function, which leaves enough degrees of freedom for the fitting of an experimental variogram by a valid variogram function. As seen in the figure, with the exception of the power variogram, all variogram functions increase with distance and level out to a maximum value called *sill* at a certain distance called *range*. The sill coincides with the overall variance, and the range is the distance beyond which data pairs are not correlated (i.e., a for the spherical variogram). Figure 3.4 shows the fitting of the variogram in direction 90° with the sum of three valid variogram functions, a nugget effect, a spherical variogram, and an exponential one:

$$\gamma(h) = \text{Nugget}(\sigma^2 = 110) + \text{Exponential}(\sigma^2 = 100, a = 65) + \text{Spherical}(\sigma^2 = 55, a = 21) \tag{3.11}$$

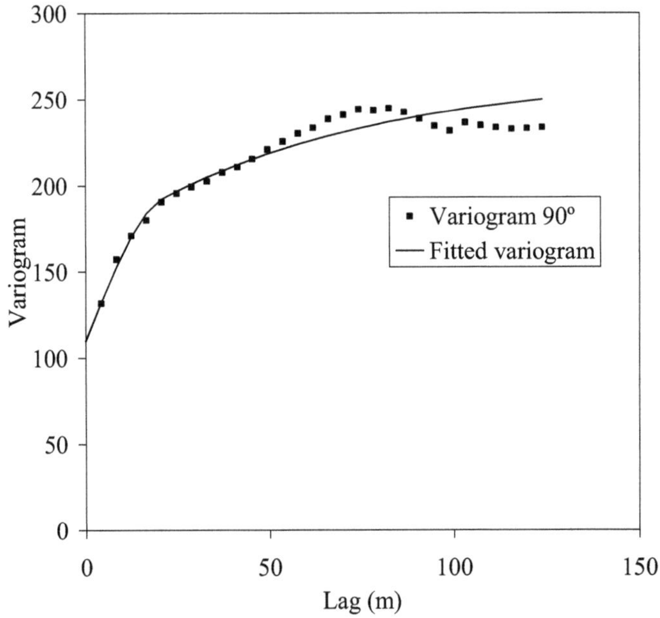

Figure 3.4. The experimental variogram for direction 90° of Figure 3.2 and its fitted variogram function

The exercise of fitting a variogram model for a given direction should be repeated for all directions in which experimental variograms have been computed. During this exercise of fitting all directions, we should keep in mind that, for the definition of a non-negative vectorial variogram function, we must use the same number and types of structures for the fitting of all directions, with the same associated variances. Thus, in the case of the sample used for Figure 3.4, all variogram models fitted to the remaining directions should be composed of a nugget effect of variance 100, plus a spherical variogram of variance 100, plus an exponential one of variance 55. If the process is isotropic, the same variogram function should fit all directions. If not, the only parameters allowed to vary between directions are the range coefficients a (except for the nugget effect). For each structure of the variogram, once these range coefficients have been determined for several directions, an ellipse function $a(\theta)$ is fitted to them.

This ellipse is defined by its major and minor semi-axes and their orientation. Figure 3.5 illustrates this step: the black dots correspond to the ranges fitted to one of the structures fitted to the experimental variograms computed along the 0°, 45°, 90° and 135° directions; the white dots are symmetric with respect to the origin of the black dots, since the definition of the variogram implies that $\gamma(\mathbf{h})=\gamma(-\mathbf{h})$

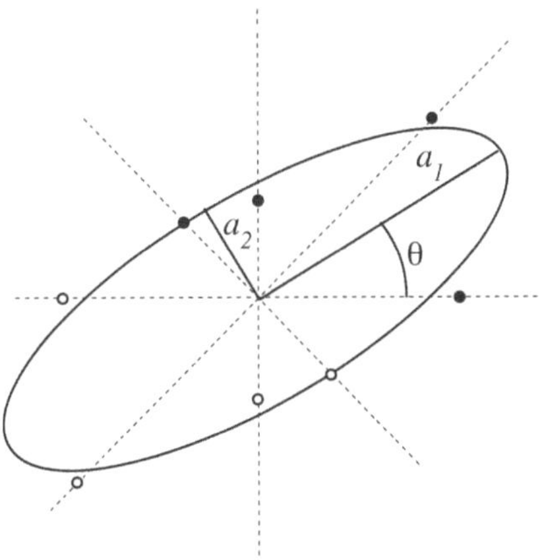

Figure 3.5. Fitting an ellipse to the ranges of one of the structures for an experimental variogram computed along the directions 0°, 45°, 90°, and 135°. This ellipse is used to model the anisotropy of the variogram function. The ellipse is defined by the major and minor semi-axes a_1 and a_2 and the angle θ.

In 3-D, the process is slightly more complex, since the variability of the range with orientation should then be fitted by an ellipsoid, defined by three semi-axes and two orientation angles. For the reader interested in variogram fitting, there is an excellent tutorial in the book by Isaaks and Srivastava (1990), which provides detailed discussions about how to construct this authorized variogram function, as well as provides answers to questions such as what to do when we cannot fit all directional variograms using the same structures and the same variances.

3.8 Estimation

After a variogram function has been constructed that is capable of providing the variogram value for any vector **h**, we can make use of the structural information (i.e., the information about the spatial correlation of the variable embedded in the variogram) to make predictions about the likely variable values at unsampled locations. Many alternatives exist when constructing an estimator for the value of a regionalized variable at an unsampled location, as a function of the sampled values. The simplest one is to consider that the estimate should be a linear combination of the sampled values. If there is no spatial correlation, the relative position of the samples with respect to each other, or with respect to the unsampled locations, does not matter. In such a case, no datum is more informative than another, and thus an equal-weighted linear combination of the data is the best prediction. However, since the variable displays spatial correlation, and this spatial correlation has been quantified through the variogram, a more elaborated estimate is warranted, taking into account the "structural" distance from the samples to the point being estimated, and also the potential redundancy of clustered sample data.

3.8.1 ORDINARY KRIGING

Given a data set $\{z(\mathbf{u}_1), z(\mathbf{u}_n), \ldots, z(\mathbf{u}_n)\}$, the problem of estimating $z(\mathbf{u}_0)$ can be cast, in the context of space random functions, as an optimization problem to find the weights λ_i in the linear combination

$$z_{SK}^{*}(\mathbf{u}_0) = \sum_{i=1}^{n} \lambda_i(\mathbf{u}_0) z(\mathbf{u}_i) \tag{3.12}$$

These weights provide an unbiased estimate satisfying the expected value of the squared difference between the random variable $Z(\mathbf{u}_0)$ and the random variable $Z_{SK}^{*}(\mathbf{u}_0)$ (obtained by the linear combination of the random variables $Z(\mathbf{u}_i)$ as in Equation (3.12). The solution to this optimization problem is given by the following:

$$\begin{Bmatrix} \lambda_1 \\ \lambda_2 \\ \vdots \\ \lambda_n \\ \mu \end{Bmatrix} = \begin{bmatrix} \gamma_{11} & \gamma_{12} & \cdots & \gamma_{1n} & 1 \\ \gamma_{21} & \gamma_{22} & \cdots & \gamma_{2n} & 1 \\ \vdots & \vdots & \ddots & \vdots & 1 \\ \gamma_{n1} & \gamma_{n2} & \cdots & \gamma_{nn} & 1 \\ 1 & 1 & 1 & 1 & 0 \end{bmatrix}^{-1} \begin{Bmatrix} \gamma_{01} \\ \gamma_{02} \\ \vdots \\ \gamma_{0n} \\ 1 \end{Bmatrix} \tag{3.13}$$

where γ_{ij} represents the variogram for vector $\mathbf{u}_i - \mathbf{u}_j$. Parameter μ is a Lagrange multiplier that appears because of the unbiasedness condition, which requires that the weights add up to 1. Notice that the matrix is symmetric since, by definition, $\gamma_{ij}=\gamma_{ji}$. Notice also that the weights, which depend on the location being estimated, are made proportional to the "structural" distance from the location being estimated and the data, γ_{0i}, and made inversely proportional to the amount of redundancy measured by the matrix containing all the "structural" interdistances between any two data.

It can be shown that kriging is an exact and nonconvex interpolator. It is exact in the sense that the kriging estimate at a location coinciding with a datum location yields the datum value, i.e., the weights are all equal to zero except for the weight associated with the co-located datum, and it is nonconvex in the sense that the kriging estimate can be outside the range of the data values.

Equation (3.12) and the weights solution of (3.13) correspond to the most popular geostatistical estimation algorithm: ordinary kriging. There are many other kriging approaches, including simple kriging, kriging with a locally varying mean, universal kriging, kriging with an external drift, disjunctive kriging, soft kriging, factorial kriging, block kriging, indicator kriging, and others; Goovaerts (1997) discusses most of them. The different variants of kriging are related to the properties of the spatial random function model chosen:

- The ordinary kriging equations—(3.13)—are deduced for a space random function stationary in both the mean and the variogram, with the particularity that the constant value of the stationary mean m is unknown.

- Simple kriging, kriging with a locally varying mean, universal kriging, and kriging with an external drift consider a space random function that is not stationary in the mean. For simple kriging, the mean is constant, m, and known; for kriging with a locally varying mean, the mean varies in space, $m(\mathbf{u})$, and is known; for universal kriging, the mean varies in space following a polynomial expression of a given order, but with unknown coefficients (i.e., in 2-D, with a first-order polynomial $m(u_1, u_2)=b_0+b_1 u_1+ b_2 u_2$, with u_1 and u_2 being the space coordinates, and b_0, b_1, b_2 being unknown coefficients). For kriging with an external drift, the mean varies in space following the expression $m(\mathbf{u})=b_0+b_1 f(\mathbf{u})$, in which $f(\mathbf{u})$ is a known drift function and b_0, b_1, are unknown coefficients.
- Disjunctive kriging gives the estimated value, not as in Equation (3.12), as a linear function of the measurements, but as the sum of unknown single-variable functions of the measurements. In this case the optimization problem is set not to find the coefficients of a linear combination, but to find the functions that, applied to the measurements, will result in a sum that minimizes the same squared difference as for the case of ordinary kriging described above.
- Soft kriging accounts for the uncertainty attached to some measurements.
- In factorial kriging, the underlying space random function is described as the linear combination of space random functions, each of which could be related to some characteristic spatial pattern of variability for the parameter.
- Block kriging deals with the problem of directly estimating block values when the block value could be expressed as a linear function of the points within the block.
- Indicator kriging is a nonparametric alternative that expands the types of space random functions that can be dealt with in geostatistics, and that can also be used for estimating the spatial distribution of categorical variables.

3.8.2 KRIGING VARIANCE

Deducing the weights in Equation (3.13) requires minimizing a squared difference, as mentioned above. The value of this minimum squared difference is known as the kriging variance. It measures the expected discrepancy between the estimate and the actual value, when the same estimation is performed numerous times within the domain for the same data configuration pattern, but with other possible measurement values. It measures the average uncertainty associated to a specific sampling configuration, but it cannot be considered a measure of the uncertainty associated with a specific data set. The expression of the ordinary kriging variance associated with the weights solution of Equation (3.13) is

$$\sigma_{OK}^2(\mathbf{u}_0) = \sigma^2 - \sum_{i=1}^{n} \lambda_i(\mathbf{u}_0)\gamma_{i0} + \mu \tag{3.14}$$

where σ^2 is the stationary variance.

3.8.3 SIMPLE KRIGING

Because of their direct relationship to the multi-Gaussian space random function, the simple kriging equations are quite instructive. These equations are derived for a space random function that is stationary on the mean and covariance. The difference with respect to ordinary kriging is that the mean m is known. In such a case, the estimation is formulated in terms of the residuals with respect to the mean:

$$z_{SK}^{*}(\mathbf{u}_0) - m = \sum_{i=1}^{n} \lambda_i(\mathbf{u}_0)(z(\mathbf{u}_i) - m) \tag{3.15}$$

and for which the kriging weights are given by

$$\begin{Bmatrix} \lambda_1 \\ \lambda_2 \\ \vdots \\ \lambda_n \end{Bmatrix} = \begin{bmatrix} \gamma_{11} & \gamma_{12} & \cdots & \gamma_{1n} \\ \gamma_{21} & \gamma_{22} & \cdots & \gamma_{2n} \\ \vdots & \vdots & \ddots & \vdots \\ \gamma_{n1} & \gamma_{n2} & \cdots & \gamma_{nn} \end{bmatrix}^{-1} \begin{Bmatrix} \gamma_{01} \\ \gamma_{02} \\ \vdots \\ \gamma_{0n} \end{Bmatrix} \tag{3.16}$$

and the associated simple kriging estimation variance is

$$\sigma_{SK}^2(\mathbf{u}_0) = \sigma^2 - \sum_{i=1}^{n} \lambda_i(\mathbf{u}_0)\gamma_{i0} \tag{3.17}$$

Notice that the Lagrange multiplier μ has disappeared from the kriging system, Equation (3.16), since the estimate is unbiased by construction. Indeed, the expected value of the right-hand side of Equation (3.15) is zero, since all of its components are zero. Therefore, the expected value of the estimate must equal the known mean m.

The particularity of the simple kriging equations (derived under the only constraints that the space random function is stationary on its mean m and on its covariance) is that, if the space random function is further assumed to be multi-Gaussian, then it can be shown that the conditional distribution of random variable $Z(\mathbf{u}_0)$—given that the variables $\{Z(\mathbf{u}_1), Z(\mathbf{u}_2), \ldots, Z(\mathbf{u}_n)\}$ take, respectively, the values $\{z(\mathbf{u}_1), z(\mathbf{u}_2), \ldots, z(\mathbf{u}_n)\}$—is a Gaussian distribution, with mean equal to the simple kriging estimate $z_{SK}^*(\mathbf{u}_0)$ and variance equal to the simple kriging variance $\sigma_{SK}^2(\mathbf{u}_0)$. This is an extremely powerful result, since it allows characterizing the full distribution of likely values at the unsampled location $\mathbf{u}_0$,. Thus, the choice of the "best" estimate is not restricted to the value minimizing a least-square criterion, but rather could be chosen in the context of a loss function in which, for instance, the cost of underestimating can be made different from that of overestimating. The concept of loss function and alternative estimates to least-squares mimima were presented by Journel (1984) and are also described by Goovaerts (1997, p. 340).

An example of ordinary kriging is displayed in Figure 3.6. The right part of the figure corresponds to the interpolation over a 40 × 40 cell grid of 64 permeability samples taken with a mini-permeameter from the slab of Berea sandstone shown on the left in the figure. The structural model used for the kriging was obtained from analyzing the sampled data and studying the variograms for directions 0°, 45°, 90°, and 135°. From this analysis, a variogram model was obtained, which included a nugget effect of 37, a spherical variogram of sill 212, and range that varied with direction as given by an ellipse of major axis at 120° (positive counter-clockwise from horizontal axis) of 70 cells and minor axis at 30° of 12 cells. The anisotropy along the 120° direction is noticeable in the kriged results (Figure 3.6, right).

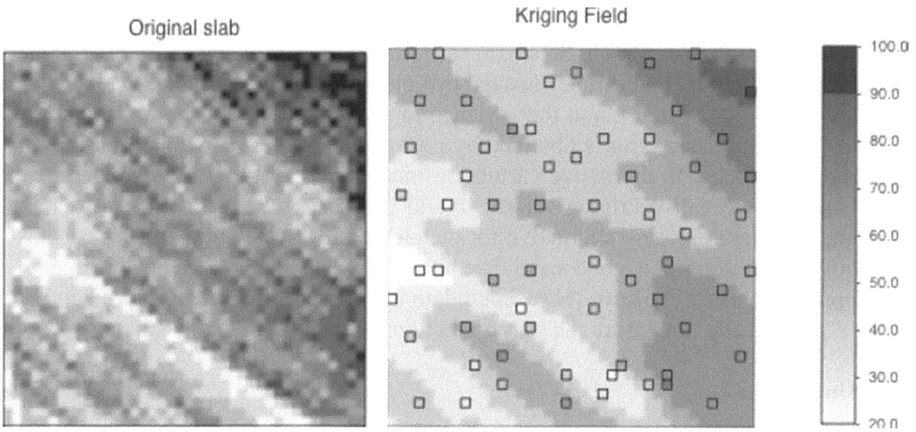

Figure 3.6. An example of ordinary kriging. A permeability map (in milidarcys) is shown on the right, which was obtained by interpolating the 64 data (in squares over a grid of 40 by 40 cells) sampled from the slab of Berea sandstone shown on the left.

3.8.4 CROSS-VALIDATION

Acknowledging the long list of alternative kriging methods, and the difficulties in fitting authorized variogram functions to experimental variograms when data are scarce, the question always remains: what choices of variogram model and kriging type are the most appropriate? As already mentioned above during the discussion of stationarity and ergodicity, the best check is the one performed outside geostatistics; that is, using the kriging maps for whatever purpose they were built, and then checking which one performs best against some control information. However, if the purpose of the kriging exercise does not go beyond the building of the maps to detect spatial trends, and no extra information exists against which to verify the performance of the alternative maps, one way of testing the validity of the option chosen is to perform cross-validation.

By cross-validation, we refer to an exercise in which kriging estimates are computed at data locations using all data except the datum co-located at the estimation location. In this way, we can compare the performance of kriging against actual measurements. The result is a set of kriging errors:

$$\{e(\mathbf{u}_i)=z(\mathbf{u}_i)-\sum_{\substack{j=1,\\ j\neq i}}^{n}\lambda_j^{CV}(\mathbf{u}_i)z(\mathbf{u}_j), i=1,\ldots,n\}, \tag{3.18}$$

in which $e(\mathbf{u}_i)$ is the cross-validation error associated with the datum $z(\mathbf{u}_i)$, and $\lambda_j^{CV}(\mathbf{u}_i)$ are the kriging weights, or the solution of a system such as (3.15) for the estimation of $z(\mathbf{u}_i)$ using the remaining n-1 $z(\boldsymbol{u}_j)$ data. These errors should satisfy some properties that can be used to rank the performance of different variogram models and kriging alternatives. Their expected values should be close to zero to meet the unbiasedness condition, their variance should be small, if plotted on a map they should be randomly distributed (something that can be checked by computing their variogram, which should be close to a nugget effect), and the variance of the so-called reduced error should be close to 1. The reduced error is obtained by normalizing the error by the square root of the kriging variance. If alternative variogram models can be fitted to the experimental variogram, or alternative kriging variants could be used in the estimation process, the variogram model or kriging alternative for which the cross-validation errors best satisfy the above conditions is a good candidate for estimating the regionalized variable at unsampled locations. The term "best" candidate is not used, since the best predictions at sample locations do not ensure the best predictions at the unsampled locations.

3.9 Simulation

An alternative to kriging interpolation is given by stochastic simulation. In simulation, the purpose is not to obtain an estimate that locally minimizes some square deviation, but rather to sample likely realizations of the attribute's spatial distribution in a manner consistent with the sample data and with the statistical properties inferred from them and used to define the space random function.

Each realization should display the same overall mean as the stationary mean of the space random function, along with the same overall variance, and the variogram inferred from any realization should match the one fitted to the data. (Note that in all rigor, this statement only holds when the space random function chosen is stationary in the mean and covariance, is ergodic, and the realization is sufficiently large.)

The above conditions are never met by a kriging map. While the stationary mean is reproduced on the kriging map, owing to the unbiasedness imposed on the derivation of the kriging equations, the variance of an estimated map is always smaller than that of the sample data. Its spatial correlation is also much stronger than that given by the variogram. In this sense, a kriging map can never represent reality, but rather an estimation of the structural component of a parameter's spatial variability.

Given a space-random-function model, characterized by its multivariate distribution function, drawing a realization is the multivariate equivalent to flipping a coin to obtain a realization of a binomial random variable with equal probabilities of each of its members. The challenge with drawing a multivariate realization (that is, generating a value on each grid node) is that even though the univariate distributions of the random variable at each grid node in the domain is known, we cannot draw independently from each univariate distribution if we wish to preserve the spatial correlation patterns, except for the rare occasion when the random variables are independent (that is, spatially uncorrelated). All values of the random variables at the grid nodes have to be drawn consistently with the spatial correlation observed in the data and modeled through the variogram function.

The simplest algorithm for generating realizations from a multivariate distribution function is sequential simulation (Gómez-Hernández and Srivastava, 1990; Gómez-Hernández and Journel, 1993; Gómez-Hernández and Cassiraga, 1994; Deutsch and Journel, 1998; Rubin and Bellin, 1998). The sequential simulation is based on the repetitive application of the conditional probability definition to a multivariate distribution, so as to express it as the product of univariate distributions. Thus, drawing a realization from the multivariate distribution is replaced by the sequential drawing from the univariate distributions. Let us consider, for example, generating a realization over n locations, with multivariate distribution of the n random variables being $F(z_1, z_2, \ldots, z_n)$). This distribution can be decomposed as follows:

$$F(z_1,z_2,\ldots,z_n) = F(z_1)F(z_2 \mid z_1)F(z_3 \mid z_2,z_1)\cdots F(z_n \mid z_1,z_2,\ldots,z_{n-1}) \tag{3.19}$$

as the product of n univariate conditional distributions. This decomposition is easy to perform when the multivariate distribution is multi-Gaussian, because then, all the conditional distributions that appear on the right-hand side of Equation (3.19) are Gaussian and can be easily computed by application of the simple kriging equations, as explained in the previous section. Thus, a realization is generated by first drawing $Z(\mathbf{u}_1)$ from $F(z_1)$, then drawing $Z(\mathbf{u}_2)$ from the conditional distribution of $Z(\mathbf{u}_2)$, given that $Z(\mathbf{u}_1)$ takes the value just drawn z_1—and so on, successively, until the last point that could be drawn from the conditional distribution of $Z(\mathbf{u}_n)$ given the values drawn at all previous nodes. In the case of the multi-Gaussian distribution, prior to the drawing we must solve a simple kriging system at the location being drawn, given the nodes previously simulated to obtain the mean and variance that define the conditional distribution.

The straightforward application of the algorithm just described has some problems related to the increasingly larger kriging system that has to be solved to determine the conditional distributions as the simulation proceeds. These problems are circumvented by approximating the conditional distributions obtained considering only the conditioning points that are "structurally" closer to the location being simulated. This approximation introduces additional implementation problems, the discussion of which is beyond the scope of this chapter, but which can be found in Gómez-Hernández and Cassiraga (1994) and Rubin and Bellin (1998).

3.9.1 CONDITIONING

Equation (3.19) assumes that simulation starts with no measured data. The realizations so generated are referred to as unconditional. If data are available, the objective is to generate conditional realizations, that is, realizations that not only reproduce the statistical patterns observed in the data, but that also honor the data at data locations. In such a case, the realizations are not drawn from the multivariate distribution in Equation (3.19), but rather from the multivariate distribution conditioned to the drawing at the data locations of the measured values. This equation transforms into

$$F(z_1,z_2,\ldots,z_n \mid (m)) = F(z_1 \mid (m))F(z_2 \mid z_1,(m))\cdots F(z_n \mid z_1,z_2,\ldots,z_{n-1},(m)) \tag{3.20}$$

in which (m) represents the set of m conditioning data at their corresponding locations. The generation procedure is identical to the unconditional case, with the only difference that there are already m conditioning values for the generation of the first random variable.

Figure 3.7 shows four conditional realizations of the same slab of Berea sandstone as in Figure 3.6. The conditioning data are indicated by squares in the first realization and coincide with those used in the production of the kriging map in Figure 3.6. All four conditional realizations honor the 64 data values and display the same patterns of variability. For their generation, a multi-Gaussian random function model was adopted.

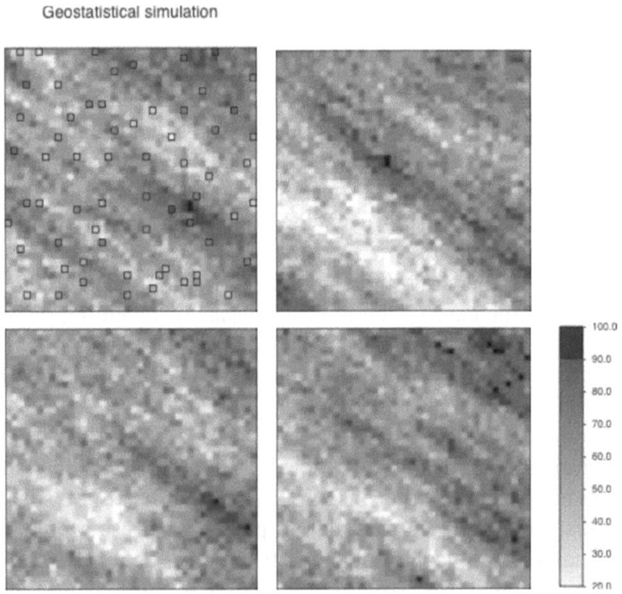

Figure 3.7. Four conditional realizations of permeability (in millidarcys). They were generated using the same 64 data and the same variogram as in Figure 3.6, and choosing a multi-Gaussian space random function.

When we compare Figures 3.6 and 3.7, it is evident that the realizations of Figure 3.7 show patterns of variability closer to those observed in the field. Indeed, not only are their overall means close to the overall mean of the conditioning data, but also their overall variability and their spatial correlation patterns are closer to those observed. The anisotropy of the variogram is clearly noticeable in the realizations, as well as the conditioning effect of the data that control the overall distribution of high and low areas.

The objective of stochastic simulation is to produce maps, conditional to the measurements, which capture the patterns of spatial variability as modeled by the chosen space random function. The differences observed among the realizations represent the uncertainty about the spatial distribution of a parameter that cannot be resolved with the available data.

3.9.2 NON-GAUSSIAN SIMULATIONS

What if we do not want to adopt a multi-Gaussian model? Are there alternatives to computing the conditional distributions in Equations (3.19) or (3.20)?

In the context of traditional geostatistics, there are several alternatives, the two most common of which are indicator simulation and direct simulation. Outside geostatistics, simulated annealing or neural networks have been used to generate realizations. These realizations impose restrictions that depart from the multi-Gaussianity inherent to the implementation of sequential simulation discussed in Section 3.9.1 above.

Indicator simulation and direct simulation both fall into the category of sequential simulation algorithms. Such algorithms offer alternative approaches for computing conditional distributions without resorting to the multi-Gaussian paradigm. In indicator simulation, the conditional distributions are computed by indicator kriging, a type of kriging that allows us to estimate the full conditional distribution nonparametrically, by transforming the data into a series of indicator variables based on whether or not the variables exceed a number of selected thresholds (Journel, 1983; Gómez-Hernández and Srivastava, 1990; Goovaerts, 1997, p. 284). (Note that indicator simulation can also be used for generating realizations of categorical variables; thus, it is especially suited for the generation of facies distributions.)

Direct simulation (Soares, 2001) is a lesser-known alternative based on a relatively unknown property of the kriging system, which is that the conditional distributions in Equations (3.19) and (3.20) do not have to be Gaussian, and that drawing from any univariate distribution (i.e., uniform, binomial, Bessel) with mean and variance—as given by the simple kriging estimate and the simple kriging variance in Equations (3.19) and (3.20)—will ensure that the resulting realization has the variogram used for the kriging estimation. The problem with direct simulation is that, except for those instances when the chosen distribution is Gaussian, the histogram of the final realization is difficult to control. If we wish to reproduce, besides the variogram, a specific non-Gaussian histogram, Soares (2001) suggests ways to choose the shape of the conditional distributions for each step of the sequential simulation to reproduce the desired univariate distribution.

3.10 Estimation or Simulation

The use of estimation or simulation to produce maps of the variable of interest will largely depend on the final use of these maps. In general, estimation is appropriate to display the major trends of variability of the attribute, and its use is advisable to evaluate linear functions of the variable. For instance, if the variable under study is rain, and we wish to estimate the annual rain in a given basin from its pluviograph measurements, then the average of the values on a kriging map provides the average rain with reasonable accuracy: the total rain is a simple linear combination of the rain values at each location in the map.

Simulation is appropriate when we need to generate maps that are plausible renditions of reality, in the sense that, taken as a whole, the map may be a representation of the unknown spatial distribution. Locally it will always be less precise than an estimated map, because the estimated map is built with the objective of being as locally precise as possible. However, the simulated realization will display spatial patterns of variability in consonance with the data. Simulations must be used when the aim is to evaluate non-linear functions of the variable under study. For instance, if we are interested in generating conductivity maps that are later to be used in predicting mass transport from a contaminant release point, we must carry out the evaluation using a Monte Carlo approach, with realizations generated by stochastic simulation. What sense does it make to perform a transport analysis with the kriging map of Figure 3.6 (in which the extreme conductivity values have been smeared out, and the short-scale variability has disappeared) when we know that what controls transport is the presence of flow channels or flow barriers, corresponding to clusters of extreme values? The mass transport phenomenon is a highly nonlinear function of the conductivities; it cannot be approximated by its evaluation in a kriging map.

When realizations are to be used in some complex, nonlinear transfer function to compute dependent variables (e.g., concentrations), the proper procedure for obtaining a probabilistic evaluation of the dependent variables is by a Monte Carlo method. This method involves using the following steps: (1) A set of conditional realizations are generated, each one being an equally likely rendition of our model of reality; (2) The transfer function is evaluated in each realization to obtain realizations of the dependent variables; and (3) The dependent variable realizations are pooled into a frequency distribution, from which an estimate of their mean values, with confidence intervals, can be made.

3.11 Recent Developments

3.11.1 MULTIPLE POINT GEOSTATISTICS

One of the main criticisms of traditional geostatistics is that it is based on only two-point statistics. It is difficult to capture curvilinear patterns of variability such as those that could be associated with certain types of geological formations, (e.g., fluvial deposits). Some attempts to address this problem were carried out by varying the orientation of the ellipse of anisotropy over the area of study, maintaining the major axis tangent to the traces of hypothetical rivers. While the resulting realizations exhibited some meandering-type patterns, the variable orientation of the anisotropy ellipse was difficult to specify, and to sustain from data. Other geological characteristics, such as preventing certain juxtapositions of facies, or enforcing a certain facies sequencing, are also difficult to impose with traditional geostatistics.

The types of patterns sought by geologists could only be captured by resorting to higher-order statistics. This avenue was successfully put into practice by Strebelle (2002), who derived a sequential simulation algorithm in which the conditional distributions in Equations (3.18) and (3.19) are computed not by kriging equations—which only take advantage of two-point statistics—but by inferring them directly for each specific data configuration from a training image.

The training image is an exhaustive realization, in which the parameter under study is known at every location. This training image could be obtained from an outcrop within the same or a similar formation, or from a digitized geologic interpretation, or from a numerical model capable of generating the spatial patterns sought, such as the code SEDSIM (Tetzlaff and Harbaugh, 1989). SEDSIM can generate realistic sedimentary patterns, but it is very difficult to condition to measurements.

The multiple-point algorithm can retrieve the probabilities associated with a given data configuration by scanning the training image, searching for repetitions of the same data configuration and the same data values, and retrieving the alternative variable values occurring at the location for which the conditional probability is sought. For example, consider a configuration in which we have sampled sand in the four corners connecting five sides of a square, and that we are interested in determining the conditional distribution of the lithology at the center of the square. A scan has to be performed in the training image, looking for facies arrangements in which the sand appears in the four corners of a square of side 5 units. Each time such a configuration is found, the lithology in the center location is recorded, and, eventually, when the scan has finished, sufficient repetitions have occurred to build a histogram. This histogram would be the conditional distribution, consistent with the training image.

The most critical point of multiple-point geostatistics is the selection of the training image. Besides this, an efficient implementation of the approach requires restricting the number of data configurations to a small number of templates. Also, in the case of continuous variables, we must bin the range of variability to sample the conditional probabilities from the training image. Moreover, to capture patterns at different scales, we must implement some type of telescoping or multigridding approach. Solutions to these problems are discussed in Strebelle (2002).

3.11.2 OBJECT-BASED ALGORITHMS

As with multiple-point geostatistics, object-based algorithms are aimed at reproducing geological structures. They originate from the so-called bombing models to define the optimal coverage of an areal bombing raid. Their fundamental characteristic is the placing of objects (resembling geological features) at random within the simulation domain. The shapes of the objects are predetermined, and the parameters controlling their sizes and orientations are chosen at random from predetermined distribution functions. Typical shapes are parallelepipeds, ellipses, ellipsoids, sinusoidal bands, spheres, or segments. For most occasions, the placing of the objects is not totally random, since there are some rules about spatial proportions, attraction, repulsion, and sequencing of the objects that have to be respected.

Typically, an object-based algorithm proceeds as follows. Each shape is associated with a certain lithology. We select a shape and randomly draw a location for its barycenter. Then, the parameters defining the size and orientation of the object are drawn from their respective distributions (for instance, the length and orientation of a segment), after which we check to see if placing the object at a certain location meets the constraints regarding proportions and relative positions to other objects. If they are met, the object is incorporated into the simulation. The process continues drawing objects with the same shape until some *a priori* proportion about the facies is reached, and then it is repeated for the other shapes (facies) until the proportions of all facies are respected. During the placing of the objects, some additional rules could be introduced, such as rules controlling erosion and accretion of objects.

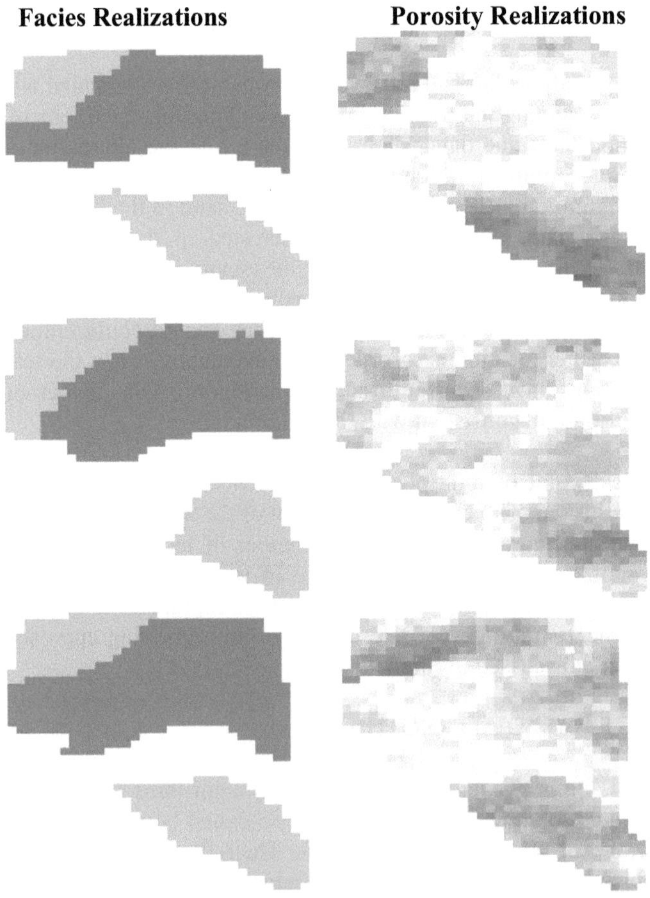

Figure 3.8. Three realizations of porosity using a hybrid method. On the left, realizations of the facies generated by sequential indicator simulation. On the right, the porosity realizations, generated by independent sequential Gaussian simulations. Light shades correspond to low porosities, dark shades to high porosities—porosity varies between 0.05 and 0.25.

From the description of the algorithm, it can be deduced that the algorithm should work when the number of objects represent a small fraction of the total volume of the area to be simulated, and the number of interaction rules are not too numerous. Otherwise, it may become very difficult to attain (by randomly choosing points within the simulation volume) the desired proportions and constraints. An additional difficulty is that of conditioning to measurement data. Still, aside from these difficulties, the final realizations are very appealing to geologists, since these realizations resemble their expectations about the architecture of the formation. For this reason, one interesting use of object simulation is for generating those training images from which to infer the higher-order statistics needed to perform multiple-point geostatistical simulations.

Some of the earlier applications of object simulation in the context of subsurface flow were carried out (in his Ph.D. thesis) by Haldorsen, which are described in the paper by Haldorsen and Damsleth (1990). A very good review of this and other methods aimed at reproducing geological patterns is given by Koltermann and Gorelick (1996). The problem of conditioning is discussed by Skorstad (1999) and Oliver (2002). A comparison between traditional geostatistics and object-based simulation for the generation of fluvial reservoir realizations is discussed by Seifert (2000).

3.11.3 HYBRID METHODS

The solution to the realistic representation of geological patterns is probably not given by any one of the methods discussed above. Each method has its advantages, and the best approach should be some clever combination of them. Under the term *hybrid methods*, we refer to a family of approaches that mixes the different algorithms. Some possibilities are given below:

- Use sequential indicator simulation to generate a realization of different facies, then within each facies simulate, by sequential Gaussian simulation, the parameter of interest within each facies, using a different Gaussian random function for each facies (Rubin, 1995; Chu et al., 1995).
- Use object simulation to generate the geological architecture of the facies, then use indicator simulation within each facies to generate the parameter, with a different nonparametric random function in each facies (Haldorsen and Damsleth, 1990).
- Use object simulation to generate an unconditional realization of the geology, then use this realization as a training image for multiple-point geostatistics and generate a conditional realization of the geology. After that, use sequential Gaussian and/or sequential indicator simulation to fill in each geological facies with a different random function model for each parameter.

Figure 3.8 shows three realizations generated by a hybrid method that combines sequential indicator and sequential Gaussian simulations. The left column contains the conditional realizations of the four facies present at the site, generated by sequential indicator simulation. The facies that are in the extremes of the simulation domain are characterized by higher porosities than the two facies towards the center of the area. The right column contains the conditional realizations of porosity obtained by running independent sequential Gaussian simulations for each facies. In this particular case, the correlation patterns, i.e., the variograms, were kept the same for the four faces, but the mean values changed according to the prior characterization.

3.12 Sources for Tools

There are some public domain codes ready to be downloaded and used for the practice of geostatistics. They include tools for variography, estimation, and simulation. Without being exhaustive, a relevant list follows (practically all of them are described in http://www.ai-geostats.org, which contains a larger list of references):

- gslib (Deutsch and Journel, 1998), Fortran codes for 2-D and 3-D geostatistics, available from ftp://geostat1.stanford.edu/gslib
- gstat (Pebesma and Wessling, 1998), C code for 2-D geostatistics, available from http://www.gstat.org
- fsstools, Linux toolbox for geostatistics, available from http://www.fssintl.com
- variowin (Pannatier, 1996), Windows code for variogram analysis, available from http://www-sst.unil.ch/research/variowin/index.html
- gcosim3d, (Gómez-Hernández and Journel, 1993), C code for the 3-D multiGaussian simulation of multiple cross-correlated variables, available from http://ttt.upv.es/~jgomez/gcosim3d.tar.gz
- isim3d, (Gómez-Hernández and Srivastava, 1990), C code for the 3-D sequential simulation of indicator variables, available from http://ttt.upv.es/~jgomez/isim3d.tar.gz
- hydrogen (Rubin and Bellin, 1998) sequential Gaussian simulation, incorporates some covariance functions specific to the stochastic hydrogeology literature, available from http://www.ing.unitn.it/~bellin/Hydro_ge.htm
- uncert (Wingle et al., 1999), geostatistical and uncertainty analysis, originally oriented to hydrogeology, although it contains many general modules, available from http://uncert.mines.edu

References

Allen-King, R. M., R.M. Halket, D.R. Gaylord, and M.J.L. Robin, Characterizing the heterogeneity and correlation of perchloroethene sorption and hydraulic conductivity using a facies-based approach, *Water Resour. Res., 34*(3), 385–396, 10.1029/97WR03496, 1998.

Bierkens, M.F.P., Modeling hydraulic conductivity of a complex confining layer at various spatial scales, *Water Resour. Res., 32*(8), 2369–2382, 10.1029/96 WR01465, 1996.

Caers, J., Geostatistical history matching under training-image based geological model constraints, SPE Paper 77429, *Proceedings of the SPE Annual Technical Conference and Exhibition*, San Antonio, Texas, 2002.

Chiles, J.-P., and P. Delfiner, *Geostatistics: Modeling Spatial Uncertainty*, John Wiley & Sons, 1999.

Chu, J., W. Xu, and A.G. Journel, 3-D Implementation of geostatistical analyses: The Amoco case study, In *Stochastic Modeling and Geostatistics: Principle, Methods and Case Studies*, Chapter 16, AAPG Computer Applications in Geology, 3, J.M. Yarus and R.L. Chambers, eds., 201–216, 1995

Deutsch, C.V., and A.G. Journel. *GSLIB, Geostatistical Software Library and User's Guide*, Oxford University Press, New York, second edition, 1998.
Deutsch, C.V., and L. Wang., Hierarchical object-based stochastic modeling of fluvial reservoirs, *Math. Geol.28*(7), 857–880, 1996.
Dreiss, S.J., and N.M. Johnson, Hydrostratigraphic interpretation using indicator geostatistics, *Water Resour. Res., 25*(12), 2501–2510, 10.1029/89WR01380, 1989.
Gómez-Hernández, J.J., and E.F. Cassiraga. Theory and practice of sequential simulation, In: *Geostatistical Simulations*, M. Armstrong and P. Dowd, eds., pp. 111–124. Kluwer Academic Publishers, 1994.
Gómez-Hernández, J.J., and A.G. Journel. Joint simulation of multiGaussian random variables. In: *Geostatistics Tróia '92*, Vol. 1, A. Soares, ed., pp. 85–94. Kluwer, 1993.
Gómez-Hernández, J.J., and R.M. Srivastava. ISIM3D: An ANSI-C three-dimensional multiple indicator conditional simulation program, *Computer and Geosciences, 16*(4), 395–440, 1990.
Gómez-Hernández, J.J., and X.-H. Wen. To be or not to be multiGaussian: A reflection in stochastic hydrogeology, *Advances in Water Resources, 21*(1), 47–61, 1998.
Goovaerts, P, *Geostatistics for Natural Resources Evaluation*, Oxford University Press, 1997.
Haldorsen, H.H., and E. Damsleth. Stochastic modeling. SPE Paper 20321. *J. of Petr. Techn (April 1990)*, 404–412, 1990.
Isaaks, E.H., and R.M. Srivastava, *An Introduction to Applied Geostatistics*, Oxford University Press, 1990.
Journel, A.G., Non-parametric estimation of spatial distributions, *Math. Geology, 5*(3), 445–468, 1983.
Journel. A.G., mAd and the conditional quantile estimators, In: *Geostatistics for Natural Resources Characterization*, G. Verly, M. David, A.G. Journel, and A. Marechal, eds., pp. 915–934, NATO Advanced Study Institute, South Lake Tahoe, California, September 6–17, D. Reidel, Dordrecht, Holland, 1984.
Journel, A.G., *Fundamentals of Geostatistics in Five Lessons*, Short Courses in Geology, Vol. 8, American Geophysical Union, 1989.
Jussel, P., F. Stauffer, and T. Dracos. Transport modeling in heterogeneous aquifer: 1. Statistical description and numerical generation of gravel deposits, *Water Resources Research, 30*(6), 1803–1817, 1994.
Krige, D.G., A statistical approach to some mine evaluations and allied problems at the Witwatersrand, Unpublished Master's Thesis, University of Witwatersrand, 1950.
Koltermann, C.E., and S.M. Gorelick, Heterogeneity in sedimentary deposits: A review of structure-imitating, process-imitating, and descriptive approaches, *Water Resour. Res., 32*(9), 2617–2658, 10.1029/96WR00025, 1996.
Matheron, G., *Traité de Géostatistique Appliquée,* Vols 1 and 2, Technip, Paris, 1962.
Olea, R., *Geostatistics for Engineers and Earth Scientists*, Kluwer, 1999.
Oliver, D., Conditioning channel meanders to well observations, *Math. Geology, 34*(2), 185–201, 2002.
Pannatier, Y., *VARIOWIN: Software for Spatial Data Analysis in 2D*, Springer-Verlag, New York, NY, 1996.
Pebesma, E. and C.G. Wesseling, Gstat: A program for geostatistical modeling, prediction and simulation, *Computers & Geosciences, 24*(1), 17–31, 1998.
Rubin, Y., Flow and transport in bimodal heterogeneous formations, *Water Resour. Res., 31*(10), 2461–2468, 1995.
Rubin, Y., and A. Bellin, Condtional simulation of geologic media with evolving scales of heterogeneity, In: *Scale Invariance in Hydrology*, G. Sposito, ed., Cambridge University Press, Cambridge, 1998.
Seifert, D., Object and pixel-based reservoir modeling of a braided fluvial reservoir, *Math. Geology, 32*(5), 581–603, 2000.
Skorstad, A., Well conditioning in a fluvial reservoir model, *Math. Geology*, *31*(7), 857–872, 1999.
Soares, A., Direct sequential simulation and cosimulation, *Math. Geology, 33*(8), 911–926, 2001.
Strebelle, S., Conditional simulation of complex geological structures using multiple-point statistics, *Math. Geology, 34*(1), 1–22, 2002.
Tetzlaff, D.A., and J.W. Harbaugh, *Simulating Clastic Sedimentation*, New York: van Nostrand Reinhold, 1989.
Wen, X.-H., and J.J. Gómez-Hernández. Numerical modeling of macrodispersion in heterogeneous media: A comparison of multi-Gaussian and non-multi-Gaussian models, *J. of Contaminant Hydrology, 30*(1998), 129–156, 1998.
Wingle, W.L., E.P. Poeter, and S.A. McKenna, UNCERT: Geostatistics, uncertainty analysis, and visualization software applied to groundwater flow and contaminant transport modeling, *Computers and Geosciences, 25*(4), 365–376, 1999.

Fundamentals of Environmental Geophysics

4 RELATIONSHIPS BETWEEN THE ELECTRICAL AND HYDROGEOLOGICAL PROPERTIES OF ROCKS AND SOILS

DAVID P. LESMES[1] and SHMULIK P. FRIEDMAN[2]

[1]*The George Washington University, Department of Earth and Environmental Sciences, Washington, D. C., U.S.A.*

[2]*Institute of Soil, Water, and Environmental Sciences, Agricultural Research Organization—The Volcanic Center, Bet Dagan, Israel*

4.1 Introduction

The ability to reliably predict the hydraulic properties of subsurface formations is one of the most important and challenging goals in hydrogeophysics. In water-saturated environments, estimation of subsurface porosity and hydraulic conductivity is often the primary objective. In partially saturated environments, characterization of the water content and the hydraulic conductivity as a function of saturation is also often required. Because the hydraulic conductivity of geologic formations varies by orders of magnitude over relatively small spatial scales, it is difficult to accurately characterize subsurface aquifer properties using just the information obtained from networks of widely spaced boreholes. A more complete and accurate characterization of the subsurface can be achieved by using an integrated exploration approach in which borehole and geophysical data sets are jointly interpreted. A key step in quantitative hydrogeophysical interpretations is the transformation of the measured geophysical properties into the desired hydrogeological parameters. This transformation typically relies on petrophysical relationships; these relationships can be developed at the field-scale using co-located hydrogeological-geophysical data, through laboratory experimentation on rock and soil samples, or by using theoretically based models. The objective of this chapter is to review the petrophysical models used to derive electrical-hydrogeological predictive relationships and to evaluate the theoretical and practical limitations of these relationships.

Electrical methods are widely used in hydrogeophysical investigations to obtain high-resolution information about subsurface conditions. These methods include electrical resistivity (ER), induced polarization (IP), electromagnetic induction (EMI), ground-penetrating radar (GPR), and time-domain reflectometery (TDR) surveys. These electrical methods, which operate at frequencies ranging from direct current (DC) to >1GHz, can be used individually or in combination to obtain information about the subsurface structure and composition. Both the structural information and the electrical property information provide important constraints for hydrogeological modeling. The structural information can be used to define hydrostratigraphic units as well as the locations of faults and fractures. The electrical property information can be used to qualitatively characterize the rock/soil type as well as the pore fluid properties. Petrophysical models can often be used to make more quantitative predictions about the

Y. Rubin and S.S. Hubbard (eds.), Hydrogeophysics, 87–128.

rock/soil properties, such as the water content, water conductivity, porosity, clay content, and hydraulic conductivity.

Petrophysical models of the electrical and hydrogeological properties of rocks and soils are formulated in terms of the intrinsic properties of the rock/soil system (i.e., the pore/grain topology and the physical and chemical properties of the pore fluids). Although the electrical and hydrogeological properties of rocks and soils depend on some common controlling factors, these relationships are often non-unique. As will be discussed in Chapter 17 of this volume, non-uniqueness can often be reduced by using multiple types of geophysical data or by applying other constraints based on a-priori information (e.g., pore fluid conductivity, porosity, and geological information). In practice, however, it is often difficult to constrain all of the model parameters required to make quantitative predictions. Therefore, theoretical models must either be simplified to reduce the number of free parameters, or empirical relationships can be established between the hydrogeological and electrical properties. These empirical relations are typically based on correlations made between electrical and hydrogeological measurements obtained from a specific site. These empirically established electrical-hydraulic relationships often work reasonably well; however, they are usually only applicable to the specific study site or to sites with similar characteristics (e.g., Huntley, 1986; Purvance and Andricevic, 2000).

We are therefore faced with the problem of the theoretically established electrical-hydrogeological petrophysical relationships being typically too complex (too many unconstrained parameters) to be applied in practice, and the empirically established relationships being valid only for specific sites. What we seek are theoretically based models that capture the intrinsic connections between the electrical and hydrogeolgoical properties, while being simple enough to be applied in the field. In this chapter, we review theoretically and empirically based models for the electrical prediction of hydrogeologic properties, and try to establish connections between these two approaches.

4.2 Hydrogeological and Electrical Properties

4.2.1 HYDROGEOLOGICAL PROPERTIES

Rocks and soils are composed of solid mineral grains and pore space. The porosity (n) is defined as the ratio of pore volume (V_p) to the total volume of the sample (V_t):

$$n = \frac{V_p}{V_t}. \tag{4.1}$$

Some materials contain isolated pores. In these cases, the effective porosity of the connected pore space is less than the total porosity.

4.2.1.1 *Saturated Flow*

The rate at which fluids flow through saturated porous materials is controlled by the saturated hydraulic conductivity (K_s). For homogeneous and isotropic materials, K_s is given by (e.g., Fetter, 2001)

$$K_s = \frac{k_s \rho_w g}{\mu}, \tag{4.2}$$

where k_s is the hydraulic permeability, μ is the fluid dynamic viscosity, ρ_w is the fluid density, and g is the acceleration of gravity. The hydraulic permeability is primarily a function of the pore size distribution, connectivity, and tortuosity of the pore network. In granular materials, the topology of the pore space is determined by the grain size distribution, the packing, and the cementation. Permeability models can either be formulated in terms of the characteristics of the pore space or in terms of the characteristics of the solid mineral grains (e.g., Nelson, 1994)

The equivalent channel model of Kozeny and Carman, or K-C (Carman, 1939), assumes that flow through a porous medium can be represented by flow through a bundle of capillaries. Each capillary is assumed to represent an independent flow path through the sample, where the effective path length (L_a) is greater than or equal to the macroscopic length of the sample (L). The tortuosity is defined as $T = (L_a/L)^2$. From considerations of laminar viscous flow through tubes, the following permeability equation is obtained (e.g., Scheidegger, 1974):

$$k_s = \frac{n r_h^2}{aT}, \tag{4.3}$$

where a is a tube shape factor (a dimensionless number between 1.7 and 3) and r_h is the hydraulic radius.

A common measure of the hydraulic radius is the reciprocal of the specific surface area (S_p), which is the ratio of pore surface area to pore volume. Furthermore, the tortuosity can be related to the electrical formation factor F (defined in Equation (4.26)) through the following relationship (e.g., Nelson, 1994)

$$T = \left(\frac{L_a}{L}\right)^2 = nF. \tag{4.4}$$

Equation (4.3) can then be expressed as

$$k_s = \frac{1}{aFS_p^2}. \tag{4.5}$$

This form of the K-C equation is more practical than Equation (4.3) because F and S_p are well-defined physical properties that can be measured, whereas T and r_h are theoretical constructs that cannot be directly measured.

Grain-based permeability models predict k_s to be dependent on the square of an effective grain size. In the Hazen model, the hydraulic conductivity is given by (e.g., Fetter, 2001)

$$K_s = Cd_{10}^2, \tag{4.6}$$

where the effective grain size d_{10} (in cm) corresponds to the grain size for which 10% of the sample is finer, and the coefficient C depends upon the grain sorting. The Hazen model is applicable to sediments where d_{10} ranges between 0.1 and 3.0 mm. More elaborate grain-based permeability models explicitly account for the effects of grain sorting and porosity (e.g., Nelson, 1994).

4.2.1.2 *Unsaturated Flow*

In the unsaturated zone, the pore space is filled with water and air, and the water is held in tension (negative pressure). The water saturation (S) is the ratio of the volume of water (V_w) to the total pore volume:

$$S = \frac{V_w}{V_p}. \tag{4.7}$$

The water content (θ) is defined as the ratio of the volume of water to the total volume of the sample:

$$\theta = \frac{V_w}{V_t}. \tag{4.8}$$

The hydraulic conductivity of partially saturated rocks and soils (K) is a function of K_s as well as the saturation level and the soil-moisture-retention properties of the material.

The soil-moisture-retention function $\theta(\psi)$ describes the relationship between the water content and the capillary head (ψ). For a capillary tube of radius r, the capillary head is given by

$$\psi = -2\gamma \cos\beta / \rho_w g r \tag{4.9}$$

where γ is the surface tension between water and air, and β is the contact angle. The shape of the soil-moisture-retention function is primarily controlled by the pore size distribution, as large pores tend to drain at low pressures and increasingly smaller pores drain at higher pressures (e.g., Fetter, 2001). Many empirical and theoretical models have been developed to fit soil-moisture-retention measurements. The most widely used function is that developed by van Genuchten (1980). The effective saturation (S_e) for the van Genuchten function is given by

$$S_e = \frac{\theta - \theta_r}{\theta_s - \theta_r} = \frac{1}{\left[1 + \alpha(\psi)^n\right]^m} \tag{4.10}$$

where θ_s and θ_r are the saturated and residual water contents, respectively, and α, m, and n are fitting parameters, which can be related to the statistics of the pore size distribution function (e.g., Kosugi, 1994).

The relative hydraulic conductivity ($K_r = K/K_s$) decreases with decreasing saturation, as under these conditions, pore water is held more tightly, and the fluid must take a more tortuous flow path as the larger pores drain. The dependence of K_r on θ is determined by the soil-moisture-retention function (e.g., Fetter, 2001). Mualem's (1976) model for predicting K_r from the soil-water-retention function is

$$K_r = K / K_s = S_e^{1/2} \left\{ \int_0^{S_e} \frac{dS_e}{|\psi|} \Big/ \int_0^1 \frac{dS_e}{|\psi|} \right\}^2 . \tag{4.11}$$

To predict K_r using Equation (4.11), one must specify $S_e(\psi)$. As described above, K_s can be predicted from measurements of F, and an effective pore or grain size. $S_e(\psi)$ can be predicted from the pore size distribution of the sample (e.g., Kosugi, 1994). Therefore, if the formation factor and the pore-size distribution of a rock or soil sample are known, it should be possible to make relatively accurate estimates of the saturated and unsaturated hydraulic properties of the sample (i.e., K_s, $S_e(\psi)$ and K_r). In the rest of this chapter, we address the feasibility of electrically estimating the key rock and soil properties required to predict flow in subsurface formations.

4.2.2 ELECTRICAL PROPERTIES

The conductive and capacitive properties of a material can be represented by a complex conductivity (σ^*), a complex resistivity (ρ^*), or a complex permittivity (ε^*), where

$$\sigma^* = \frac{1}{\rho^*} = i\omega\varepsilon^* \tag{4.12}$$

and ω is the angular frequency ($\omega = 2\pi f$) and $i = \sqrt{-1}$ (e.g., Schön, 1996). The complex electrical parameters can be expressed either in polar or rectangular form. For example, the complex conductivity can be expressed in terms of a magnitude ($|\sigma|$) and phase (ϕ), or by real (σ') and imaginary (σ'') components:

$$\sigma^* = |\sigma| e^{i\phi} = \sigma' + i\sigma'' . \tag{4.13}$$

The complex conductivity response of Berea sandstone saturated with 0.01M KCl and measured over the frequency range of 10^{-3} Hz to 10^{6} Hz is plotted in Figure 4.1. Also plotted in Figure 4.1 is the real part of the relative permittivity or dielectric constant. The dielectric constant (κ') is the ratio of the permittivity of the sample (ε') to the permittivity of vacuum (ε_0), and it is proportional to the imaginary component of the complex conductivity

$$\kappa' = \frac{\varepsilon'}{\varepsilon_0} = \frac{\sigma''}{\omega\varepsilon_0} . \tag{4.14}$$

As the dielectric response approaches the low-frequency limit (κ'_{static}), the imaginary conductivity goes to zero. As the dielectric response approaches the high-frequency limit (κ'_{∞}), the imaginary conductivity becomes large.

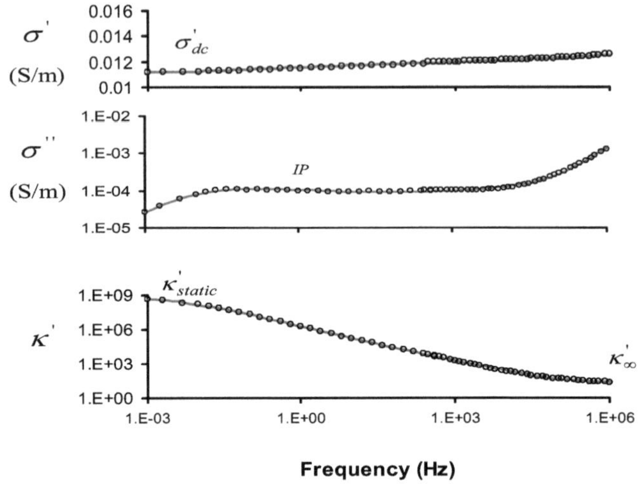

Figure 4.1. Complex responses of Berea sandstone saturated with 0.01 M NaCl (Lesmes and Frye, 2001): (top) real component of complex conductivity, (middle) imaginary component of complex conductivity, (bottom) dielectric constant or permittivity

In the laboratory, three different measurement systems are needed to measure the broadband electrical impedance response from 10^{-3} Hz to 10^{9} Hz. Four electrode systems are used to measure the impedance response from 10^{-3} Hz to 10^{3} Hz; two-electrode systems are used to measure the impedance response from 10^{2} Hz to 10^{7} Hz; and transmission line systems (e.g., TDR) are used to measure the impedance response from 10^{6} Hz to 10^{10} Hz (Olhoeft, 1985, 1986). The effective frequency range of each of these measurement systems is limited by systematic errors intrinsic to the measurement configuration. For each measurement system, it is possible to have large systematic errors that are stable and repeatable but give erroneous results. It is important, therefore, to test and calibrate these systems with known standards (e.g., standard resistors and standard brines) to determine the overall accuracy and precision of the measurements (Olhoeft, 1985; VanHalla and Soinenen, 1995; Lesmes and Frye, 2001). It is not possible to measure the complete broadband impedance response in the field with currently available instrumentation. As with laboratory measurement systems, theoretical and practical considerations limit the effective frequency ranges of the various electrical geophysical instruments and methods. The most commonly measured electrical parameters in the field are the high-frequency permittivity or dielectric constant (κ'_{∞}), the low-frequency conductivity (σ'_{dc}), and a low-frequency capacitance, commonly referred to as the induced polarization (IP) or complex resistivity (CR) response. These electrical parameters can be directly measured with point sensors or indirectly estimated by inverting noninvasive geophysical measurements for the subsurface electrical properties. Dielectric measurements are typically made using TDR (refer to Chapter 15 of this volume) or GPR methods (Chapter 7), and conductivity measurements are typically made using electrical resistivity (Chapter 5) or electromagnetic induction methods (Chapter 6). The IP and CR methods are similar to the electrical resistivity method, but they measure both the low-frequency conductive and capacitive properties of a material (see Chapter 5).

The three electrical parameters—κ'_∞, σ'_{dc}, and an IP parameter (there are several different but equivalent measures of the IP response)—contain complementary information about a material. Petrophysical models can be used to interpret these electrical measurements in terms of the physical and chemical properties of a material. Models for the high-frequency dielectric response (κ'_∞) and the low-frequency conductivity response (σ'_{dc}) of rocks and soils are reviewed below in Sections 4.3 and 4.4, respectively. Models for the induced polarization response and the frequency-dependent complex conductivity response of rocks and soils are reviewed below in Sections 4.5 and 4.6, respectively.

4.3 Permittivity Models

In this section, we review models for the frequency-independent permittivity responses of rocks and soils. These models are generally applicable to permittivity measurements made in the frequency range of 100 MHz to 10 GHz. At lower frequencies, polarization mechanisms cause the permittivity responses of rocks and soils to increase with decreasing frequency (see Figure 4.1). At very high frequencies (f >10 GHz), the permittivity begins to decrease as the relaxation frequency of water molecules is approached. The high-frequency limit of the permittivity response, indicated as κ'_∞ in Figure 4.1, is referred to as κ_{eff} in the models presented in this section. The high-frequency permittivity response of rocks and soils is primarily sensitive to the water content, as the relative permittivity of water, $\kappa_W \cong 80$, is much higher than the relative permittivity of dry mineral grains, κ_s = 4 to 8, or air, κ_a = 1 (e.g., Olhoeft, 1981). Secondary factors affecting the permittivity responses of rocks and soils include the effective shapes of the pores and grains, fine-scale laminations, temperature, and to a lesser degree the salinity of the saturating solution (e.g., Olhoeft, 1981; Schön, 1996). Models of the high-frequency permittivity responses of rocks and soils are generally not dependent upon the pore or grain size. As discussed at the end of this section, some attempts have been made to predict the hydraulic conductivity of earth materials from high-frequency permittivity measurements. However, it is not possible to theoretically derive a predictive relationship between permittivity and hydraulic conductivity, so the empirically obtained relationships are quite site specific.

4.3.1 WATER-SATURATED MEDIA

The refractive index (RI) model (Birchak et al., 1974) is the most widely used relationship to predict the relative volume fractions of two-phase mixtures. For water-saturated porous materials, the RI model is given by

$$\sqrt{\kappa_{eff}} = n\sqrt{\kappa_W} + (1-n)\sqrt{\kappa_S} \quad . \tag{4.15}$$

By specifying κ_W and κ_S, the RI model can be used to predict the porosity of water - saturated materials from their measured permittivity. The physical assumption behind the RI model is that the time for an EM wave to travel through a porous material is equal to the sum of the travel times for it to pass through the separate phases of the material (solid grains and pore water) as if they were arranged in series. Equation (4.15) is also referred to as the CRIM equation or the time propagation equation (Wharton et al., 1980). The slight decrease in κ_w with increasing temperature and salinity can be computed using Equations 4.68 and 4.69, respectively, given in the Appendix.

The dielectric constant κ_s of common rock and soil-forming minerals is much smaller that the dielectric constant of water—for example: quartz $\cong 4.5$, calcite $\cong 9$, and clay $\cong 5.5$ (Robinson and Friedman, 2003; Robinson, 2004). For mixtures of different types of mineral grains, the dielectric constant of the solid matrix can be estimated as the weighted arithmetic mean of the dielectric constants for the different mineral constituents (Robinson and Friedman, 2002).

Theoretically based models can be used to predict the dielectric properties of heterogeneous materials in terms of the permittivities and volume fractions of the individual phases and their microgeometrical configurations. Continuum mean field theories are widely used to predict the dielectric and conductive properties of rocks and soils. The following "universal" mixing formula can be used to represent several dielectric mixing models derived using different mean-field theory approaches (Sihvola and Kong, 1988, 1989):

$$\kappa_{eff} = \kappa_w + \frac{\left[\dfrac{(1-n)(\kappa_s-\kappa_w)\left[\kappa_w + a(\kappa_{eff}-\kappa_w)\right]}{\left[\kappa_w + a(\kappa_{eff}-\kappa_w) + \frac{1}{3}(\kappa_s-\kappa_w)\right]}\right]}{\left[1 - \dfrac{\frac{1}{3}(1-n)(\kappa_s-\kappa_w)}{\left[\kappa_w + a(\kappa_{eff}-\kappa_w) + \frac{1}{3}(\kappa_s-\kappa_w)\right]}\right]}. \tag{4.16}$$

Equation (4.16) expresses the effective dielectric constant as a function of a background phase of the pore solution (with relative permittivity κ_w and volume fraction n), with spherical inclusions of the mineral grains (relative permittivity κ_S and volume fraction 1-n). The heuristic parameter a in Equation (4.16), which ranges from 0 to 1, accounts for the effect of neighboring particles on the internal electrical field of a reference particle. The term ($\kappa_w + a(\kappa_{eff} - \kappa_w)$) in this equation represents the apparent permittivity of the background as felt by a reference particle. When there are no interactions between the embedded particles, the heuristic parameter $a = 0$ and the universal mixing formula reduces to the Maxwell-Garnett (MG) model (Maxwell and Garnett, 1904), which can be expressed as

$$\kappa_{eff} = \kappa_w + 3(1-n)\kappa_w \frac{\kappa_s - \kappa_w}{\kappa_s + 2\kappa_w - (1-n)(\kappa_s - \kappa_w)}. \tag{4.17}$$

The MG model results in an upper bound for the permittivity response of a water-saturated sample. When $a = 2/3$, the universal mixing formula results in the symmetric effective-medium approximation (Polder and van Santen, 1946) (PVS model); and when a=1, it results in the coherent potential (CP) mixing formula (Tsang et al., 1985). The PVS and CP mixing formulas account for interactions between the embedded particles; they usually underpredict the permittivity responses of high-porosity granular materials such as soils (Jones and Friedman, 2000). To account for modest interactions between the embedded particles Friedman and Robinson (2002) set a = 0.2 in Equation (4.16) and obtained good fits to permittivity measurements made on water-saturated packings of glass beads, quartz sand, and tuff grains.

The mean-field theories expressed in Equations (4.16) and (4.17) are for spherical inclusions; however, pore/grain shape significantly affects the permittivity response of porous materials. In effective media theories, the particle shape is typically represented by an ellipsoid of revolution (spheroid) for which analytical expressions can be derived (e.g., Sihvola and Kong, 1988; Jones and Friedman, 2000). By extending or contracting the b and c axes while keeping a constant, a sphere can be transformed into either disk-like (oblate) particles or needle-shaped (prolate) particles. The effect of particle shape and orientation on the permittivity response is characterized by the depolarization factors: N^a, N^b, and N^c. For ellipsoids of revolution, where $a \neq b = c$, Jones and Friedman (2000) found that $N^a(a/b)$ is well approximated by a single empirical expression (r^2 = 0.9999):

$$N^a = \frac{1}{1+1.6(a/b)+0.4(a/b)^2} \quad ; \quad N^b = N^c = \frac{1}{2}(1-N^a)\,. \tag{4.18}$$

The depolarization factors for a sphere (a/b=1) are $N^{a,b,c}$ = 1/3, 1/3, 1/3; for a thin disk (a/b<<1) $N^{a,b,c}$ =1,0,0; and for a long needle (a/b>>1) $N^{a,b,c}$=0, 1/2, 1/2.

Nonspherical particles can form either anisotropic or isotropic packings, depending on whether the particles are aligned or randomly oriented. The effective permittivity of a material composed of nonspherical particles, aligned to make a uniaxial-anisotropic medium, is described by a second-order tensor with diagonal components given by $\kappa_{eff}{}^a \neq \kappa_{eff}{}^b = \kappa_{eff}{}^c$. The effective permittivity in the ith direction takes the form (Sihvola and Kong, 1988):

$$\kappa_{eff}{}^i = \kappa_w + \frac{\dfrac{(1-n)[\kappa_w + a(\kappa_{eff}{}^i - \kappa_w)](\kappa_s - \kappa_w)}{\kappa_w + a(\kappa_{eff}{}^i - \kappa_w) + N^i(\kappa_s - \kappa_w)}}{1 - \dfrac{(1-n)N^i(\kappa_s - \kappa_w)}{\kappa_w + a(\kappa_{eff}{}^i - \kappa_w) + N^i(\kappa_s - \kappa_w)}} \tag{4.19}$$

where the depolarization factors (N^i) are defined by the particle aspect ratio according to Equation (4.18). Nonspherical particles that are randomly oriented form an isotropic medium with a scalar effective permittivity given by (Sihvola and Kong,1988):

$$\kappa_{eff} = \kappa_w + \frac{\sum\limits_{i=a,b,c} \frac{(1-n)[\kappa_w + a(\kappa_{eff} - \kappa_w)](\kappa_s - \kappa_w)}{3[\kappa_w + a(\kappa_{eff} - \kappa_w) + N^i(\kappa_s - \kappa_w)]}}{1 - \sum\limits_{i=a,b,c} \frac{(1-n)N^i(\kappa_s - \kappa_w)}{3[\kappa_w + a(\kappa_{eff} - \kappa_w) + N^i(\kappa_s - \kappa_w)]}} \tag{4.20}$$

The MG model formed the basis of a self-similar (SS) model derived by Bruggeman (1935) for spheres and generalized to ellipsoids by Sen et al. (1981), in which the solid-phase inclusions were sequentially added to the host water phase, while the background pore system remained intact to low values of porosity. The resulting formula is impressively simple (Sen et al., 1981):

$$\left(\frac{\kappa_s - \kappa^i_{eff}}{\kappa_s - \kappa_w}\right)\left(\frac{\kappa_w}{\kappa^i_{eff}}\right)^{N^i} = n\,, \tag{4.21}$$

where the depolarization factors can again be computed using Equation (4.18). This model applies, in principle, to a fractal medium of infinitely wide particle size distribution and therefore forms a lower bound for the estimate of κ^i_{eff} of water-saturated porous media (Robinson and Friedman, 2001). An isotropic form of the SS model can be obtained by averaging the depolarization factors (N^i) over all possible particle orientations (Mendelson and Cohen, 1982). The isotropic form of the SS model can be expressed as

$$\kappa_{eff} = \kappa_w n^m \left(\frac{1-\kappa_s/\kappa_w}{1-\kappa_s/\kappa_{eff}}\right)^m, \tag{4.22}$$

where the effective cementation exponent m is a function of the particle shape (Mendelson and Cohen, 1982), with an arithmatic correction by Sen (1984):

$$m = \left\langle \frac{(5-3N)}{3(1-N^2)} \right\rangle. \tag{4.23}$$

As will be described in Section 4.4, this cementation exponent m can also be used to describe the effects of cementation and grain shape on the conductivity response of rocks and soils (e.g., Equations (4.26) and (4.35)). For high-porosity granular materials such as soils, we typically set m=1.5, and for consolidated rocks with lower porosity, we typically use a cementation exponent of m=2.0 (Sen et al., 1981; Mendelson and Cohen, 1982; Sen, 1984; Robinson and Friedman, 2001).

The relative permittivity responses of isotropic materials computed using the RI, MG, and SS models with κ_W=80 and κ_S = 5 are plotted as a function of porosity in Figure 4.2. The MG and SS model responses are computed for both spherical particles (a/b=

1.0, N^a=1/3, m=1.5) and oblate particles with random orientations (a/b= 0.22, N^a=0.73, m=2.0). As shown in Figure 4.2, the permittivity responses of all the models increase with increasing porosity. As previously discussed, the MG and SS models give realistic upper and lower bounds, respectively, for the permittivity responses of earth materials. The MG and SS models with randomly oriented oblate grains give smaller permittivity responses than the spherical grain models. The response of the RI model is similar to the responses predicted by the SS model with spherical grains and the MG model with oblate grains. Porosity predictions made using these three models differ by less than 0.03 porosity units over the porosity range of 0 to 0.5, thus illustrating the practicality of the RI model.

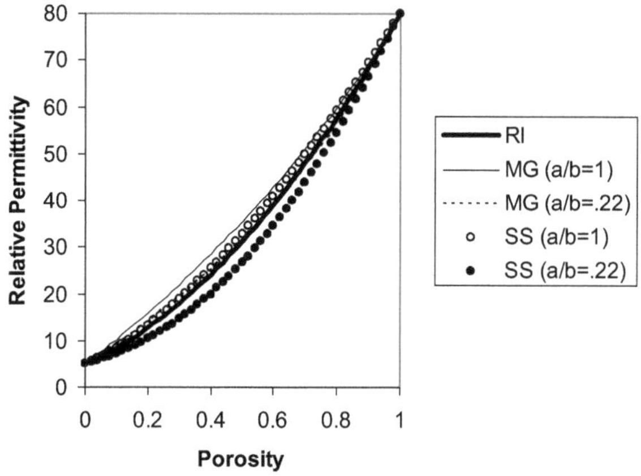

Figure 4.2. Relative permittivity responses of a water-saturated media ($\kappa_w = 80$; $\kappa_s = 5$) predicted by the refractive index (RI) model, the Maxwell-Garnett (MG) model, and the self-similar (SS) model. The aspect ratios of the spherical (a/b=1) and oblate (a/b=0.22) particle inclusions used in the MG and SS model calculations are indicated in the figure legend. An aspect ratio of a/b=0.22 corresponds to a cementation exponent of *m*=2.0. Note that the MG (a/b=0.22) curve is not visible because it is very similar to, and overlaid by, the SS (a/b=1) and the RI responses.

The effects of particle shape on the relative permittivity responses of the MG and SS models are further illustrated in Figure 4.3. In this plot, the relative permittivity is computed for *n*=0.3 and plotted as a function of the aspect ratio (*a*/*b*) for the MG and SS models. The solid lines refer to the MG model computations and the dashed lines refer to the SS model computations for κ_{eff}^{a}, κ_{eff}^{b}, and the isotropic mixture of randomly oriented particles κ_{eff}. The κ_{eff}^{a} and κ_{eff}^{b} responses for the anisotropic MG and SS models approach the same limiting values for very oblate and very prolate particle shapes. The κ_{eff}^{a} responses are maximum for long needle-like particles ($a/b=10^3$), and the κ_{eff}^{b} responses are maximum for thin disk-like particles ($a/b=10^{-3}$). The MG and SS model responses for randomly oriented particles have similar dependencies upon the particle shape, except that the SS model response, which takes into account particle interactions, is shifted approximately 3 permittivity units lower than the response of the MG model.

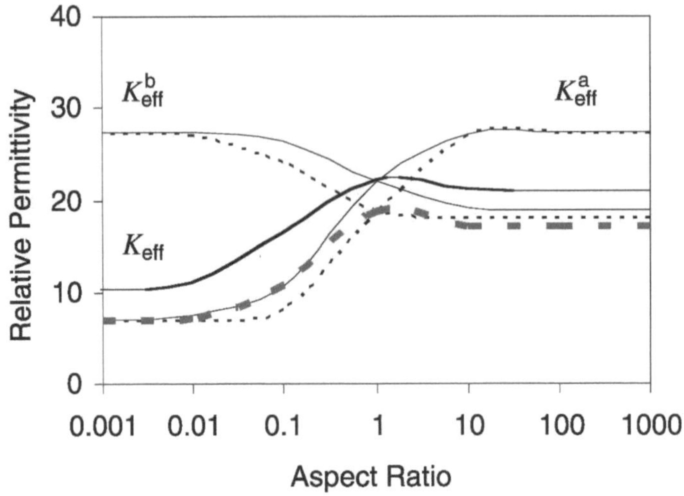

Figure 4.3. Relative permittivity of water saturated media ($\kappa_w = 80$; $\kappa_s = 5$; n= 0.3) computed using the MG model (solid lines) and the SS model (dashed lines). The orientations of the particles relative to the applied fields (κ_{eff}^{a} and κ_{eff}^{b}) are indicated on the plot. The bold solid and bold dashed lines are the MG and SS model responses, respectively, for randomly oriented particles κ_{eff} .

4.3.2 UNSATURATED MEDIA

The refractive index model (Equation 4.15) can be extended to model the effective permittivity of three-phase mixtures (e.g., Alharthi and Lange, 1987)

$$\sqrt{\kappa_{eff}} = \theta\sqrt{\kappa_w} + (n-\theta)\sqrt{\kappa_a} + (1-n)\sqrt{\kappa_s} \quad . \tag{4.24}$$

Several further modifications of Equation (4.24) have been proposed, such as splitting the solid phase into sand and clay phases of different permittivities (Wharton et al., 1980; Knoll et al., 1995), or replacing the ½ exponent with different powers (Loyenga, 1965; Roth et al., 1990). However, to predict water content from permittivity measurements using Equation (4.24), the porosity of the material must be specified in addition to the permittivities of the individual phases. Since the porosity of soil and rock formations is variable and often unknown, several empirical equations have been developed to directly predict the relationship between κ_{eff} and θ (Topp et al., 1980; Wang and Schmugge, 1980; Roth et al., 1992). Most of these empirical relationships use polynomial functions, which in some cases account for soil textural parameters such as the clay and sand percentage (Dobson et al., 1984) or the bulk density and clay and organic matter contents (Jacobsen and Schjonning, 1993). The most reliable and widely used $\kappa_{eff}(\theta)$ relationship for soils is the empirical formula of Topp et al. (1980):

$$\kappa_{eff} = 3.03 + 9.3\theta + 146\theta^2 - 76.7\theta^3 , \tag{4.25}$$

which was best-fitted to careful measurements made on four soil types and was suggested to hold approximately for all types of mineral soils. This relationship was found to give good predictions for coarse and medium-textured soils (Dirksen and

Dasberg, 1993; Friedman, 1998a), and gives similar predictions for several theoretically derived models (Friedman, 1997, 1998a).

The permittivity responses of the Topp model and the three-phase RI model with n=0.5 are plotted in Figure 4.4. The Topp and R.I. equations work well for coarse and medium-textured soils (Dirksen and Dasberg, 1993; Friedman, 1998a). However, they do not work as well for fine textured soils. The presence of soil minerals affects the dielectric properties of the water molecules adjacent to their surfaces by restricting their rotational movements, thus reducing their polarizability and permittivity (e.g., Dobson et al., 1985; Dirksen and Dasberg, 1993; Heimovaara et al., 1994; Or and Wraith, 1999). In these cases, the Topp and RI equations underpredict the water content of fine-textured soils. This effect can be accounted for in $\kappa_{eff}(\theta)$ models by adding a fourth, bound water phase with a permittivity similar to that of ice κ_{bw}=3.5 (Dobson et al., 1985; Dirksen and Dasberg, 1993), or by varying continuously κ_w as a function of the mean thickness of the water films surrounding the soil particles (Friedman, 1998a).

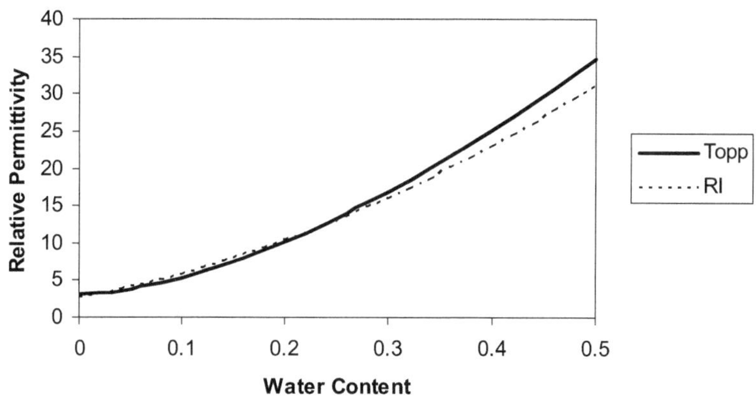

Figure 4.4. Relative permittivity of an unsaturated medium (n=0.5) computed using the the three-phase refractive index model ($\kappa_w = 80$; $\kappa_s = 5$) and the Topp et al (1980) equation

For permittivity measurements made at frequencies <100 MHz, the surface polarization of fine-textured soils can cause large enhancements in the permittivity response. In these cases, the Topp and RI equations overpredict the water content of fine-textured rocks and soils. Four-phase mixing formulae can also be used to account for clay polarization effects (Knoll et al., 1995); however, the permittivity responses of clay can be much larger than the permittivity response of water, and the clay response varies with clay type, water conductivity, and temperature. Therefore, it is difficult to accurately account for the clay polarization effects in lower-frequency permittivity estimates of water content.

Theoretically based effective medium models can be used to better understand the geometric factors controlling the permittivity responses of partially saturated rocks and soils. For example, Friedman (1998a) developed a MG-type three-phase mixing model consisting of concentric spheres of grains, water, and air. There are six possible concentric arrangements of the three phases, and because the permittivity of water is much higher than that of the other phases, it is primarily the location of the water phase that dictates the resulting κ_{eff}. In order to use theoretically derived models to predict water content from permittivity measurements, the model parameters must be measured or their values assumed to be known. For this reason, most practical applications use models calibrated with site-specific data or (more commonly) empirical formulae such as the Topp equation, which are based on fits to laboratory measurements.

4.3.3 PERMEABILITY PREDICTION

Both k_s and κ_{eff} depend on porosity and the effective pore/grain shapes of earth materials. Therefore, permittivity measurements could possibly be used to constrain these parameters in a permeability prediction formula (e.g., Equation 4.3). However, k_s is also strongly dependent upon the pore/grain size, which cannot be determined from high-frequency permittivity measurements. Therefore, it is not possible to derive a general predictive relationship between κ_{eff} and k_s. Lower-frequency permittivity measurements ($f<100$ MHz), which are sensitive to surface polarization effects, can be used to predict the sample surface area, and therefore an effective pore size (e.g., Knight and Nur, 1987a; Knoll et al., 1995). However, as explained in Sections 4.5 and 4.6, it is more practical to measure these polarization effects and parameters using induced polarization methods.

In some cases, site-specific predictive relationships between κ_{eff} and k_s can be empirically developed. For example, at sites where porosity is well correlated with formation permeability, κ_{eff} can potentially be used to estimate the formation permeability (e.g., Hubbard et al., 1997). Alternatively, in partially saturated environments where the water content is often controlled by soil water-retention properties, the net water content as measured by κ_{eff} could also be a predictive measure of the formation's hydraulic properties. However, whenever possible, we recommend using κ_{eff} information in conjunction with other geophysical data such as electrical conductivity and induced polarization measurements, and advanced geostatistical interpretation tools. The site-specific empirical relationships should be based on collocated, core-scale geophysical and hydrogeologic measurements, e.g., the linear κ_{eff} ($\log(k_s)$) relationship used by Hubbard et al. (1999). These relationships, depending on their statistical significances, will probably contribute to evaluating the absolute permeabilities; if not, they will at least provide relative k_s information on an aquifer's structural heterogeneity and anisotropy at larger scales (e.g., Chen et al., 2001).

4.4 Electrical Conductivity Models

In this section, we review models for the frequency-independent conductivity response of rocks and soils. Strictly speaking, these models are applicable to low-frequency conductivity measurements where $\sigma'(\omega)$ approaches σ'_{dc} (e.g., Figure 4.1). However, because the dispersion in the conductivity response is generally much smaller than σ'_{dc}, the dispersion effects can be neglected when $f < 1$ MHz. The frequency dependence in the conductivity (and permittivity) response is discussed below in Section 4.6.

The conductivity response of rocks and soils ($f < 1$ MHz) is primarily a function of the water content, the conductivity of the saturating solution, and the sample lithology. The conductivity of aqueous solutions generally increases with the concentration, mobility, and electronic charge of the ions in the solution, as well as the temperature of the solution (see Appendix). The effective shape of the grains/pores also affects the conductivity response. Surface conductivity at the grain/solution interface can also be significant in fine-grained materials, especially if the solution conductivity is low. Surface conductivity models are dependent on the amount of surface area and the surface chemical properties of the solid/liquid interface. If the solution conductivity is known, electrical conductivity measurements can be used to estimate the effective porosity of water-saturated formations or the water content of partially saturated formations. Alternatively, soil scientists often use electrical conductivity measurements to estimate soil water salinity. In some cases, electrical conductivity measurements can be used to estimate the permeability of subsurface formations. However, as described at the end of this section, these empirical correlations are quite site specific.

4.4.1 WATER-SATURATED MEDIA

Archie's empirical law (Archie, 1942) is the most widely used relationship to predict the effective electrical conductivity responses of water-saturated geological materials:

$$\sigma_{eff} = \frac{\sigma_w}{F} = \sigma_w n^m . \tag{4.26}$$

The electrical formation factor F is an intrinsic measure of material microgeometry

$$F = n^{-m} , \tag{4.27}$$

and it is often assumed to be an indicator of the hydraulic tortuosity (Equation 4.4). Archie found that the exponent m ranged from 1.3 for unconsolidated sands to approximately 2.0 for consolidated sandstones (see Table 4.1 and Figure 4.5). As m increases with cementation, Archie termed it the cementation index. Theoretically derived petrophysical models, which are discussed later in this section, relate the cementation index to the effective grain shape. Jackson et al. (1978) made electrical conductivity measurements on natural and artificial sand samples. They determined that the cementation index increased as the grains became less spherical (see Figure 4.5) while variations in grain size and sorting had little effect on m. Archie's law implicitly assumes that the effective porosity (n_e) is equal to the total porosity (n) of the sample, and that all electrical conduction in a water-saturated rock or soil results from the migration of ions in the bulk pore-solution. If there are isolated pores through which ions cannot migrate, then $n_e < n$, and Archie's law will overpredict sample conductivity.

Table 4.1. Archie's law exponents (m) of different consolidated and nonconsolidated media

I. Medium	Porosity Range	m, Archie's Exponent	*Reference*
clean sand	0.12-0.40	1.3	Archie (1942)
consolidated sandstones	0.12-0.35	1.8-2.0	
glass spheres	0.37-0.40	1.38	Wyllie and Gregory (1955)
binary sphere mixtures	0.147-0.29	1.31	
cylinders	0.33-0.43	1.47	
disks	0.34-0.45	1.46	
cubes	0.19-0.43	1.47	
prisms	0.36-0.52	1.63	
8 marine sands	0.35-0.50	1.39-1.58	Jackson et al. (1978)
glass beads (spheres)	0.33-0.37	1.20	
quartz sand	0.32-0.44	1.43	
rounded quartz sand	0.36-0.44	1.40	
shaley sand	0.41-0.48	1.52	
shell fragments	0.62-0.72	1.85	
fused glass beads	0.02-0.38	1.50	Sen et al. (1981)
fused glass beads	0.10-0.40	1.7	Schwartz and Kimminau (1987)
sandstone	0.05-0.22	1.9-3.7	Doyen (1988)
polydisperse glass beads	0.13-0.40	1.28-1.40	de Kuijper et al. (1996)
fused glass beads	0.10-0.30	1.6-1.8	Pengra and Wong (1999)
sandstones	0.07-0.22	1.6-2.0	
limestones	0.15-0.29	1.9-2.3	
Syporex®	0.80	3.8	Revil and Cathles III (1999)
Bulgarian altered tuff	0.15-0.39*	2.4-3.3	Revil et al. (2002)
Mexican altered tuff	0.50*	4.4	
glass beads	0.38-0.40	1.35	Friedman and Robinson (2002)
quartz sand	0.40-0.44	1.45	
tuff particles	0.60-0.64	1.66	

*connected (inter-granular) porosity

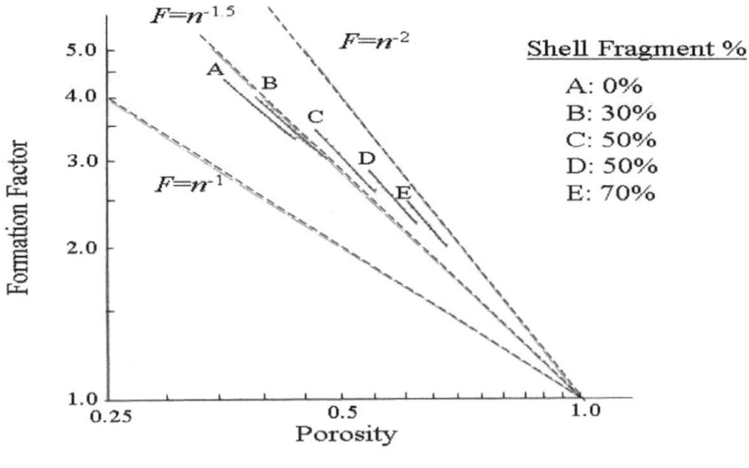

Figure 4.5. A log-log plot of formation factor versus porosity for granular mixtures consisting of rounded quartz grains and platey shell fragments (after Jackson et al., 1978, Figure 8). The dashed lines are Archie's law predictions for cementation indicies of 1.0, 1.5, and 2.0.

In fine-grained materials or in materials saturated with resistive pore solutions, surface conduction can be significant, causing Archie's law to underpredict the electrical conductivity of the sample. The conductivities of two sandstone cores measured as a function of solution conductivity are plotted in Figure 4.6 (Waxman and Smits, 1968). For the clean sandstone sample C1, the surface conductivity effects are negligible, and Archie's law is valid. The formation factor can simply be computed as the ratio of σ_w / σ_{eff}, and if the sample porosity is known, the cementation index can be computed using Equation (4.27). For the shaly sandstone sample C26, however, significant surface conductivity effects occur that result in a nonlinear relationship between σ_{eff} and σ_w. Archie's law can be modified to include a surface conduction term in parallel with the bulk conduction term, which results from the migration of ions through the bulk pore solution (e.g., Schön, 1996):

$$\sigma_{eff} = \frac{\sigma_w}{F} + \sigma_{surface}. \tag{4.28}$$

Although the bulk and surface conduction mechanisms do not strictly act in parallel (e.g., Friedman, 1998b), this simple parallel-conduction model has several practical advantages. One major advantage is that F and $\sigma_{surface}$ can be easily estimated by plotting on a linear scale σ_{eff} versus σ_w. The formation factor can be estimated from the slope of the linear portion of the σ_{eff} versus σ_w plot at high solution conductivity, and $\sigma_{surface}$ can be estimated from the extrapolated y-intercept at $\sigma_w = 0$ (e.g., Waxman and Smits, 1968; Nadler, 1982).

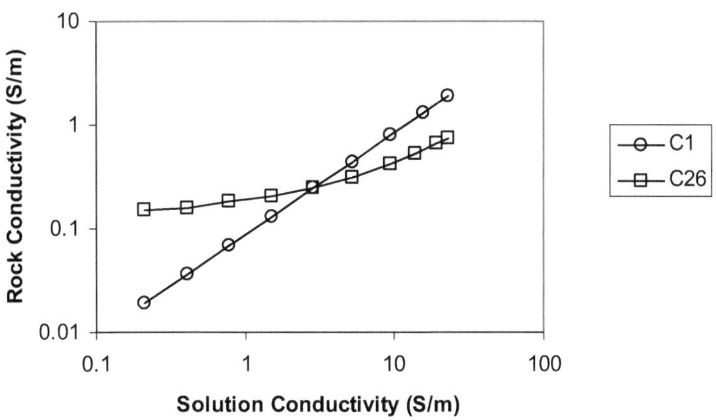

Figure 4.6. Conductivity of two sandstone cores measured as a function of the NaCl solution conductivity (Waxman and Smits, 1968). Surface conductivity effects are negligible in core C1 (n=0.239, Q_v=0.017 meq/cm^3) and significant in core C2 (n=0.229, Q_v=1.47meq/cm^3), which has a much higher cation exchange capacity.

The effects of lithology and solution chemistry on the surface conductivity term in Equation (4.28) were accounted for by Waxman and Smits (1968) in the following model:

$$\sigma_{surface} = \frac{BQ_v}{F}. \tag{4.29}$$

The cation exchange capacity per unit pore volume, Q_v, is a measure of the effective clay content, and B is the equivalent ionic conductance of the clay exchange cations. Waxman and Smits empirically obtained the following formula for the dependence of B on the solution conductivity:

$$B = \alpha[1 - \beta \exp(-\gamma\sigma_w)], \tag{4.30}$$

where the fitting parameters α, β, and γ depend upon the solution type. Sen et al. (1988) developed a modified version of the Waxman and Smits model, in which the empirical parameter B is a function of the solution conductivity (σ_w in S/m) as well as the cementation index m

$$B = \frac{1.93m}{1 + 0.7/\sigma_w}. \tag{4.31}$$

The advantage of the Sen et al. model is that it accounts for lithologic effects on the surface conductivity response; however, the empirical parameters in the equation are still dependent upon the solution type. This equation provided a good fit to conductivity measurements made on a suite of 140 shaly sandstone cores (Sen et al., 1988).

More sophisticated surface conductivity models have been developed in terms of the electrical double layer (EDL) that forms between the mineral grains and the bulk pore solution (e.g., Rink and Schopper, 1974; Johnson, et al., 1986; Schwartz et al., 1989; Revil and Glover, 1997). These models can also be represented by a bulk conduction term in parallel with a surface conduction term (Equation 4.28), where

$$\sigma_{surface} \cong \frac{\Sigma_s S_p}{f}, \tag{4.32}$$

with Σ_s the specific surface conductance, S_p the specific surface area, and f a parameter characterizing the "tortuosity" of the surface, which is similar to, but not necessarily equal to, the formation factor (e.g., Johnson, et al.; 1986). The specific surface conductance (Σ_s) represents the conduction in the fixed and diffuse parts of the electric double-layer (EDL). In general, Σ_s can be expressed as a function of the mineral surface charge density (Ω_0) and the surface ionic mobility (μ_S), such that

$$\sigma_{surface} \cong \frac{e\mu_s \Omega_0 S_p}{f}, \tag{4.33}$$

where e is the electronic charge. The charge within the EDL (Ω_0) is partitioned between the fixed layer (Ω_f) and the diffuse layer (Ω_d). The specific mobility of each ion within the EDL is a function of the ionic radius, valence, and the ion's distance from the mineral surface. Therefore, μ_S is an effective or average surface ionic mobility for the ensemble of counter-ions within the fixed and diffuse parts of the EDL (Revil and Glover, 1997). Surface complexation models can be used to predict the surface conductivity response as a function of the geochemical parameters Ω_0 and μ_S, which vary with solution type, concentration, and mineralogy. In practical applications, however, it is usually impossible to constrain all of these parameters for a formation. Therefore, semi-empirical equations such as those developed by Waxman and Smits (1968) and Sen et al. (1988) are typically used to interpret electrical conductivity measurements of shaly sand formations or clay rich soils. For simplicity, it is assumed in Equation 4.28 that the bulk and surface conduction mechanisms are independent of each other, and that they act in parallel. Three-dimensional networks of conductors can be used to more accurately model the interactions between the bulk and surface conduction mechanisms (e.g., Bernabe and Revil, 1995; Friedman, 1998b). However, these network models are not very applicable in practical applications.

The mean field theories introduced in Section 4.3 above can also be used to model the effective conductivities of two- and three-phase mixtures. For example, the "universal" mixing formula for conductivity is obtained by replacing the relative permittivity parameters in Equation (4.19) (κ^i_{eff}, κ_w, κ_s) with the corresponding conductivity parameters (σ^i_{eff}, σ_w, σ_s). If the solid-phase conductivity is negligible, then σ_s=0 and the Maxwell-Garnet formula for electrical conductivity (a=0) is given by

$$\sigma^i_{eff} = \sigma_w \left(1 - \frac{(1-n)}{1+(1-n)N^i} \right). \tag{4.34}$$

The conductivity of an isotropic mixture of spheroids with random orientations can similarly be obtained from Equation (4.20). This model assumes that there are no interactions between the embedded particles; therefore, this Maxwell-Garnet formula results in a maximum estimate for the conductivity of a mixture, and it is most appropriate for high-porosity soils and granular materials.

The self-similar model (Equation 4.22) for the effective conductivity of a mixture of randomly oriented spheroidal grains is given by

$$\sigma_{eff} = \sigma_w n^m \left(\frac{1-\sigma_s/\sigma_w}{1-\sigma_s/\sigma_{eff}} \right)^m . \tag{4.35}$$

The self-similar model accounts for strong interactions between the embedded particles and can be used to predict the conductivity of rocks and lower porosity soils. The cementation exponent m is determined by the effective grain shape as expressed by Equation (4.23). If $\sigma_s \cong 0$, which is true for most nonmetallic minerals, the self-similar model reduces to Archie's law (Sen et al., 1981). Bussian (1983) used Equation (4.35) with a non-zero σ_s term to model the shaly sandstone conductivity measurements of Waxman and Smits (1968). In this approach, the σ_s term represents both the conductivity of the dry mineral grain (essentially zero) and the surface conduction effects. As the surface conductivity increases with increasing specific surface area (e.g., Equation 4.33), $\sigma_{surface}$ will increase with decreasing grain size. It will also be dependent upon the surface charge density and surface ionic mobility, which vary with the solution conductivity. Although the surface conductivity mechanisms are not explicitly accounted for in Bussian's model, the interaction between the surface and bulk conduction mechanisms is treated more realistically than in the parallel conduction model (Equation 4.28). Furthermore, as is discussed in Section 4.6 below, a similar approach can be used to explicitly account for the surface conductivity mechanisms and to simultaneously predict the conductive and dielectric properties of rocks and soils.

4.4.2 UNSATURATED MEDIA

In his seminal paper, Archie (1942) also addressed the effects of saturation on the conductivity responses of consolidated and unconsolidated geological materials. Archie observed that the formation conductivity increased with saturation (*S*) according to the following power law:

$$\sigma_{eff}(S) = \sigma_{sat} S^d , \tag{4.36}$$

where σ_{sat} is the conductivity of the fully saturated sample. The saturation index *d* was observed to be about 2 for consolidated rocks and to range from 1.3 to 2 for unconsolidated sands (e.g., Schön, 1996). This power law was observed to hold down to saturations of about 0.15 to 0.20. At lower saturations, the power law breaks down, especially in fine-textured materials, as surface effects become dominant. In the absence of surface conductivity, the extended form of Archie's law can be used to predict the conductivity of partially saturated rocks and soils:

$$\sigma_{eff} = \sigma_w n^m S^d . \tag{4.37}$$

The saturation index is usually larger than the cementation index ($d>m$), because as saturation decreases, the water films surrounding the grains become thinner and the conducting paths become more tortuous. For coarse-textured sands, for example, the semi-empirical model of Mualem and Friedman (1991) predicts that $m=1+\gamma$ and $d=2+\gamma$, where the tortuosity exponent γ can be taken as 0.5 (Equation 4.11), making $m=1.5$ and $d=2.5$. Waxman and Smits (1968) also studied the effects of saturation on the electrical conductivity of oil-bearing shaly sandstones. They proposed the equation

$$\sigma_{eff} = \frac{S^d}{F}\left(\sigma_w + BQ_v / S\right). \tag{4.38}$$

The surface conductivity term (BQ_v/S) increases with decreasing saturation, which Waxman and Smits attributed to an increase in the volume concentration of clay exchange cations at low saturation.

Soil scientists often express the electrical conductivity in terms of water saturation (Mualem and Friedman, 1991; Weerts et al., 1999). If we assume for simplicity that $m=d$ in Archie's model, then the conductivity is given by

$$\sigma_{eff} = \sigma_w \theta^m . \tag{4.39}$$

This approach was used by Amente et al. (2000), who obtained an average exponent of $m=1.58$ by fitting Equation (4.39) to conductivity measurements of sandy loam soils. If the water content is independently determined, for example by using TDR measurements of permittivity and a model such as that given by (4.25), then the soil water conductivity can be estimated using Equation (4.39), with an average m determined for the field site. The most commonly used empirical equation to predict the electrical conductivity of soils in terms of the water content is that of Rhoades et al. (1976), who assumed that the surface conductivity term ($\sigma_{surface}$) is independent of θ and σ_w:

$$\sigma_{eff} = \sigma_w \theta T_c(\theta) + \sigma_{surface}; \qquad T_c(\theta) = a\theta + b. \tag{4.40}$$

The transmission coefficient $T_c(\theta)$ is assumed to be a linear function of θ, and the empirical coefficients a and b are a function of the soil type. For clay soils, a=2.1 and b=-0.25; for loam soils, $1.3 \le a \le 1.4$ and $-0.11 \le b \le -0.06$ (Rhoades et al., 1976).

Effective media theories and network models can also be used to model the effects of partial saturation on the electrical conductivity responses of rocks and soils (e.g., Feng and Sen, 1985; Man and Jing, 2001). These models can be used to better understand the various structural factors controlling the conductivity responses of rocks and soils, and to investigate important phenomena such as hysteresis. In practical applications, however, there is usually never enough information to constrain all of the parameters in the theoretically based models. Therefore, more empirically based approaches are typically used to interpret field measurements.

4.4.3 PERMEABILITY ESTIMATION

The Kozeny-Carman (K-C) permeability model (Equation 4.3) forms the basis of many permeability prediction equations (e.g., Nelson, 1994). In Equation (4.5), obtained from the K-C model, the formation factor (F) and the specific surface area (S_p) are the two key parameters needed to predict permeability. Because the formation factor can be estimated from electrical conductivity measurements, it is not surprising that many investigators have tried to use this approach for *in situ* permeability estimation (e.g., Nelson, 1994). In fact, in his seminal paper, Archie (1942) showed good correlations between k_s and $1/F$ for his suite of consolidated sandstone cores. Anisotropy measurements for the electrical conductivity of cores can also be used to estimate anisotropy in the permeability of the samples (Friedman and Jones, 2001). These approaches are limited, however, because S_p, which is a measure of the effective pore size, is much more variable than F for rock and soil formations. Therefore, in practice, F is of secondary importance in predicting the permeability of rock and soil formations. Electrical estimation of the effective pore/grain size using induced polarization methods is the focus of Sections 4.5 and 4.6 below.

Several empirically based equations have been proposed to electrically predict the permeability (or hydraulic conductivity) of rock and soil formations (e.g., Heigold et al., 1979; Huntley, 1986; Kosinski and Kelly, 1981; and Purvance and Andricevic, 2000). These equations, which are site specific, can be categorized into two groups. For clay-free formations and coarse-textured soils, the permeability and conductivity tend to be positively correlated, because they both increase with increasing porosity (or decreasing F), as is predicted by the K-C model (e.g., Heigold et al., 1979). For shaly sand formations and fine-textured soils, where surface conductivity effects are significant, the permeability and electrical conductivity tend to be inversely correlated (e.g., Kosinski and Kelly, 1981). Increasing clay content causes the permeability to decrease as the effective size of the pores is decreased, but it causes the electrical conductivity to increase because of increasing surface conductivity effects. This inverse correlation between k_s and σ_{eff} is often enhanced in partially saturated formations, where high-permeability materials tend to drain faster and have lower water-retention properties (i.e., lower field capacity) than low-permeability materials. Since the electrical conductivity is also a function of water content, high permeability soils tend to be drier and have lower conductivities than lower permeability soils.

4.5 Induced Polarization Models

Historically, IP methods have been used mostly to explore for metallic ore deposits. Recently, however, they are increasingly being used in a wide variety of environmental and engineering applications (e.g., Slater and Lesmes, 2002a; Chapter 5 of this volume). The IP response of non-metal-bearing earth materials is primarily controlled by the effective clay content or internal surface area of the sample (e.g., Vinegar and Waxman, 1984; Börner and Schön, 1991). Because the permeability of rocks and soils is also strongly dependent upon these "lithologic" parameters it is possible to develop permeability prediction equations in terms of the measured IP response of the sample (e.g., Börner et al., 1996; Slater and Lesmes, 2002b). The effectiveness of IP-permeability prediction formulae is complicated by the dependence of the IP response on secondary factors, such as the type of clay and its distribution, the solution conductivity and composition, the organic matter content, and water saturation. Because of these complicating factors, IP-permeability prediction formulae must be calibrated for the conditions at specific sites. However, the IP-permeability predictions are generally more accurate and robust than those made using high-frequency permittivity measurements or low-frequency conductivity measurements, as discussed in Sections 4.3 and 4.4 above. The IP (or complex conductivity) response is also a function of frequency; but over the frequency range of typical field instruments (10^{-2} Hz to 10^{2} Hz) the IP responses of many earth materials are relatively constant. In this section, we review models for the frequency-independent IP parameters typically measured in the field. In the next section, we review more comprehensive models for the broadband complex-conductivity responses of rocks and soils as measured in the laboratory. In the remainder of this chapter, we will refer to the effective complex conductivity σ^*_{eff} as simply σ^* and to the effective complex permittivity as ε^*_{eff} as ε^*. The relationship between these complex electrical parameters is defined by Equation (4.12).

4.5.1 INDUCED POLARIZATION PARAMETERS

Field IP surveys can be conducted using complex resistivity (CR), frequency-domain IP, or time-domain IP measurement systems (e.g., Ward, 1990). All of these systems are operationally similar to the DC resistivity method. However, in addition to measuring the conductive properties of the media, IP systems also measure, either directly or indirectly, the low-frequency capacitive properties of the media. The proportionality between the CR phase (θ), percent frequency effect (PFE), and chargeability (M) is both theoretically and experimentally well established (e.g., Marshall and Madden, 1959; Vinegar and Waxman, 1984). These field IP parameters, defined in Table 4.2, effectively measure the ratio of the capacitive to conductive properties of the material at low frequencies. The low-frequency capacitive component (σ'') is primarily controlled by electrochemical polarization mechanisms, whereas the low-frequency conductive component (σ') is primarily controlled by electrolytic conduction in the bulk pore solution and can be modeled using all the relationships presented in the previous section. Therefore, the field IP parameters are sensitive to the ratio of surface conductivity to bulk conductivity effects (e.g., Lesmes and Frye, 2001; Slater and Lesmes, 2002a).

Dividing the field IP parameters by the formation resistivity, or multiplication by the formation conductivity, yields the following normalized IP parameters: quadrature conductivity or imaginary conductivity (σ''), metal factor (MF), and a normalized chargeability (MN). These normalized IP parameters, defined in Table 4.2, are more directly related to the surface chemical properties of the material. Normalized IP parameters are therefore useful for characterizing lithological and geochemical variability (e.g., Lesmes and Frye, 2001; Slater and Lesmes, 2002a). In the rest of this section, the IP data and models are expressed in terms of the real and imaginary components of the complex conductivity, where the imaginary conductivity is directly proportional to the other normalized IP parameters, MN and MF.

Table 4.2. Commonly measured field IP parameters (θ, PFE*, and M) are equivalent measures of the IP response. The normalized IP parameters (σ'', MF, and MN) are more directly related to the surface properties of the rock/soil sample.*

Field IP Parameters		Normalized IP Parameters
$\theta = \tan^{-1}\left(\sigma'' / \sigma'\right) \cong \sigma'' / \sigma'$		$\sigma'' = \sigma' \tan(\theta) \cong \sigma' \theta$
$PFE = 100 * \dfrac{\sigma(\omega_1) - \sigma(\omega_0)}{\sigma(\omega_0)}$.	F	$M = a(\sigma(\omega_1) - \sigma(\omega_0))$
$M = \dfrac{1}{V_{max}(t_1 - t_0)} \displaystyle\int_{t_0}^{t_1} V(t) dt$.		$MN = M\sigma'$

*PFE is the relative dispersion in the conductivity response measured between a low frequency (ω_0) and a higher frequency (ω_1); a is a dimensionless constant; $V(t)$ is the potential difference measured at a time t after the current is shut off, V_{max} is the maximum potential difference measured during current transmission, and t_0 and t_1 define the time window over which the voltage decay curve is integrated.

4.5.2 WATER-SATURATED MEDIA

In a classic study, Vinegar and Waxman (1984) measured the complex conductivity response of a suite of 21 shaly sandstone cores as a function of pore water conductivity. They used this data set to develop the following complex form of the Waxman and Smits (1968) surface conductivity model

$$\sigma^* = \frac{1}{F}\left(\sigma_w + BQ_v\right) + i\frac{\lambda Q_v}{Fn}. \tag{4.41}$$

The real part of Equation (4.41) is Waxman and Smits (1968) model, in which the bulk and surface conduction terms are assumed to add in parallel. The imaginary (or quadrature) conductivity results from displacement currents that are 90 degrees out-of-phase with the applied field. Vinegar and Waxman assumed that the displacement currents were caused by the following two polarization mechanisms: (1) the blockage of ions by clay minerals at pore throats (membrane polarization) and (2) the accumulation of counter-ions migrating along grain/pore surfaces. Although these polarization mechanisms are intrinsically frequency dependent, Vinegar and Waxman showed that over the low frequency range of their measurements (3 Hz to 1 kHz), the quadrature conductivity response was essentially independent of frequency. Vinegar and Waxman assumed that both the membrane and the counter-ion polarization mechanisms were proportional to the effective clay content or specific surface area, represented by the cation exchange capacity of the rock per unit pore volume (Q_v). The parameter λ in Equation (4.41) represents an effective quadrature conductance for these surface polarization mechanisms, and it is analogous to the specific surface conductance term (Σ_s) in Equation (4.32). Vinegar and Waxman empirically determined λ to be slightly dependent on salinity. The polarization was also assumed to increase with decreasing porosity, as more pores become blocked.

4.5.3 UNSATURATED MEDIA

Vinegar and Waxman also measured the complex conductivity responses of their samples as a function of the water/oil saturation. Following the approach of Waxman and Smits (1968), they assumed that the effective cation exchange capacity would increase with decreasing saturation, such that $Q_v' = Q_v / S$. The imaginary conductivity therefore had the following dependence upon saturation: $\sigma''(S) = \sigma''_{sat} S^q$, where σ''_{sat} is the imaginary conductivity of the fully saturated sample. They observed that $q \cong d - 1$ where d is the saturation index for the real conductivity (Equation 4.36). Their measurements confirmed the validity of this expression for their samples. Ulrich and Slater (2004) measured the complex conductivity responses of unconsolidated sediments as a function of the water/air saturation. They observed a similar power-law dependence of the imaginary conductivity on water saturation, with the saturation exponent of the imaginary conductivity being less than the saturation index for the real conductivity ($q<d$). They also observed that the imaginary conductivity was a function of the saturation history (i.e., hysteretic effects). Effective media theories, discussed in

more detail in Section 4.6 below, can be used to model the effects of saturation on the complex conductivity response (e.g., Endres and Knight, 1992; Samstag, 1992).

4.5.4 PERMEABILITY ESTIMATION

Börner and Schön (1991) established the following linear relationship between S_p and σ'', based on complex conductivity measurements made on unconsolidated sediments at a frequency of 1 Hz (see Figure 4.7):

$$S_{p[el]} = b(\sigma''_{1Hz}), \tag{4.42}$$

where $S_{p[el]}$ is in μm^{-1}, σ''_{1Hz} is in S/m, and $b = 10^{-11}$ S^{-1}. Using this relationship in a KC-type model, they obtained the following expression for the hydraulic conductivity:

$$K_s = \frac{a}{FS^{c}_{p[el]}} = \frac{a}{F\left(10^5\,\sigma''_{1Hz}\right)^c} \tag{4.43}$$

where K_s is in m/s, a=10^{-5} and c ranged between 2.8 and 4.6, depending upon the material type and the method used to measure K_s.

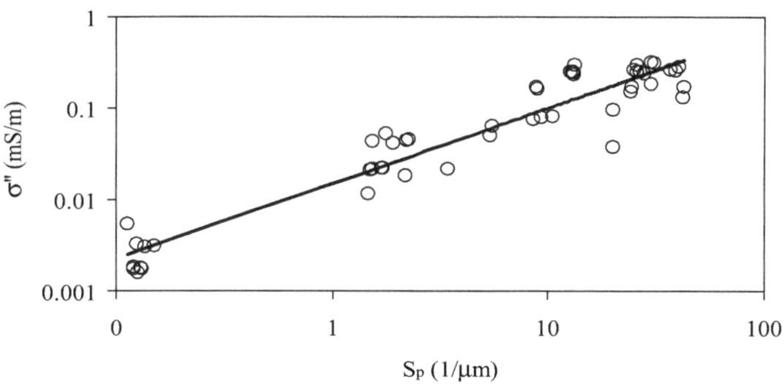

Figure 4.7. Börner and Schön (1991) observed a linear correlation between the imaginary conductivity of their saturated sandstone samples and the specific surface area. The samples were saturated with 0.01 M NaCl and measured at a frequency of 1 Hz.

Slater and Lesmes (2002b) measured the complex conductivity response and hydraulic properties of sand-clay mixtures and glacial tills. They observed a power law relationship between S_p and σ''_{1Hz}:

$$S_{p[el]} = 2000\left(\sigma''_{1Hz}\right)^p, \tag{4.44}$$

where p=0.5±0.2 (R^2=0.53, CI=95%). This relationship also appeared to be a function of the material type. They found for their samples, however, that σ'' was better correlated with the effective grain size d_{10} (see Figure 4.8):

$$\sigma''_{1Hz} = 0.0005\left(d_{10}\right)^{-b}, \tag{4.45}$$

where d_{10} is in μm, σ'' is in S/m, and b=1.0±0.1 (R^2=0.83, CI=95%). Using a Hazen type of grain permeability model (Equation 4.6), they obtained the following expression for the hydraulic conductivity in terms of σ'':

$$K_s = a\left(\sigma''_{1Hz}\right)^b, \tag{4.46}$$

where a=800±1200 and b=1.1±0.2 (σ'' is in S/m, K_s in m/s, R^2=0.7).

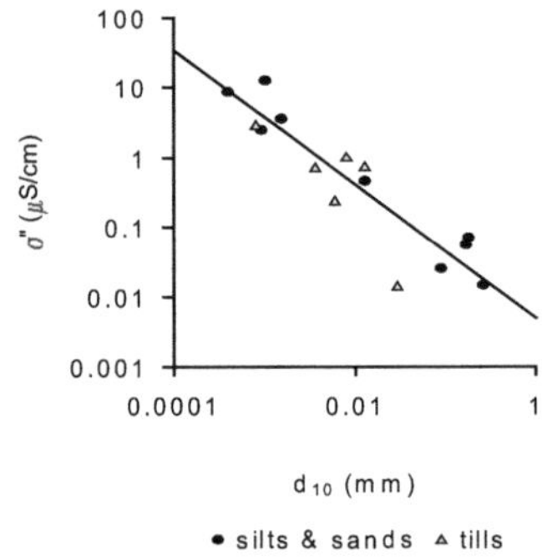

Figure 4.8. Imaginary conductivity of unconsolidated sand samples plotted as a function of the effective grain size d_{10} (Slater and Lesmes, 2002b). The complex conductivity was measured at 1 Hz, and the samples were saturated with 0.01 M NaCl.

A cross plot of the predicted-versus-measured hydraulic conductivity measurements is shown in Figure 4.9. The samples indicated by the + symbol in Figure 4.9 were not part of the data set used to establish the relationship between σ''and d_{10} in Equation (4.45), so they serve as an independent check on the robustness of this predictive equation.

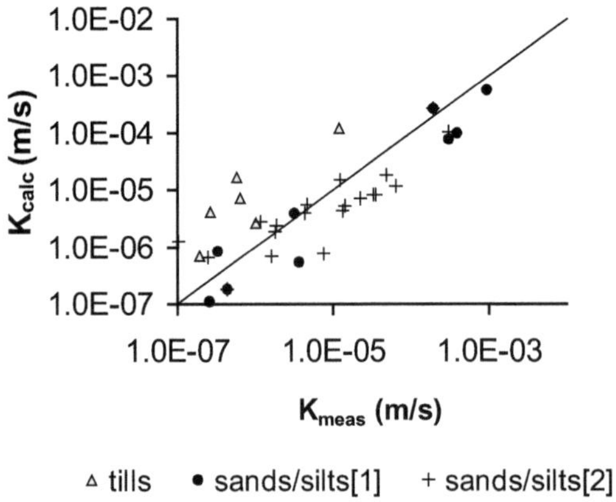

Figure 4.9. Cross plot of the predicted and measured hydraulic conductivity of the unconsolidated samples plotted in Figure 4.8. The hydraulic conductivity was predicted from the measured imaginary conductivity using Equation (4.46). The samples indicated by the + symbol were not used to establish the relationship in Equation (4.46), and they therefore are an independent check on the robustness of this permeability predication equation (Slater and Lesmes, 2000b).

The model consistently overpredicts the K_s values for the glacial till samples, indicated by the triangles in Figure 4.9. These samples are highly heterogeneous and have broad grain-size distributions. More accurate permeability predictions could perhaps be achieved by taking into account additional factors such as the grain sorting. As is discussed in Section 4.6, this additional information may possibly be obtained by measuring the frequency dependence in the imaginary conductivity response.

The development of permeability prediction equations using normalized IP parameters such as σ'' seems promising. However, the normalized IP parameters can also vary with solution chemistry, which can complicate the predictive relationships between the IP parameters and the desired lithologic variables. For example, the imaginary conductivity response of three shaly sandstone samples measured by Vinegar and Waxman (1984) are plotted versus solution conductivity in Figure 4.10. This plot shows that σ'' increases with increasing clay content, as expected, but there is a non-monotonic dependence upon solution conductivity, which increases with increasing clay content. Using a more sophisticated shaly sand model, de Lima and Niwas (2000) were able to account for the salinity dependence of σ'' and obtained a good prediction of the permeabilities of the Vinegar and Waxman samples. However, the model was significantly more complicated than, for example, Equation (4.46), and it would be difficult to constrain all of the model parameters without having complex conductivity measurements of the core samples made as a function of salinity. Therefore, this approach is difficult to apply in the practical interpretation of field data sets.

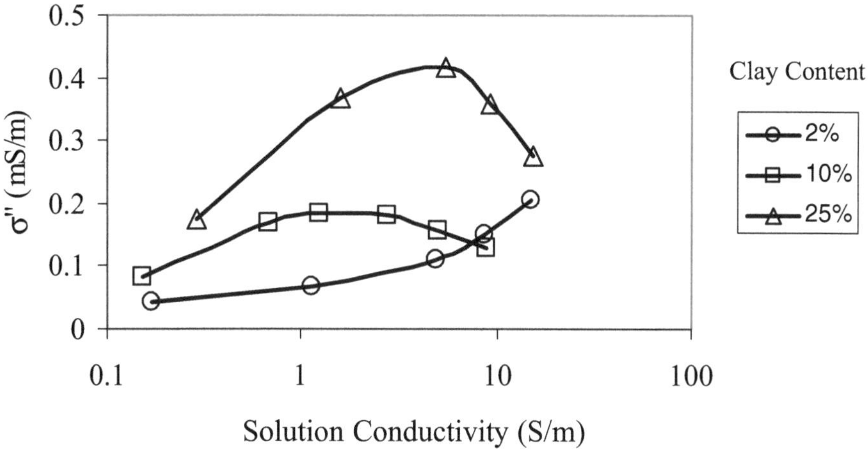

Figure 4.10. Imaginary conductivity of shaly sandstone samples plotted as a function of the solution conductivity (Vinegar and Waxman, 1984). Although the imaginary conductivity generally increases with increasing clay content, it has a nonmonotonic dependence upon the solution conductivity.

4.6 Complex Conductivity Models

The dispersion in the conductivity and dielectric responses of rocks and soils is controlled by physical and chemical polarization mechanisms that result in a broad distribution of relaxation times. To characterize the electrical properties of a sample fully requires that the impedance response be measured over a very wide range of frequencies (e.g., 10^{-3} Hz to 10^{9} Hz), so that the entire distribution of relaxation times can be captured (e.g., Olhoeft, 1985, 1986; Lesmes and Morgan, 2001). To span this frequency range requires using three different measurement systems with different sample configurations, which is both costly and time consuming. In most laboratory investigations, a single type of measurement system is used to characterize the impedance response over a limited range of frequencies while varying the sample lithology, solution chemistry, and water saturation. These data can be used to partially test models for the frequency-dependent complex conductivity responses of rocks and soils, but broadband data, collected from the low- to high- frequency limits, are required to fully constrain all of the model parameters. In this section, we review both empirical and theoretical models for the broadband complex conductivity responses of rocks and soils, and try to establish connections between the empirical and theoretical model parameters. We also discuss how broadband complex conductivity measurements and models can perhaps be used to more accurately predict the hydraulic properties of rocks and soils in the laboratory and in the field.

4.6.1 EMPIRICAL MODELS

Relatively simple empirical models can be used to fit the frequency-dependent permittivity and conductivity responses of rocks and soils. Havriliak and Negami (1967) introduced the following empirical dielectric response function, which is a generalized form of the widely used Cole-Cole (1941) and Cole-Davidson (1950) dielectric models:

$$\varepsilon^*_{HN}(\omega) = \frac{\Delta\varepsilon}{(1+(i\omega\tau_o)^{1-\alpha})^{\beta}}. \tag{4.47}$$

Equation (4.47) only predicts the frequency dependence in the permittivity and conductivity responses. It does not account for the high-frequency permittivity response (ε_∞) or the low-frequency conductivity response (σ_{dc}) of the sample. Specific models for ε_∞ and σ_{dc} were discussed above in Sections 4.3 and 4.4, respectively. One approach to modeling the "complete" response (the frequency-dependent and independent properties of a sample) is to assume that the ε_∞ and σ_{dc} terms act in parallel with the dispersive term $\varepsilon^*_{HN}(\omega)$ in Equation (4.47), such that the "complete" complex permittivity response is given by:

$$\varepsilon^*(\omega) = \left(\varepsilon_\infty + \frac{\sigma_{dc}}{i\omega}\right) + \frac{\Delta\varepsilon}{(1+(i\omega\tau_o)^{1-\alpha})^{\beta}}. \tag{4.48}$$

The Havriliak-Negami (HN) equation simplifies to the Cole-Cole (CC) expression when $\beta = 1$, it reduces to the Cole-Davidson (CD) expression when $\alpha = 0$, and it simplifies to the Debye expression when $\alpha = 0$ and $\beta = 1$. The dielectric increment $\Delta\varepsilon = \varepsilon_{static} - \varepsilon_\infty$ determines the amount of dispersion in the dielectric and conductivity responses, the central relaxation time τ_0 determines the characteristic angular frequency ($\omega_0 = 2\pi f_0 = 1/\tau_0$) of the relaxation, and the exponents α and β determine the shape of the relaxation (e.g., Bottcher and Bordewijk, 1978). To accurately estimate the model parameters, the dielectric response must be measured from the low- to high-frequency limits. These broadband measurements are difficult to make in the laboratory and essentially impossible to make in the field. Therefore, models such as the Cole-Cole response are often not fully constrained by the available data, and the model parameters obtained by inversion are non-unique.

If the relaxation in the permittivity and conductivity responses cannot be fully captured, then it is better to use models with fewer parameters to more uniquely fit the available data. For this reason, the constant phase-angle (CPA) model is widely used to model the dispersion in the complex conductivity responses of rocks and soils:

$$\sigma^*_{CPA}(\omega) = \sigma_0 (i\omega/\omega_0)^p = \sigma_0 (\omega/\omega_0)^p e^{ip\pi/2}. \tag{4.49}$$

In the CPA model, σ_0 is the magnitude of the complex conductivity response measured at an arbitrary angular frequency of ω_0. The dispersion parameter p determines both the phase angle $\theta = p\pi/2$ and the rate of change for the conductivity amplitude with increasing frequency. Adding the frequency-independent terms ε_∞ and σ_{dc} in parallel with the dispersive CPA response gives:

$$\sigma^*(\omega) = (\sigma_{dc} + i\omega\varepsilon_\infty) + \sigma_0 (\omega/\omega_0)^p e^{ip\pi/2}. \tag{4.50}$$

To fit Equation (4.50) to complex conductivity measurements, the reference frequency ω_0 must be arbitrarily defined (e.g., f_0=1Hz or $\omega_0 = 2\pi$ radians/sec). Furthermore, unless very broadband data are used, the estimates of σ_{dc} and ε_∞ will change with the frequency range of the measurement, such that: $\sigma_{dc} \cong \sigma'(\omega_{\min})$ and $\varepsilon_\infty \cong \varepsilon'(\omega_{\max})$. Although these parameters are not uniquely defined, it is important to include them in the inversion to obtain a good fit to the data and to accurately estimate σ_0 and p.

It is instructive to compare the CPA model to the high-frequency limit $(\omega >> 1/\tau_0)$ of the Cole-Cole model, for which the complex conductivity response is given by

$$\sigma^*_{CC}(\omega) \cong \frac{\Delta\varepsilon}{\tau_0}(i\omega\tau_0)^\alpha = \frac{\Delta\varepsilon}{\tau_0}(\omega\tau_0)^\alpha e^{i\alpha\pi/2}. \tag{4.51}$$

The relationships between the CPA and Cole-Cole model parameters is as follows: $\sigma_0 = \Delta\varepsilon/\tau_0$, $\omega_0 = 1/\tau_0$, and $p = \alpha$. This comparison illustrates the utility of the CPA model when fitting narrow band data. In the rest of this chapter, we use theoretically based models and published data to try to establish the physical and chemical significance of the empirical model parameters from the Cole-Cole and CPA models.

4.6.2 THEORETICAL MODELS

Theoretically based effective medium theories can be used to simultaneously model the conductivity and permittivity responses of rocks and soils. The Bruggeman-Hanai-Sen (BHS) effective medium model (Bruggeman, 1935; Hanai, 1960; Sen et al., 1981) is a complex form of the self-similar, asymmetric, effective media theories used to model the high-frequency permittivity (Equation 4.22) and the low-frequency conductivity (Equation 4.35) responses of rocks and soils:

$$\varepsilon^* = \varepsilon_w^* n^m \left(\frac{1 - \varepsilon_s^* / \varepsilon_w^*}{1 - \varepsilon_s^* / \varepsilon^*} \right)^m . \tag{4.52}$$

The cementation index *m* is a function of the grain shape, and for isotropic mixtures it is defined by Equation (4.23). The complex permittivity of the pore water is a function of frequency:

$$\varepsilon_w^* = \varepsilon_0 \kappa_w + i\omega\sigma_w . \tag{4.53}$$

If surface conductivity and polarization effects are neglected, then the complex permittivity of the mineral grains is purely real and frequency independent: $\varepsilon_s^* = \varepsilon_s'$. The conductivity and permittivity responses predicted by the BHS model have relaxations at relatively high frequencies (MHz range), and the relaxation frequency increases with increasing solution conductivity (e.g., Kenyon, 1984). The magnitude of the dispersions in the permittivity and conductivity responses increases slightly with increasing *m*, but the predicted dispersions are much smaller than observed permittivity and conductivity responses of rocks and soils (e.g., Lesmes and Morgan, 2001).

The BHS model represents the bulk properties of the rock/soil system. However, the surface phase that forms between the mineral grains and the pore solution can significantly affect both the conductive and capacitive properties of rocks and soils. Excess conduction and polarization at the solid/liquid interface leads to enhanced conductivity and permittivity responses that are frequency dependent. When ions migrate through water-bearing rocks and soils, they can accumulate at pore throat constrictions, at blockages caused by clay minerals, or on rough grain/pore surfaces. The re-equilibration of these charge accumulations is a diffusion-controlled process. Therefore, the time that it takes the charges to re-equilibrate is dependent upon the distances over which they are polarized. Complex conductivity measurements (such as those for Berea sandstone shown in Figure 4.1) can be inverted for a distribution of diffusive relaxation times, which can be transformed into a distribution of diffusion lengths. The diffusion-length distribution can then be equated to an effective distribution of grain or pore sizes, depending upon the specific geometries used to model the diffusive polarization mechanisms (e.g., Morgan and Lesmes, 1994; Chelidze and Guegen, 1999; Chelidze et al, 1999; Lesmes and Morgan, 2001).

Theoretical models have been derived for a variety of diffusive polarization mechanisms that can contribute to the permittivity and conductivity responses of water-bearing rocks and soils (Marshall and Madden, 1959; Schwarz, 1962; Wong,

1979; Chew and Sen, 1982, de Lima and Sharma, 1992). All of these models are similar in that the relaxation time distribution, which controls the dispersions in the conductivity and permittivity responses, is determined by the distribution of diffusion lengths. The characteristic lengths in the models can be grain size, pore size, the spacing between clay blockages of pore throats, or surface roughness. For simplicity, we consider just the EDL polarization model of Schwarz (1962) for a fixed-layer of charge surrounding a spherical grain. In this model, the "fixed" charge in the EDL (Ω_f) is allowed to migrate tangentially to the grain surface, but is restricted from migrating radially away from the grain. This model results in a single Debye response, where the relaxation time is a function of the particle radius (*R*) and the surface ionic mobility (μ_s) (or the surface diffusion coefficient: $D_s = \mu_s kT$):

$$\tau_0 = \frac{R^2}{2\mu_s kT}, \tag{4.54}$$

where *k* is Boltzman's constant and *T* is absolute temperature. The dielectric increment is given by

$$\Delta\varepsilon = \frac{eR\Omega_f}{kT}. \tag{4.55}$$

The polarization of a grain-size distribution, which can represent the water-wet rock matrix, will result in a distribution of relaxation times. If we use the Cole-Cole model to represent the complex permittivity response of the water-wet rock matrix (ε_m^*), then the Cole-Cole parameters can be defined in terms of the microgeometrical and surface chemical properties of the granular mixture (Lesmes et al., 2000). In this case, the central relaxation time τ_0 is approximately proportional to the grain radius (R_0) corresponding to the peak in the grain volume distribution:

$$\tau_0 \cong \frac{R_0^2}{2\mu_s kT}. \tag{4.56}$$

The dispersion parameter α is related to the width of the grain-size distribution. For fractal grain-size distributions, α can be related to the fractal dimension (*d*) of the grain-size distribution (e.g., Lesmes et al., 2000; Lesmes and Morgan, 2001):

$$\alpha \cong \frac{d-2}{2}. \tag{4.57}$$

The conductivity increment ($\Delta\sigma = \sigma_0 = \Delta\varepsilon / \tau_0$) is given by:

$$\Delta\sigma \cong \frac{e\mu_f \Omega_f S_p}{f}, \tag{4.58}$$

where f=3/2; and the dielectric increment is given by

$$\Delta\varepsilon \cong \frac{e\Omega_f S_p R_0^2}{3kT}. \tag{4.59}$$

The total response of the water-wet rock matrix will be the combination of all of the polarization mechanisms (e.g., fixed layer polarization, diffuse layer polarization, membrane polarization, etc.) and their interactions. The details of the physical and chemical interactions in the polarization models may be very complex, but laboratory experiments indicate that the net response of all of these interactions will result in a distribution of relaxation times that has a rather simple Cole-Cole type of response.

The complex electrical response of the complete sample is obtained by incorporating the response of the water-wet rock matrix into an effective media model for the entire rock or soil sample (Knight and Endres, 1990; Samstag, 1992; de Lima and Sharma, 1992). In terms of the BHS model, the complex permittivity response is given by

$$\varepsilon^* = \varepsilon_w^* n^m \left(\frac{1 - \varepsilon_m^* / \varepsilon_w^*}{1 - \varepsilon_m^* / \varepsilon^*} \right)^m , \tag{4.60}$$

and the complex conductivity response is given by

$$\sigma^* = \sigma_w^* n^m \left(\frac{1 - \sigma_m^* / \sigma_w^*}{1 - \sigma_m^* / \sigma^*} \right)^m . \tag{4.61}$$

The model for ε_m^* (or σ_m^*) will depend upon the specific polarization mechanisms that are assumed to control the complex electrical response of the water-wet rock matrix. If we assume for simplicity that ε_m^* results from fixed-layer polarization of the EDL as described above, then

$$\varepsilon_m^* = \frac{\Delta\varepsilon}{1 + (i\omega\tau_0)^{1-\alpha}} , \tag{4.62}$$

where τ_0, α, and $\Delta\varepsilon$ are given by Equations (4.56), (4.57), and (4.59), respectively. Equation (4.62) accounts for the frequency dependence in the EDL polarization as described by the Schwartz model. This model assumes that no interaction occurs between the grains, and therefore counterions in the EDL cannot migrate across multiple grain lengths. Therefore, the conductivity of the surface phase is predicted to go to zero in the low-frequency limit of Equation (4.62). To account for surface conductivity effects, Lesmes and Morgan (2001) added a surface conductivity term to the EDL polarization response. Since $\Delta\sigma$ in Equation (4.58) has the same functional form as $\sigma_{surface}$ in Equation (4.32), we can use $\Delta\sigma$ to represent the surface conductivity. The complex permittivity response of the water-wet matrix is then given by

$$\varepsilon_m^* = \frac{\Delta\varepsilon}{1 + (i\omega\tau_0)^{1-\alpha}} + \frac{\Delta\varepsilon / \tau_0}{i\omega} , \tag{4.63}$$

where $\Delta\sigma = \Delta\varepsilon / \tau_0$. This model can also be expressed as a complex conductivity and written in term of $\Delta\sigma$:

$$\sigma_m^* = \frac{i\omega\tau_0\Delta\sigma}{1+(i\omega\tau_0)^{1-\alpha}} + \Delta\sigma . \tag{4.64}$$

This model can be used to invert complex permittivity data for an effective grain-size distribution (Lesmes et al., 2000; Lesmes and Morgan, 2001). If the porosity, cementation index, and solution conductivity are known then Equation (4.60) can be explicitly solved for the complex permittivity of the water-wet rock matrix. Equation (4.63) can then be fit to $\varepsilon_m^*(\omega)$ in order to estimate the Cole-Cole model parameters, which can in turn be used to estimate the effective grain-size distribution of the sample (Lesmes and Morgan, 2001). For samples with significant surface polarization effects, the measured response of the sample $\varepsilon^*(\omega)$ is dominated by $\varepsilon_m^*(\omega)$, especially at lower frequencies (f<1 MHz). Therefore, fits of the Cole-Cole model to $\varepsilon^*(\omega)$ and $\varepsilon_m^*(\omega)$ will yield essentially the same estimates for τ_0 and α (Lesmes and Morgan, 2001). This may not be true, however, at higher frequencies, where the interfacial polarization mechanisms can be comparable to or greater than the electrochemical polarization mechanisms.

4.6.3 WATER-SATURATED MEDIA

In this section, we briefly review published experimental studies on the complex electrical properties of rocks and soils to illustrate how the Cole-Cole parameters τ_0 and α depend upon the sample lithology and pore solution chemistry. (For a more complete review of experimental studies, refer to Chelidze et al., 1999; and Schön, 1996.) Diffusive polarization models predict that $\tau_0 \propto R_0^2$. Complex conductivity measurements of samples with metallic grains are generally observed to follow this trend (e.g., Olhoeft, 1985); however, Pelton et al. (1978) showed that τ_0 was also a function of the particle concentration. Titov et al. (2002) measured the differential polarizability (a form of spectral induced polarization) of sieved sand samples saturated with tap water. They observed $\tau_0 \propto R_0^{1.7}$, where R_0 is the radius of the sand grains. They interpreted their results using a short-narrow-pore model, which is similar to the membrane polarization mechanism of Marshall and Madden (1959). Scott and Barker (2003) measured the complex impedance response (10^{-3} Hz to 10^3 Hz) of 18 sandstone cores and showed that there was a good correlation between the peak in the phase spectrum, which is often taken to be a measure of τ_0, and the characteristic size of the pore-throats determined by mercury injection measurements. Their data indicate that the relaxation frequency ($\omega_0 = 1/\tau_0$) varied by five orders of magnitude for approximately one order of magnitude change in characteristic pore size. This would roughly correspond to a power-law relationship of $\tau_0 \propto R_0^5$. They did not, however, provide a physical or chemical model to explain this interesting observation.

Klein and Sill (1982) measured the electrical-impedance response (10^{-3} Hz to 10^3 Hz) of glass-bead and clay mixtures, with varying pore solution chemistries. They used a form of the Cole-Cole model to invert their data for a central relaxation time and a chargeability parameter. They showed that the relaxation time generally increased with the size of the glass beads: $\tau_0 \propto R_0^a$ where $0.5 \leq a \leq 1.1$. The power law exponent a was less than 2 and dependent upon the solution conductivity and the amount of clay in the mixture. These experiments indicate that the simple model in Equation (4.56) may work relatively well for a mixture of single-sized particles (e.g., Titov et al., 2002), but it seems to break down for more complicated sand-clay mixtures and lithified rocks. One of the biggest limitations of the model expressed by Equation (4.56) is that it does not take into account the interactions between the grains; it simply assumes that all of the grain polarizations are superimposed (Lesmes and Morgan, 2001).

In the Cole-Cole model, the dispersion parameter α determines the width of the relaxation time distribution (e.g., Bottcher and Bordewijk, 1978). Several investigators have derived relationships between α and the fractal dimension d (e.g., Ruffet et al., 1991). Nearly all of these models are similar to Equation (4.57) in that they predict α to increase with increasing d. Dielectric measurements of sandstone cores by Ruffet et al. (1991) and Knight and Nur (1987a) showed α to increase with the specific surface area of the sample, which is consistent with the idea that α is related to the surface roughness or fractal dimension. Glover et al. (1994) measured the dielectric response of Berea and Darley Dale sandstone cores as a function of the solution conductivity. They showed in their study that α increases with increasing solution conductivity, which is not explicitly predicted by the fractal models. One possible explanation for this observation is that the effective surface roughness increases as the EDL compresses with increasing solution concentration. This may imply that the diffuse part of the EDL is indeed contributing to the net electrochemical polarization response.

4.6.4 UNSATURATED MEDIA

Three-phase effective medium models can be used to predict the conductivity and permittivity responses of partially saturated samples (Feng and Sen, 1985; Endres and Knight, 1992; Samstag, 1992). The BHS model can be used to simulate the effects of partial saturation (e.g., water and air mixtures) by replacing ε_w^* in Equation (4.53) with the effective complex permittivity response of the water/air mixture. In this case, the BHS equation can be used to first compute the complex permittivity of the water/air mixture. Then, a second embedding is performed to mix the water-wet mineral grains into the water/air background (e.g., Samstag, 1992). In this double-embedding procedure, it is assumed that the response of the surface phase ε_m^* does not change with saturation. Knight and Nur (1987a) demonstrated that this is generally true as long as the saturation is greater than the critical saturation, which roughly corresponds to one monolayer coverage of water molecules on the pore/grain surface. They also showed that the critical saturation increased linearly with the internal surface area of the sample, and that the dispersion parameter α is relatively independent of saturation for saturations greater than the critical saturation. Below the critical saturation, the permittivity response and the dispersion parameter α rapidly decrease. The permittivity and conductivity responses are also dependent upon the saturation history: the permittivity and conductivity responses measured during imbibition are consistently greater than the responses measured during drainage (Knight and Nur, 1987b; Knight, 1991; Roberts and Lin, 1997). Roberts and Lin (1997) showed that the hysteretic effects increased with the surface area of the sample and the solution resistivity (i.e., increasing thickness of the EDL), indicating that these effects are at least partially affected by the surface properties of the sample.

4.6.5 PERMEABILITY PREDICTION

Petrophysical models predict the permeability of rocks and soils to be proportional to the square of a characteristic pore or grain size (Friedman and Seaton, 1998): $k \propto R_0^2$ (e.g., Equations 4.3 and 4.5). Diffusion polarization mechanisms for the complex conductivity response predict that $\tau_0 \propto R_0^2$ (e.g., Olhoeft, 1985; Chelidze and Gueguen, 1999). The K-C equation (4.3) can then be expressed in terms of the electrical parameters τ_0 and F, such that (e.g., Kemna, 2000)

$$k \propto \frac{\tau_0}{F} \tag{4.65}$$

Complex conductivity or spectral IP measurements (made in the time or frequency domains) can be used to estimate τ_0 if there is sufficient bandwidth (e.g., Pelton et al., 1978; Johnson, 1984). If the pore solution conductivity is known, then the formation factor (F) can be estimated from the measured formation conductivity. Furthermore, the dispersion parameter α can be related to the width of the pore or grain-size distribution. As the permeability increases with increased sorting (e.g., Nelson, 1994), Equation (4.65) can be modified to include α as a sorting parameter (e.g., Kemna, 2000):

$$k \propto \frac{\tau_0}{F^b} e^{-a\alpha}, \tag{4.66}$$

where a and b are fitting parameters. Sturrock et al (1999) tried to use this approach to predict the permeability of eight sandstone cores with permeability ranging over three orders of magnitude. They found that τ_0 was not very sensitive to the mean pore/grain size for these samples, and that the permeability prediction was mainly controlled by the variations in F and α. More experiments on well-characterized samples are required to better test this approach to permeability prediction.

It is interesting to note the mathematical similarity between the Havriliak-Negami dielectric model (Equation 4.47) and the van Genuchten model for soil-moisture-retention curves (Equation 4.10). Both of these empirically derived functions can be theoretically related to the statistics of the grain- or pore-size distributions of the sample. It is possible that broadband dielectric measurements could eventually be used to predict the van Genuchten parameters for at least some types of materials. However, the dielectric response is also dependent upon the surface chemical properties of the sample (e.g., clay type and solution chemistry), so any predictive relationships that are established will have to account for these effects. Even if it is not possible to predict van Genuchten parameters from dielectric spectra, IP estimates of the effective pore or grain size, as discussed in Section 4.5, could be helpful for characterizing the water-retention properties of rock and soil formations, especially if some core measurements are available to establish site-specific relationships.

4.7 Conclusions

Electrical methods are commonly used in hydrogeophysical investigations to characterize the lithologic properties of subsurface formations and to characterize and monitor the pore fluid saturation and salinity. The most commonly measured electrical parameters are the high-frequency permittivity and the low-frequency conductivity. Permittivity measurements can be used to estimate the porosity of fully saturated formations and the water content of partially saturated formations. Electrical conductivity measurements are sensitive to the water content, the water conductivity, and the lithology. Petrophysical relationships can be used to predict the electrical conductivity responses of rocks and soils in terms of these controlling parameters. The hydrological properties of subsurface formations can be estimated by performing dynamic experiments, such as permittivity monitoring of water infiltration tests in the vadose zone or conductivity monitoring of saline tracer tests in the saturated zone. However, direct estimates of the subsurface permeability structure from permittivity and conductivity measurements are generally not possible. The correlations used in the literature are very site specific, and there is a wide degree of scatter in the permeability predictions. The IP response, however, is strongly correlated with the effective clay content and the specific surface area of rocks and soils. Because the permeability is largely controlled by these lithologic parameters, IP measurements can be used in many cases to estimate subsurface permeability variations to within an order of magnitude. More accurate and robust IP-permeability prediction equations will need to account for the dependence of the IP response on the solution conductivity and saturation degree. This will require jointly measuring the IP, permittivity, and conductivity responses of subsurface formations and developing the appropriate petrophysical models.

The permittivity and conductivity responses of water-bearing rocks and soils also vary with frequency. The empirical Cole-Cole model can be used to fit the frequency-dependent permittivity and conductivity responses of rocks and soils with only five model parameters. Theoretically based effective media models can be used to interpret the Cole-Cole model parameters in terms of fundamental physical and chemical properties of the porous media. These models and laboratory data illustrate the important affect that the surface phase has on the frequency-dependent permittivity and conductivity responses of rocks and soils. More laboratory experiments are needed on well-characterized samples made as a function of frequency, solution chemistry, and saturation to better understand the polarization mechanisms operating within rocks and soils. These laboratory experiments and theoretical analyses should lead to a better understanding of the Cole-Cole type of empirical functions widely used to interpret and model electrical field surveys.

Lastly, the electrical methods and petrophysical models discussed in this chapter will be most effective when combined with other geophysical, geological, and hydrological measurements and integrated using appropriate geostatistical and geological models, as is illustrated by other chapters in this volume.

4.8 Appendix: Electrical Properties of Aqueous Solutions

This appendix contains equations for computing the dielectric constant and electrical conductivity of aqueous solutions as a function of temperature and ionic concentration.

The dielectric constant of aqueous solutions decreases with increasing temperature and to a lesser degree with increasing ionic concentration. The temperature dependence of pure water is given by (Weast, 1983):

$$\kappa_w(t) = 78.54\left[1 - 4.759\times10^{-3}(t-25) + 1.19\times10^{-5}(t-25)^2 - 2.8\times10^{-8}(t-25)^3\right], \quad (4.67)$$

where t is in °C. The concentration dependence for mono valent aqueous solutions (e.g., LiCl and NaCl) is approximately given by Olhoeft (1981):

$$\kappa_w(c) = \kappa_w(t) - 13.00c + 1.065c^2 - 0.03006c^3, \quad (4.68)$$

where c is the effective salt concentration in moles per liter of solution.

Pure water has a very low electrical conductivity of $\sim 4\times10^{-6}$ S/m (Weast, 1983). The conductivity of natural waters containing dissolved salts and other ionic components is much larger with pore water conductivities ranging from ~0.01 S/m for freshwater aquifers to 20 S/m for oil field brines (e.g., Schön, 1996). The conductivity of an aqueous solution containing n ionic components is given by:

$$\sigma_w \cong \sum_{i=1}^{n} \alpha_i c_i z_i \mu_i , \quad (4.69)$$

where the parameters are ionic concentration (c_i), valence (z_i), ionic mobility (μ_i), and the degree of dissociation (α_i) (e.g., Schön, 1996; Weast, 1983). Empirical relationships can be used to relate the electrical conductivity of aqueous solutions to the total dissolved solids, or *TDS* (Fishman and Friedman, 1989):

$$\sigma_w(S/m) \cong a \times TDS(mg/L), \quad (4.70)$$

where, depending upon the ionic composition and temperature of the solution, the constant a can range from $\sim 1.2\text{x}10^{-4}$ to $\sim 2.0\text{x}10^{-4}$. An average value of a=$1.5\text{x}10^{-4}$ is typically used to predict the conductivity of natural waters at 25 degrees C (Fishman and Friedman, 1989)). The solution conductivity also increases with increasing temperature as the ionic mobility and the degree of dissociation are temperature dependent. The solution conductivity increases by ~2% per degree C increase in temperature for temperatures ranging between 20 and 30 degrees C. The temperature dependence of NaCl solutions is described by the following empirical equation (Arps, 1953):

$$\sigma_w(t_2) = \sigma_w(t_1)\frac{t_2 + 2\ .5}{t_1 + 2\ .5}, \tag{4.71}$$

where the temperature is in °C. A comprehensive set of equations for the electrical conductivity of NaCl solutions as a function of concentration and temperature was published by Worthington et al. (1990).

Acknowledgments

We thank Lee Slater for his helpful advice about the content and organization of this chapter and his insightful review of the manuscript. We also thank Susan Hubbard for her advice, encouragement and careful review of this chapter.

References

Alharthi, A., and J. Lange, Soil water saturation: Dielectric determination, *Water Resour. Res.*, *23*, 591–595, 1987.

Amente, G., J.M. Baker, and C.F. Reece, Estimation of soil solution electrical conductivity from bulk soil electrical conductivity in sandy soils, *Soil. Sci. Soc. Am. J.*, *64,* 1931–1939, 2000.

Archie, G.E., The electrical resistivity log as an aid in determining some reservoir characteristics, *Trans. Amer. Inst. Mining Metallurgical and Petroleum Engineers*, *146*, 54–62, 1942.

Arps, J.J., The effect of temperature on the density and electrical resistivity of sodium chloride solutions, *Trans. AIME*, *198*, 327, 1953.

Bernabé, Y., and A. Revil, Pore-scale heterogeneity, energy dissipation and the transport properties of rocks, *Geophys. Res. Lett.*, *22,* 1529–1232,1995.

Birchack, J.R., C.G.Z. Gardner, J.E. Hipp, and J.M. Victor, High dielectric constant microwave probes for sensing soil moisture, *Proc. IEEE*, *62,* 93–98, 1974.

Börner, F.D., and J.H. Schön, A relation between the quadrature component of electrical conductivity and the specific surface area of sedimentary rocks, *The Log Analyst*, *32,* 612–613, 1991.

Börner, F.D., J.R. Schopper, and A. Weller, Evaluation of transport and storage properties in the soil and groundwater zone from induced polarization measurements, *Geophysical Prospecting*, *44,* 583–602, 1996.

Böttcher, C.J.F., and P. Bordewijk, *Theory of Electric Polarization, Vol II*, Elsevier Sci., New York, 1978.

Bruggeman, D.A.G., The calculation of various physical constants of heterogeneous substances: I. The dielectric constants and conductivities of mixtures composed of isotropic substances, *Ann. Phys.*, *24,* 636–664, 1935.

Bussian, A.E., Electrical conductance in a porous medium, *Geophysics*, *49,* 1258–1268, 1983.

Carman, P.C., Permeability of saturated sands, soils and clays, *J. Agric. Sci.*, *29,* 262–273, 1939.

Chelidze, T.L., and Y. Gueguen, Electrical spectroscopy of porous rocks: a review- I. Theoretical models, *Geophys. J. Int.*, *137,* 1–15, 1999.

Chelidze, T.L., Y. Gueguen, and C. Ruffet, C., 1999, Electrical spectroscopy of porous rocks: a review- II. Experimental results and interpretation, *Geophys. J. Int.*, *137,* 16–34, 1999.

Chen, J., S. Hubbard, and Y. Rubin, Estimating the hydraulic conductivity at the South Oyster Site from geophysical tomographic data using Bayesian techniques based on the normal linear regression model, *Water Resour. Res.*, *37,* 1603–1613, 2001.

Chew, W.C., and P.N. Sen, Dielectric enhancement due to electrochemical double layers: Thin double layer approximation, *J. Chem. Phys.*, *77,* 4683–4693, 1982.

Cole, K.S., and R.H. Cole, Dispersion and adsorption in dielectrics, I, Alternating current characteristics, *J. Chem. Phys.*, *9,* 341–351, 1941.

Davidson, D. W., and R.H. Cole, Dielectric relaxation in glycerol, propylene glycol and n-propanol, *J. Chem. Phys.*, *19,* 1484–1490, 1951.

de Kuijper, A., R.K.J. Sandor, J.P. Hofman, and J.A. de Waal, Conductivity of two-component systems. *Geophysics*, *61,* 162–168, 1996.

de Lima, O.A.L., and S. Niwas, Estimation of hydraulic parameters of shaly sandstone aquifers from geoelectrical measurements, *Journal of Hydrology, 235*, 12–26, 2000.

de Lima, O.A.L., and M.M. Sharma, A generalized Maxwell-Wagner theory for membrane polarization in shaly sands, *Geophysics*, *57*, 431–440, 1992.

Dirksen, C., and S. Dasberg, Improved calibration of time domain reflectometry soil water content measurements, *Soil Sci. Soc. Am. J.*, *57*, 660–667, 1993.

Dobson, M.C., F. Kouyate, and F.T. Ulaby, A reexamination of soil textural effects on microwave emission and backscattering, *IEEE Trans. Geosc. Remote Sensing, GE-22*, 530–536, 1984.

Dobson, M.C., F.T. Ulaby, M.T. Hallikainen, and M.A. El-Rayes, Microwave dielectric behavior of wet soil: II. Dielectric mixing models, *IEEE Trans. Geosc. Remote Sensing, GE-23*, 35–46, 1985.

Doyen, P.M., Permeability, conductivity, and pore geometry of sandstones, *J. Geophys. Res.*, *93*, 7729–7740, 1988.

Endres, A.L., and R.J. Knight, A theoretical assessment of the effect of microscopic fluid distribution on the dielectric response of partially saturated rocks, *Geophysical Prospecting*, *40,* 307–324, 1992.

Feng, S., and P.N. Sen, Geometrical model of conductive and dielectric properties of partially saturated rocks, *J. Appl. Phys.*, *58*, 3236–3243, 1985.

Fetter, C.W., *Applied Hydrogeology*, 4th ed., Prentice Hall, Upper Saddle River, NJ, 2001.

Fishman, M. J., and Friedman, L. C., Methods of determination of organic substances in water and fluvial sediments, *Techniques of Water-Resource Investigations of the U.S. Geological Survey, book 5, chapter A1.,* 1989.

Friedman, S.P., Statistical mixing model for the apparent dielectric constant of unsaturated porous media, *Soil Sci. Soc. Am. J., 61,* 742–745, 1997.

Friedman, S.P., A saturation degree-dependent composite spheres model for describing the effective dielectric constant of unsaturated porous media, *Water Resour. Res.*, *34,* 2949–2961, 1998a.

Friedman, S.P., Simulation of a potential error in determining soil salinity from measured apparent electrical conductivity, *Soil Sci. Soc. Am. J., 62,* 593–599, 1998b.

Friedman, S.P., and D.A. Robinson, Particle shape characterization using angle of repose measurements for predicting the effective permittivity and electrical conductivity of saturated granular media, *Water Resour. Res., 38,* doi:10.1029/2001WR000746, 2002.

Friedman, S.P., and N.A. Seaton, Critical path analysis of the relationship between permeability and electrical conductivity of 3-dimensional pore networks, *Water Resour. Res., 34,* 1703–1710, 1998.

Glover, P.W.J., P.G. Meredith, P.R. Sammonds, and S.A.F. Murrel, Ionic surface electrical conductivity in sandstone, *J. Geophys. Res., 99,* 21635–21650, 1994.

Hanai, T., Theory of the dielectric dispersion due to the interfacial polarization and its applications to emulsions, *Kolloid Z., 171,* 23–31, 1960.

Havriliak, S., and S. Negami, A complex plane analysis of α—dispersion in some polymer systems in transition and relaxations in polymers, *J. Polymer Sci.,* Part C Polym. Lett., *14,* 99–117, 1966.

Heigold, P.C., R.H. Gilkeson, K. Cartwright, and P.C. Reed, Aquifer transmissivity from surficial electrical methods, *Ground Water*, *17,* 338–345, 1979.

Heimovaara, T. J., Frequency domain analysis of time domain reflectometry waveforms, 1, Measurements of complex dielectric permittivity of soils, *Water Resour. Res., 30,* 189–199, 1994.

Hubbard, S.S., Y. Rubin, and E. Majer, Ground-penetrating-radar assisted saturation and permeability estimation in bimodal systems. *Water Resour. Res. 33***,** 971–990, 1997.

Hubbard, S.S., Y. Rubin, and E. Majer, Spatial correlation structure estimation using geophysical and hydrological data. *Water Resour. Res., 35,* 1809–1825, 1999.

Huntley, D., 1986, Relations between permeability and electrical resistivity in granular aquifers, *Ground Water*, *24,* 466–474, 1986.

Jackson, P.D., D.T. Smith, and P.N. Stanford, Resistivity-porosity-particle shape relationships for marine sands, *Geophysics*, *43,* 1250–1268, 1978.

Jacobsen, O.H., and P. Schjonning, A laboratory calibration of time domain reflectometry for soil water measurement including effects of bulk density and texture, *J. Hydrol., 151,* 147–157, 1993.

Johnson, D.L., J. Koplik, and L.M. Schwartz, New pore-size parameter characterizing transport in porous media, *Phys. Rev. Lett.*, *57,* 2564–2567, 1986.

Johnson, I.A., Spectral induced polarization parameters as determined through time-domain measurements, Geophysics, *49,* 1993–2003, 1984.

Jones, S.B., and S.P. Friedman, Particle shape effects on the effective permittivity of anisotropic or isotropic media consisting of aligned or randomly oriented ellipsoidal particles. *Water Resour. Res.*, *36,* 2821–2833, 2000.

Kemna, A., Tomographic inversion of complex resistivity – Theory and application, Ph.D. thesis, Bochm Univ. (published by: Der Andere Verlag, Osanabruck, Germany), 2000.
Kenyon, W.E., Texture effects on the megahertz dielectric properties of rock samples, *J. Appl. Phys., 55,* 3153–3159, 1984.
Klein, J.D., and W.R. Sill, Electrical properties of artificial clay-bearing sandstone, *Geophysics*, *47,* 1593–1605, 1982.
Kosugi, K., Three-parameter lognormal distribution model for soil water retention, *Water Resour. Res.*, *30,* 891–901, 1994.
Knight, R.J., and A. Nur, The dielectric constant of sandstones, 60 kHz to 4 MHz, *Geophysics*, *52,* 644–654, 1987a.
Knight, R. J., and A. Nur, Geometrical effects in the dielectric response of partially saturated sandstones, *The Log Analyst*, *28,* 513–519, 1987b.
Knight, R.J., Hysteresis in the electrical resistivity of partially saturated sandstones, *Geophsyics*, *56*, 2139–2147, 1991.
Knight, R. J., and A.L. Endres, A new concept in modeling the dielectric response of sandstones: Defining a wetted rock and bulk water system, *Geophysics*, *55,* 586–594, 1990.
Knoll, M.D., R. Knight, and E. Brown, Can accurate estimates of permeability be obtained from measurement of dielectric properties? Paper Presented at Symposium on the Application of Geophysics to Environmental and Engineering Problems, *Environ. and Eng. Geophys. Soc.*, Orlando, FL, April 23–26, 1995.
Kosinski, W.K. and W.E. Kelly, Geoelectric soundings for predicting aquifer properties, *Ground Water*, *19,* 163–171, 1981.
Lesmes, D.P., and K.M. Frye, The influence of pore fluid chemistry on the complex conductivity and induced polarization responses of Berea sandstone, *J. Geophys. Res., 106,* 4079–4090, 2001.
Lesmes, D.P., and F.D. Morgan, Dielectric spectroscopy of sedimentary rocks, *J. Geophy. Res.*, *106,* 13,329–13,346, 2001.
Lesmes, D.P., J. Sturrock, and K.M. Frye, A physicochemical interpretation of the Cole-Cole dielectric model, *Proceedings of the Symposium on the Application of Geophysics to Environmental and Engineering Problems (SAGEEP),* Environ. and Eng. Geophys. Soc., Arlington, VA, 2000.
Looyenga, H., Dielectric constants of mixtures, *Physica*, *31,* 401–406, 1965.
Man, H.N., and X.D. Jing, Network modeling of strong and intermediate wettability on electrical resistivity and capillary pressure, *Adv. Water Resour.*, *24,* 345–363, 2001.
Marshall, D.J., and T.R. Madden, Induced polarization: A study of its causes, *Geophysics, 24,* 790–816, 1959.
Maxwell-Garnett, J.C., Colours in metal glasses and in metal films, *Trans. R. Soc. London*, *203,* 385–420, 1904.
Mendelson, K.S., and M.H. Cohen, The effect of grain anisotropy on the electrical properties of sedimentary rocks, *Geophysics 47,* 257–263, 1982.
Morgan, F.D., and D.P. Lesmes, Inversion for dielectric relaxation spectra, *J. Chem. Phys*., *100,* 671–681, 1994.
Mualem, Y., A new model for predicting the hydraulic conductivity of unsaturated porous media, *Water Resour. Res.*, *12,* 513–522, 1976.
Mualem, Y., and S.P. Friedman, Theoretical prediction of electrical conductivity in saturated and unsaturated soil, *Water Resour. Res.*, *27,* 2771–2777, 1991.
Nadler, A., Estimating the soil water dependence of the electrical conductivity soil solution/electrical conductivity bulk soil ratio, *Soil Sci. Soc. Am. J.*, *46,* 722–726, 1982.
Nelson, P.H., Permeability-porosity relationships in sedimentary rocks, *The Log Analyst, 35,* 38–61, 1994.
Olhoeft, G.R., Electrical properties of rocks. In *Physical Properties of Rocks and Minerals*, Y.S. Touloukian and C.Y. Ho, eds., McGraw-Hill / CINDAS Data Series on Material Properties, Volume II-2, pp. 298–329, McGraw-Hill book company, New York, 1981.
Olhoeft, G.R., Low-frequency electrical properties, *Geophysics, 50,* 2492–2503, 1985.
Olhoeft, G.R., Electrical properties from 10^{-3} to 10^{9} Hz- physics and chemistry, in *Physics and Chemistry of Porous Media, II,* J.R. Banavar, J. Koplik, and K. W. Winkler, eds., *AIP Conf. Proc., 154*, pp. 775–786, 1986.
Or, D., and J.M. Wraith, Temperature effects on soil bulk dielectric permittivity measured by time domain reflectometry: A physical model, *Water Resour. Res.*, *35,* 371–383, 1999.
Pelton, W.H., S.H. Ward, P.G. Hallof, W.R. Sill, and P.H. Nelson, Mineral discrimination and removal of inductive coupling with multifrequency IP, *Geophysics*, *43,* 588–609, 1978.

Pengra, D.B., and P.Z. Wong, Low-frequency AC electrokinetics, *Colloids & Surf. A: Physicochem. Eng. Aspects, 159,* 283–292, 1999.
Polder, D., and J.H. van Santen, The effective permeability of mixtures of solids, *Physica, 12,* 257–271, 1946.
Purvance, D.T., and R. Andricevic, On the electrical-hydraulic conductivity correlation in aquifers, *Water Resour. Res., 36,* 2905–2913, 2000.
Revil, A., and L.M. Cathles, Permeability of shaly sands, *Water Resour. Res., 35,* 651–662, 1999.
Revil, A., and P.W.J. Glover, Theory of ionic-surface electrical conduction in porous media, *Phys. Rev. B., 55,* 1757–1773, 1997.
Revil, A., D. Hermitte, E. Spangenberg, and J.J. Cocheme, Electrical properties of zeolitized volcaniclastic materials. *J. Geophys. Res., 107,* doi:10.1029/2001JB000599, 2002.
Rhoades, J.D., P.A.C. Ratts, and R.J. Prather, Effects of liquid-phase electrical conductivity, water content, and surface conductivity on bulk soil electrical conductivity, *Soil Sci. Soc. Am. J., 40,* 651–655, 1976.
Rink, M., and J.R. Schopper, Interface conductivity and its implications to electric logging, *Trans. 15th Ann. Logging Symp., Soc. Prof. Well Log Analysts,* 1–15, 1974.
Roberts, J.J., and W. Lin, Electrical properties of partially saturated Topopah Spring tuff: Water distribution as a function of saturation, *Water Resour. Res., 33,* 577–587, 1997.
Robinson, D.A., Measurement of the solid dielectric permittivity of clay minerals and granular samples using time domain reflectometry immersion method, *Vadose Zone J., 3,* 705–713, 2004.
Robinson, D.A., and S.P. Friedman, The effect of particle size distribution on the effective dielectric permittivity of saturated granular media, *Water Resour. Res., 37,* 33–40, 2001.
Robinson, D.A., and S.P. Friedman, The effective permittivity of dense packings of glass beads, quartz sand and their mixtures immersed in different dielectric background, *J. Non-Crystaline Solids, 305,* 261–267, 2002.
Robinson, D.A., and S.P. Friedman, A method for measuring the solid particle permittivity or electrical conductivity of rocks, sediments, and granular materials, *J. Geophys. Res., 108*(B2), 2075, doi:10.1029/2001JB000691, 2003.
Robinson, D.A., C.M.K. Gardner, and J.D. Cooper, Measurement of relative permittivity in sandy soils using TDR, capacitance and theta probes: comparison, including the effects of bulk soil electrical conductivity, *J. Hydrol., 223,* 198–211, 1999.
Roth, C. H., M.A. Malicki, and R. Plagge, Empirical evaluation of the relationship between soil dielectric constant and volumetric water content as the basis for calibrating soil moisture measurements by TDR, *J. Soil Sci., 43,*1–13, 1992.
Roth, K., R. Schulin, H. Fluhler, and W. Attinger, Calibration of time domain reflectometry for water content measurement using a composite dielectric approach, *Water Resour. Res., 26,* 2267–2273, 1990.
Ruffet, C., Y Gueguen, and M. Darot, Complex measurements and fractal nature of porosity, *Geophysics, 56,* 758–768, 1991.
Samstag, F. J., and F.D. Morgan, Induced polarization of shaly sands: Salinity domain modeling by double embedding of the effective medium theory, Geophysics, *56,* 1749–1756, 1991.
Samstag, F. J., An effective-medium model for complex conductivity of shaly sands in the salinity, frequency, and saturation domains, Ph.D. Thesis, Texas A&M Univ., College Station, TX, 1992.
Scheidegger, A.E., *The Physics of Flow through Porous Media*, Toronto: University of Toronto Press, 1974.
Schön, J.H., Physical properties of rocks – fundamentals and principles of petrophysics, Elsevier Science Ltd., *Handbook of Geophysical Exploration, Seismic Exploration, 18,* 379–478, 1996.
Schwartz, L.M., and S. Kimminau, Analysis of electrical conduction in the grain consolidated model, *Geophysics, 52,* 1402–1411, 1987.
Schwartz, L.M., P.N. Sen, and D.L. Johnson, Influence of rough surfaces on electrolyte conduction in porous media, *Phys. Rev. B., 40,* 2450–2458, 1989.
Schwarz, G., A theory of the low-frequency dielectric dispersion of colloidal particles in electrolyte solution, *J. Phys. Chem., 66,* 2636–2642, 1962.
Scott, J.B.T., and R.D. Barker, Determining pore-throat size in Permo-Triassic sandstones from low-frequency electrical spectroscopy, *Geophys. Res. Lett., 30*(9), 1450, doi:10.1029/2003GL016951, 2003.
Sen, P.N., C. Scala, and M.H. Cohen, A self-similar model for sedimentary rocks with application to the dielectric constant of fused glass beads. *Geophysics, 46,* 781–795, 1981.
Sen, P.N., Grain shape effects on dielectric and electrical properties of rocks, *Geophysics, 49,* 586–587, 1984.
Sen, P.N., P.A. Goode, and A. Sibbit, Electrical conduction in clay bearing sandstones at low and high salinities, *J. Appl. Phys., 63,* 4832–4840, 1988.
Sihvola, A., and J.A. Kong, Effective permittivity of dielectric mixtures, *IEEE Trans. Geosci. Remote Sens.,*

26, 420–429, 1988.

Sihvola, A., and J.A. Kong, Correction to "Effective permittivity of dielectric mixtures," *IEEE Trans. Geosci. Remote Sens., 21,* 101–102, 1989.

Slater, L., and D.P. Lesmes, IP Interpretation in environmental investigations, *Geophysics*, *67,* 77–88, 2002a.

Slater, L., and D.P. Lesmes, Electrical-hydraulic relationships observed for unconsolidated sediments, *Water Resour. Res*., *38*(10), 1213, doi:10.1029/2001WR001075, 2002b.

Sturrock, J.T., D.P. Lesmes, and F.D. Morgan, Permeability estimation using spectral induced polarization measurements, Paper presented at the Symposium on the Application of Geophysics to Environmental and Engineering Problems (SAGEEP*), Environ. Eng. Geophys. Soc.*, Oakland, CA, 1999.

Titov, K., V. Komarov, V. Tarasov, and A. Levitski, Theoretical and experimental study of time domain-induced polarization in water-saturated sands, *J. Appl. Geophys*., *50,* 417–433, 2002.

Topp, G.C., J.L. Davis, and A.P. Annan, Electromagnetic determination of soil water content: Measurements in coaxial transmission lines, *Water Resour. Res.*, *16,* 574–582, 1980.

Tsang, L., J.A. Kong, and R.T. Shin, *Theory of Microwave Remote Sensing*, John Wiley, New York, 1985.

Ulrich, C., and L. Slater, Induced polarization measurements on unsaturated sands, *Geophysics*, *69,* 762–771, 2004.

Van Genuchten, M.T., A closed-form equation for predicting the hydraulic conductivity of unsaturated soils, *Soil Sci. Soc. Am. J.*, *44,* 892–898, 1980.

Vanhala, H., and H. Soininen, Laboratory technique for measurement of spectral induced polarization soil samples, *Geophysical Prospecting*, *43,* 655–676, 1995.

Vinegar, H.J., and M.H. Waxman, Induced polarization of shaly sands, *Geophysics, 49,* 1267–1287, 1984.

Ward, S.H., Resistivity and Induced Polarization Methods, in *Geotechnical and Environmental Geophysics*, Vol. 1, *Review and Tutorial*, *Invest. Geophys. 5,* S.H. Ward, ed., pp. 141–190, Soc. of Explor. Geophys., Tulsa, OK, 1980.

Wang, J.R., and T.J. Schmugge, An empirical model for the complex dielectric constant of soils as a function of water content, *IEEE Trans. Geosci. Remote Sensing*, *GE-18*, 288–295, 1980.

Waxman, M.H., and L.J.M. Smits, Electrical conductivities in oil-bearing shaly sands, *Soc. Pet. Eng. J.*, *8,* 107–122, 1968.

Weast, R.C., ed., *CRC Handbook of Chemistry and Physics* (63rd edition), CRC Press, Inc., Boca Raton, Florida, 1983.

Weerts, A. H., W. Bouten, and J.M. Verstraten, Simultaneous measurement of water retention and electrical conductivity in soils: Testing the Mualem-Friedman tortuosity model, *Water Resour. Res.*, *35,* 1781–1787, 1999.

Wharton, R.P., G.A. Hazen, R.N. Rau, and D.L. Best, Electromagnetic propagation logging: Advances in technique and interpretation, *Soc. Pet. Eng.*, *Pap. No. 9267,* 1980.

Wong, J., An electrochemical model of the induced-polarization phenomenon in disseminated sulfide ores, *Geophysics*, *44,* 1245–1265, 1979.

Worthington, A.E., J.H. Hedges, and N. Pallatt, SCA guidelines for sample preparation and porosity measurement of electrical resistivity samples: Part I–Guidelines for preparation of brine and determination of brine resistivity for use in electrical resistivity measurements, *The Log Analyst, 31,* 20, 1990.

Wyllie, M.R.J., and A.R. Gregory, Fluid flow through unconsolidated porous aggregates: Effect of porosity and particle shape on Kozeny-Carman constants, *Indust. Eng. Chem*, *47,* 1379–1388, 1955.

5 DC RESISTIVITY AND INDUCED POLARIZATION METHODS

ANDREW BINLEY[1] and ANDREAS KEMNA[2]

[1]*Lancaster University, Department of Environmental Science, Lancaster, LA1 4YQ, UK*
[2]*Agrosphere Institute, ICG IV, Forschungszentrum Jülich GmbH, 52425 Jülich, Germany*

5.1 Introduction and Background

Direct current (DC) resistivity (here referred to as *resistivity*) and induced polarization (IP) methods allow, respectively, the determination of the spatial distribution of the low-frequency resistive and capacitive characteristics of soil. Since both properties are affected by lithology, pore fluid chemistry, and water content (see Chapter 4 of this volume), these methods have significant potential for hydrogeophysical applications. The methods can be applied at a wide range of laboratory and field scales, and surveys may be made in arbitrary geometrical configurations (e.g., on the soil surface and down boreholes). In fact, resistivity methods are one of the most widely used sets of geophysical techniques in hydrogeophysics. These surveys are relatively easy to carry out, instrumentation is inexpensive, data processing tools are widely available, and the relationships between resistivity and hydrological properties, such as porosity and moisture content, are reasonably well established. In contrast, applications of induced polarization methods in hydrogeophysics have been limited. As noted by Slater and Lesmes (2002), this is partly because of the more complex procedure for data acquisition, but also because the physicochemical interpretation of induced polarization parameters is not fully understood.

Resistivity and IP methods are deployed in a similar manner and are therefore considered jointly here. Unlike electromagnetic and radar methods, resistivity and IP require galvanic contact with the soil. Consequently, this may limit application of the methods to appropriate sites and also require longer survey times, in comparison with electromagnetic and radar techniques (although recent advances have been made in continuous survey methods, as will be discussed later).

Modern resistivity and IP methods have emerged from early exploration tools developed in the early 1900s. During the late 1970s, increased computing power led to the emergence of new modeling tools (for example, Dey and Morrison, 1979a,b) and the development of electrical imaging concepts (for example, Lytle and Dines, 1978), which emerged in parallel with advances in biomedical imaging. Despite the apparent complexity of present-day methodologies and algorithms, the resistivity and IP imaging tools used today are very closely related to these pioneering developments in the 1970s.

Most resistivity and IP methods still adopt the four-electrode measurement approach traditionally used in exploration geophysics. Two electrodes act as current source and

Y. Rubin and S.S. Hubbard (eds.), Hydrogeophysics, 129–156.

sink, while the other two electrodes are used for potential difference (voltage) measurement. The emergence of imaging concepts prompted the development of multi-electrode measurement systems in the 1980s. By the mid- to late 1990s, most instrument manufacturers were able to offer multiplexed instruments, and in some cases multi-measurement channel hardware, thus enabling reduced survey time. For hydrogeophysical applications, this also meant the ability to study time-dependent processes.

Modern resistivity and IP methods both have powerful imaging capability. Terms such as electrical resistance (or resistivity) tomography (ERT), electrical impedance tomography (EIT), and applied potential tomography (APT) have been widely used in geophysics, biomedicine, and process engineering to describe imaging of electrical properties. In geophysics, the term "tomography" was originally used for investigations in which sensors bounded the region under investigation. Eventually, the term became synonymous with any form of imaging, even using linear arrays of sensors, as in surface deployed surveys. As imaging techniques have developed to study 3-D and even 4-D systems, the term "tomography" is not appropriate, and we prefer here to simply use the term "imaging."

Conventional DC resistivity and IP methods for geophysical exploration are covered well in many texts, for example Telford et al. (1990) and Reynolds (1998). However, these texts, and others, do not address more advanced developments of these methods (for example, cross-borehole imaging) or approaches for small-scale characterization (related to those used in the biomedical or process engineering fields). All of these approaches now provide the hydrogeophysicist with more flexible investigative tools. Our aim here is to review a wide range of resistivity and IP tools available for hydrogeophysical investigation. In Section 5.2, we outline the basic measurement principles for these methods. This is followed, in Section 5.3, by a presentation of the range of survey configurations. Data processing is required to transform measured data to a 1-D, 2-D, or 3-D spatial distribution of electrical properties, and details of the methods commonly used are covered in Section 5.4. In Section 5.5, we discuss important practical issues that must be addressed before the techniques covered here can be applied. Finally, in Section 5.6, we summarize future challenges for developers and users of resistivity and IP methods for hydrogeophysical surveys.

5.2 Basic Measurement Principles

5.2.1 RESISTIVITY

In the resistivity method, the spatial variation of resistivity ρ (or conductivity σ, the inverse) in the field is determined using four-electrode measurements. Two (transmitter) electrodes are deployed to create an electrical circuit. Measurement of the potential difference (voltage) between the two other electrodes permits determination of an apparent resistivity (i.e., the resistivity a homogenous halfspace should have to give the actual measurement). Inverse methods may be applied to such measurements to determine an image of the subsurface structure, as illustrated later. Electrodes may be placed on the ground surface and/or in boreholes.

Stainless steel is the most widely used electrode material for field measurements, although others, such as copper or brass, are also used. To avoid polarization at the electrodes, an alternating power source is utilized. A switched square wave (see Figure 5.1) is the most common current waveform; it is generally applied at frequencies of about 0.5 to 2 Hz. As shown in Figure 5.1, a background (self-potential) voltage, V_{sp}, may be observed. Note that the level of this may change over time, but such drift is easily removed owing to the shape of the injected waveform. The measured transfer resistance is given by $R = V_p / I$, where V_p is the primary (peak) voltage and I is the injected current, as shown in Figure 5.1. Note that the voltage series in Figure 5.1 is idealized since no capacitive (electrical charge storage) effects are observed. In practice, such effects will exist, as outlined in Section 5.2.2.

For a 3-D, isotropic electrical-conductivity distribution, $\sigma(\mathbf{r})$, the electric potential, $V(\mathbf{r})$, at a point $\mathbf{r}$ due to a single current electrode, idealized as a point source at the origin with strength I, is defined by the Poisson equation

$$\nabla \cdot (\sigma \nabla V) = -I\delta(\mathbf{r}), \tag{5.1}$$

subject to the condition

$$\frac{\partial V}{\partial n} = 0 \tag{5.2a}$$

at the ground surface and the condition

$$V = 0 \tag{5.2b}$$

at other, infinite boundaries; with δ the Dirac delta function and n the outward normal.

As illustrated in Figure 5.2, current flow in a homogenous earth from an electrode placed on the ground surface will follow

$$V = \frac{\rho I}{2\pi} \frac{1}{r}, \tag{5.3}$$

where ρ is the resistivity and r the distance from the electrode. Since the apparent resistivity, ρ_a, is defined as the resistivity of a homogenous earth to which the measured transfer resistance is equivalent, this equation may be used with the superposition principle to derive expressions for the apparent resistivity of specific electrode arrangements.

A number of electrode configurations are commonly used for ground-surface surveys. Figure 5.3 illustrates the Wenner, dipole-dipole, and Schlumberger surveys. Others (for example, pole-pole, pole-dipole) are also deployed (see, for example, Telford et al., 1990). As an example, for the Wenner array, the apparent resistivity, ρ_a, is given by

$$\rho_a = 2\pi a R, \tag{5.4}$$

where a is electrode spacing (as shown in Figure 5.3).

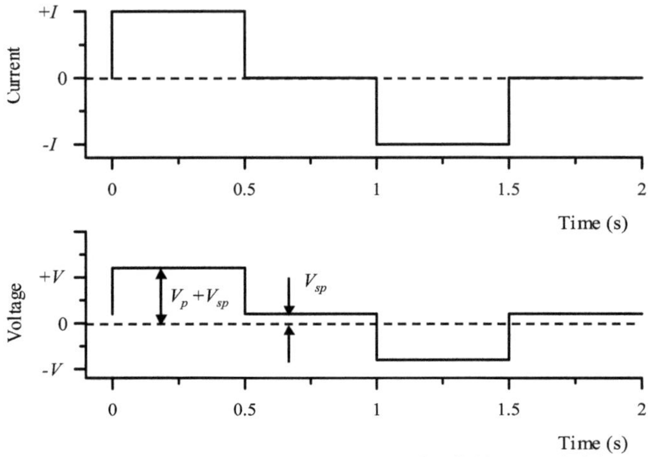

Figure 5.1. Typical current and idealized voltage waveforms for field DC resistivity surveys. V_p is the primary voltage, V_{sp} is the observed self potential voltage.

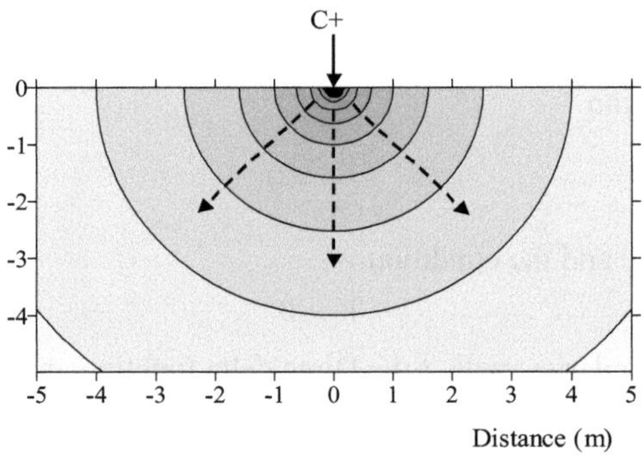

Figure 5.2. Potential variation in a half space with uniform resistivity distribution due to current injection at the ground surface

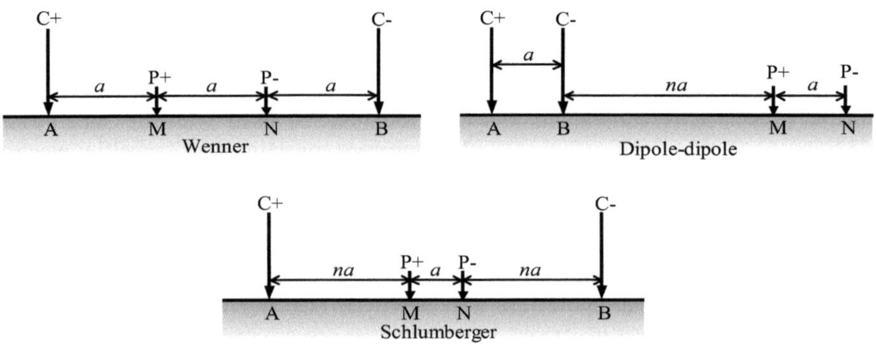

Figure 5.3. Example surface electrode configurations

5.2.2 INDUCED POLARIZATION

Whereas soil resistivity is controlled mainly by electrical conduction within the pore fluid, IP is strongly affected by processes at the fluid-grain interface (see Chapter 4 of this volume). In an IP survey, both resistive and capacitive properties of the soil are measured. As a result, IP surveys, at least in theory, permit additional information about the spatial variation in lithology and grain-surface chemistry to be determined.

IP measurements are made in the field using a four-electrode arrangement. Measurements can be made in time-domain or frequency-domain mode. For the former, the voltage decay with time is measured after the current injection is stopped (see Figure 5.4). The gradual (rather than abrupt) decrease in measured voltage is a complex function of the electrical charge polarization at the fluid-grain interface, and the conduction within the pore fluid and along the grain boundaries.

Seigel (1959) defined the apparent chargeability (m_a) as

$$m_a = \frac{V_s}{V_p}, \tag{5.5}$$

where V_s is the secondary voltage (voltage immediately after the current is shut off) and V_p is the primary voltage. The secondary voltage is difficult to measure accurately in the field and, as a result, an integral measure of apparent chargeability,

$$m_a = \frac{1}{(t_2 - t_1)} \frac{1}{V_p} \int_{t_1}^{t_2} V(t)dt, \tag{5.6}$$

is used, as illustrated in Figure 5.4. Units of chargeability are typically millivolts per volt (mV/V).

In frequency-domain mode, a phase-shifted voltage relative to an injected alternating current is measured. Traditionally, the percent frequency effect (PFE) has been used as an IP measure in the frequency-domain. Here, a comparison of impedance magnitudes is made at different injection frequencies (see for example, Reynolds, 1998). Alternatively, the impedance (in terms of magnitude and phase angle) may be used as a measure of IP. This is commonly referred to as complex resistivity. By applying the injected current at different injection frequencies, a spectrum of impedances results. This is commonly referred to as spectral IP (SIP).

While traditional IP time-domain instruments measure a number of points on a decay curve, modern instruments digitize the signal at relatively high sampling rates and permit Fourier analysis in the frequency-domain. In doing so, time-domain instruments can also provide a measure of IP in terms of complex resistivity, although the frequency range will be constrained by the sampling rate of the instrument.

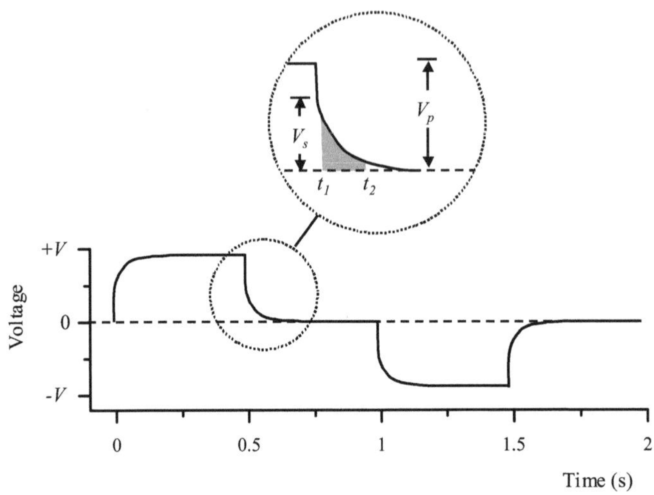

Figure 5.4. Measurement of time-domain induced polarization

IP can be measured in the field using a similar approach to that for DC resistivity. The potential electrodes should, ideally, be nonpolarizing (for example, copper-copper sulfate), although conventional DC resistivity electrodes have been used with some success (see Dahlin et al., 2002). To avoid electromagnetic coupling effects, the cable used for current injection should be short and isolated, as much as possible, from cables connected to the potential electrodes. Dipole-dipole arrays for ground-surface surveys are often preferred because of their minimal coupling effects and safer operating conditions (particularly for long survey transects). For IP surveys, injection currents often need to be much higher than those used for DC resistivity to ensure good signal-to-noise ratios. This is particularly important when using a dipole-dipole electrode configuration.

There are clearly a number of measures of IP. The measured apparent chargeability will depend upon the time window (t_1 to t_2) used, PFE will depend upon the two selected injection frequencies, and the phase angle of the impedance will depend upon the injection frequency. Furthermore, all of these measured properties will be a function of the pore-fluid conductivity and the grain-surface polarization. Slater and Lesmes (2002) demonstrate how these measurements may be scaled to derive IP parameters dependent only on polarization at the grain surface. In doing so, an IP survey can be used to provide information about the conductivity of the pore space (i.e., fluid chemistry and/or pore water content), as in a conventional resistivity survey, together with information about the grain-surface chemistry (lithology). Slater and Lesmes (2002) illustrate this approach using two field studies, showing how zones of high and low clay content may be delineated irrespective of pore-fluid chemistry. Kemna et al. (2004a) show further evidence of the value of such an approach, in this case through cross-borehole IP imaging.

5.3. Survey Configurations

Electrical methods may be applied in a range of survey configurations. The approach adopted will depend on: (1) the objectives of the survey, (2) the expected spatial variability of electrical properties, (3) access to suitable electrode sites, (4) equipment availability, and (5) data processing capabilities. Resistivity and IP methods are commonly applied on the ground surface using multiple four-electrode sites. This allows a reasonably low-cost way of mapping lateral and vertical variability in resistivity. Deployment of electrodes in boreholes will permit improved resolution at depth and, relatively recently, borehole-to-borehole or cross-borehole methods have been developed for this purpose. We list here the main configuration types and summarize the strengths and weaknesses for each when applied for hydrogeophysical investigations. In some cases, significant postprocessing of data (using modeling techniques discussed below) is required. In others, minimal effort is required to achieve the survey objectives. The survey configurations discussed apply to both resistivity and IP methods.

5.3.1 SURFACE PROFILING

The term "profiling" is used here to describe surveys of the lateral variability (i.e., at a constant depth) of electrical properties. Profiling is typically carried out along ground-surface transects. Using a particular electrode array (Wenner, dipole-dipole, etc. [see Figure 5.3]), measurements are made at different positions while keeping the electrode spacing constant. Each four-electrode position will provide an integrated measure of the electrical property. The spatial variability of signal contribution will depend on the array type (see for example, Barker, 1979) and the (unknown) variability of the resistivity. Typically, a single position is designated for the measurement at one four-electrode site. The lateral position will be the center of the four-electrode array; the depth position will depend on the configuration and spacing. For example, using a Wenner array, half the electrode spacing ($a/2$) is often assigned as the approximate depth zone of influence.

Profiling is useful for relatively rapid spatial coverage and does not require specialized multicore cables, multiplexing units, or lengthy setting out. The method is routinely used by archeological geophysicists for single depth surveys over large areas, often using a twin-array (mobile pole-pole) configuration (see for example, Mussett and Khan, 2000). Equipment to permit continuous profiling has been developed in Denmark (Sørensen, 1996) allowing surveys of 10 km in one day by taking 1 m spaced measurements using a string of electrodes towed by a vehicle. Hydrogeophysical uses of such surveys may include near-surface soil moisture, salinity, or textural mapping. However, because profiling does not provide information about variability with depth, electromagnetic methods may be preferred owing to their improved survey speed. By using large electrode spacing, deeper investigations may allow mapping of vertical hydrogeological boundaries or fracture zones (for example, Bernard and Valla, 1991)

.

5.3.2 VERTICAL ELECTRICAL SOUNDING

A vertical electrical sounding (VES) is used to determine the variation of electrical properties with depth at a single location in space, assuming no lateral variation. Four-electrode measurements are made with progressively larger spacing, centered at the same position. This provides a sounding curve as shown in Figure 5.5, which may be interpreted, using methods described below, to delineate horizontal layers. Such an approach has been used widely to differentiate hydrogeological units and, in some cases, estimate hydrological properties (e.g., Kosinski and Kelly, 1981; Börner et al., 1996). It has also been adopted in a number of studies of dynamic hydrological processes (e.g., Kean et al., 1987; Frohlich and Parke, 1989) and to estimate travel times of pollutants moving through the vadose zone (Kalinski et al., 1993).

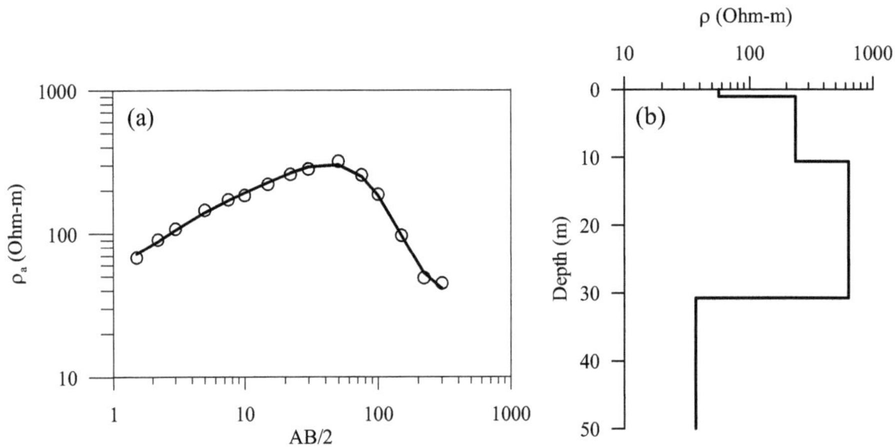

Figure 5.5. (a) Example of a Schlumberger array sounding curve from VES survey. Circles indicate measured data and solid line shows modeled curve. AB/2 (in m) is the half distance between the current electrodes (see Figure 5.3). (b) Interpreted model of resistivity as a function of depth at a single location (data and inversion available at www.geol.msu.ru/deps/geophys/rec_labe.htm).

The VES method suffers from the necessary, but often unrealistic, assumption that 1-D variations in electrical properties exist. Most modern DC resistivity and IP instruments offer the capability of addressing electrodes using multicore cables. When using such an arrangement, information about the lateral variability of electrical properties can be easily determined using combined sounding and profiling (see Section 5.3.4). VES surveys are only advantageous when such multiplexing of electrode configurations is not available. Simms and Morgan (1992) illustrate how VES survey interpretation can be ambiguous or non-unique.

5.3.3 AZIMUTHAL SURVEYS

Azimuthal surveys are conducted by carrying out four-electrode measurements at different horizontal angles, centered on a single point. This provides information about the azimuthal anisotropy of the electrical property. Such surveys can be useful for determining the orientation of fracture planes (see for example, Taylor and Fleming, 1988).

5.3.4 SURFACE IMAGING

Surface imaging combines surface profiling with vertical sounding to produce a 2-D or 3-D image of the subsurface resistivity or IP properties. Although the method has been widely used for decades, the relatively recent development of computer-controlled DC resistivity and IP instruments, coupled with advances in modeling techniques, have led to a widespread use of this approach for hydrogeological applications.

A surface imaging survey is carried out by acquiring profiles along transects using different electrode spacings. Three-dimensional surveys may be carried out as a sequence of 2-D surveys in parallel to each other (for example, Dahlin and Loke, 1997) or as a true 3-D survey using a 2-D grid of electrodes on the ground surface (see for example, Loke and Barker, 1996). A pseudo-section of electrical properties is normally constructed: for each measurement, a survey level and lateral position (midpoint of the four electrodes) is determined, and the apparent resistivity assigned to that position. Figure 5.6 illustrates the concept of building up the pseudo-section for a 2-D survey using a Wenner electrode configuration. Here, each survey level corresponds to a different electrode spacing. As the spacing increases, the effective depth of sensitivity increases. Note, however, that the pseudo-section does not necessarily show an accurate image of the subsurface distribution of resistivity; it merely serves as a means of plotting measured data. Figure 5.7 illustrates the pseudo-sections obtained using Wenner and dipole-dipole configurations associated with the synthetic model (shown in Figure 5.7 [top]). Note that the two configurations yield a different response, since a pseudo-section is not a true vertical cross section. This figure also reveals that each configuration type has different signal sensitivity patterns: Wenner pseudo-sections will tend to exaggerate lateral layering, whereas dipole-dipole surveys will reveal better vertical features.

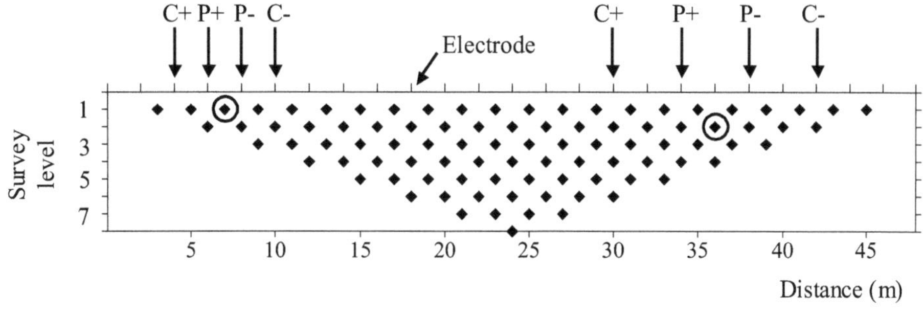

Figure 5.6. Surface electrical imaging: building a pseudo-section. Circles identify the location assignment for the two measurement configurations shown. Each survey level corresponds to a different electrode spacing.

Modeling of electrical imaging data, using methods described in Section 5.4, can provide an image of modeled resistivity within the zone of investigation. Figure 5.8 illustrates this for the synthetic model example in Figure 5.7. The modeled sections show greater similarity to the model itself than the pseudo-sections. However, note again that the Wenner model tends to reveal the horizontal element of the target, whereas the vertical structure is better resolved using the dipole-dipole model. Note also that the modeling process does not recover the true structure of the target. This is partly due to the variation in signal sensitivity within the region (particularly towards the edges) but also due to the smoothing of the model inversion process (discussed below).

Numerous examples of surface electrical imaging applied to hydrogeological problems exist, a number of which are discussed in Chapter 14 of this volume. The method is now widely used for shallow investigations (for example, to depths of 30 m). An entire survey along a transect 100 to 200 m long, covering depths to 20 to 30 m, can be carried out in under two hours using multicore cables and computer-controlled instrumentation. Longer surveys, while maintaining the same depth of survey, can be achieved using "roll along" extensions—see for example, Dahlin (1993). Panissod et al. (1997) and Christensen and Sørensen (2001) both describe methods for continuous electrical soundings using vehicle-towed electrode arrays.

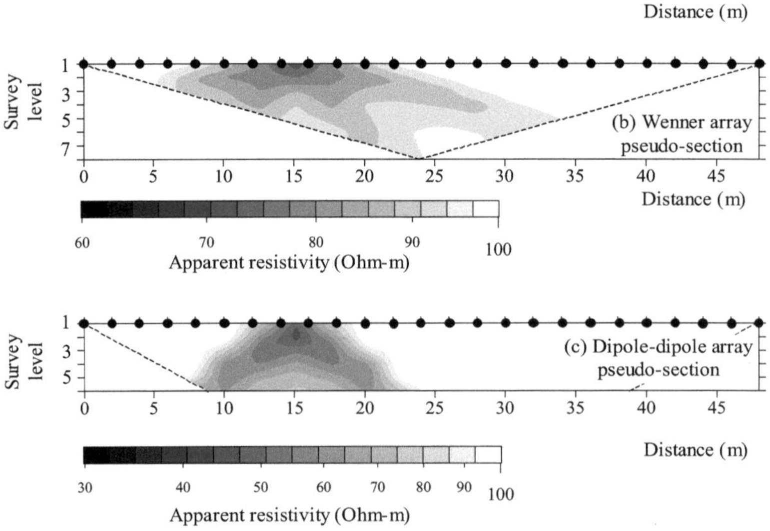

Figure 5.7. Surface electrical imaging example pseudo-sections: (top) synthetic model, (middle) Wenner pseudo-section, (bottom) dipole-dipole pseudo-section

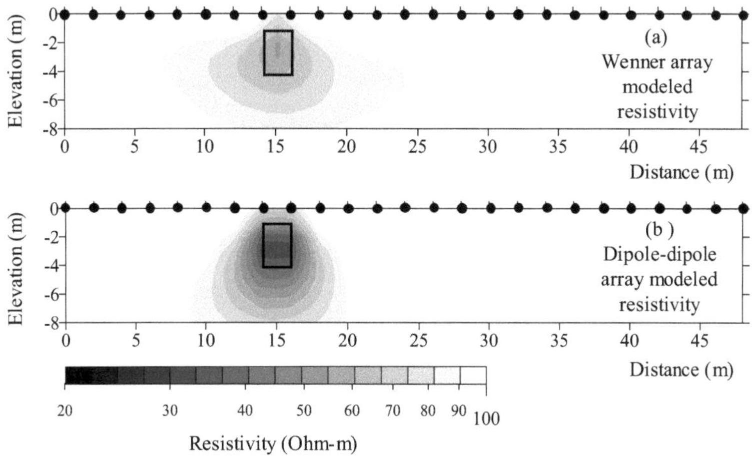

Figure 5.8. Modeled resistivity sections for the synthetic model in Figure 5.7 (inversion tools available at http://www.es.lancs.ac.uk/es/people/teach/amb)

5.3.5 SINGLE BOREHOLE SURVEYS

Electrical surveys can be conducted in several modes using electrodes deployed in a borehole. Electrodes may be permanently installed in a borehole, using suitable backfill material (ideally the native material, i.e. drill returns) in the vadose zone. In the saturated zone, open (uncased) boreholes can be used as electrode sites and, relatively recently, arrays have been manufactured for this purpose, some with isolation packers to minimize current flow along the borehole water column. Slotted plastic cased wells may also be used as electrode sites below the water table, although the well screen will add to the resistance between electrode and host material, and channeling along the borehole water column will occur unless isolation packers are used. In theory, however, the effect of the channeling can be accounted for in any data modeling by incorporating measurements of borehole water electrical conductivity and borehole geometry.

A common borehole-based approach used in mineral exploration is the mise-à-la-masse method. In such a survey, one of the current-carrying electrodes is installed in a borehole at depth, while the other current electrode is installed on the surface at a significant distance from the borehole. Potential measurements are then made at several sites on the surface, relative to a remote potential electrode, as shown in Figure 5.9. The measurements may be compared to modeled voltages for a homogenous resistivity. In mineral exploration, such measurements can help delineate electrically conductive ore bodies. In hydrological studies, the method may be used to identify orientation of fracture planes or the migration of an electrically conductive tracer injected in the well used for current injection (see, for example, Osiensky, 1997). Improved characterization may be achieved by supplementing the ground-surface measurements with electrodes installed in other boreholes (for example, Nimmer and Osiensky, 2001). In all cases, interpretation may be subjective unless modeling techniques are used (see, for example, Bevc and Morrison, 1991).

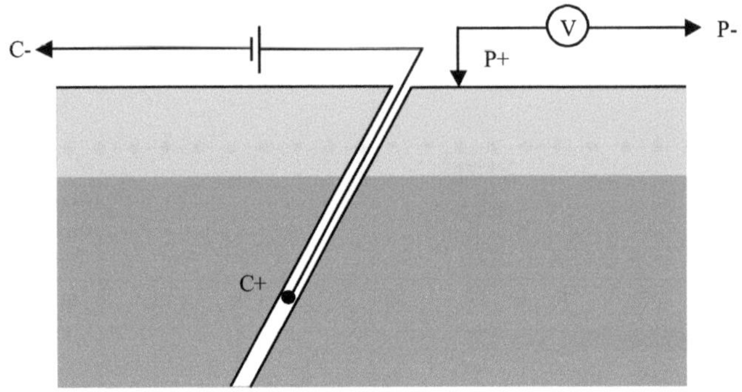

Figure 5.9. Mise-à-la-masse survey layout

Arrays of electrodes in a borehole can be used to obtain profiles in the same manner as a surface electrical profile. Such an approach provides greater support volume of measurements than conventional borehole-water electrical conductivity sampling. These methods also offer improved sensitivity to variation in electrical properties with depth compared to surface-applied electrical surveys. Binley et al. (2002a), for example, used a 32-electrode array installed in the vadose zone of a sandstone aquifer to assess changes in moisture content in unsaturated sandstone over a two-year period and detected changes in resistivity associated with a 2 % absolute variation in moisture content (see also Chapter 14 of this volume).

Imaging using a borehole electrode array is a straightforward extension of surface electrical imaging, although applications have been rare. These surveys may be supplemented with surface electrodes to form a similar geometrical arrangement to vertical seismic profiling (see Chapter 8 of this volume); however, for electrical surveys, the signal sensitivity will be confined to areas close to the borehole and surface electrode sites.

5.3.6 CROSS-BOREHOLE SURVEYS

Cross-borehole electrical resistivity surveying is an extension of conventional surface resistivity imaging. Using measurements from electrodes placed in at least two boreholes, sometimes supplemented by surface electrodes, an image of the resistivity between the boreholes is obtained. One of the earliest attempts to use such an arrangement is outlined by Daily and Owen (1991). Daily et al. (1992) demonstrated how this technique can be applied in hydrogeophysics, in their study of monitoring moisture movement in the vadose zone. Following the development of robust inversion routines and suitable data acquisition systems, cross-borehole resistivity has been applied to a wide range of hydrogeophysical problems, including vadose zone studies (Binley et al., 2002b; French et al., 2002), characterizing the transport of tracers in the subsurface (Slater et al., 2000; Kemna et al., 2002), monitoring remediation technologies (Ramirez et al., 1993; Schima et al., 1996) and monitoring leakage from underground storage tanks (Ramirez et al., 1996). A case study illustrating the use of cross-borehole resistivity imaging to monitor vadose zone flow is described in Chapter 14 of this volume.

The main advantages of cross-borehole imaging compared to surface imaging are that: (1) high resolution at depth is possible and (2) investigations can be made without the need for surface access (for example, surveys under buildings are possible). The disadvantages are that: (1) boreholes are required, (2) data sensitivity is constrained to the region between the boreholes, (3) data acquisition may require more sophisticated instrumentation, (4) for vadose zone surveys, noise levels may be much higher than those using surface electrodes, owing to weaker electrical contact (increased contact resistance), and (5) data processing is more complex.

Using borehole-deployed electrodes, a very large number of measurement schemes are possible. In Figure 5.10, two example bipole-bipole schemes are shown (here we use the term bipole-bipole to differentiate from the conventional surface dipole-dipole array). Other arrangements using remote (pole) electrodes are also possible. In the AM-BN scheme shown in Figure 5.10a, current is injected between the two boreholes and a potential difference is measured between the boreholes. In the AB-MN scheme in Figure 5.10b, current is injected between electrodes in the same borehole, and the potential difference is measured between electrodes in the other borehole. The AM-BN scheme has good signal-to-noise characteristics, in comparison to AB-MN, because of the dipole length. Bing and Greenhalgh (2000) have studied these schemes, and others, in a synthetic modeling exercise, and demonstrate the impact of poor signal-to-noise characteristics on resistivity images.

Although such synthetic studies are useful, the concept of a "universal" measurement scheme for cross-borehole imaging is probably not achievable. Conditions for cross-borehole surveys are extremely variable because of: (1) contrasts in electrode contact and influence of backfill or any borehole water column, (2) different borehole separations, and (3) instrumentation resolution and measurement rate. Moreover, the assessment of the true measurement errors must be considered in any analysis of cross-borehole electrical measurements, because such errors will be strongly dependent on the environment to which the method is applied. It is thus extremely useful to consider the acquisition geometry of cross-borehole methods on a case-by-case basis.

To illustrate the resolution of cross-borehole resistivity imaging, Figure 5.11 shows an example of a synthetic model and the associated resistivity distribution. For this example, the AB-MN scheme with an electrode spacing of 8 m was used. The "target" is well resolved, but the final image is smoothed, owing to the inverse modeling scheme adopted. In Figure 5.12, the effect of increased borehole spacing is shown. Here, the final image reveals a more smeared effect caused by the reduced sensitivity towards the center of the image. Clearly, care should be taken to design the position of borehole arrays when using cross-borehole techniques. This is discussed further in Section 5.5.

Schenkel (1995) has demonstrated how steel-cased boreholes can be used to determine electrical resistivity images between boreholes. This approach is no different from the conventional cross-borehole arrangement, except that the highly conductive well casing is accounted for in the model analysis. Although such an approach has reduced sensitivity, given the widespread availability of existing steel cased wells, this method offers potential for further development.

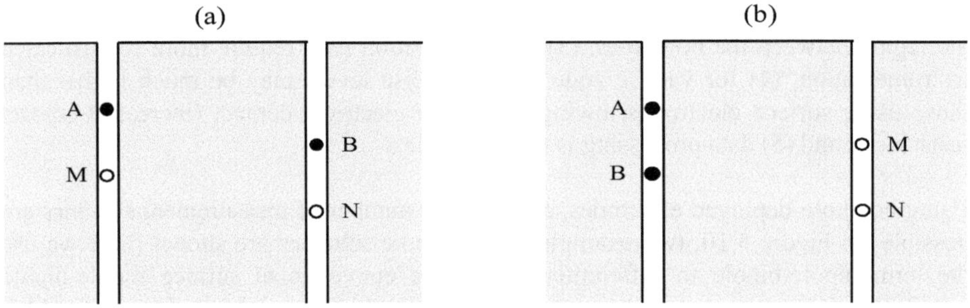

Figure 5.10. Example measurement configuration for cross-borehole resistivity imaging. Electrodes A and B are for current injection, M and N are for voltage measurement. Scheme (a) is AM-BN and Scheme (b) is AB-MN.

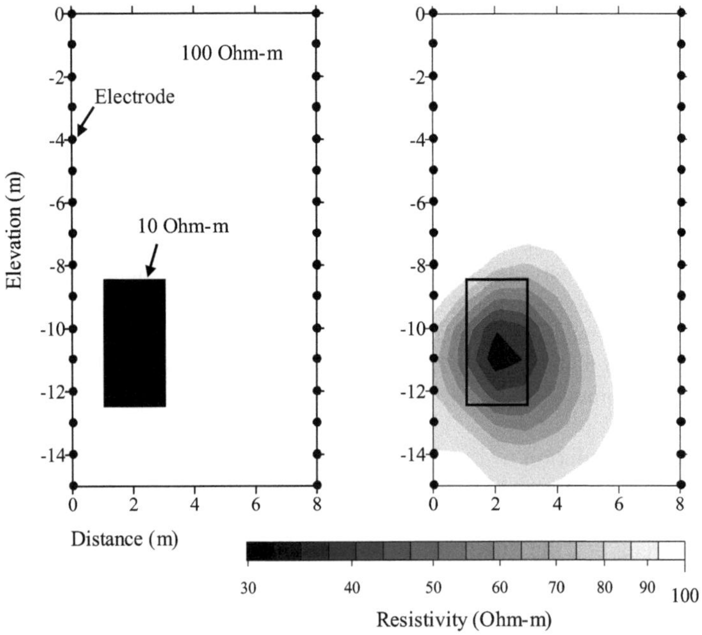

Figure 5.11. Cross-borehole resistivity imaging applied to a synthetic model

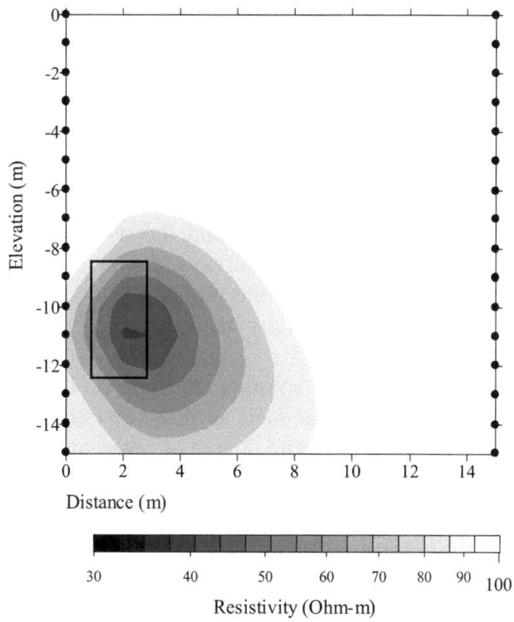

Figure 5.12. Cross-borehole resistivity imaging applied to a synthetic model to illustrate reduced resolution with increased borehole spacing. The 10 ohm-m target outline is indicated in the image. The background resistivity of the model is 100 ohm-m, as in Figure 5.11.

5.3.7 CORE AND BLOCK SCALE IMAGING

Electrical methods have also been applied to image the variability of electrical properties at the core or block scale. In fact, one of the early applications of electrical imaging was the study of rock cores by Daily et al. (1987) using circular electrode arrays. Binley et al. (1996) also used circular arrays in their investigation of preferential flow in 0.3 m diameter undisturbed soil cores, and Ramirez and Daily (2001) demonstrate how electrical imaging can be used to study changes in moisture distribution in a 3 m × 3 m × 4.5 m block of welded tuff. Because the electric potential field can be modeled for any arbitrary geometry, there are no constraints on the size and shape of the object under investigation. In Chapter 15 of this volume, examples of core-scale imaging for hydrogeophysical studies are shown.

5.4 Modeling and Data Inversion

5.4.1 GENERAL CONCEPTS

The ultimate goal of electrical methods is to derive the distribution of (low-frequency) electrical properties inside an object, here generally the subsurface, from a set of measurements conducted on the boundary of the object, or at least outside the region of interest, according to the principles outlined in the previous section. The theoretical outcome of such a measurement can be mathematically determined (modeled) for given electrical properties from the governing physical law, the Poisson equation, subject to given boundary conditions (Equations (5.1) and (5.2)). This exercise defines the so-called "forward problem." For the purpose of subsurface investigations, however, the "inverse problem" needs to be solved, i.e., given a set of measurements (data), the distribution of electrical properties (model) is sought that explains the observations to an acceptable degree (Figure 5.13). For resistivity surveys, data will be in the form of transfer resistances or apparent resistivities, and the model will be parameterized in terms of resistivity or conductivity. For IP surveys, data will be in the form of apparent chargeability or transfer impedance, and the model will be parameterized in terms of intrinsic chargeability or complex resistivity, respectively.

Unfortunately, there is no unique solution to this problem (this has promoted development of stochastic approaches such as those discussed in Chapter 17 of this volume). Electrical methods bear a certain degree of inherent non-uniqueness, i.e., there typically exists a variety of different models that effectively produce the same response. In addition, because of practical limitations, data are neither complete nor perfectly accurate, but mostly insufficient and inconsistent. Therefore, in principle, an infinite number of models fit the data within a given level of uncertainty. However, by systematically restricting the model search in the inversion process, for instance by claiming predefined model characteristics, a "unique" solution with practical relevance can be obtained. This is usually accomplished by formulating the inverse problem as a regularized optimization problem, which involves minimization of an objective function comprising both data misfit (measured vs. modeled) and a penalty term accounting for deviations from the desired model attributes.

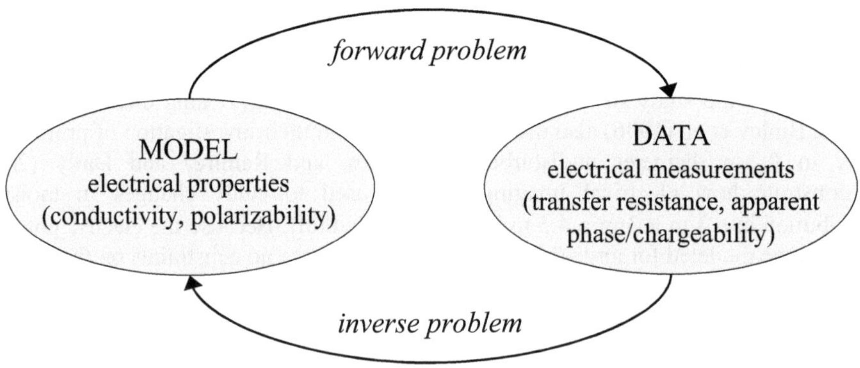

Figure 5.13. Definition of electrical forward and inverse problem

Starting with the formulation of the electrical forward problem, the following discussion focuses on the principles of appropriate inversion strategies. Moreover, different approaches for appraisal of the final inversion result are addressed; these are essential tools for a critical interpretation in real applications. We include details of the approaches here, since this material is not covered in existing exploration geophysical texts. The reader who is not concerned with details of the modeling approaches may wish to bypass the following and go directly to Section 5.5. It is, however, useful for any user of resistivity and IP methods to be aware of the limitations and assumptions of the modeling tools used for data processing. Inverse models provide an image of electrical properties, but the image is subject to uncertainty. All data will be subject to error, and the user should be aware of these errors in order to prevent overfitting of any inverse model. The model itself will be an approximation and thus acts as another source of error. Data sensitivity will vary considerably in the image and an appreciation of how to explore such variability will allow greater confidence in assessing the reliability of the image values. Many inverse methods for resistivity and IP methods employ some form of smoothing to ensure that a solution is reached. In some cases, such smoothing may be inappropriate—for example, if it is known that sharp contrasts exist, perhaps resulting from lithological boundaries.

5.4.2 FORWARD PROBLEM

5.4.2.1 DC Resistivity

If the conductivity distribution is 2-D, for instance, constant in the y direction, application of the Fourier transform

$$\tilde{V}(x,k,z) = \int_0^\infty V(x,y,z)\cos(ky)\,\mathrm{d}y \tag{5.7}$$

to Equation (5.1) yields a simplified, 2-D problem (for example, Hohmann, 1988), that is,

$$\frac{\partial}{\partial x}\left(\sigma\frac{\partial\tilde{V}}{\partial x}\right)+\frac{\partial}{\partial z}\left(\sigma\frac{\partial\tilde{V}}{\partial z}\right)-k^2\sigma\tilde{V} = -\frac{I}{2}\delta(x)\delta(z), \tag{5.8}$$

with k being the wave number. Once this differential equation is solved for the transformed potential $\tilde{V}(x,k,z)$, an inverse Fourier transform must be applied to obtain $V(\mathbf{r})$ (for example, LaBrecque et al., 1996; Kemna, 2000).

For a horizontally layered earth model, i.e., σ only varying with z, a compact solution of the Poisson equation can still be formulated. Assuming for instance a Schlumberger electrode configuration (see Figure 5.3), which is typically adopted for vertical electrical soundings (see Section 5.3.2), the apparent resistivity is given by the so-called Stefanescu integral (for example, Koefoed, 1979):

$$\rho_\mathrm{a}(r) = r^2\int_0^\infty T(\lambda)J_1(\lambda r)\,\lambda\,\mathrm{d}\lambda\,, \tag{5.9}$$

where r is the half distance between the current electrodes, J_1 is the Bessel function of order one, and the kernel $T(\lambda)$ is a function of the given layer thicknesses and resistivities, with λ being an integration variable (wavenumber).

Whereas the integral in Equation (5.9) can be readily calculated, for instance using fast filtering techniques (for example, Anderson, 1979; Christensen, 1990), for arbitrary 2-D or 3-D conductivity distributions, computationally more expensive numerical approaches are required to solve Equation (5.8) or (5.1). In these cases, finite element (FE) or finite difference (FD) methods are typically used, where the continuous conductivity distribution is approximated by a mesh of individual elements or cells, each with constant conductivity. The potential is then calculated at discrete points (nodes of the mesh) by solving a linear system of equations derived from the discretized differential equation and boundary conditions. Hohmann (1988) provides an overview of the approach. Details on the use of FE and FD methods to solve the DC electrical problem are well documented in the literature (for example, Coggon, 1971; Dey and Morrison, 1979a,b; Pridmore et al., 1981; Lowry et al., 1989; Spitzer, 1995; Bing and Greenhalgh, 2001). A direct comparison of the performance of both methods was recently presented by Li and Spitzer (2002). However, in general, FE methods are preferred if greater flexibility with respect to mesh geometry is desired, for instance to account for irregular electrode positions or to incorporate topography.

5.4.2.2 Induced Polarization

Two different approaches are commonly used to model IP data. According to Seigel (1959), the time-domain voltage response of a polarizable medium with conductivity σ_{dc} and intrinsic chargeability m can be interpreted as the response of a nonpolarizable medium with decreased conductivity $\sigma_{ip} = (1-m)\sigma_{dc}$. This results from the fact that any present polarization dipoles effectively reduce the overall current density. Consequently, the apparent chargeability m_a (see Equation (5.5)) can be modeled by carrying out DC resistivity forward models for σ_{ip} and σ_{dc}, according to (for example, Oldenburg and Li, 1994)

$$m_a = \frac{f[(1-m)\sigma_{dc}] - f[\sigma_{dc}]}{f[(1-m)\sigma_{dc}]}. \tag{5.10}$$

In Equation (5.10), f represents the forward modeling operator implicitly defined by the equations stated in the previous paragraph for 1-D, 2-D, and 3-D conductivity distributions, respectively. An alternative, frequency-domain approach is based on the combined description of conduction and polarization properties by means of a complex conductivity, σ^*. In the quasi-static frequency limit (typically < 10 Hz), a complex-valued Poisson equation,

$$\nabla \cdot (\sigma^* \nabla V^*) = -I\delta(\mathbf{r}), \tag{5.11}$$

may be solved for a complex potential, V^*, analogous to the DC case outlined above. From V^*, both magnitude and phase, as the relevant DC resistivity and IP data, are obtained. Weller et al. (1996), for example, followed this approach using a finite-difference framework.

5.4.3 INVERSE PROBLEM

To formulate the inverse problem, the considered distribution of electrical properties is discretized into a set of parameters defining a model vector **m**. While for 1-D problems,

m normally contains the conductivities and thicknesses of a multilayer model, for arbitrary 2-D and 3-D distributions its elements generally correspond to the conductivities of the individual elements or cells (or lumped blocks thereof) of the FE or FD mesh used in the forward modeling. That is,

$$m_j = \ln \sigma_j \quad (j = 1, ..., M). \tag{5.12}$$

Here, the logarithm accounts for the large possible range in earth conductivity. Analogously, the given set of measured transfer resistances, R_i, is assembled in a data vector **d** according to

$$d_i = -\ln R_i \quad (i = 1, ..., N). \tag{5.13}$$

Again, log-transformed data are normally used on account of the wide range in observed resistances for arbitrary electrode configurations, while the minus sign in Equation (5.13) accounts for a physical dimension consistent with Equation (5.12).

The inverse problem now is to find a model **m** which, using the forward mapping according to Equations (5.1), (5.8), or (5.9), reproduces data **d** to the specified level of uncertainty. However, since inherent non-uniqueness of the resistivity inverse problem, together with the presence of data errors, can effectively lead to an extremely ill-posed numerical problem, additional constraints must be imposed on the inversion. This is normally accomplished by solving the inverse problem as a regularized optimization problem (Tikhonov and Arsenin, 1977), where an objective function of the form

$$\Psi(\mathbf{m}) = \Psi_{\mathrm{d}}(\mathbf{m}) + \alpha\, \Psi_{\mathrm{m}}(\mathbf{m}) \tag{5.14}$$

is sought to be minimized. Here,

$$\Psi_{\mathrm{d}}(\mathbf{m}) = \left\| \mathbf{W}_{\mathrm{d}}\, [\mathbf{d} - \mathbf{f}(\mathbf{m})] \right\|^2 \tag{5.15}$$

is a measure of the data misfit, with **f** denoting the forward operator and $\mathbf{W}_{\mathrm{d}} = \mathrm{diag}(1/\varepsilon_1, ..., 1/\varepsilon_N)$ representing a data weighting matrix associated with the individual (uncorrelated) data errors ε_i. In addition, $\Psi(\mathbf{m})$ contains a stabilizing model objective function usually expressed as

$$\Psi_{\mathrm{m}}(\mathbf{m}) = \left\| \mathbf{W}_{\mathrm{m}}\, (\mathbf{m} - \mathbf{m}_{\mathrm{ref}}) \right\|^2, \tag{5.16}$$

which is used to incorporate certain model constraints relative to a reference model $\mathbf{m}_{\mathrm{ref}}$ by appropriate choice of a model weighting matrix $\mathbf{W}_{\mathrm{m}}$. The regularization parameter, α, in Equation (5.14) controls the tradeoff between influence of data misfit and model objective function in the inversion. Usually a smoothness constraint is imposed on the model, i.e., $\mathbf{W}_{\mathrm{m}}$ is chosen such as to evaluate the roughness of $\mathbf{m} - \mathbf{m}_{\mathrm{ref}}$. Using directional weights in the evaluation of model roughness, anisotropic smoothing may be applied (for example, Ellis and Oldenburg, 1994), for instance to account for a layered subsurface environment. The reference model $\mathbf{m}_{\mathrm{ref}}$ may contain expected parameter values as well as the result of a previous inversion in monitoring (time-lapse) applications, or just be assigned to a homogenous halfspace or the null vector if no additional information is available.

Minimization of Equation (5.14) can be achieved through application of gradient search methods. Adopting the Gauss-Newton approach, an iterative scheme results, where at each step, k, the linear system of equations

$$(\mathbf{J}_k^T \mathbf{W}_d^T \mathbf{W}_d \mathbf{J}_k + \alpha \mathbf{W}_m^T \mathbf{W}_m)\Delta\mathbf{m}_k = \mathbf{J}_k^T \mathbf{W}_d^T \mathbf{W}_d[\mathbf{d} - \mathbf{f}(\mathbf{m}_k)] - \alpha \mathbf{W}_m^T \mathbf{W}_m(\mathbf{m}_k - \mathbf{m}_{ref}) \quad (5.17)$$

is solved for a model update $\Delta\mathbf{m}_k$. Here, $\mathbf{J}_k$ is the Jacobian (sensitivity) matrix; that is, $J_{ij} = \partial d_i / \partial m_j$, evaluated for the current model $\mathbf{m}_k$. Starting from a model $\mathbf{m}_0$ (for instance, homogenous or equal to $\mathbf{m}_{ref}$, if available), the iteration process $\mathbf{m}_{k+1} = \mathbf{m}_k + \Delta\mathbf{m}_k$ according to Equation (5.17) is continued for an optimum choice of α (see, for example, deGroot-Hedlin and Constable, 1990) until $\Psi_d(\mathbf{m}_k)$ matches the desired data misfit target value.

Numerous DC resistivity inversion schemes have been proposed that are, apart from implementation details, in principle based on the above approach, including 2-D and 3-D approaches for both surface (for example, Loke and Barker, 1995) and cross-borehole (for example, LaBrecque et al., 1996) data. For the interpretation of time-lapse data, the scheme may be slightly modified to directly invert temporal changes in the data for relative model variations (Daily et al., 1992; LaBrecque and Yang, 2000; Kemna et al., 2002). Importantly, the approach can also be straightforwardly extended to include IP data in the inversion process, either in terms of time-domain chargeability according to Equation (5.10) (LaBrecque, 1991; Oldenburg and Li, 1994) or in terms of frequency-domain phase angle according to Equation (5.11) (Kemna and Binley, 1996; Kemna et al., 2000).

5.4.4 MODEL APPRAISAL

To reliably interpret resistivity or IP inversion results, some knowledge about the final model resolution is often desired. Generally, model resolution is a complicated function of numerous factors, including electrode layout, measurement scheme, data signal-to-noise ratio, and resistivity distribution, as well as parameterization and regularization used in the inversion. According to inverse theory (for example, Menke, 1989), the model resolution matrix

$$\mathbf{R} = (\mathbf{J}_k^T \mathbf{W}_d^T \mathbf{W}_d \mathbf{J}_k + \alpha \mathbf{W}_m^T \mathbf{W}_m)^{-1} \mathbf{J}_k^T \mathbf{W}_d^T \mathbf{W}_d \mathbf{J}_k \quad (5.18)$$

may be computed for the final inversion iteration k to assess how individual parameters are resolved in multidimensional imaging approaches. $\mathbf{R}$ may be defined as

$$\mathbf{m} = \mathbf{R}\mathbf{m}_{true}, \quad (5.19)$$

where $\mathbf{m}$ is the vector of parameters obtained by the inversion and $\mathbf{m}_{true}$ is the vector of true (unknown) parameters. Any deviations of $\mathbf{R}$ from the identity matrix indicate that inverted parameter values result from an averaging process in the inversion. Off-diagonal entries in $\mathbf{R}$ may be used to display the effect of regularization (for instance smoothing) on each parameter (for example, Alumbaugh and Newman, 2000). Use of only the diagonal of $\mathbf{R}$ has been shown to be insightful for image appraisal in cross-borehole applications (Ramirez et al., 1995).

Since the actual calculation of **R** in large-scale inverse problems is cumbersome, an indirect approach based on a simple accumulated sensitivity map, **s**, according to (Park and Van, 1991; Kemna, 2000)

$$s_j = (\mathbf{J}_k^{\mathrm{T}} \mathbf{W}_{\mathrm{d}}^{\mathrm{T}} \mathbf{W}_{\mathrm{d}} \mathbf{J}_k)_{jj} \tag{5.20}$$

may be used as a computationally inexpensive alternative for image appraisal. Obviously, resolution is supposed to be low in model regions where sensitivity of the measurements is poor (i.e., **s** shows low values) and correspondingly regularization is more influential. Figure 5.14 shows a comparison of the two approaches, Equations (5.18) and (5.20), for the 2-D cross-borehole example in Figure 5.11. Both images in Figure 5.14 illustrate the decrease in sensitivity away from the boreholes. In addition, the resolution matrix in Figure 5.14a shows clearly the asymmetry in the image sensitivity caused by the distribution of resistivity. Any application that requires accurate estimation of electrical properties away from the boreholes must take into account the rapidly decreasing sensitivity. Imaging a tracer plume, for example, will result in an effective mass-balance error or, if the boreholes are spaced too far apart, no sensitivity to the change in pore-fluid concentration.

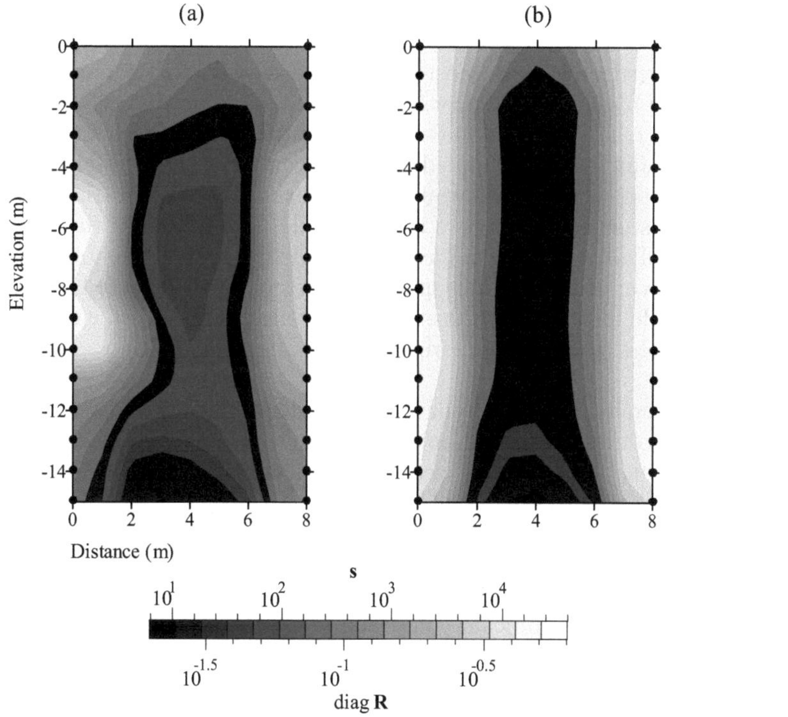

Figure 5.14. Illustration of resolution estimation for the 2-D cross-borehole survey in Figure 5.11 (dark areas in images indicate poor resolution): (a) diagonal of resolution matrix R in Equation (5.18), and (b) sensitivity map according to Equation (5.20). Note that (b) exhibits characteristics similar to (a).

An alternative procedure for assessing image resolution was proposed by Oldenburg and Li (1999). In their approach, two independent inversions for different reference models, $\mathbf{m}_{\text{ref}}^{(1)}$ and $\mathbf{m}_{\text{ref}}^{(2)}$, are computed, and the deviation of the respective inversion results, $\mathbf{m}^{(1)}$ and $\mathbf{m}^{(2)}$, is assessed in terms of the so-called depth of investigation index:

$$DOI_j = \frac{m_j^{(1)} - m_j^{(2)}}{m_{\text{ref }j}^{(1)} - m_{\text{ref }j}^{(2)}}. \tag{5.21}$$

In areas of the image where sensitivity is low, the *DOI* index will be close to unity, revealing regions where the model is weakly constrained by the data. Similarly, low *DOI* values will indicate high sensitivity and consequently allow greater faith in inverted values.

All of these model appraisal techniques are useful in assessing the uncertainty in areas of any image produced by an inverse method. These methods should be used, along with other factors discussed in the next section, to design an appropriate survey.

5.5 Survey Design and Implementation

For any investigation, the most appropriate survey configuration (surface, cross-borehole, etc.) will strongly depend on the specific objectives of the project. Surface arrays allow relatively rapid surveys to be carried out in a noninvasive manner, but depth of sensitivity will be limited. Galvanic contact is required for the electrodes, and consequently in some environments surface cover may limit the use of surface surveys. Investigations under areas where electrodes cannot be installed (under a building, for example) will also pose difficulties. In such environments, borehole arrays may prove more effective. However, the spacing between boreholes used for cross-borehole surveys should ideally be less than half the shortest length of an electrode array in any of the boreholes (as illustrated earlier).

The use of boreholes for electrode sites will often depend on the availability of existing boreholes or drilling budgets. Below the water table, electrodes can be lowered into open boreholes or slotted cased (non-metallic) boreholes. For vadose zone investigations, the electrodes will need to be sacrificially installed with an appropriate backfill to ensure good contact with the host material. For IP surveys, the backfill may significantly affect the measured polarization, and thus an appropriate backfill material must be selected. Bentonite, which is widely used for sealing sections of borehole backfill, will create significant polarization, which, even for DC resistivity, will cause problems in data quality.

The adopted measurement scheme (for example, Wenner, dipole-dipole, etc., for surface arrays) should be selected based on the expected target, but will also depend on environmental conditions and instrument availability. Dipole-dipole surveys, for example, are ideal for resolving localized targets (such as a tracer plume), but suffer from poor signal-to-noise ratios. Thus, high input voltages are required for dipole pairs that are widely spaced. For closely spaced dipoles, high voltage input will often result in "over voltages" (saturation of the receiver), and thus the dynamic range of the instrument needs to be considered before carrying out the survey. For surface DC resistivity surveys, the Wenner array is popular because of the good signal-to-noise ratio and reasonably narrow range of resistances measured in a typical survey. These factors allow the use of relatively inexpensive field instruments. In addition, a Wenner survey will enhance horizontal structures in the final image. While these may be appealing in many cases and give an indication of realism, in cases where the resistivity contrast is not horizontally layered, the Wenner array may be inappropriate. Ideally, synthetic model studies should be carried out before any survey to design the most appropriate configuration. Forward models are useful for computing the expected signals; these can be compared with the instrument resolution to assess the likely signal resolution. Inverse models, applied to synthetic targets, will allow the user to determine the expected resolution. Model appraisal methods, such as those discussed earlier, will enhance the user's assessment of the appropriateness of the survey configuration selected.

To illustrate the use of synthetic studies, Figure 5.15 shows the results from three inverse models applied to a synthetic model representing a groundwater plume. The plume model is shown in Figure 5.15a. In Figure 5.15b the inversion of surface electrode data using a dipole-dipole scheme is shown. The target is reasonably well resolved, but is improved by using a reduced set of surface electrodes, combined with two widely spaced boreholes, as shown in Figure 5.15c. Using three boreholes, one of which intercepts the plume, results in a vast improvement in plume delineation as shown in Figure 5.15d. In all these models, a dipole-dipole scheme was used, and 2% Gaussian noise was added to the forward model response prior to inversion. In addition, all forward model "measurements" with a transfer resistance lower than 10^{-3} ohm were ignored in the inversion to simulate instrument resolution.

For many hydrogeophysical studies of dynamic processes, the measurement scheme used will be constrained by data collection speed and the rate at which changes occur in the subsurface. Slow instrument relay switching results in data collection rates of two to four hundred measurements per hour. Many modern instruments offer multi-measurement channel capability, but each measurement "frame" may still take several hours for 3-D configurations. Data collection speed should therefore also be recognized at the planning stage of any investigation.

If inverse methods are used, as is the case in many electrical surveys, the user should be aware of what is a satisfactory misfit between the model prediction and the data. This will depend on the error level in the data; however, for many resistivity surveys, users often fail to recognize the significance of data error in any investigation. Measurement errors in conventional surface resistivity surveys are traditionally assessed using repeatability checks. Such checks prove to be unreliable, particularly for cross-borehole surveys, and measurements of reciprocity are far superior in assessing true error levels. A reciprocity check is made by interchanging the electrode pair used for voltage measurement with the electrode pair used for current injection. The transfer impedance for these two cases will be identical if the system is responding linearly (i.e., according to Ohm's law) and there is no measurement error. Data errors should be used to weight the data (see Equation (5.15)). Over-fitting in data inversion results in "noisy" images and artifacts, which can be avoided by appropriate attention during data collection (see, for example, LaBrecque et al., 1996).

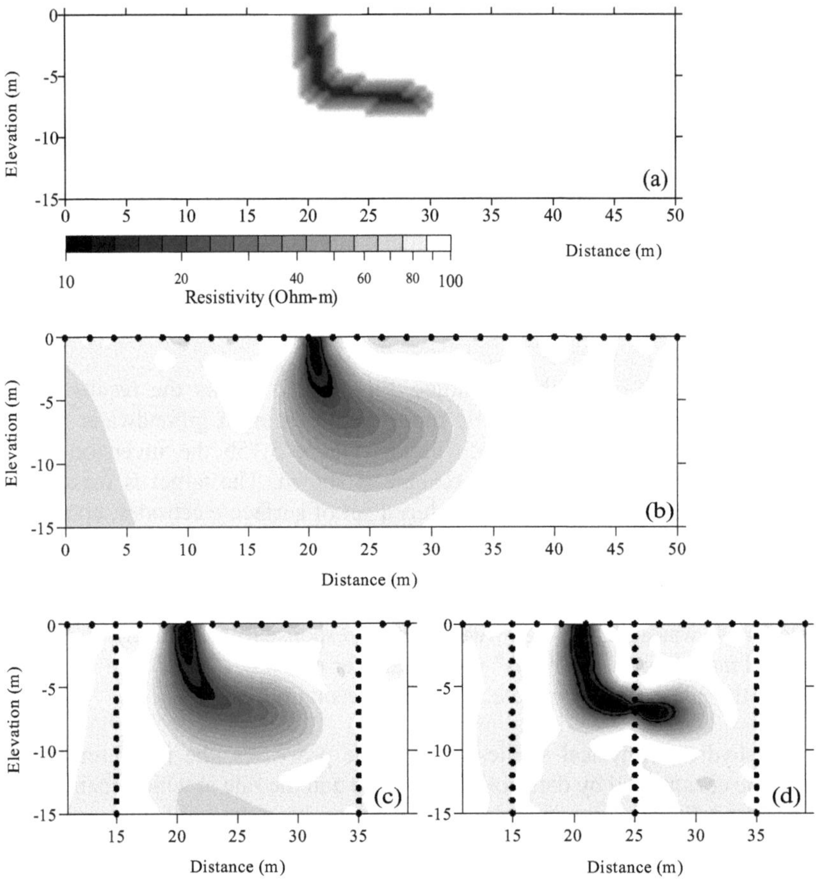

Figure 5.15. Synthetic plume model application: (a) synthetic model resistivity, (b) inverted model using only surface electrodes, (c) inverted model using surface electrodes and two borehole arrays, and (d) inverted model using surface electrodes and three borehole arrays

The survey selected will also be influenced by the assumed spatial distribution of the electrical properties. Soundings are clearly not appropriate if significant 2-D or 3-D variability exists. Similarly, 2-D (surface or cross-borehole) surveys may suffer from 3-D effects; for cross-borehole surveys, these 3-D effects may be result of the boreholes, or the backfill.

In summary, the hydrogeophysicist should consider a wide range of issues in selecting an appropriate survey. When appropriate, models should be used to ensure optimal subsurface characterization, and particular attention should be addressed to data quality prior to using any inverse method.

5.6 Future Challenges

Despite their long history, electrical methods are still developing, yet many hydrogeophysical applications are currently constrained by the availability of appropriate hardware and software tools. Data collection speed is improving, but still remains a serious constraint for time-lapse 3-D investigations. Autonomous data acquisition systems are emerging and offer great potential for remote monitoring of dynamic processes. Recently, Daily et al. (2004) demonstrated the use of such a system for remote monitoring of leakage from an underground storage tank.

Many inverse methods are based on a general smoothness constraint (see Equation (5.16)). However, for some cases assumptions of smoothness are not appropriate: sharp contrasts in the subsurface may exist, for example, at the edges of a contaminant or tracer plume or a lithological boundary. Alternative regularization approaches are being developed to address this problem (for example, Loke et al., 2001), and these offer promise for some future hydrogeophysical applications.

Linking electrical properties derived from electrical images to hydrological parameters, through relationships such as those covered in Chapter 4 of this volume, may permit estimation of hydraulic variables such as water content (for field examples, see Chapter 14 of this volume) or solute concentration (see for example, Kemna et al., 2002). Recently, Kemna et al. (2004b) proved in a synthetic experiment that time-lapse electrical imaging in conjunction with equivalent transport model analysis can actually be used to characterize solute transport in a heterogeneous aquifer in a quantitative manner. However, the effects of reduced sensitivity with distance from electrodes, coupled with smoothing effects from common regularization methods, will lead to poor mass balance. This limits the amount of quantitative information available to the hydrologist directly from an image. New approaches are needed that constrain image reconstruction, for instance based on the incorporation of flow models, such as proposed by Seppänen et al. (2001) for state estimation in process tomography, or by the incorporation of hydrologic-electrical parameter models (Liu and Yeh, 2004). The computation of geostatistical characteristics from electrical images is also likely to be an area of future research in hydrogeophysics, because such information can be of immense value to hydrological modelers. Joint hydraulic-geophysical inversion methods are also likely to develop. Dam and Christensen (2003) illustrate how hydraulic head and resistivity data may be used jointly to constrain groundwater model calibration, albeit through (perhaps limiting) assumptions of a relationship between hydraulic and electrical conductivity.

With rapidly developing hardware and software, some of which will be guided by hydrogeophysicists, we expect that there will be an increasing number of appropriate applications of electrical methods to an increasing range of hydrological problems, as well as greater integration of electrically derived information within hydrological modeling studies.

Acknowledgments

We would like to acknowledge the help of Lee Slater and Niels Christensen, whose excellent reviews greatly improved this chapter.

References

Alumbaugh, D.L., and G.A. Newman, Image appraisal for 2-D and 3-D electromagnetic inversion, *Geophysics*, *65*, 1455–1467, 2000.

Anderson, W.L., Numerical integration of related Hankel transforms of orders 0 and 1 by adaptive digital filtering, *Geophysics*, *44*, 1287–1305, 1979.

Barker, R.D., Signal contributions and their use in resistivity studies, *Geophys. J. Royal Astr. Soc.*, *59*, 123–129, 1979.

Bernard, J., and P. Valla, Groundwater exploration in fissured media with electrical and VLF methods, *Geoexploration*, *27*, 81–91, 1991.

Bevc, D., and H.F. Morrison, Borehole-to-surface electrical resistivity monitoring of a salt water injection experiment, *Geophysics*, *56*, 769–777, 1991.

Bing, Z., and S.A. Greenhalgh, Cross-hole resistivity tomography using different electrode configurations, *Geophys. Prosp.*, *48*, 887–912, 2000.

Bing, Z., and S.A. Greenhalgh, Finite element three-dimensional direct current resistivity modelling: Accuracy and efficiency considerations, *Geophys. J. Internat.*, *145*, 679–688, 2001.

Binley, A., S. Henry-Poulter, and B. Shaw, Examination of solute transport in an undisturbed soil column using electrical resistance tomography, *Water Resour. Res.*, *32*, 763–769, 1996.

Binley, A., P. Winship, L.J. West, M. Pokar, and R. Middleton, Seasonal variation of moisture content in unsaturated sandstone inferred from borehole radar and resistivity profiles, *J. Hydrol.*, *267*, 160–172, 2002a.

Binley, A., G. Cassiani, R. Middleton, and P. Winship, Vadose zone model parameterisation using cross-borehole radar and resistivity imaging, *J. Hydrol.*, *267*, 147–159, 2002b.

Börner, F.D., J.R. Schopper, and A. Weller, Evaluation of transport and storage properties in the soil and groundwater zone from induced polarization measurements, *Geophy. Prosp.*, *44*, 583–601, 1996.

Christensen, N.B., Optimized fast Hankel transform filters, *Geophys. Prosp.*, *27,* 876–901, 1990.

Christensen, N.B., and K. Sørensen, Pulled array continuous electrical sounding with an additional inductive source: An experimental design study, *Geophys. Prosp.*, *49*, 241–254, 2001.

Coggon, J.H., Electromagnetic and electrical modeling by the finite element method, *Geophysics*, *36*, 132–155, 1971.

Dahlin, T., On the automation of 2-D resistivity surveying for engineering and environmental applications, PhD Thesis, Lund Univ., Sweden, 1993.

Dahlin, T., and M.H. Loke, Quasi-3-D resistivity imaging-mapping of three-dimensional structures using 2-D resistivity techniques, *Proc. 3rd Mtg. Environmental and Engineering Geophysics*, Environ. Eng. Geophys. Soc., Eur. Section, 143–146, 1997.

Dahlin, T., V. Leroux, and J. Nissen, Measuring techniques in induced polarisation imaging, *J. Appl. Geophys.*, *50*, 279–298, 2002.

Dam, D., and S. Christensen, Including geophysical data in ground water model calibration, *Ground Water, 41*, 178–189, 2003.

Daily, W., and E. Owen, Cross borehole resistivity tomography, *Geophysics*, *56*, 1228–1235, 1991.

Daily, W.D., W. Lin, and T. Buscheck, Hydrological properties of Topopah Spring tuff: Laboratory measurements, *J. Geophys. Res.*, *92*, 7854–7864, 1987.

Daily, W.D., A.L. Ramirez, D.J. LaBrecque, and J. Nitao, Electrical resistivity tomography of vadose water movement, *Water Resour. Res.*, *28*, 1429–1442, 1992.

Daily, W., A. Ramirez, and A. Binley, Remote monitoring of leaks in storage tanks using electrical resistance tomography: Application at the Hanford Site, *J. Environ. Eng. Geophys.*, *9*, 11–24, 2004.

deGroot-Hedlin, C., and S.C. Constable, Occam's inversion to generate smooth, two-dimensional models from magnetotelluric data, *Geophysics*, *55*, 1613–1624, 1990.

Dey, A., and H.F. Morrison, Resistivity modeling for arbitrarily shaped three-dimensional structures, *Geophysics*, *44*, 753–780, 1979a.

Dey, A., and H.F. Morrison, Resistivity modelling for arbitrarily shaped two-dimensional structures, *Geophys. Prosp.*, *27*, 106–136, 1979b.

Ellis, R.G., and D.W. Oldenburg, Applied geophysical inversion, *Geophys. J. Internat.*, *116*, 5–11, 1994.

French, H.K., C. Hardbattle, A. Binley, P. Winship, and L. Jakobsen, Monitoring snowmelt induced unsaturated flow and transport using electrical resistivity tomography, *J. Hydrol.*, *267*, 273–284, 2002.

Frohlich, R.K., and C.D. Parke, The electrical resistivity of the vadose zone—Field survey, *Ground Water*, *27*, 524–530, 1989.

Hohmann, G.W., Numerical modeling for electromagnetic methods of geophysics, in *Electromagnetic Methods in Applied Geophysics, Vol. 1, Theory*, M.N. Nabighian, ed., Soc. Expl. Geophys., pp. 313–363, 1988.

Kalinski, R.J., W.E. Kelly, I. Bogardi, and I. Pesti, Electrical resistivity measurements to estimate travel times through unsaturated ground water protective layers, *J. Appl. Geophys.*, *30*, 161–173, 1993.

Kean, W.F., M.J. Waller, and H.R. Layson, Monitoring moisture migration in the vadose zone with resistivity, *Ground Water*, *27*, 562–571, 1987.

Kemna, A., Tomographic inversion of complex resistivity—Theory and application, PhD Thesis, Bochum Ruhr-Univ., Germany (published by: Der Andere Verlag, Osnabrück, Germany), 2000.

Kemna, A., and A. Binley, Complex electrical resistivity tomography for contaminant plume delineation, *Proc. 2nd Mtg. Environmental and Engineering Geophysics*, Environ. Eng. Geophys. Soc., Eur. Section, 196–199, 1996.

Kemna, A., A. Binley, A.L. Ramirez, and W.D. Daily, Complex resistivity tomography for environmental applications, *Chem. Eng. J.*, *77*, 11–18, 2000.

Kemna, A., J. Vanderborght, B. Kulessa, and H. Vereecken, Imaging and characterisation of subsurface solute transport using electrical resistivity tomography (ERT) and equivalent transport models, *J. Hydrol.*, *267*, 125–146, 2002.

Kemna, A., A. Binley, and L. Slater, Cross-borehole IP imaging for engineering and environmental applications, *Geophysics*, *69*, 97–107, 2004a.

Kemna, A., J. Vanderborght, H. Hardelauf, and H. Vereecken, Quantitative imaging of 3-D solute transport using 2-D time-lapse ERT: A synthetic feasibility study, *Proc. Symp. Application of Geophysics to Engineering and Environmental Problems*, Environ. Eng. Geophys. Soc., 342–353, 2004b.

Koefoed, O., 1979, *Geosounding Principles, Vol. 1: Resistivity Sounding Measurements*, Elsevier Science Publ. Co., Inc., 1979.

Kosinski, W.K., and W.E. Kelly, Geoelectric soundings for predicting aquifer properties, *Ground Water*, *19*, 163–171, 1981.

LaBrecque, D.J., IP tomography, *61st Ann. Internat. Mtg., Expanded Abstracts*, Soc. Expl. Geophys., 413–416, 1991.

LaBrecque, D.J., M. Miletto, W. Daily, A. Ramirez, and E. Owen, The effects of noise on Occam's inversion of resistivity tomography data, *Geophysics*, *61*, 538–548, 1996.

LaBrecque, D.J., and X. Yang, Difference inversion of ERT data: A fast inversion method for 3-D in-situ monitoring, *Proc. Symp. Application of Geophysics to Engineering and Environmental Problems*, Environ. Eng. Geophys. Soc., 723–732, 2002.

Li, Y., and K. Spitzer, Three-dimensional DC resistivity forward modelling using finite elements in comparison with finite-difference solutions, *Geophys. J. Internat.*, *151*, 924–934, 2002.

Liu, S., and T.-C.J. Yeh, An integrative approach for monitoring water movement in the vadose zone, *Vadose Zone J.*, in press, 2004.

Loke, M.H., and R.D. Barker, Least-squares deconvolution of apparent resistivity pseudosections, *Geophysics*, *60*, 1682–1690, 1995.

Loke, M.H., and R.D. Barker, Practical techniques for 3-D resistivity surveys and data inversion, *Geophys. Prosp.*, *44*, 499–523, 1996.

Loke, M.H, I. Acworth, and T. Dahlin, A comparison of smooth and blocky inversion methods in 2-D electrical imaging surveys, *Proc. 15th Geophysical Conference and Exhibition*, Austr. Soc. Expl. Geophys., 2001.

Lowry, T., M.B. Allen, and P.N. Shive, Singularity removal: A refinement of resistivity modeling techniques, *Geophysics*, *54*, 766–774, 1989.

Lytle, R.J., and K.A. Dines, An impedance camera: A system for determining the spatial variation of electrical conductivity, *Lawrence Livermore National Laboratory Report UCRL-52413*, Livermore, California, USA, 1978.

Menke, W., *Geophysical Data Analysis: Discrete Inverse Theory*, Academic Press, Inc., 1989.

Mussett, A.E., and M.A. Khan, *Looking into the Earth. An Introduction to Geological Geophysics*, Cambridge University Press, 2000.

Nimmer, R.E., and J.L Osiensky, Direct current and self-potential monitoring of an evolving plume in partially saturtated fractured rock, *J. Hydrol.*, *267*, 258–272, 2001.

Oldenburg, D.W., and Y. Li, Inversion of induced polarization data, *Geophysics*, *59*, 1327–1341, 1994.

Oldenburg, D.W., and Y. Li, Estimating depth of investigation in DC resistivity and IP surveys, *Geophysics*, *64*, 403–416, 1999.

Osiensky, J.L., Ground water modeling of mise-a-la-masse delineation of contaminated ground water plumes, *J. Hydrol.*, *197*, 146–165, 1997.

Panissod, C., M. Lajarthe, and A. Tabbagh, Potential focusing: A new multielectrode array concept, simulation study and field tests in archaeological prospecting, *J. Appl. Geophys.*, *38*, 1–23, 1997.

Park, S.K., and G.P. Van, Inversion of pole-pole data for 3-D resistivity structure beneath arrays of electrodes, *Geophysics*, *56*, 951–960, 1991.

Pridmore, D.F., G.W. Hohmann, S.H. Ward, and W.R. Sill, An investigation of finite-element modeling for electrical and electromagnetic data in three dimensions: *Geophysics*, *46*, 1009–1024, 1981.

Ramirez, A., and W. Daily, Electrical imaging at the large block test—Yucca Mountain, Nevada, *J. Appl. Geophys.*, *46*, 85–100, 2001.

Ramirez, A., W. Daily, D. LaBrecque, E. Owen, and D. Chesnut, Monitoring an underground steam injection process using electrical resistance tomography, *Water Resour. Res.*, *29*, 73–87, 1993.

Ramirez, A., W.D. Daily, and R.L. Newmark, Electrical resistance tomography for steam injection monitoring and process control, *J. Environ. Eng. Geophys.*, *0*, 39–51, 1995.

Ramirez, A., W. Daily, A. Binley, D. LaBrecque, and D. Roelant, Detection of leaks in underground storage tanks using electrical resistance methods, *J. Environ. Eng. Geophys.*, *1*, 189–203, 1996.

Reynolds, J.M., *An Introduction to Applied and Environmental Geophysics*, Wiley, 1998.

Seigel, H.O., Mathematical formulation and type curves for induced polarization, *Geophysics*, *24*, 547–565, 1959.

Schima, S., D.J. LaBrecque, P.D. Lundegard, Using resistivity tomography to monitor air sparging, *Ground Water Monitoring and Remediation*, *16*, 131–138, 1996.

Schenkel, C.J., Resistivity imaging using a steel cased well, *Lawrence Livermore National Laboratory Report UCRL-JC-121653*, Livermore, California, USA, 1995.

Seppänen, A., M. Vauhkonen, P.J. Vauhkonen, E. Somersalo, and J.P. Kaipio, State estimation with fluid dynamical evolution models in process tomography: An application to impedance tomography, *Inverse Problems*, *17*, 467-483, 2001.

Simms, J.E. and F.D. Morgan, Comparison of four least-squares inversion schemes for studying equivalence in one-dimensional resistivity interpretation, *Geophysics*, *57*, 1282–1293, 1992.

Slater, L., and D. Lesmes, IP interpretation in environmental investigations, *Geophysics*, *67*, 77–88, 2002.

Slater, L., A. Binley, W. Daily, and R. Johnson, Cross-hole electrical imaging of a controlled saline tracer injection, *J. Appl. Geophys.*, *44*, 85–102, 2000.

Sørensen, K., Pulled array continuous electrical profiling, *First Break*, *14*, 85–90, 1996.

Spitzer, K., A 3-D finite-difference algorithm for DC resistivity modelling using conjugate gradient methods, *Geophys. J. Internat.*, *123*, 903–914, 1995.

Taylor, R.W., and A.H. Fleming, Characterizing jointed systems by azimuthal resistivity surveys, *Ground Water*, *26*, 1988.

Telford, W.M., L.P. Geldart, and R.E. Sheriff, *Applied Geophysics*, 2nd ed., Cambridge Univ. Press, 1990.

Tikhonov, A.N., and V.Y. Arsenin, *Solutions of Ill-Posed Problems*, W.H. Winston and Sons, 1977.

Weller, A., M. Seichter, and A. Kampke, Induced-polarization modelling using complex electrical conductivities, *Geophys. J. Internat.*, *127*, 387–398, 1996.

6 NEAR-SURFACE CONTROLLED-SOURCE ELECTROMAGNETIC INDUCTION: *BACKGROUND AND RECENT ADVANCES*

MARK E. EVERETT
Dept. of Geology and Geophysics, Texas A&M University
College Station TX 77843 U.S.A.

MAX A. MEJU
Dept. of Environmental Science, Lancaster University
Lancaster LA1 4YQ, U.K.

6.1 Introduction

The controlled-source electromagnetic (CSEM) induction method is emerging as a leading geophysical technique in hydrogeological studies. However, the technique is quite often misunderstood compared to other common techniques of applied geophysics: namely, seismic reflection and refraction, magnetics, gravity, and ground-penetrating radar (GPR). In this chapter we review the fundamental physical principles behind the CSEM prospecting technique, with emphasis on near-surface applications, and present some recent advances in this field that have been made by the authors. CSEM methods are defined here to be those in which the experimenter has knowledge of and control over the electromagnetic field transmitted into the ground and hence excludes magnetotellurics, related natural-source methods, and the various uncontrolled-source methods involving, for example, radio transmissions.

CSEM methods for investigating subsurface geology began in earnest in the 1950s and 1960s with the advent of airborne systems mainly for mining applications. Later, with the development of portable and inexpensive ground-based instruments, CSEM systems were applied to groundwater prospecting in arid or hard-rock environments. Due to the upsurge in interest in environmental applications in the past one or two decades, the CSEM method is currently experiencing a rapid growth in use by hydrogeologists, civil and geotechnical engineers, engineering geologists, and others not trained specifically in the technique.

The CSEM method is sensitive to electrical conductivity averaged over the volume of ground in which induced electric currents are caused to flow. Amongst the surface-based geophysical methods that sense bulk electrical properties of the ground, CSEM offers deeper penetration capability than GPR techniques and greater resolving power than DC resistivity methods. The CSEM method performs well in conductive soils or high radar reflectivity zones where GPR often encounters difficulties. The CSEM method also performs well in highly resistive terrains where establishing good electrode contact with the ground, as required for most DC methods, often is problematic.

Y. Rubin and S.S. Hubbard (eds.), Hydrogeophysics, 157–183.

CSEM techniques play an important role in hydrogeophysical investigations since the sensed physical property, electrical conductivity, is linked petrophysically to hydrological variables of interest such as moisture content, hydraulic conductivity, and porosity. This chapter focuses on the relationship between CSEM measurements and spatially averaged subsurface electrical conductivity; it does not address the correspondence between electrical conductivity and hydrological variables. Information on this important topic can be found in Chapter 3 of this book. CSEM case histories relevant to hydrogeophysics are found in Chapter 12.

The CSEM should be attractive to hydrogeologists also because it is simple to operate in the field with the current generation of inexpensive commercial instruments. A variety of data processing options are available, ranging from construction of apparent conductivity curves based on simple asymptotic formulas for rapid subsurface evaluation, all the way to 1-D, 2-D, and advanced 3-D forward modeling and inversion for more detailed analyses.

Previous reviews of CSEM studies for near-surface applications have been carried out by Nobes (1996) and Tezkan (1999). Tutorial articles specifically concerned with the CSEM method have been written by McNeill (1980*a*,*b*) and West and Macnae (1991), while theoretical treatments are available in the books by Grant and West (1965), Ward and Hohmann (1987), Wait (1982), and Zhdanov and Keller (1994). Senior- or graduate-level exploration textbooks such as Telford et al. (1990), Sharma (1997), and Kearey et al. (2002) provide a brief overview of the CSEM method along with case studies. Standard physics textbooks such as Wangsness (1986) and Jackson (1998) are useful for review of the underlying laws of classical electromagnetism. However, these texts emphasize wave propagation in dielectric media over electromagnetic induction in conducting media. The older books by Jones (1964), Stratton (1941), and Smythe (1967) treat induction in more detail and are recommended for advanced study.

In order to use CSEM data for hydrogeological applications, it is important for potential practitioners to understand the basic physics of the technique and how factors such as noise, scaling, and inversion impact the interpretation. This chapter is organized as follows. After an overview of the CSEM method, the basic physics behind frequency and time-domain CSEM prospecting systems is discussed, followed by a discussion on CSEM applications to hydrogeophysical problems and a brief account of 3-D forward modeling. Then, inversion of CSEM data is treated. The paper concludes with an outlook and discussion section. The emphasis in this paper is on inductively coupled loop-loop CSEM systems, although other configurations, including those that employ a directly coupled source, will be discussed. Loop-loop systems are chosen to simplify the discussion and because they are easy to use in the field and widely employed in hydrogeophysical applications.

6.2 CSEM Overview

The CSEM method of geophysical prospecting is founded on Maxwell's equations that govern electromagnetic phenomena. Combining Ohm, Ampere, and Faraday laws (Wangsness, 1986; Jackson, 1998) results in the damped wave equation

$$\nabla^2 \mathbf{B} - \mu_0 \sigma \frac{\partial \mathbf{B}}{\partial t} - \mu_0 \epsilon \frac{\partial^2 \mathbf{B}}{\partial t^2} = \mu_0 \nabla \times \mathbf{J}_S, \tag{6.1}$$

where $\mathbf{B}$ is magnetic field, μ_0 is magnetic permeability, σ is electrical conductivity, ϵ is dielectric permittivity, $\mathbf{J}_S$ is the source current distribution, and t is time. In the CSEM method for most hydrogeophysical applications, the earth is generally considered to be nonmagnetic, such that μ_0=4π $\times 10^{-7}$ H/m, the magnetic permeability of free space. This assumption can break down if highly magnetic volcanic soils or ferrous metal objects are encountered, in which case the ground or target magnetic permeability influences the CSEM response and should be taken into account. The electrical conductivity of most near-surface geological materials lies in the range σ=0.0001-0.1 S/m (Palacky, 1987).

The third term on the left hand side of Equation (6.1) is the energy storage term describing wave propagation (Powers, 1997). The second term on the left hand side of Equation (6.1) is the energy dissipation term describing electromagnetic diffusion. The CSEM method, which is the topic of our consideration, operates at low frequencies ($\sim$100 Hz to 1 MHz), or alternatively, slow ($\sim$1 μs to 10 ms) disturbances of the source currents. In this case, $|\sigma \partial \mathbf{B}/\partial t|$ is larger by several orders of magnitude than $|\epsilon \partial^2 \mathbf{B}/\partial t^2|$, so that the wave propagation term is safely ignored and dielectric permittivity ϵ plays no further role in the discussion. The result is that CSEM is a purely diffusive phenomenon. If the electrical conductivity is low enough and the frequency high enough that $\sigma \sim \epsilon\omega$, the electromagnetic response of the earth is properly described by Equation (6.1) with all terms kept.

6.3 Basic Physics, Time Domain

A physical understanding of CSEM responses can be obtained by recognizing that the induction process is equivalent to the diffusion of an image of the transmitter (TX) loop into a conducting medium. The similarity of the equations governing electromagnetic induction and hydrodynamic vortex motion, first noticed by Helmholtz, leads directly to the association of the image current with a smoke ring (Lamb, 1945; p. 210). The latter is not "blown," as commonly thought, but instead moves by self-induction with a velocity that is generated by the smoke ring's own vorticity and described by the familiar Biot-Savart law (Arms and Hama, 1965). An electromagnetic smoke ring dissipates in a conducting medium, much as the strength of a hydrodynamic eddy is attenuated by the viscosity of its host fluid (Taylor, 1958; pp. 96-101). The medium property that dissipates the electromagnetic smoke ring is electrical conductivity.

An inductively coupled time domain CSEM (TDEM) system is shown in Figure 6.1*a*. A typical TX current waveform $I(t)$ is a slow rise to a steady-on value I_0 followed by a rapid shut-off, as exemplified by the linear ramp in Figure 6.1*b* (*top*). Passing a disturbance through the TX loop generates a primary magnetic field that is in-phase with, or proportional to, the TX current. According to Faraday's law of induction, an impulsive electromagnetic force (emf) that scales with the time rate of change of the primary magnetic field is also generated. The emf drives electromagnetic eddy currents in the conductive earth, notably in this case during the ramp-off interval, as shown in Figure 6.1*b* (*middle*). After the ramp is terminated, the emf vanishes and the eddy currents start to decay via ohmic dissipation of heat. A weak, secondary magnetic field is produced in proportion to the waning strength of the eddy currents. The receiver (RX) coil voltage measures the time rate of change of the decaying secondary magnetic field, Figure 6.1*b* (*bottom*). In many TDEM systems, RX voltage measurements are made during the TX off-time when the primary field is absent. The off-time advantage is that the relatively weak secondary signal is not swamped by the much stronger primary signal. A good tutorial article on TDEM has been written by Nabighian and Macnae (1991).

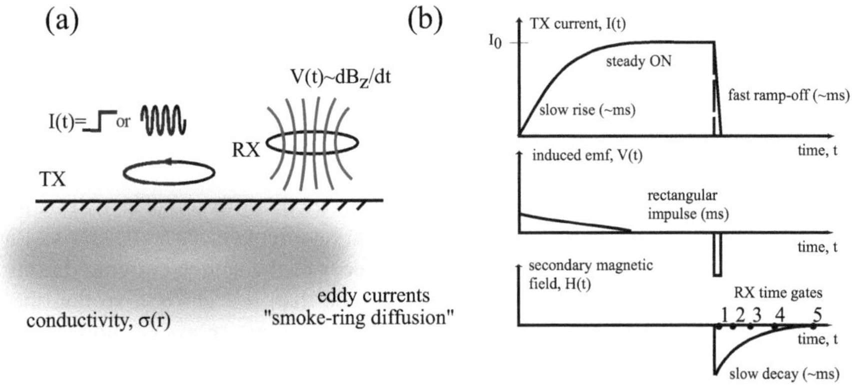

Figure 6.1. (*a*) TX loop lying on the surface of an isotropic uniform halfspace of electrical conductivity, σ. A disturbance $I(t)$ in the source current immediately generates an electromagnetic eddy current in the ground just beneath the TX loop. A RX coil measures the induced voltage $V(t)$, which is a measure of the time-derivative of the magnetic flux generated by the diffusing eddy current. (*b*) Typical TX current waveform $I(t)$ with slow rise time and fast ramp-off; induced emf $V(t)$, proportional to the time rate of change of the primary magnetic field; decaying secondary magnetic field $H(t)$ due to the dissipation of currents induced in the ground.

During the ramp-off, the induced current assumes the shape of the horizontal projection of the TX loop onto the surface of the conducting ground. The sense of the circulating induced currents is such that the secondary magnetic field they create tends to maintain the total magnetic field at its original steady-on value prior to the TX ramp-off. In this case, therefore, the induced currents flow in the same direction as the TX current, i.e., opposing the TX current decrease that served as the emf source. The image current then diffuses downward and outward while diminishing in amplitude.

In TDEM offset-loop soundings, the TX and RX loops are separated by some distance L. As indicated in Figure 6.2a, at a fixed instant in time t, the vertical magnetic field $H_z(L)$ due to the underground circuit exhibits a sign change from positive to negative as distance L increases. Alternatively, the vertical magnetic field $H_z(t)$ changes sign from positive to negative as the filament passes beneath a fixed measurement location. The "normal moveout" of the sign reversal with increasing TX-RX separation distance is shown in Figure 6.2b, Field examples from Utah and Texas showing sign reversal in the TDEM offset-loop response are shown in Figure 6.3.

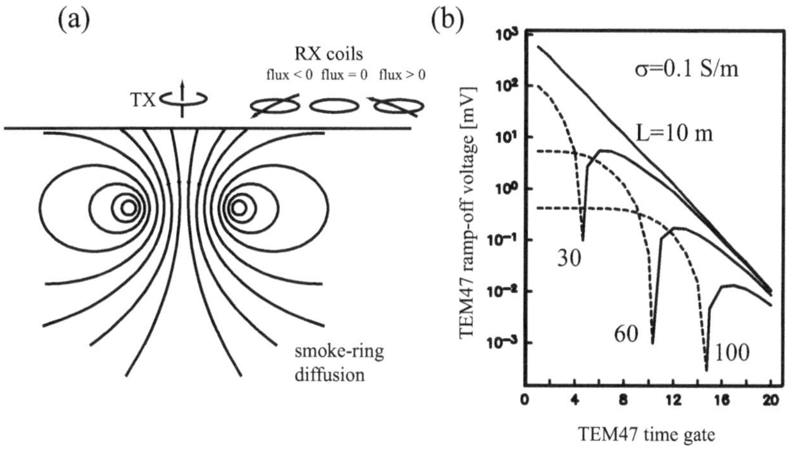

Figure 6.2. (a) The electromagnetic smoke ring may be viewed as a system of equivalent current filaments diffusing downward and outward from the TX loop (Nabighian, 1979). The equivalent filament concept can be used to understand the spatiotemporal behavior of the voltage induced in a horizontal RX coil. The figure shows that, at a fixed instant in time t, the flux-proportional vertical magnetic field $H_z(x)$ due to the underground circuit exhibits a sign change from positive to negative as distance x increases from the TX loop. Similarly, at a fixed location in space, the vertical magnetic field $H_z(t)$ also changes sign from positive to negative as the filament passes beneath a RX coil. The apparent conductivity of the ground is determined by analyzing the times at which the sign reversals occur in the various RX coils. (b) Transient decays for a loop-loop CSEM system (Geonics PROTEM47) over a uniform halfspace. The time of the sign reversal increases with TX-RX separation distance L, in meters. Dashed line: negative voltage; solid line: positive voltage.

A mathematical treatment of the electromagnetic smoke-ring phenomenon appears in Nabighian (1979). Hoversten and Morrison (1982) and Reid and Macnae (1998) have further explored smoke-ring diffusion into a layered conducting earth. These papers provide physical insight which can greatly assist hydrogeophysicists to understand the TDEM response in idealized situations. The description of a directly coupled TDEM system is similar to the inductively coupled case, but requires additional physics due to the presence of ground-contacting electrodes.

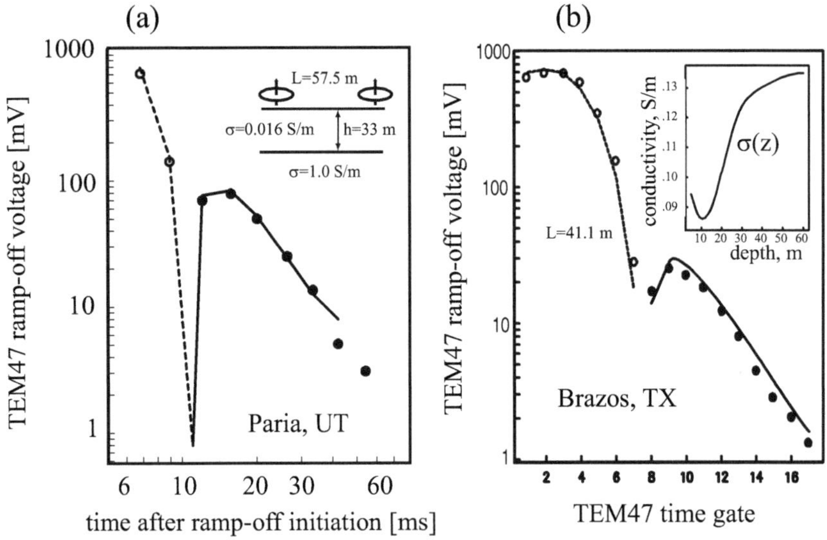

Figure 6.3. (*a*) A transient sounding using the Geonics PROTEM47 loop-loop configuration (TX loop radius, 2 m) observed atop an eolian Page Sandstone outcrop near Paria Campground, UT. The response of the 2-layer model in the insert is also shown. (*b*) A similar PROTEM47 sounding observed over Brazos, TX, clay-rich floodplain alluvium, after Sananikone (1998). The response of the smooth model $\sigma(z)$ in the insert is also shown. The low conductivity at depths ~9–12 m owes to a sand and gravel aquifer overlying basement shale.

6.3.1 TDEM APPARENT CONDUCTIVITY

The conductivity of the ground can be estimated from TDEM offset-loop sounding responses by analyzing the times at which sign reversals occur in the various RX coils. It is sometimes convenient, on the other hand, to transform TDEM sounding data into an apparent conductivity curve. Apparent conductivity is defined (Spies and Frischknecht, 1991) as the conductivity of the uniform half-space that would generate the observed response at each discrete measurement time after the TX ramp-off. Caution is required, however, since for some electrical conductivity structures at certain time gates, the apparent conductivity can be nonexistent or multivalued.

Examples of apparent conductivity curves for two synthetic and one actual field TDEM offset-loop response are given in Figure 6.4. The synthetic responses were generated by forward modeling of transient EM induction in a two-layer conductivity structure. The apparent conductivity at each time gate was calculated by matching the response at that time gate to the synthetic response of a uniform half-space. The trend of an apparent conductivity curve with increasing time is roughly indicative of the trend of the subsurface electrical conductivity with increasing depth, assuming that a layered structure is a good approximation to the electrical conductivity of the underlying geological formations.

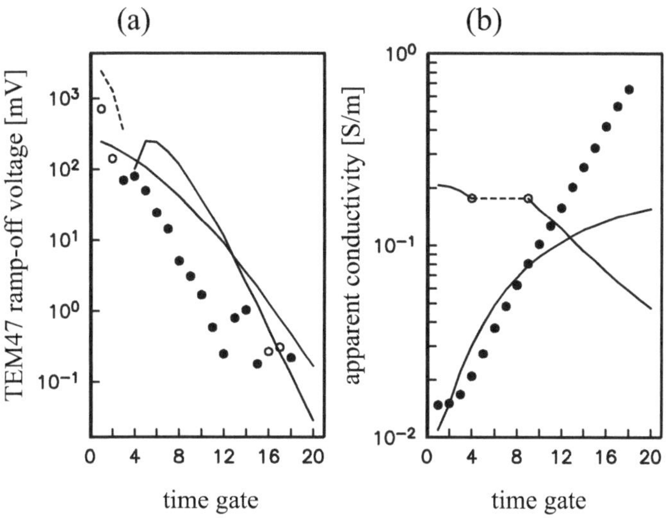

Figure 6.4. (*a*) Geonics PROTEM47 responses. Symbols represent the Paria, UT response shown in the previous figure. The heavy line is the synthetic response of a resistive layer (0.02 S/m, 10 m thick) overlying a conductive halfspace (0.2 S/m). The lighter dashed/solid line shows the negative/positive portions of the synthetic response of a conductive layer (0.2 S/m, 10 m thick) overlying a resistive halfspace (0.02 S/m.) (*b*) The corresponding apparent conductivity curves, showing trends with increasing time that reflect trends of electrical conductivity with increasing depth. The gap in the apparent conductivity curve for the conductor-over-resistor model indicates that no uniform conductor can be found with a response that exactly matches the two-layer response for PROTEM47 time gates 5–8.

Various asymptotic formulas for apparent conductivity that are valid at either early or late times after TX ramp-off have been developed (Spies and Frischknecht, 1991). They yield estimates of shallow ($z \ll L$) and deep ($z \gg L$) electrical conductivity, respectively. Other TDEM sounding configurations also may be analyzed using apparent conductivity. In the popular central loop configuration, a RX coil is placed in the center of the TX loop. No sign reversal occurs. In this type of TDEM sounding, the shape and strength of the decaying transient are used to determine apparent conductivity, either by exact forward modeling or using asymptotic formulas. In the coincident loop method, the TX and RX coils are collocated. Apparent conductivity in this case is often computed using the method of Spies and Raiche (1980).

6.4 Basic Physics, Frequency Domain

In the frequency domain methods (FDEM), the TX current oscillates at a given angular frequency, ω. The emf, being proportional to the rate of change of primary magnetic field, is out of phase with the TX current. The emf drives eddy currents in the conductive earth. The secondary magnetic field due to the induction of eddy currents contains both in-phase and out-of-phase (quadrature) components, owing to the complex impedance of the ground. Excellent, rigorous treatments of the frequency domain theory of CSEM induction may be found in Grant and West (1965), Wait (1982), and Zhdanov and Keller (1994).

In hydrogeophysical applications, FDEM is typically applied at sufficiently low frequencies that the response is equivalent to that of a system of uncoupled electric circuits containing only passive inductive and resistive elements. This approximation is equivalent to the low-induction-number (LIN) principle upon which terrain conductivity meters such as the Geonics EM31 and EM34 instruments operate (McNeill, 1980b). The approximation is good, provided that the frequency is low enough that the effective penetration depth, or skin depth δ given by

$$\delta = \sqrt{2/\mu_0 \sigma \omega}, \tag{6.2}$$

is much greater than the TX-RX intercoil spacing L.

At LIN frequencies, each layer in the ground can be modeled to first-order as a separate wire loop. The LIN principle is equivalent to the Doll approximation (Doll, 1949) which has long formed the conceptual basis for analysis of induction logs in the petroleum industry, and assumes that the mutual inductance between each pair of underground loops is negligible. The response at the RX loop, in such case, is the sum of the primary magnetic flux from the TX loop, plus the sum of the secondary magnetic fluxes generated by the induced currents flowing in the circuits.

A representation of the LIN circuit approximation for a 2-layer earth is shown in Figure 6.5a. The mutual inductance M_{12}, defined as the magnetic flux through Circuit 1 caused by a unit current flow in Circuit 2 (Wangsness, 1986, p. 312), is assumed to be negligibly small. The primary magnetic field $H_p(\omega)$ caused by time-harmonic excitation of the TX loop at frequency ω generates an out-of-phase, or quadrature, emf in each of the underground circuits, according to Faraday's law. The emf's drive induced currents in the circuits. These currents each generate a secondary magnetic field that is flux-linked to the RX loop. The total secondary field $H_s(\omega)$, along with the primary magnetic field $H_p(\omega)$, is measured by the RX loop. The primary field is known precisely, since the TX and RX loops are under control of the experimenter.

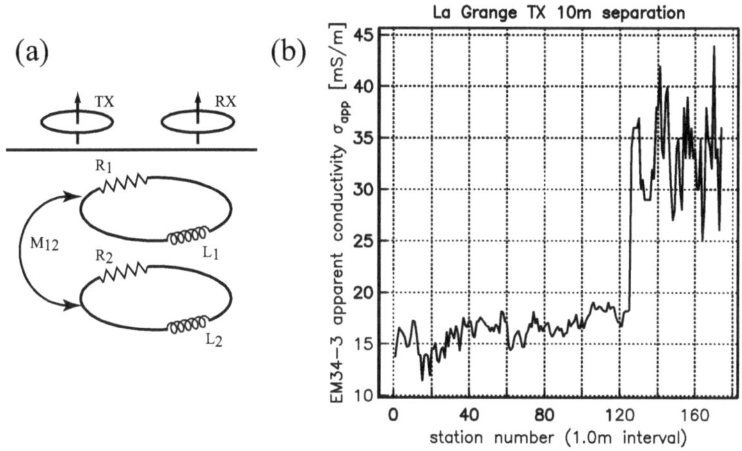

Figure 6.5. (*a*) A conceptual view of low-frequency CSEM induction as the interaction between the time-harmonic primary magnetic flux from a TX loop and underground circuits (resistance R_i, self-inductance L_i) with vanishing mutual inductance, M_{12}=0. The secondary flux due to the induced currents flowing in the underground circuits is measured at the RX loop (after McNeill, 1980b). (*b*) Field example from a gravel deposit near La Grange TX, illustrating the utility of EM34 $\sigma_{app}(x)$ profiles for lateral reconnaissance. The sharp increase in σ_{app} near Station 125 marks an abrupt transition from electrically resistive gravel of economic value to conductive, interbedded sands and clays.

The ratio of secondary to primary magnetic fields over a homogeneous halfspace, in the Doll or LIN approximation ($\delta \gg L$), is derived by McNeill (1980b) as

$$Hs/Hp = i\omega\mu_0 L^2/4, \tag{6.3}$$

which can be rearranged to define the apparent conductivity σ_{app} of the ground as

$$\sigma_{app} = 4\,[Hs/Hp]^Q\,/\omega\mu_0 L^2, \tag{6.4}$$

where the notation $[\ ]^Q$ refers to quadrature component.

The Geonics EM34 instrument can be operated in either vertical or horizontal dipole modes. In vertical dipole mode, the TX and RX coils are coplanar horizontal. As illustrated by curves in McNeill (1980b), the operating frequency (10–100 kHz range) is such that the apparent conductivity σ_{app} in this orientation is sensitive to electrical conductivity variations in the depth range $0.3L<z<0.6L$, with zero sensitivity at the surface $z=0$ and at depths beyond $z>2L$. In the horizontal dipole mode, the TX and

RX coils are arrayed in a common vertical plane. The maximum sensitivity to electrical conductivity is at the surface z=0, with sensitivity decreasing smoothly to zero beyond $z>2L$.

Interpretation of FDEM responses is typically performed by analyzing the behavior of apparent conductivity $\sigma_{app}(x)$ as a function of the TX-RX midpoint position x along a profile, while the TX-RX intercoil spacing L and operating frequency ω are kept fixed. Terrain conductivity meters such as EM34 provide information to a few tens of meters in depth, but with limited resolving power, since they operate at just a single frequency. The GEM3 system from Geophex, Ltd. operates at multiple frequencies spanning 2–3 decades in the range 100 Hz–50 kHz, but depth resolution remains modest, since earth's FDEM response is a slowly varying function of frequency in this band (Huang and Won, 2003).

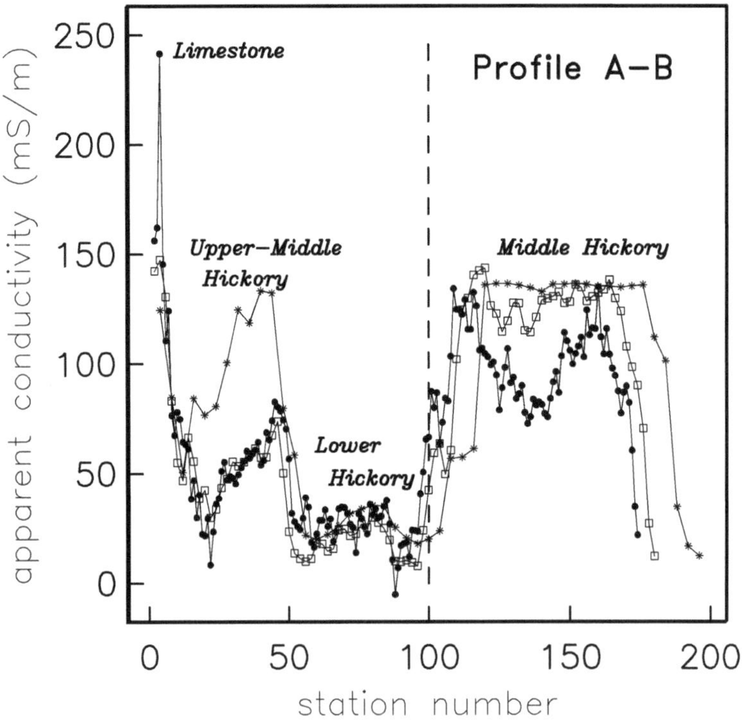

Figure 6.6. EM34 mapping of a buried near-surface fault in a Cambrian sandstone aquifer near Mason, TX, after Gorman (1998). The electrically resistive, coarse-grained Lower Hickory unit is juxtaposed to the more conductive, fine-grained Middle Hickory unit. The inferred fault location is shown by the dashed line at Station 100. The various symbols indicate different TX-RX separations: L=10 m, solid dots; L=20 m, squares; L=40 m, stars. The station interval is 5.0 m.

Terrain conductivity meters are much better suited for rapid, noninvasive reconnaissance mapping of lateral changes in near-surface electrical conductivity. Figure 6.5*b* illustrates the use of EM34 terrain conductivity mapping to determine the lateral extent of a subsurface economic gravel deposit. This example demonstrates the use of the FDEM technique to map paleochannels comprised of coarse sediments. Roughly speaking, fine-grained material with abundant clay is conductive, while clean, coarse-grained sediment is resistive. In Figure 6.6, the EM34 method at three different frequencies is used to map faults and stratigraphic contacts in a consolidated sandstone aquifer. The faults and contacts act as barriers and conduits compartmentalizing groundwater flow in the aquifer. In both EM34 field examples, the lateral contrasts in apparent electrical conductivity are caused by sharp lateral contrasts in texture.

6.5 CSEM Hydrogeophysics

6.5.1 HYDROGEOPHYSICAL TARGETS

The CSEM method was originally developed and continues to be refined for natural resource prospecting and reserve evaluation applications (Meju, 2002) such as mining geophysics, in which conductive ore bodies form excellent, compact targets within resistive host crystalline rocks. The CSEM method is also able to identify conductive fracture zones in crystalline bedrock aquifers for groundwater prospecting applications (Meju et al., 2001). In FDEM applications, fracture zones are typically identified by the presence of anomalously high apparent conductivity readings along a measurement profile. Typical length scales in mining and groundwater problems range from tens of meters to tens of kilometers.

As Figure 6.7*a* dramatically illustrates, a generic hydrogeophysical site characterization problem is very difficult. The subsurface geology may contain quasi-localized features such as weathered mantle, bedrock, bedding planes, faults, joints, and fracture zones, in addition to continuously distributed textural and compositional variations. In fact, because of the evident multiscale complexity, a geological medium is often viewed as a discrete hierarchy (Vogel et al., 2002), in which different length scales possess different patterns of spatial variability and correlation; or a fractal (Mandelbrot, 1988), in which each length scale possesses a similar pattern of spatial variability and correlation.

The spatiotemporal CSEM response of a discrete hierarchial or fractal medium is likely to be very complicated, characterized by a broadband, power-law spatial wavenumber spectrum (Everett and Weiss, 2002). It is the difficult task of the EM hydrogeophysicist to interpret such CSEM responses in terms of the subsurface geology, with its attendant spatial complexity. The presence of man-made conductors in the subsurface adds to the difficulty. However, man-made objects tend to be geometrically regular, more or less, rather than hierarchial or fractal. This offers the possibility that such objects may be detected by their band-limited wavenumber spectrum, or localized by spatial/wavenumber analysis using, for example, the continuous wavelet transform (Benavides and Everett, 2004).

6.5.2 THE SCALE EFFECT

Much has been written in the hydrological literature on the effect of measurement scale on hydraulic conductivity (Neuman, 1994; Winter and Tartakovsky, 2001). For example, owing to the increased likelihood of a hydraulic test being influenced by a connected pathway from injection to monitoring well, hydraulic conductivity estimates tend to increase with increasing measurement scale, as shown in Figure 6.7*b*. Basin-scale hydraulic conductivity estimates are often 1-2 orders of magnitude higher than aquifer pump-test determinations, which in turn are larger than laboratory measurements on core samples of the same aquifer material. Hydraulic and electric conductivity are both transport properties of geological media that depend on geometric parameters at many length scales, such as the connectivity of fracture networks and the tortuous microstructure of the pore space (see Chapter 5 of this volume. Thus, electrical conductivity should exhibit a similar scale effect, but field-based studies are rare (Purvance and Andricevic, 2000).

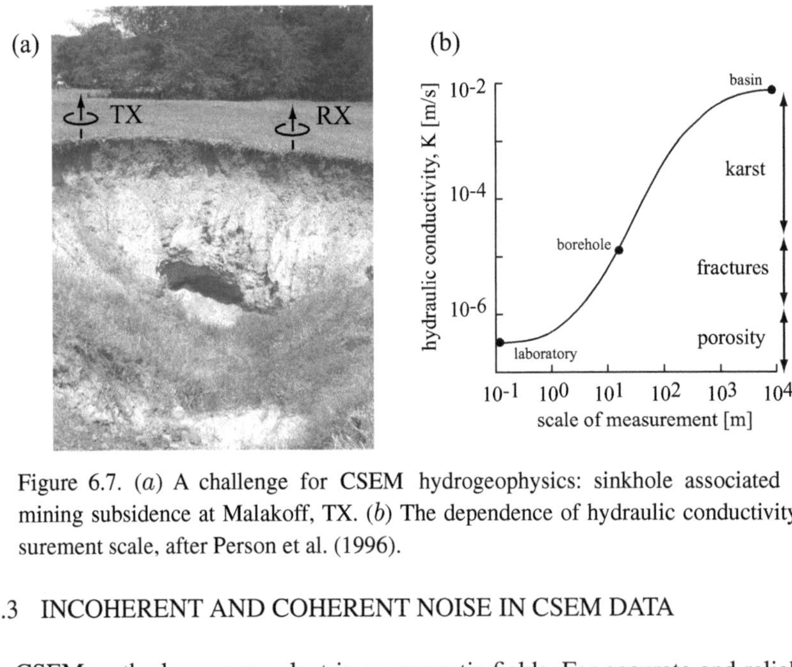

Figure 6.7. (*a*) A challenge for CSEM hydrogeophysics: sinkhole associated with coal mining subsidence at Malakoff, TX. (*b*) The dependence of hydraulic conductivity on measurement scale, after Person et al. (1996).

6.5.3 INCOHERENT AND COHERENT NOISE IN CSEM DATA

The CSEM method measures electric or magnetic fields. For accurate and reliable interpretation of the responses in terms of subsurface geology, it is important to recognize the various sources of electromagnetic noise that may be present in CSEM responses.

Foremost, it is essential to differentiate between coherent and incoherent noise. Incoherent noise is temporally uncorrelated with the transmitted source current. In Figure 6.8, the main sources of environmental incoherent noise are shown. In the frequency range of interest for hydrogeophysical studies, 100 Hz-1 MHz, the predominant environmental noise sources (sferics) are caused by impulsive lightning discharges whose EM radiation can propagate over thousands of kilometers within the earth-ionosphere waveguide. The labels TEM and TM refer to different waveguide modes of propagation (Wait, 1962), whereas the shape of the background continuum within the sferics band is determined by the cutoff frequencies of the various modes (Porrat et al., 2001). The lower frequency continuum below 1 Hz is caused by large-scale geomagnetic pulsations that arise in response to the solar wind interaction with earth's magnetosphere.

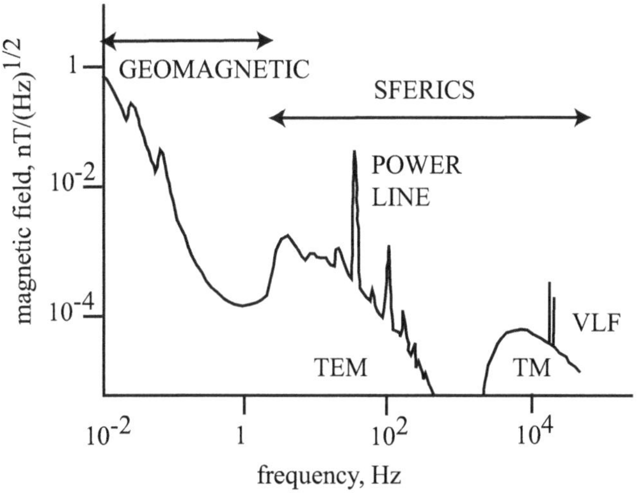

Figure 6.8. Electromagnetic noise spectrum in the frequency range of interest to CSEM hydrogeophysical investigations, after Palacky and West (1991).

The spectral lines superimposed on the background continuum in Figure 6.8 are caused by 50/60 Hz power line harmonics and radio signals from very low frequency (VLF) transmitters used primarily for military communication. Newman et al. (2003) describe the application of a method that utilizes VLF radio signals for hydrogeophysical investigations of the near-surface, but for the most part these transmissions are a source of noise. Other types of incoherent noise include motional noise induced in the RX coil and electronic and thermal noise in the RX amplifiers. Incoherent noise in TDEM data generally is treated by filtering and stacking the received signal. Sferics appear as irregularly spaced, high amplitude spikes of brief (ms) duration in RX voltage time series. Robust statistical methods are required for efficient sferics reduction (Buselli and Cameron, 1996).

While incoherent noise sources pose a modest challenge to accurate CSEM data interpretation, much of the data processing is performed internal to the commercial instrument and hence is transparent to the user. Coherent noise, however, is much more problematic. Such noise is temporally correlated with the transmitted source current and can include transmitter or receiver loop misalignment or instrumental drift. These effects are reduced by careful experimental and data processing procedures. Other sources of coherent noise are currents induced in conductive structures that are not the primary target of the hydrogeophysical investigation. The two major classifications are geological and cultural noise.

Geological noise (Everett and Weiss, 2002) is defined as the electromagnetic field generated by induced currents flowing in fine-scale geological heterogeneities that are too small, numerous, or complicated to be described in detail and hence accounted for in numerical simulations (see later section on 3-D forward modeling). Cultural noise (Qian and Boerner, 1995) is defined as the electromagnetic field generated by currents induced in man-made conductors such as pipelines, buried tanks, steel fences, buried metal drums, or other objects. These conductors must be included in numerical simulations, if possible, so that their distorting effects on the CSEM response can be properly evaluated. As an example, the effect on measured TDEM responses of a small steel sphere buried in alluvium soil is shown in Figure 6.9. Geological and cultural noise cannot be filtered or stacked out of CSEM data because of the mutual induction between the causative structures and the host geology.

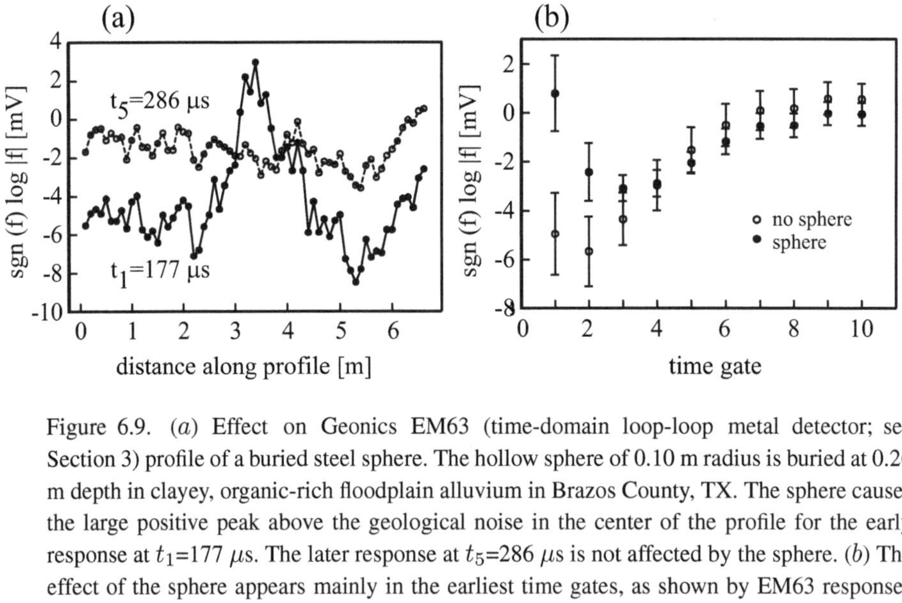

Figure 6.9. (*a*) Effect on Geonics EM63 (time-domain loop-loop metal detector; see Section 3) profile of a buried steel sphere. The hollow sphere of 0.10 m radius is buried at 0.26 m depth in clayey, organic-rich floodplain alluvium in Brazos County, TX. The sphere causes the large positive peak above the geological noise in the center of the profile for the early response at t_1=177 μs. The later response at t_5=286 μs is not affected by the sphere. (*b*) The effect of the sphere appears mainly in the earliest time gates, as shown by EM63 responses averaged over the 8 stations nearest the sphere (solid symbols) and the 60 other stations (open symbols) along the profile. The EM63 voltage signal $f(t)$ [mV] is plotted in the form of $\mathtt{sgn}(f)\,\mathtt{log}|f|$ for convenience.

6.6 Forward Modeling

Calculation of the electromagnetic field generated by a magnetic or electric dipole source situated on, above, or within a layered conducting medium is a well-known boundary value problem of classical physics. CSEM operates at low frequencies such that the energy storage term proportional to $|\epsilon\partial^2\mathbf{B}/\partial t^2|$ in Equation (6.1) can be ignored. One standard solution technique for the resulting vector diffusion equation expresses the electric and magnetic fields in terms of Hertz vector potentials (Sommerfeld, 1964; p. 236 ff). The Sommerfeld technique definitely should be studied in detail by students interested in mastering the theoretical foundations of the subject. Other analytic approaches are possible; for example, Chave and Cox (1982) derive seafloor CSEM fields using a poloidal/toroidal mode decomposition.

The closed-form expressions for the CSEM field components are Hankel transforms, or integrals over Bessel functions. A variety of special-purpose quadrature algorithms are available (e.g., Chave, 1983; Guptasarma and Singh, 1997) for evaluating these slowly convergent integrals. A 1-D forward modeling code based on a fast, reliable Hankel transform subroutine should be an essential part of the EM induction geophysicist's computational toolbox, since it permits rapid, quantitative layered-earth interpretations of measured CSEM responses.

An exact solution for FDEM is available and shown here for illustrative purposes in the specific case of a horizontal loop transmitter of radius a carrying current I and situated on the surface of a uniform halfspace of electrical conductivity σ. The vertical magnetic field $B_z(r)$ on the surface at distance r from the transmitter is given by the formula (Ward and Hohmann, 1987)

$$B_z(r) = \mu_0 I a \int_0^\infty \frac{\lambda^2}{\lambda + i\gamma} J_1(\lambda a)\, J_0(\lambda r)\, d\lambda, \tag{6.5}$$

where $\gamma^2 \equiv -i\mu_0\sigma\omega - \lambda^2$ and J_0, J_1 are Bessel functions. Similar analytic solutions are available for layered earths excited by other TX/RX configurations.

There exist specialized 1-D analytic solutions for certain anisotropic structures of relevance to hydrogeophysics. For example, an exact expression for the EM34 response of a fractured rock formation is derived by Al-Garni and Everett (2003) under the assumption that the electrical conductivity is represented by a uniaxial tensor. The effect of crossbedding anisotropy on CSEM responses has been calculated by Anderson et al. (1998) in the context of well logging, but their methodology readily applies to surface geophysical investigations.

The CSEM response to 3-D subsurface conductivity distributions requires the application of numerical methods for solving the governing Maxwell partial differential equations. A good introduction to integral equation, finite element, and finite difference techniques is provided by Hohmann (1987). Fully 3-D numerical simulations are now feasible on the current generation of personal computers featuring $\sim$2.5 GHz processors and $\sim$1 GB memory. These simulations involve a complete description of CSEM induction physics, including galvanic and vortex contributions. The galvanic term appears when induced currents encounter spatial gradients in electrical conductivity. Electric charges accumulate across interfaces in electrical conductivity. In 1-D layered media excited by a purely inductive CSEM source such as a loop, induced currents flow horizontally and do not cross layer boundaries. In that case, only the vortex contribution is present.

The following is a very brief description of some recent advances in 3-D numerical modeling. An integral equation solution of the CSEM response to a buried plate is provided by Walker and West (1991). Badea et al. (2001) have developed a 3-D finite element approach to computing CSEM responses for arbitrary subsurface conductivity distributions. Stalnaker and Everett (2004) have made modifications to the algorithm so as to calculate 3-D loop-loop CSEM responses of direct relevance to hydrogeophysical investigations. Zhdanov et al. (2000) have emphasized fast 3-D quasi-analytic forward solutions to cut down the cost of forward solutions, so that large CSEM data sets may be rapidly interpreted. Additional developments in forward modeling may be found in the book by Zhdanov and Wannamaker (2002).

6.7 Inversion of CSEM Data

An important goal in CSEM exploration is to determine the electrical conductivity $\sigma(\mathbf{r})$, or its reciprocal, the electrical resistivity $\rho(\mathbf{r})$ structure of the subsurface from measurements taken on, above, or below the earth's surface. The recorded data bear the conductivity signature of the subsurface and permit inferences to be made about the structure and physico-chemical state of geological formations and anthropogenic targets. Mathematical modeling methods are used to relate these responses to hypothetical earth structures that could have generated them. In the general 3-D case, hypothetical models of the subsurface describe the lateral and vertical distribution of electrical conductivity beneath the CSEM survey area. For a stratified geological environment (i.e., the 1-D case), the subsurface is simply parameterized into a succession of plane horizontal layers of varying conductivity and thickness beneath each sounding site.

The computation of the observable or theoretical responses (referred to as predicted or synthetic data) for a given susbsurface model is facilitated using the methods of the previous section. This constitutes the forward problem or direct approach in CSEM data interpretation and is formally described as: "Given the parameters of a hypothetical earth model, determine the observable responses for a given experimental setup." The inverse problem is stated: "Given a finite collection of observations, find a conductivity model whose theoretical responses satisfactorily describe or match them." For example, the determination of the conductivity structure of aquiferous targets and their confining formations based on CSEM measurements is the central inverse problem in hydrogeophysics. The underlying theory of inverse problems is well described in the geophysical literature (e.g., Jackson, 1972, 1979; Tarantola and Valette, 1982; Constable et al., 1987; Meju, 1994a; Newman and Alumbaugh, 1997), and there are several books available to the interested reader (e.g., Twomey, 1977; Menke, 1984; Meju, 1994*b*).

The CSEM field observations and their associated measurement errors are respectively denoted by the vectors $\mathbf{d}$ and $\mathbf{e}$, while the parameters of the sought subsurface model are grouped into another vector $\mathbf{m}$. The forward model for predicting the theoretical responses of any given earth model is a nonlinear function of earth conductivity and thickness parameters, e.g., Equation (6.5), and is denoted by the functional $\mathbf{f}(\mathbf{m})$. The inverse problem is thus the search for a model $\mathbf{m}$ whose responses best replicate the field observations in the least-squares sense, but within bounds defined by the data errors. It is also desirable for the sought model to be in accord with any available and realistic geophysical, geological, or well log data (dubbed *a priori* information or extraneous data and denoted by the vector $\mathbf{h}$). The inverse problem in mathematical terms can be stated as (e.g., Meju, 1994a): minimize the objective function

$$\phi = (\mathbf{Wd} - \mathbf{Wf}(\mathbf{m}))^T(\mathbf{Wd} - \mathbf{Wf}(\mathbf{m})) + (\beta[\mathbf{Dm} - \mathbf{h}])^T(\beta[\mathbf{Dm} - \mathbf{h}]) \qquad (6.6)$$

where the superscript T denotes the transposition. The first term in the right side of Equation (6.6) describes the differences between the predicted and actual field data. The second term in the right side imposes constraints on the desired solution, namely that it should be in accord with the *a priori* information $\mathbf{h}$ unless justified by the field data. The parameter β is an undetermined multiplier that helps force the solution into conformity with $\mathbf{h}$. Typically, the vector $\mathbf{h}$ contains the model parameters (resistivities and boundary positions) towards which we wish to bias the sought solution. Large values of β will keep $\mathbf{m}$ close to $\mathbf{h}$, with the optimal value of β determined using a simple line search. For numerical stability and to prevent undue importance being given to poorly estimated field data, each field datum is weighted by its error. The diagonal matrix $\mathbf{W}$ contains the reciprocals of the standard observational errors $\mathbf{e}$ so as to emphasize precise data in the inversion process.

The second term in the right side of Equation (6.6) helps to regularize the otherwise ill-posed inverse problem. Without this term, there are an infinite number of statistically acceptable solutions. The regularization matrix $\mathbf{D}$ (Constable et al., 1987) may be the identity matrix or a first or second difference operator that allows model parameters to vary either smoothly or sharply in the vertical and lateral directions. There is an extensive literature on the use of *a priori* or extraneous information to reduce non-uniqueness (Jackson, 1979) in CSEM inversion (see e.g., Meju, 1994a, 1996, 2004; Meju et al., 2000).

The solution to Equation (6.6) yields the sought parameter estimates and is (cf. Meju, 1994a)

$$\mathbf{m}_{est} = [(\mathbf{WA})^T(\mathbf{WA}) + \beta^2\mathbf{D}^T\mathbf{D}]^{-1}[(\mathbf{WA})^T\mathbf{d}_c + \beta\mathbf{D}^T\mathbf{h}] \qquad (6.7)$$

where $\mathbf{d}_c$=$\mathbf{Wy} + \mathbf{WAm}_0$. In general, $\mathbf{y} = [\mathbf{d} - \mathbf{f}(\mathbf{m}_0)]$ is called the discrepancy vector and $\mathbf{A} = \partial\mathbf{f}(\mathbf{m}_0)/\partial\mathbf{m}_0$ is the matrix of partial derivatives evaluated at an initial model, $\mathbf{m}_0$. To ensure positivity, and hence always a physical solution to the inverse problem (e.g., Meju, 1996), the logarithms of the apparent resistivity data are used, i.e., $\mathbf{y} = ln\mathbf{d} - ln\mathbf{f}(\mathbf{m}_0)$ so that $\mathbf{A} = \partial ln\mathbf{f}(\mathbf{m}_0)/\partial\mathbf{m}_0$. The components of $\mathbf{m}$ are taken to be the logarithms of the resistivities and interface depths or layer thicknesses.

The first term in square brackets on the right side of Equation (6.7) is referred to as the generalized inverse. It operates on the CSEM data and any prior constraints (i.e., the second term in square brackets on the right side) to yield the desired constrained least-squares solution ($\mathbf{m}_{est}$) to the above-stated inverse problem. Because of its nonlinear nature, the solution to the inverse problem cannot be obtained in one step, but instead is found by a series of steps in which Equation (6.7) is applied successively to improve an initial model, $\mathbf{m}_0$. The latter is typically an informed guess of the subsurface conductivity distribution. There are simple direct data transformation or imaging schemes (e.g., Meju, 1998) that can serve for generating $\mathbf{m}_0$ in a first-pass interpretation process.

The inversion of CSEM survey data is a time-consuming process. It is fast for idealized 1-D problems (e.g., Christensen 1995, 2000; Meju et al., 2000) but in reality the subsurface is heterogeneous and requires a 3-D approach. Fully 3-D CSEM inversion is computationally expensive. New advances in multidimensional numerical modeling (e.g., Newman and Alumbaugh, 1997; Sasaki, 2001) and instrumentation (e.g., Sorensen, 1997) have led to improved data acquisition (such as efficient "continuous array profiling") and interpretation techniques. The key challenge currently facing CSEM inversionists is the development of software for fast 3-D inversion on portable computers. This will permit leaps forward in our ability to characterize hydrogeophysical targets.

6.8 CSEM and Other Methods of Conductivity Depth Sounding

Owing to current technical limitations, no single conductivity-depth-sounding technique furnishes complete, consistent, and sufficient data to fully characterize the subsurface. The effectiveness of each technique varies from one geological environment to another. Electrical (i.e., DC resistivity and induced polarization) methods are widely used for soundings to image the resistivity structure of near-surface targets (i.e., within a few tens of meters of the ground surface) due to ease of operation and low cost.

High-frequency CSEM tools such as the Geonics EM31, EM38 and EM34 terrain conductivity meters play important roles in shallow-depth conductivity mapping, but have limited frequency bandwidth and hence poor depth sounding capability. DC resistivity methods require large electrode spacings relative to the maximum depth at which useful information can be obtained (about 5-6 times the target depth of interest) and thus are difficult to operate in some environments. In addition, large electrode array dimensions can result in significant interpretation problems caused by lateral variation in resistivity. For deeper soundings, $\geq$500 m, magnetotelluric (MT) sounding is the mainstay of the geoelectromagnetic community, being particularly appropriate for regional hydrogeophysical studies (e.g., Meju et al., 1999; Bai et al., 2001; Mohamed et al., 2002; see also Chapter 12 of this volume).

DC resistivity and IP methods (see Chapter 7 of this volume) remain popular tools for resistivity characterization of the near-surface (ca. 20 m depth). To cover the depth range 20 to 500 m, the TDEM method is best in terms of (a) the potential to give vertical and lateral resistivity information, and, (b) fewer problems in terms of ambiguity and lack of resolution. The TDEM method has the best resolution for subsurface conductivity targets (σ>2 S/m) and offers a high rate of productivity.

The TDEM method is an essential tool for hydrogeophysical studies in past-glaciated terrains or areas characterized by lateral changes in near-surface conductivity (e.g., an irregular weathered layer or buried glacial channels). This is because the presence of small-scale 3-D heterogeneities distorts MT and DC/IP depth sounding measurements. The resulting problem of "electrical static shift" are best corrected using TDEM data (e.g., Sternberg et al., 1988; Pellerin and Hohmann, 1990; Meju, 1996, 2004; Meju et al., 1999; Mohamed et al., 2002). Electrical static shift is caused by accumulation of charges around small-size 3-D bodies and manifests itself as a vertical shift of log-apparent resistivity sounding curves for MT (see Sternberg et al., 1988) and DC resistivity (see Spitzer, 2001; Meju, 2004).

The inversion of distorted MT sounding curves leads to erroneous resistivities and depths (Sternberg et al., 1988; Pellerin and Hohmann, 1990; Meju, 1996) and incorrect resistivities in DC geometric soundings (Meju, 2004). This has negative implications for groundwater quality or contaminant hydrogeophysical studies employing these methods. To effectively identify and remove static shift in DC and MT soundings for hydrogeophysical investigations, collocated TDEM measurements are required. It is useful to display the data in a common scale using relative space-time relations for electrical and EM depth sounding arrays (Meju et al., 2003; Meju, 2004).

The apparent resistivity data from TDEM and in-line four-electrode DC (Schlumberger, Wenner and dipole-dipole) arrays may be compared using the relation (Meju, 2004)

$$t = \frac{\pi}{2}\mu\sigma L^2 \tag{6.8}$$

or equivalently, $L = 711.8\sqrt{t\rho}$ [m] where time t is in seconds (s), μ is the magnetic permeability (taken to be that of free-space, $\mu_0 = 4\pi\times10^{-7}$ H/m), L is one half the electrode array length (i.e., the distance from the center of the array to an outermost electrode), and $\rho=1/\sigma$ is the homogeneous subsurface resistivity [Ω·m], which is estimated by the measured apparent resistivity ρ_a. Since it is shown semi-analytically and empirically (Sternberg et al., 1988; Meju, 1998) that the equivalent MT period (T) for a given transient time in seconds is $T{\sim}4t$, Meju (2004) defines the scaling relation for MT and DC resistivity as

$$T = 2\pi\mu\sigma L^2 \tag{6.9}$$

or L=355.9$\sqrt{T\rho}$ [m]. For a given depth sounding (apparent resistivity versus time or frequency) from a TDEM or FDEM experiment, one may estimate the half-electrode array length for the appropriate in-line four-electrode configuration that will yield the equivalent relative information and vice versa. This permits ready comparison of data from CSEM, MT, and electrical resistivity methods (Meju, 2004).

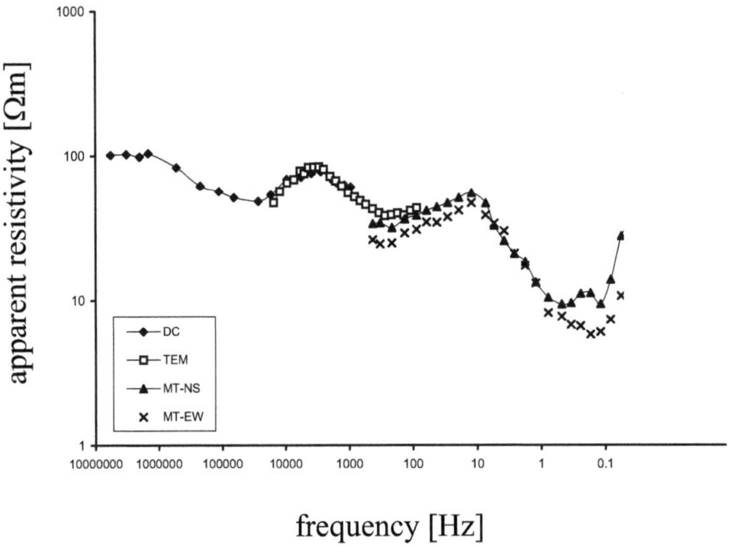

Figure 6.10. Examples of relative time-space scaling of TDEM, DC resistivity and MT soundings (Meju 2004). Shown are Schlumberger DC resistivity, central-loop TEM, and MT soundings at a deep borehole site in Parnaiba basin, Brazil (Meju et al. 1999). The DC and TDEM data are presented as a function of the equivalent MT frequencies. The DC sounding position is offset 40 m from the TEM and MT sounding point. The symbols NS and EW denote north-south and east-west sounding directions, respectively.

Figure 6.10 shows Schlumberger DC, central loop TDEM, and bidirectional MT apparent resistivity data from a single sounding acquired during a regional groundwater survey in the Parnaiba basin, NE Brazil (Meju et al., 1999). The MT soundings were carried out in magnetic north-south and east-west directions (see Chapter 12 of this volume). The MT and TDEM sounding positions are coincident, but the DC sounding point was offset by 40 m for logistical reasons at this site. All data are presented as a function of MT frequency (inverse of period) using Equations (6.8) and (6.9). Notice that all methods furnish concordant sounding curves, but the MT sounding curves in both directions are shifted down by different amounts relative to the overlapping TDEM curve, i.e., they are affected by static shift. The MT curves thus require static shift correction before inversion.

Figure 6.11 shows collocated DC resistivity and TDEM soundings from a site in midland England where glacial deposits overlie the Mercian Mudstone group (Meju, 2004). The DC and IP apparent resistivity sounding curves show marked parallelism to, and vertical displacement from, the overlapping TDEM curve and thus require correction before reliable inversion.

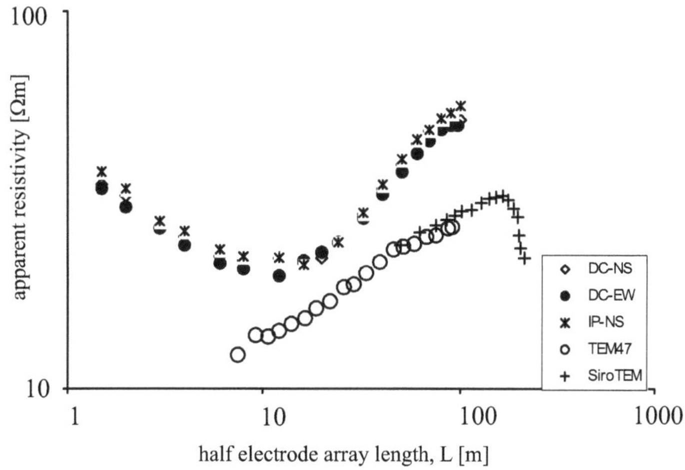

Figure 6.11. Comparison of collocated bi-directional DC resistivity and central-loop TDEM soundings at a glacial-covered test site in Leicester, England (after Meju, 2004). The DC electrode arrays were deployed in both the north-south (NS) and east-west (EW) directions at this station. The TDEM soundings employed a loop size of 100 m×100 m (with the SiroTEM equipment) and 50 m×50 m (with the Geonics TEM47 equipment).

Meju (2004) presents joint DC resistivity and TDEM inversion that corrects for a shift in the DC field curves to produce an isotropic model. Previously, it would have been necessary to incorporate anisotropy (Christensen, 2000) to explain these data. The model resulting from the joint inversion is shown in Figure 6.12*a*, and the fit of the model responses is shown in Figure 6.12*b*. Numerous examples of joint inversion of TDEM and distorted DC resistivity soundings from borehole sites in past glaciated terrains all point towards improved hydrogeophysical characterization of the subsurface (Meju, 2004).

6.9 Outlook for the Future

Technical, logistical, and conceptual difficulties often arise in the acquisition, interpretation and inversion of CSEM data. For example, better methods for treating distortion caused by induction in cultural conductors are required. Most hydrogeophysical surveys are located at or near developed sites that contain metal objects and other artifacts. The occurrence of man-made features often impacts survey design. Establishment of portable, user-friendly, multiple TX-multiple RX configurations with on-board real-time navigation, processing, and modeling software packages is required to seriously attack many hydrogeophysical problems and should be made a high priority. The paper by Nelson and McDonald (2001) provides a useful perspective on the possibilities.

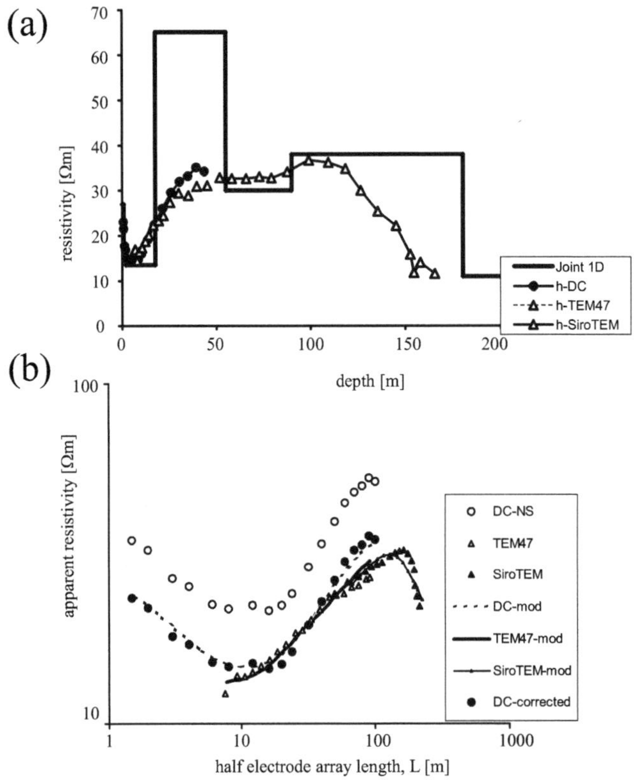

Figure 6.12. Summary of the result of 1-D joint inversion of the DC and TEM data from the Leicester test site (after Meju, 2004). The north-south DC resistivity, TEM47 and SiroTEM data were jointly inverted. (*a*) Shown are the 7-layer model for the most-squares plus solution (blocky structure) and the resistivity-depth transform of the apparent resistivity data (Meju, 1995, 1998) that served as **h** term in Equation (6.7) after accounting for static shift in DC sounding data. From top to bottom in the 1-D inversion model, the layer resistivities are: 27, 18, 13.5, 65, 30, 38, and 11 m; and the depths-to-base of the layers are: 0.9, 2, 18, 55, 90, and 181 m. (*b*) The fit between field and model responses. The model responses (-mod), the observed DC curve (DC-NS), and the resulting DC curve from joint inversion (DC-corrected) are shown for comparison.

Other challenges confront hydrogeophysicists interested in using CSEM techniques to address hydrological problems. Foremost is the need to closely examine the applicability of conventional 3-D forward modeling techniques (finite element, finite difference, and integral equation) that are based on piecewise constant electrical conductivity models of the subsurface. Field studies and theoretical considerations continue to indicate that geological media exhibit heterogeneity arranged in hierarchial patterns spanning wide ranges of spatial scales, reflecting the geological processes that generated the rock formations over vast periods of time. This type of spatial structure is not well represented by piecewise constant functions. New forward modeling strategies are required to solve electromagnetic diffusion problems in fractal media. One promising line of inquiry involves continuous-time random walk methods (e.g., Metzler and Nonnenmacher, 2002.)

It is also worthwhile to reconsider the role of conventional inversion in CSEM data analysis. It is well known that conductivity models obtained by inversion are non-unique, and *ad hoc* constraints must be imposed to provide stability. Further, field-scale mappings between electrical conductivity and the hydrological variables of interest remain tenuous, at best. The role of pattern recognition methods such as machine learning and target feature extraction, as an alternative to inversion, is gaining rapid acceptance in areas such as unexploded ordnance (UXO) and landmine detection. An important early paper in classification of buried spheroids by their CSEM response was written by Chesney et al. (1984). It would be interesting to explore whether such concepts can be applied to hydrogeophysical settings in which the subsurface target is not necessarily an isolated, well-defined man-made object but instead could be a subtle, finely distributed, and irregular variation in the subsurface electrical conductivity distribution.

References

Al-Garni, M., and M.E. Everett, The paradox of anisotropy in electromagnetic loop-loop responses over a uniaxial halfspace, Geophysics 68, 892–899, 2003.

Anderson, B., T.D. Barber, and S.C. Gianzero, The effect of crossbedding anisotropy on induction tool response, Transactions Soc. Prof. Well Logging Assoc. 39th Annual Meeting, Paper B, 14pp., 1998.

Arms, R.J., and F.R. Hama, Localized-induction concept on a curved vortex and motion of an elliptic vortex ring, Physics of Fluids 8, 553–559, 1965.

Benavides, A., and M.E. Everett, Target signal enhancement in near-surface controlled source electromagnetic data, Geophysics, to appear, 2004.

Badea, E.A., M.E. Everett, G.A. Newman, and O. Biro, Finite element analysis of controlled-source electromagnetic induction using Coulomb gauged potentials, Geophysics 66, 786–799, 2001.

Buselli, G., and M. Cameron, Robust statistical methods for reducing sferics noise contaminating transient electrmagnetic measurements, Geophysics 61, 1633–1646, 1996.

Chave, A.D., and C.S. Cox, Controlled electromagnetic sources for measuring electrical conductivity beneath the oceans. 1. Forward problem and model study, Journal of Geophysical Research 87, 5327-5338, 1982.

Chave, A.D., Numerical integration of related Hankel transforms by quadrature and continued fraction expansion, Geophysics 48, 1671–1686, 1983.

Chesney, R.H., Y. Das, J.E. McFee, and M.R. Ito, Identification of metallic spheroids by classification of their electromagnetic induction responses, I.E.E.E. Transactions on Pattern Analysis 6, 809–820, 1984.

Christensen, N. B., 1D imaging of central loop transient electromagnetic soundings, J. Eng. Environ. Geophys. 2, 53–66, 1995.

Christensen, N. B., Difficulties in determining electrical anisotropy in subsurface investigations, Geophys. Prospecting 48, 1–19, 2000.

Constable, S.C., R.L. Parker, and C.G. Constable, Occam's inversion: a practical algorithm for generating smooth models from electromagnetic sounding data. Geophysics 52, 289–300, 1987.

deGroot-Hedlin, C., and S. Constable, Occam's inversion to generate smooth two dimensional models from magnetotelluric data. Geophysics 55, 1613–1624, 1990.

Doll, H.G., Introduction to induction logging and application to logging of wells drilled with oil base mud, Petroleum Transactions AIME 186, 148–162, 1949.

Everett, M.E., and C.J. Weiss, Geological noise in near-surface electromagnetic induction data, Geophysical Research Letters 29, 2001GL014049, 2002.

Gorman, E., Controlled-source electromagnetic mapping of a faulted sandstone aquifer in central Texas, MS Thesis, Texas A&M University, 1998.

Grant, F.S., and G.F. West, Interpretation Theory in Applied Geophysics, McGraw-Hill, 583pp., 1965.

Guptasarma, D., and B. Singh, New digital linear filters for Hankel J_0 and J_1 transforms, Geophysical Prospecting 45, 745–762., 1997

Hohmann, G.W., Numerical modeling for electromagnetic methods of geophysics, in M.N. Nabighian (ed.), Electromagnetic Methods in Applied Geophysics, vol.1, Society of Exploration Geophysics, 313–363, 1987.

Hoversten, G.M., and H.F. Morrison, Transient fields of a current loop source above a layered Earth, Geophysics 47, 1068–1077, 1982.

Huang, H., and I.J. Won, Real-time resistivity sounding using a hand-held broadband electromagnetic sensor, Geophysics 68, 1224–1231, 2003..

Jackson, D.D., Interpretation of inaccurate, insufficient, and inconsistent data, Geophys. Journal of the Royal Astron. Soc. 28, 97–109, 1972.

Jackson, D.D., The use of a priori information to resolve non-uniqueness in linear inversion, Geophys. Journal of the Royal Astron. Soc. 57, 137–157, 1979.

Jackson, J.D., Classical Electrodynamics, 3rd Edition, John Wiley & Sons, 808pp., 1998.

Jones, D.S., Theory of Electromagnetism, Macmillan, 807pp., 1964.

Kearey, P., M. Brooks, and I. Hill, An Introduction to Geophysical Exploration, 3rd Edition, Blackwell Science Ltd., 262 pp., 2002.

Lamb, H., Hydrodynamics, by Sir Horace Lamb, 6th Edition, Dover Publications, 738pp., 1945.

Mandelbrot, B.B., The Fractal Geometry of Nature, W.H. Freeman, 468 pp., 1998.

McNeill, J.D., Electromagnetic terrain conductivity measurement at low induction number, Geonics Ltd. Technical Note TN-6, 1980*a*.

McNeill, J.D., Applications of transient electromagnetic techniques, Geonics Ltd. Technical Note

TN-17, 1980*b*.

Meju, M.A, S.L. Fontes, E.U. Ulugergerli, E.F. La Terra, C.R. Germano, and R.M. Carvalho, A joint TEM-HLEM geophysical approach to borehole siting in deeply weathered granitic terrains, Ground Water 39, 554-567, 2001.

Meju, M.A., and V.R.S. Hutton. Iterative most-squares inversion: application to magnetotelluric data. Geophysical Journal International 108, 758-766, 1992.

Meju, M.A., Biased estimation: a simple framework for parameter estimation and uncertainty analysis with prior data. Geophysical Journal International 119, 521-528, 1994*a*.

Meju, M.A., Geophysical Data Analysis: Understanding Inverse Problem Theory and Practice. Society of Exploration Geophysicists Course Notes Series, Vol. 6, SEG Publishers, Tulsa, Oklahoma, 296pp., 1994*b*.

Meju, M.A., P.J. Fenning, and T.R.W. Hawkins, Evaluation of small-loop transient electromagnetic soundings to locate the Sherwood Sandstone aquifer and confining formations at well sites in the Vale of York, England. J. Applied Geophysics 44, 217-236, 2000.

Meju, M.A., P. Denton, and P. Fenning, Surface NMR sounding and inversion to detect groundwater in key aquifers in England: comparisons with VES-TEM methods. J. Applied Geophysics 50, 95-112, 2002.

Meju, M.A., L.A. Gallardo, and A.K. Mohamed, Evidence for correlation of electrical resistivity and seismic velocity in heterogeneous near-surface materials. Geophys. Res. Lett. 30, 1373-1376, 2003.

Meju, M.A., Joint inversion of TEM and distorted MT soundings: Some effective practical considerations. Geophysics 61, 56-65, 1996.

Meju, M.A., A simple method of transient electromagnetic data analysis. Geophysics 63, 405-410, 1998.

Meju, M.A., Geoelectromagnetic exploration for natural resources: Models, case studies and challenges, Surveys of Geophysics 23, 133-205, 2002.

Meju, M.A., Simple relative space-time scaling of electrical and electromagnetic depth sounding arrays. Geophysical Prospecting, submitted, 2004.

Menke, W., Geophysical Data Analysis: Discrete Inverse Theory, Academic Press, 1984.

Metzler, R., and T.F. Nonnenmacher, Space and time fractional diffusion and wave equations, fractional Fokker-Planck equations, and physical motivation, Chemical Physics 284, 67-90, 2002.

Mohamed, A.K., M.A. Meju, and S.L. Fontes, Deep structure of the northeastern margin of Parnaiba basin, Brazil, from magnetotelluric imaging, Geophysical Prospecting 50, 589-602, 2002.

Nabighian, M.N., Quasi-static transient response of a conducting half-space : an approximate representation, Geophysics 44, 1700-1705, 1979.

Nabighian, M.N., and J.C. Macnae, Time domain electromagnetic prospecting methods, in Nabighian, M.N. (editor), Electromagnetic Methods in Applied Geophysics, vol.2A, Society of Exploration Geophysics, 427-520, 1991.

Nelson, H.H., and J.R. McDonald, Multisensor towed array detection system for UXO detection system, I.E.E.E. Transactions on Geoscience and Remote Sensing 39, 1139-1145, 2001.

Neuman, S.P., Generalized scaling of permeabilities - validation and effect of support scale, Geophysical Research Letters 21, 349-352, 1994.

Newman, G.A., S. Recher, B. Tezkan, and F.M. Neubauer, 3-D inversion of a scalar radio magnetotelluric field data set, Geophysics 68, 791-802, 2003.

Newman G.A., and D.L. Alumbaugh, Three-dimensional massively parallel electromagnetic inversion 1, Theory. Geophys. J. Int. 128, 345-354, 1997.

Nobes, D.C., Troubled waters: environmental applications of electrical and electromagnetic methods, Surveys of Geophysics 17, 393-454, 1996.

Palacky, G.J., Resistivity characteristics of geological targets, in Nabighian, M.N. (editor), Electromagnetic Methods in Applied Geophysics, vol.1, Society of Exploration Geophysics, 53-129, 1987.

Palacky, G.J., and G.F. West, Airborne electromagnetic methods, in Nabighian, M.N. (editor), Electromagnetic Methods in Applied Geophysics, vol.2B, Society of Exploration Geophysics, 811-879, 1991.

Pellerin, L., and G.W. Hohmann, Transient electromagnetic inversion: A remedy for magnetotelluric static shift. Geophysics 55, 1242-1250, 1990.

Porrat, D., P.R. Bannister, and A.C. Fraser-Smith, Modal phenomena in the natural electromagnetic spectrum below 5 kHz, Radio Science 36, 499-506, 2001.

Person, M., J.P. Raffensperger, S. Ge, and G. Garven, Basin-scale hydrogeologic modeling, Reviews of Geophysics 34, 61-87, 1996.

Powers, M.H., Modeling frequency-dependent GPR, The Leading Edge 16, 1657-1662, 1997.

Purvance, D.T., and R. Andricevic, Geoelectric characterization of the hydraulic conductivity field and its spatial structure at variable scales, Water Resources Research 36, 2915-2924, 2000.

Qian, W., and D.E. Boerner, Electromagnetic modeling of buried line conductors using an integral equation, Geophysical Journal International 121, 203-214, 1995.

Reid, J.E., and J.C. Macnae, Comments on the electromagnetic "smoke ring" concept, Geophysics 63, 1908-1913, 1998.

Sananikone, K., Subsurface characterization using time-domain electromagnetics at the Texas A&M University Brazos River hydrological field site, Burleson County, Texas, MS Thesis, Texas A&M University, 1998.

Sasaki, Y., Full 3-D inversion of electromagnetic data on PC, J. Applied Geophys. 46, 45-54, 2001.

Sharma, P.V., Environmental and Engineering Geophysics, Cambridge University Press, 475pp., 1997.

Smythe, W.R., Static and Dynamic Electricity, McGraw-Hill, 623pp., 1967.

Sommerfeld, A., Partial Differential Equations in Physics, Academic Press, 335pp., 1964.

Sorensen K.I., The pulled array transient electromagnetic method. Proc. 3rd Meeting of EEGS-ES, Aarhus, Denmark, 135-138, 1997.

Spies, B.R., and A.P. Raiche , Calculation of apparent conductivity for the transient electromagnetic (coincident loop) method using an HP-67 calculator, Geophysics 45, 1197-1204, 1980.

Spies, B.R., and F.C. Frischknecht, Electromagnetic sounding, in Nabighian, M.N. (editor), Electromagnetic Methods in Applied Geophysics, vol.2A, Society of Exploration Geophysics, 285-425, 1991.

Spitzer K., Magnetotelluric static shift and direct current sensitivity. Geophys. J. Int. 144, 289-299, 2001.

Stalnaker, J., and M.E. Everett, Finite element analysis of controlled-source electromagnetic induction for near-surface geophysical prospecting, Geophysics, to appear., 2004.

Sternberg, B. K., J.C. Washburne, and L. Pellerin, Correction for the static shift in magnetotellurics using transient electromagnetic soundings. Geophysics 53, 1459-1468, 1988.

Stratton, J.A., Electromagnetic Theory, McGraw-Hill, 615pp., 1941.

Tarantola, A., and B. Valette, Generalized nonlinear inverse problems solved using least squares criterion, Reviews of Geophysics & Space Physics 20, 219-232, 1982.

Taylor, G.I., On the dissipation of eddies, in G.K. Batchelor (editor), Scientific Papers. Edited by G.K. Batchelor, vol.2, Cambridge University Press, 96-101, 1958.

Telford, W.M., L.P. Geldart, and R.E. Sheriff, Applied Geophysics, 2nd Edition, Cambridge University Press, 770pp., 1990.

Tezkan, B., A review of environmental quasi-stationary electromagnetic techniques, Surveys of Geophysics 20, 279-308, 1999.

Twomey, S., An introduction to the mathematics of inversion in remote sensing and indirect measurements, Elsevier Scientific Publishing Company, 1977.

Vogel, H.J., I. Cousin, and K. Roth, Quantification of pore structure and gas diffusion as a function of scale, European Journal of Soil Science 53, 465-473, 2002.

Wait, J.R., Electromagnetic Waves in Stratified Media, Pergamon, 1962.

Wait, J.R., Geo-electromagnetism, Academic Press, 268pp., 1982.

Walker, P.W., and G.F. West, A robust integral equation solution for electromagnetic scattering by a thin plate in conductive media, Geophysics 56, 1140-1152, 1991.

Wangsness, R.K., Electromagnetic Fields, 2nd Edition, John Wiley & Sons, 608pp, 1986.

Ward, S.H., and G.W. Hohmann, Electromagnetic theory for geophysical applications, in M.N. Nabighian (ed.), Electromagnetic Methods in Applied Geophysics, vol.1, Society of Exploration Geophysics, 131-311, 1987.

West, G.F., and J.C. Macnae, Physics of the electromagnetic induction exploration method, in Nabighian, M.N. (editor), Electromagnetic Methods in Applied Geophysics, vol.2A, Society of Exploration Geophysics, 1-45 1991.

Winter, C.L., and D.M. Tartakovksy, Theoretical foundation for conductivity scaling, Geophysical Research Letters 28, 4367-4369, 2001.

Zhdanov, M.S., V.I. Dmitriev, S. Fang, and G. Hursan, Quasi-analytical approximations and series in electromagnetic modeling, Geophysics 65, 1746-1757, 2000.

Zhdanov, M.S., and G.V. Keller, The Geoelectrical Methods in Geophysical Exploration, Elsevier, 884pp., 1994.

Zhdanov, M.S., and P.E. Wannamaker, Three-dimensional Electromagnetics, Elsevier, 304 pp., 2002.

7 GPR METHODS FOR HYDROGEOLOGICAL STUDIES

A. PETER ANNAN

Sensors & Software Inc., 1040 Stacey Court, Mississauga, ON L4W 2X8 Canada

7.1 Introduction

Use of ground-penetrating radar (GPR) for geologic applications grew considerably in the 1970s. In recent years, use of GPR has increased in hydrogeological investigations, since the presence or absence of water dominates GPR responses.

Use of radiowaves for groundwater was first reported by El Said (1953). The same fundamental concepts used for subsurface lunar investigations (Annan, 1973) underpin modern GPR concepts. Annan (2002) provides a brief history of GPR over the past 50 years; the GPR conference proceedings (see the special reference list at the end of this chapter) also provide insight into the evolution of GPR.

The fundamentals of GPR for hydrogeological studies are the focus of this chapter. Discussions of applications of GPR for hydrogeological investigations can be found in Chapters 13 and 14 of this volume. These applications grew after it became abundantly clear that GPR signal propagation was strongly controlled by water content (Davis and Annan, 1989; Topp et al., 1980).

GPR can be employed to address numerous hydrogeological questions, ranging from geological structure to material properties. One of the earliest recognized benefits of GPR was its ability to delineate fine-scale depositional stratigraphy, which has a significant impact on groundwater flow. Excellent examples of this can be found in Bristow and Jol (2002). Related to the stratigraphic information are the texture and spatial scales of variations in the subsurface, which again are important in terms of understanding correlation lengths and the scale of heterogeneity for hydraulic conductivity and porosity.

GPR performs best in coarse-grained materials, such as sands and gravels, which are transparent to radiowave signals. Its use is limited in finer-grained soils such as clays and silts, or in saline groundwater, all of which strongly attenuate signals. Fine-grained soils usually exhibit low hydraulic conductivity, forming barriers to groundwater flow. GPR can thus be effective at delineating zones that impede water movement. Excellent examples are given by van Overmeeren (1998).

GPR sensitivity to water content provides a technique for mapping water table and perched water tables. Subtle variations in water table and coupling between zones of groundwater flow can be determined from the GPR response at the water table horizon (van Overmeeren, 1998); other examples are presented in Chapters 13 and 14 of this volume.

Y. Rubin and S.S. Hubbard (eds.), Hydrogeophysics, 185–213.

In some situations, GPR can be used for groundwater contaminant detection. Detectable contaminants either displace water or dissolve in water, thereby modifying pore-fluid properties. Generally, GPR is sensitive to pore fluid only at the few percent level rather than the part per million level. GPR detects how contaminants impact the distribution and properties of pore fluid, not the actual contaminant itself (Brewster and Annan, 1994; Redman et al., 1994).

GPR can monitor changes nonintrusively and quickly; thus it lends itself to monitoring changes in the subsurface versus time. Several examples of using GPR this way are described by Brewster and Annan (1994), Redman et al. (1994, 2000), and in Chapters 13 and 14 of this volume.

Building on the history of time-domain reflectometry (Topp et al., 1980), it is possible to estimate water content based on the GPR velocity (Hubbard et al., 1997; Parkin et al., 2002). Similarly, estimates of conductivity can be derived from attenuation which, in controlled circumstances, can be used to estimate salinity or total dissolved solids (TDS) in the groundwater. The quantitative extraction of these measurable quantities is the subject of current research. All these analyses require mixing relationships to enable the conversion of GPR information to material property.

In the following, the basic principles of GPR are discussed, with particular focus on hydrogeologic applications.

7.2 GPR Basic Principles

7.2.1 OVERVIEW

The foundations of GPR lie in electromagnetic (EM) theory. The history of this field spans more than two centuries and is the subject of numerous texts such as Jackson (1962) and Smythe (1989). This cursory overview outlines the basic building blocks needed to work quantitatively with GPR.

Maxwell's equations mathematically describe the physics of EM fields, while constitutive relationships quantify material properties. Combining the two provides the foundations for quantitatively describing GPR signals.

7.2.2 MAXWELL'S EQUATIONS

In mathematical terms, EM fields and related properties are expressed as:

$$\overline{\nabla} \times \overline{E} = -\frac{\partial \overline{B}}{\partial t} \tag{7.1}$$

$$\overline{\nabla} \times \overline{H} = \overline{J} + \frac{\partial \overline{D}}{\partial t} \quad (7.2)$$

$$\overline{\nabla} \bullet \overline{D} = q \quad (7.3)$$

$$\overline{\nabla} \bullet \overline{B} = 0 \quad (7.4)$$

where:

$\overline{E}$ - electric field strength vector (V/m)

$\overline{B}$ - magnetic flux density vector (T)

$\overline{D}$ - electric displacement vector (C/m^2)

$\overline{H}$ - magnetic field intensity (A/m)

q - electric charge density (C/m^3)

$\overline{J}$ - electric current density vector (A/m^2)

t - time (s)

Maxwell succinctly summarized the work of numerous researchers in this compact form. From these relationships, all classic electromagnetics (induction, radio waves, resistivity, circuit theory, etc.) can be derived when combined with formalism to characterize material electrical properties.

7.2.3 CONSTITUTIVE EQUATIONS

Constitutive relationships are the means of relating the material physical properties to the EM fields. For GPR, the electric and magnetic properties are of importance. Constitutive equations—Equations (7.5), (7.6), and (7.7)—provide a macroscopic (or average behavior) description of how electrons, atoms, molecules, and ions respond *en masse* to the application of a field.

$$\overline{J} = \tilde{\sigma}\overline{E} \quad (7.5)$$

$$\overline{D} = \tilde{\varepsilon}\overline{E} \quad (7.6)$$

$$\overline{B} = \tilde{\mu}\overline{H} \quad (7.7)$$

Electrical conductivity $\tilde{\sigma}$ describes how free charges flow to form a current when an electric field is present. Dielectric permittivity $\tilde{\varepsilon}$ describes how constrained charges are displaced in response to an electric field. Magnetic permeability $\tilde{\mu}$ describes how intrinsic atomic and molecular magnetic moments respond to a magnetic field.

These descriptions, $\tilde{\sigma}$, $\tilde{\varepsilon}$, and $\tilde{\mu}$, are tensor quantities and can also be nonlinear (i.e., $\tilde{\sigma} = \tilde{\sigma}(E)$). For virtually all practical GPR issues, these quantities are treated as field-independent scalar qualities. (In other words, the response is in the same direction as the exciting field and independent of field strength.) While these assumptions are seldom fully valid, they are typically reasonable for practical applications.

Material properties can also depend on the history of the incident field. Time dependence most often manifests itself when the electrical charges in a structure have a finite response time, making them appear fixed for slow rates of field change and free for faster rates of field change. To be fully correct, Equations (7.5), (7.6), and (7.7) should be expressed in the following form (only Equation (7.5) is written for brevity)

$$\bar{J}(t) = \int_0^{\infty} \tilde{\sigma}(\beta) \cdot \bar{E}(t-\beta)d\beta \tag{7.8}$$

This more complex form of the constitutive equations must be used when physical properties are frequency dependent (dispersive).

For most GPR applications, assuming the scalar constant form of ε, μ, σ suffices. For GPR, ε and σ are of most importance.

7.2.4 WAVE NATURE OF EM FIELDS

GPR exploits the wave character of EM. Maxwell's equations—Equation (7.1) through Equation (7.4)—describe a coupled set of electric and magnetic fields when the fields vary with time. Changing electric currents create magnetic fields that, in turn, induce electric fields, which drive new currents (as depicted in Figure 7.1). This continuing succession of one field driving another results in fields that move through the medium. Depending on the relative magnitude of losses, the fields may diffuse or propagate as waves. With GPR, we are most concerned with conditions that have a wave-like response.

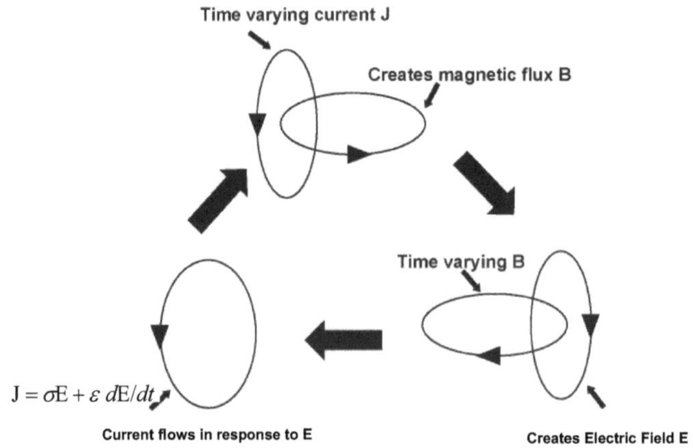

Figure 7.1. Maxwell's equations describe self-perpetuating field sequences that evolve in space and time. A moving charge or current J creates a magnetic field B, which induces an electric field E, which in turn causes electric charge to move, and so forth.

The wave character becomes evident when Maxwell's equations are rewritten to eliminate either the electric or magnetic field. The result is a transverse vector wave equation:

$$\underset{\mathbf{A}}{\underbrace{\overline{\nabla}\times\overline{\nabla}\times\overline{E}}}+\underset{\mathbf{B}}{\underbrace{\mu\sigma\cdot\frac{\partial\overline{E}}{\partial t}}}+\underset{\mathbf{C}}{\underbrace{\mu\varepsilon\cdot\frac{\partial^2\overline{E}}{\partial t^2}}}=0 \tag{7.9}$$

GPR is effective in low-loss materials in which energy dissipation, **B**, is small compared to energy storage, **C**.

Solutions to the transverse wave equation, Equation (7.9), take the form depicted in Figure 7.2. The electric field and the magnetic field vectors are orthogonal to each other and to the spatial direction of the field movement, $\hat{k}$.

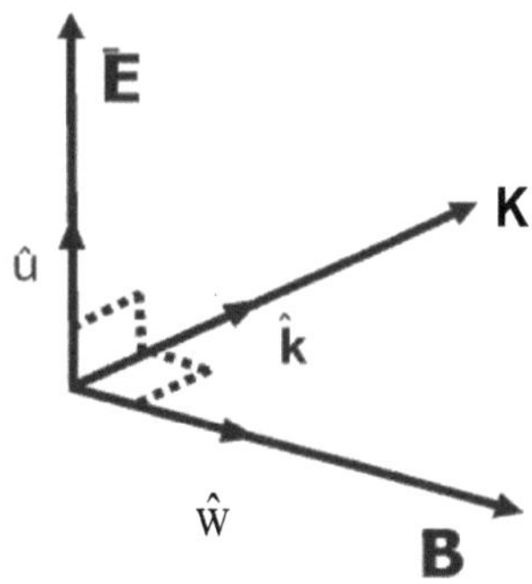

Figure 7.2. The electric field E, magnetic field B, and the propagation directions k, are orthogonal; $\hat{u}$, $\hat{w}$ and $\hat{k}$ are orthogonal unit vectors

Such solutions are referred to as plane-wave solutions to Maxwell's equations. With GPR, the electric field is the field normally measured. It has the form

$$\overline{E}=f\left(\overline{r}\cdot\overline{k},t\right)\hat{u} \tag{7.10}$$

where $\overline{r}$ is a vector describing spatial position and $f\left(\overline{r}\cdot\hat{k},t\right)$ satisfies the equation

$$\frac{\partial^2}{\partial\beta^2}f(\beta,t)-\mu\sigma\frac{\partial}{\partial t}f(\beta,t)-\mu\varepsilon\frac{\partial^2}{\partial t^2}f(\beta,t)\equiv 0, \tag{7.11}$$

which is the damped scalar-wave equation. $\beta = \overline{r} \cdot \hat{k}$ is the distance in the propagation direction.

In low loss conditions:

$$f(\beta, t) \approx f(\beta \pm vt) e^{\mp\alpha\beta} \tag{7.12}$$

where

$$v = \frac{1}{\sqrt{\varepsilon\mu}} \qquad \alpha = \frac{1}{2}\sigma\sqrt{\frac{\mu}{\varepsilon}} \tag{7.13}$$

are velocity and attenuation.

The wave nature is indicated by the fact that the spatial distribution of the fields translates in the β direction between observation times, as depicted in Figure 7.3.

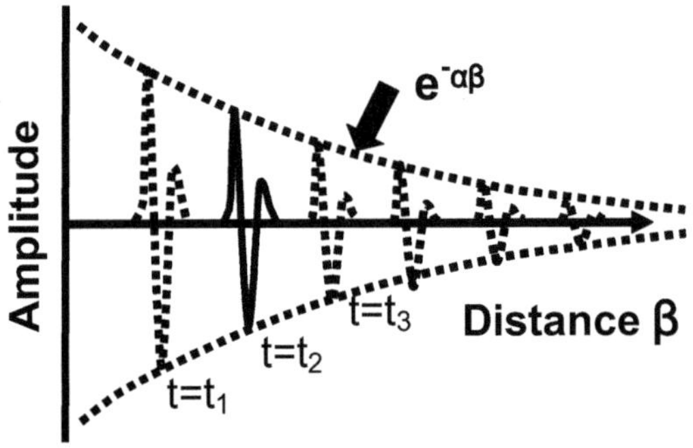

Figure 7.3. EM fields in low loss conditions move through space at finite velocity and decrease in amplitude as energy is dissipated. Time sequences t_1, t_2, t_3 indicate how the field moves through space.

7.2.5 WAVE PROPERTIES

GPR responses indicate the wave field properties, namely phase velocity (v), attenuation (α), and electromagnetic impedance (Z) (Annan, 2003). The behavior of these wave properties for a simple medium with fixed permittivity, conductivity, and permeability are most easily expressed if a sinusoidal time variation is assumed. The variation of v and α versus sinusoidal frequency, f, are shown in Figure 7.4 (note $\omega = 2\pi f$).

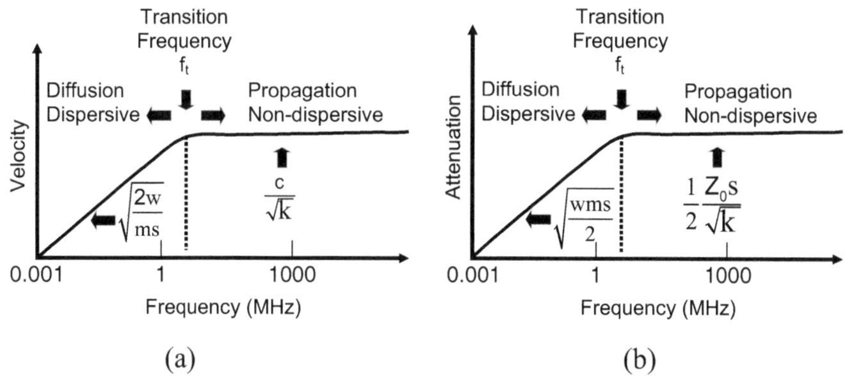

Figure 7.4. Variation of velocity and attenuation in a simple medium with nondispersive physical properties, with c and Z_0 the velocity and impedance of free space (i.e., a vacuum)

The wave properties all exhibit similar behavior. At low frequencies, all properties depend on $\sqrt{\omega}$, which are indicative of diffusive field behavior. At high frequencies, the properties become frequency independent if ε, μ, σ are constant. The high-frequency behavior is the feature most important to GPR.

The transition from diffusion to propagation behavior occurs when the electrical currents change from conduction (free charge) dominant to displacement (constrained charge) current dominant behavior. For a simple material, the transition frequency is defined as

$$f_t = \frac{\sigma}{2 \cdot \pi\varepsilon} \tag{7.14}$$

For GPR, dielectric permittivity is an important parameter. More often, the terms "relative permittivity" or "dielectric constant" are commonly used and defined as

$$\kappa = \frac{\varepsilon}{\varepsilon_0} \tag{7.15}$$

ε_0 is the permittivity of a vacuum, 8.89×10^{-12} F/m.

In the high-frequency plateau above f_t in Figure 7.4, all frequency components travel at the same velocity and suffer the same attenuation. An impulsive signal will travel with its shape intact, which is propagation without dispersion (Annan, 1996). In this case, the velocity, attenuation, and impedance can be expressed as

$$v = \frac{1}{\sqrt{\varepsilon \cdot \mu}} = \frac{c}{\sqrt{\kappa}} \tag{7.16}$$

$$\alpha = \sqrt{\frac{\mu}{\varepsilon}} \cdot \frac{\sigma}{2} = Z_0 \cdot \frac{\sigma}{2 \cdot \sqrt{\kappa}} \tag{7.17}$$

$$Z = \sqrt{\frac{\mu}{\varepsilon}} = \frac{Z_0}{\sqrt{\kappa}} \tag{7.18}$$

with the right-most expression being valid when magnetic property variations are assumed negligible, making $\mu = \mu_o$ where $\mu_o = 1.25 \times 10^{-6}\,\mathrm{H/m}$ is the free space magnetic permeability. In the above, c is the speed of light and Z_0 is the impedance of free space.

$$Z_0 = \sqrt{\frac{\mu_0}{\varepsilon_0}} = 377 \cdot \mathrm{ohms} \tag{7.19}$$

The "GPR plateau" normally exhibits a gradual increase in velocity and attenuation with frequency. Two primary factors cause this increase. First, water starts to absorb energy more and more strongly as frequency increases toward the water relaxation frequency, in the 10 to 20 GHz range (Hasted, 1972). Even at 500 MHz, water losses can start to be seen in otherwise low-loss materials. Second, scattering losses are extremely frequency dependent and can become more important than electrical losses (Annan, 2003).

7.2.6 REFLECTION, REFRACTION, AND TRANSMISSION AT INTERFACES

GPR methods depend on detecting reflected or scattered signal. Planar boundaries provide the simplest model for qualifying behavior. The fresnel reflection (and transmission) coefficients (Jackson, 1962; Born and Wolf, 1980) quantify how the amplitudes of the electromagnetic fields vary across an interface between two materials. When an EM wave impinges on a boundary, the incident wave, I, is partially transmitted, TI, and partially reflected, RI, as shown in Figure 7.5.

Travel direction also changes (i.e., the wavefront is refracted) in accordance with Snell's law:

$$\frac{\sin \theta_1}{v_1} = \frac{\sin \theta_2}{v_2} \tag{7.20}$$

When $v_1 > v_2$, medium 2 has a critical angle beyond which energy cannot propagate from medium 1 to 2. The critical angle is determined by setting $\theta_1 = 90^\circ$. (This angle plays a role in many GPR responses.)

An incident wave is composed of two independent components whose vector elements have a compatible orientation with respect to the boundary. These waves are referred to as the TE (transverse electric field) and TM (transverse magnetic field) waves, and are depicted in Figure 7.5. By decomposing the incident wave into TE and TM components, the specific mathematical forms for R and T can be derived.

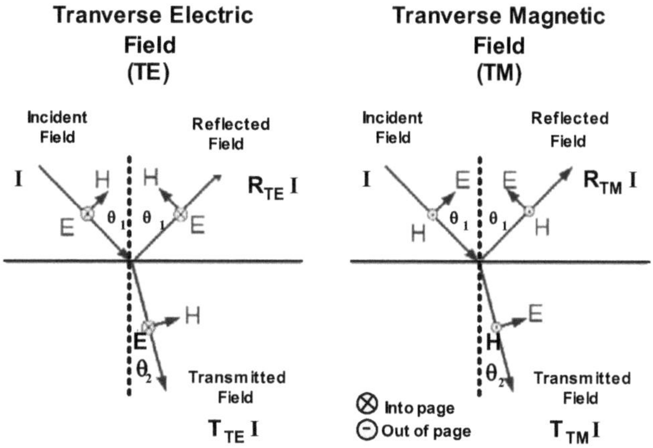

Figure 7.5. EM waves are transverse vector wave fields. For any given propagation direction, two independent fields exist. For planar interfaces, it is traditional to discuss the two waves, one with the electric field in the interface plane, called traverse electric (TE), and one with the magnetic field vector in the interface plane, called traverse magnetic (TM).

The incident reflected and transmitted field strengths are related by

$$I + R \cdot I = T \cdot I. \tag{7.21}$$

R and I are determined by requiring Snell's law to be satisfied, the electric and magnetic fields in the plane of the interface to be continuous, and the electric current and magnetic flux density crossing the interface to be equal. The result is

$$R_{TE} = \frac{Y_1 \cdot \cos\theta_1 - Y_2 \cdot \cos\theta_2}{Y_1 \cdot \cos\theta_1 + Y_2 \cdot \cos\theta_2} \tag{7.22}$$

$$R_{TM} = \frac{Z_1 \cdot \cos\theta_1 - Z_2 \cdot \cos\theta_2}{Z_1 \cdot \cos\theta_1 + Z_2 \cdot \cos\theta_2} \tag{7.23}$$

and

$$T_{TE} = 1 + R_{TE} = \frac{2 \cdot Y_1 \cdot \cos\theta_1}{Y_1 \cdot \cos\theta_1 + Y_2 \cdot \cos\theta_2} \tag{7.24}$$

$$T_{TM} = 1 + R_{TM} = \frac{2 \cdot Z_1 \cdot \cos\theta_1}{Z_1 \cdot \cos\theta_1 + Z_2 \cdot \cos\theta_2} \tag{7.25}$$

where Z_i and Y_i are the impedances and admittances ($Y_i = 1/Z_i$) of the ith material, respectively. The critical factor is that an electromagnetic impedance contrast must exist for there to be a response.

When the EM wave is vertically incident on the interface ($\theta_1 = \theta_2 = 0°$), no distinction exists between a TE and TM wave, and the TE and TM reflection coefficients become identical (for the field components).

Note that increasing reflectivity occurs with increasing angle of incidence, that total reflection occurs for incidence angles beyond the critical angle when moving from a low to high velocity material, and that reflected signals can be positive or negative, depending on whether impedance decreases or increases at an interface.

7.2.7 GPR SOURCE NEAR AN INTERFACE

Only very simple forms of EM fields have been discussed to this point. In practice, field sources and how fields spread from a finite-sized transmitter are important issues.

GPR sources are normally deployed close to the ground. The signals extend outward as depicted in Figure 7.6 (Annan, 2003). When the wavefront impinges on the ground, the field at any point along the ground interface can be visualized locally as a planar wave impinging on the boundary at a specific incidence angle defined by geometry (source height and lateral distance). Locally the signal is reflected and refracted according to Snell's law and the fresnel coefficients.

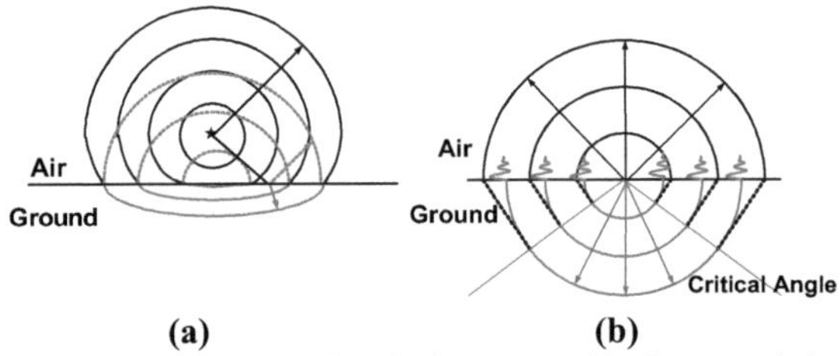

Figure 7.6. Wavefronts spreading out from a localized source. In (a), the source is located above the ground, with the dotted lines indicating the reflected signal. In (b), the source is located on the air-ground interface, with the dashed lines indicating refracted waves. The oscillating lines indicate evanescent waves.

The wavefront in the air is spherical, while in the ground it is no longer spherical, because bending occurs at the interface unless the source is right on the surface. To understand what is happening in the ground and near the interface, the limiting case of the source right at the interface is informative (see Figure 7.6b). The incident and reflected waves in the air coalesce into an upgoing spherical wave. In the ground, the transmitted signal divides into two parts, a spherical wave and a planar wavefront traveling at the critical angle linking the direct spherical air wave and the spherical ground wave. Near the interface, the spherical ground wave extends into the air as an evanescent field.

References such as Sommerfield (1949), Wait (1962), Brekhovskikh (1960), and Annan (1973) provide more detailed discussion. The various wave fields are clearly visible at large distances from the source and/or very short wavelengths. For short distances from the source or long wavelengths, the separation of events becomes blurred, but the essential concepts remain valid.

7.3 Water and GPR

7.3.1 PHYSICAL PROPERTIES

Physical properties are very important in GPR, since low electrical loss conditions are not prevalent. Mineralogical clay-rich environments or areas of saline groundwater can create conditions in which GPR signal penetration is very limited.

The electrical properties (ε, μ, σ) of materials are a wide-ranging topic. Background can be found in Olhoeft (1981, 1987), Santamarina et al (2001), and in Chapter 4. Discussion here is limited to key issues related to GPR and hydrogeologic applications.

Earth materials are invariably composites of many other materials or components. Water and ice represent the few cases where only a single component is present. Understanding the physical properties of mixtures is thus a key factor in interpretation of a GPR response.

7.3.2 PROPERTIES OF MIXTURES OF MATERIALS

Earth environments are always complex and mixtures of many components. Table 7.1 illustrates some idealized materials that might be encountered. A simple quartz sand can be visualized as a mixture of soil grains, air, water, and ions dissolved in water. Soil grains will typically occupy 60 to 80% of the available volume.

Table 7.1. Examples of Composite Materials

Quartz Sand	Contaminated Sand	Concrete
quartz grains	quartz grains	aggregate
air	air	cement hydrated
water	water	cement partially hydrated
dissolved ions	dissolved ions	air
	liquid hydrocarbon	water
	biodegradation products	dissolved ions
	hydrocarbon vapor	

A hydrocarbon-contaminated quartz sand will additionally contain hydrocarbons in both liquid and vapor form and, if time has passed, biodegraded hydrocarbon derivatives as well.

The third example is concrete, which consists of cement and aggregate. Cement is often used to create grout barriers to modify groundwater flow. The aggregate is typically broken rock with variable mineralogy. The cement can be fully hydrated or partially hydrated. An optimal mix will have very little pore space containing water and/or air; a less than optimal mix will have substantial pore space, which may be connected or disconnected, and which can contain air and/or water.

Mixtures of materials seldom exhibit properties directly proportional to the volume fraction of the constitute components. In many respects, this complexity can make quantitative analysis of GPR data impossible without ancillary information.

While the subject of mixtures is complex, the big picture GPR perspective is simpler. In the 10–1000 MHz frequency range, the presence or absence of water in the material dominates behavior, with the general picture as follows:

- Bulk minerals and aggregates in mixtures generally are good dielectric insulators. They typically have a permittivity in the range of 3 to 8 (depending on mineralogy and compaction) and are usually insulating with virtually zero conductivity.

- Soils, rocks, and construction materials such as concrete and asphalt have empty space between the grains (pore space) available to be filled with air, water, or other material.

- Water is by far the most polarizable naturally occurring material (in other words, it has a high permittivity with $\kappa \approx 80$).

- Water in pore space normally contains ions, and the water electrical conductivity associated with ion mobility is the dominant factor in determining bulk-material electrical conductivity.

- Since water is invariably present in the pore space of natural (geologic) materials, except in such unique situations where vacuum drying or some other mechanism assures the total absence of water, it has a dominant effect on electrical properties.

Empirically derived forms such as the Topp relationship (Topp et al., 1980) and variations of Archie's law (Archie, 1942) have long demonstrated the relationship between permittivity, electrical conductivity, and volumetric water content for soils.

The Topp relationship expresses apparent relative permittivity, κ_a, as a function of volume fraction water content, θ_v :

$$\kappa_a = 3.03 + 9.3\theta_v + 146.0\theta_v^2 - 76.6\theta_v^3 \tag{7.26}$$

Archie's law, although derived for low frequencies, is qualitatively useful at GPR frequencies and relates bulk electrical conductivity to pore-water electrical conductivity:

$$\sigma = a\phi^m s^n \sigma_w + \sigma_c \tag{7.27}$$

where

ϕ - porosity

m - constant 1.3 to 2.5

a - constant 0.4 to 2

s - pore space fraction water filled

n - constant about = 2

σ_w - pore water electrical conductivity

σ_c - soil grain surface electrical conductivity

Permittivity at zero volumetric water content is in the 3 to 4 range in soils, and electrical conductivity is very small. As water is added to the mix, the permittivity and conductivity rise until no more water can be squeezed into the available pore space. Obviously porosity of the material dictates the maximum limit on the volume of water that can be placed in the material, and ultimately that in turn dictates the maximum permittivity and conductivity (for a fixed pore water conductivity) of the mix.

Another heuristic mixing formula commonly used is the complex refractive index model (CRIM) (Wharton et al., 1980)—which expresses a mixture's relative permittivity as a function of the volume fraction and the square root of the complex permittivity of the components with

$$\sqrt{\kappa_{mix}} = \sum_i \sqrt{\kappa_i \theta_i} \tag{7.28}$$

where κ_i and θ_i are the respective component permittivities and volume fractions. This formulation can be quite helpful for a first-order analysis of a problem. A more advanced relationship is the Bruggeman- Hanai-Sen, or BHS model, (Sen et al., 1981) which uses effective media theory to derive a composite material property from constituents.

A wide variety of theoretical and empirical mixing relationships exist for mixtures with varying degrees for scope of applicability, depending on the nature of the problem to be addressed (Sihvola, 2002). Site- or material-specific versions of the various relationships are sometimes developed. Refer also to Chapter 4 of this volume for more discussion of these very important concepts.

7.4 GPR Survey Methodology

GPR measurements fall into two categories, reflection and transillumination, as depicted in Figure 7.7. Surveys using a single transmitter and a single receiver are the most common, although multiple source and receiver configurations are starting to appear for specialized applications. Survey design details can be found in Annan and Cosway (1992) whereas instrumentational details are discussed by Annan (2003) and frequency selection is addressed by Annan and Cosway (1994).

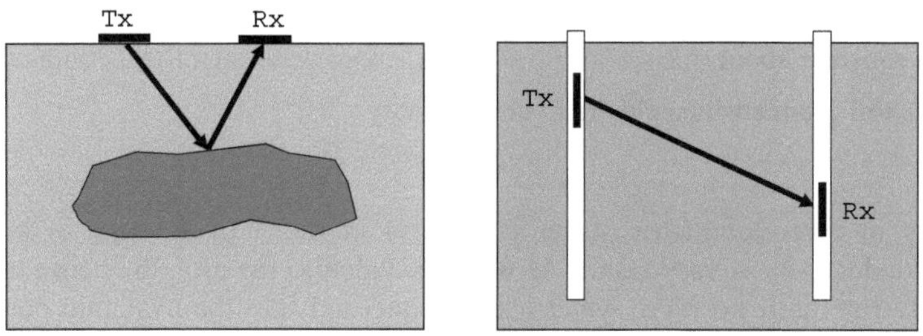

Figure 7.7. Example of a borehole transillumination measurement

7.4.1 COMMON OFFSET REFLECTION SURVEY

With reflection surveys, the objective is to map subsurface reflectivity versus spatial position. Variations in reflection amplitude and time delay indicate variations in v, α, and Z. GPR reflection surveys are traditionally conducted on "straight" survey lines, and systems are designed to operate primarily in this fashion. Area coverage most often entails data acquisition on a rectilinear grid of lines that cover the area, such as depicted in Figure 7.8.

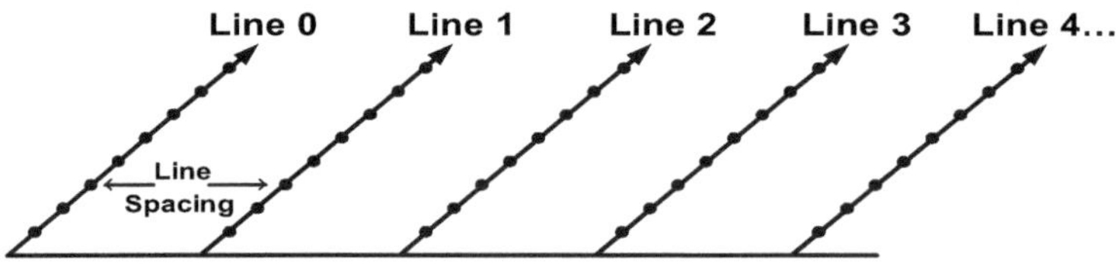

Figure 7.8. A survey area spanned by a number of survey lines. The ground response is measured at discrete points along the survey line. While field practice may be more erratic, data of this format are key to most systematic data processing and visualization.

Common offset surveys usually deploy a single transmitter and receiver, with a fixed offset or spacing between the units at each measurement location. The transmitting and receiving antennas have a specific polarization character for the field generated and detected. The antennas are placed in a fixed geometry and measurements made at regular fixed station intervals, as depicted in Figure 7.9. Data on regular grids at fixed spacing are normally needed if advanced data processing and visualization techniques are to be applied.

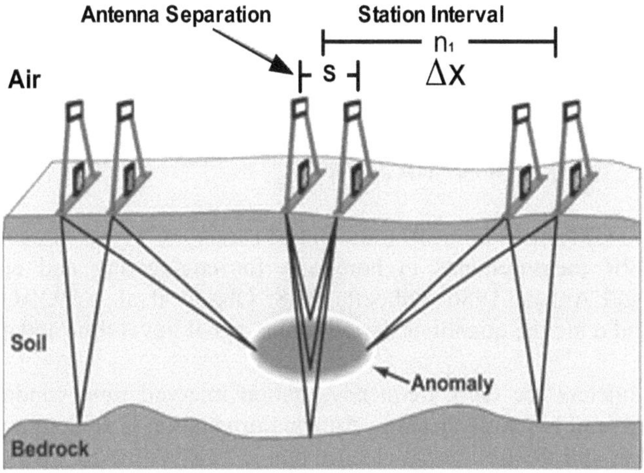

Figure 7.9. Schematic illustration of common-offset single-fold profiling along a line showing major survey specification parameters

The parameters defining a common offset survey are GPR center frequency, the recording time window, the time-sampling interval, the station spacing, the antenna spacing, the line separation spacing, and the antenna orientation.

7.4.2 MULTI OFFSET CMP/WARR VELOCITY SOUNDING DESIGN

The common midpoint (CMP) or wide-angle reflection and refraction (WARR) sounding mode of operation are the electromagnetic equivalent to seismic refraction and wide angle reflection (see Chapter 8 of this volume). CMP soundings are primarily used to estimate the radar signal velocity versus depth in the ground, by varying the antenna spacing and measuring the change of the two-way travel time (as illustrated in Figure 7.10).

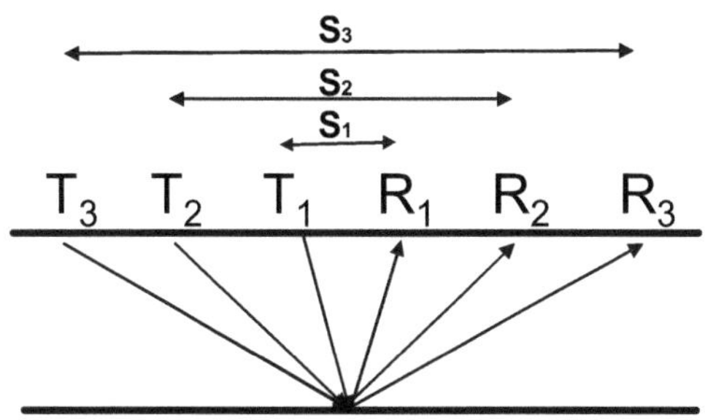

Figure 7.10. Procedure for conducting a CMP velocity sounding

CMP/WARR measurements can be performed at every station of a survey, resulting in a multifold reflection survey, which is the norm in most seismic surveys. Two benefits are that CMP stacking can improve signal-to-noise (Fisher et al, 1992), and that a full velocity cross section can be derived (Greaves et al., 1996). Multifold GPR surveys are seldom performed because they are time consuming, more complex to analyze, and not cost-effective (most of the cost benefit is obtained with well-designed single-fold surveys).

7.4.3 TRANSILLUMINATION SURVEYS

Transillumination GPR measurements (Annan and Davis, 1978) are less common. Most uses involve GPR measurements in boreholes for engineering and environmental studies (Davis and Annan, 1986; Olhoeft, 1988; Olsson et al., 1992; Owen, 1981). Estimates of v and α are the quantities derived from signal travel time and amplitude.

The survey parameters are GPR frequency, station interval, time window, temporal sampling interval, and borehole spacing. Antenna orientation is seldom an issue, since boreholes are slim and the traditional electric dipole axis is aligned with the borehole. Some specialty borehole systems for larger diameter holes have been designed to have source and receiver directionality.

The GPR frequency selection is usually dictated by spatial scale, and the survey design logic for reflection surveys applies. Most often, attenuation is an issue, so frequencies are kept as low as possible to maximize penetration. For closely spaced boreholes, higher-frequency antennas may be required to obtain sufficient travel-time resolution.

Deviations in geometry will introduce errors into values of v and α derived from GPR time and amplitude observations. A critical part of transillumination surveys involves controlling and/or measuring geometry as auxiliary data to the GPR measurement.

7.4.3.1 Forms of Transillumination Profiling

Zero offset profiling (ZOP) is a quick and simple survey method to locate velocity anomalies or attenuation (shadow) zones (Gilson et al., 1996). This technique is a quick way for assessing variations in water content. The methodology is depicted in Figure 7.11a: The transmitter and receiver are moved from station-to-station in synchronization. In a uniform environment, the received signal should be the same at each location.

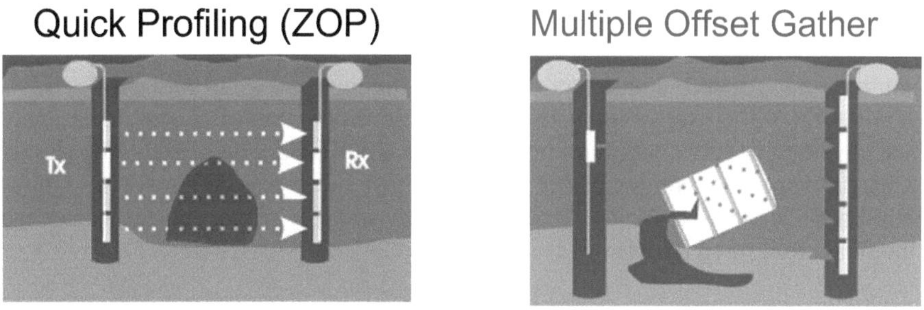

(a) **(b)**

Figure 7.11. (a) Illustration of a transillumination zero offset profiling (ZOP); (b) Illustration of a transillumination multi-offset gather (MOG). Combining MOGs for each transmitter station provides the data for tomographic imaging.

7.4.3.2 Multi-Offset Transillumination

Multi-offset gather (MOG) surveying provides the basis of tomographic imaging. The objective is to measure a large number of angles passing through the volume between the boreholes, as depicted in Figure 7.11b. As with ZOP measurements, borehole geometry plays a critical role in data analysis.

Critical factors in survey design are borehole depth, D, to borehole separation, S. Ideally, D/S > 2 is desirable for several reasons. First, keeping D/S large maximizes the range of view angles through the ground. Second, refracted wave events can mask direct arrivals (see Figure 7.12).

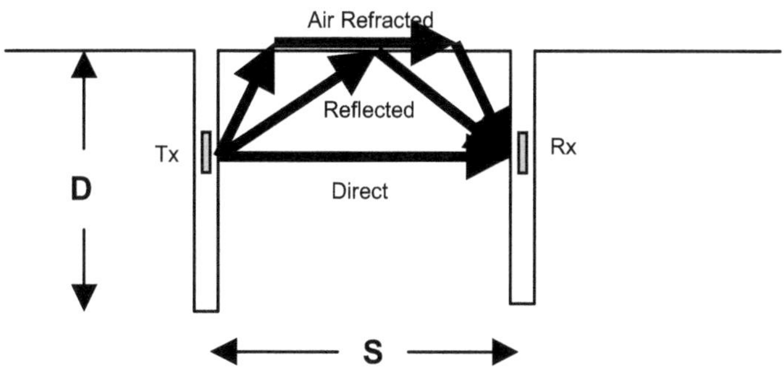

Figure 7.12. For borehole transillumination surveys, D/S should be kept as large as practical. Direct signals may be masked by faster refracted airwave arrivals.

While maximizing D/S, S should be kept sufficiently large to be in the far-field of the antenna. In other words,

$$S > \frac{v_g}{f_c} \tag{7.29}$$

where v_g is the ground velocity and f_c is the GPR center frequency.

7.5 Data Processing and Display

Ubiquitous inexpensive computers have made digital processing and display of GPR data common. GPR users exploit many of the developments of seismic data analysis, which have evolved to a very high level (Yilmaz, 2000) and which are briefly discussed in Chapter 8 of this volume. GPR data are most often treated as scalar, while the electromagnetic fields are vector quantities. Uses of GPR respecting the vector nature of the field are limited but growing.

Transforming GPR information into hydrogeological information follows two paths. The first is common to all geophysical methods, in that the GPR response measured is presented in a section, plan, or volume form to indicate spatial change, texture, or target location. The second is to extract quantifiable variables, such as velocity, attenuation, and impedance, and then use these wave properties to derive hydrogeological quantities such as water content, porosity, and salinity, Applications of GPR in both modes are presented in Chapter 14 of this volume.

The typical processing flow for GPR data is depicted in Figure 7.13. Data processing focuses on the highlighted areas: data editing, basic processing, advanced processing, and visualization/ interpretation processing. Processing is usually an iterative activity. Data will flow through the processing loop several times, with the results visually monitored. Batch processing with limited interactive control may be applied on large data sets, after initial iterative testing on selected data samples has been performed.

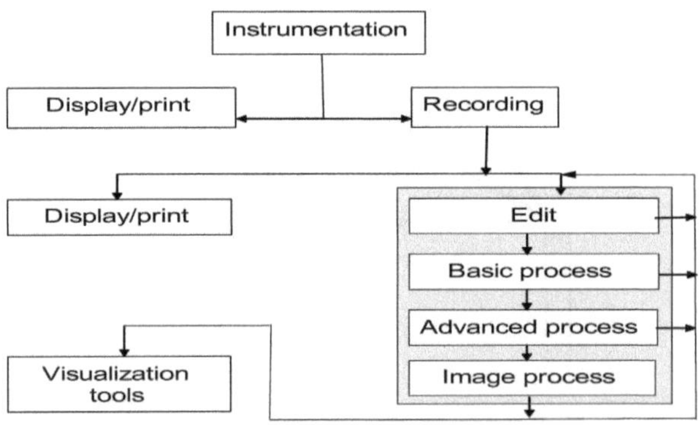

Figure 7.13. Overview of GPR data flow. The data processing subdivisions indicated are meant to reflect that the data manipulation becomes more interpreted and application dependent, and hence more subjective (or biased), because a particular end result is sought as processing blurs into interpretation (Annan, 1993).

7.5.1 DATA EDITING AND STORAGE

Field acquisition is seldom so routine that no errors, omissions, or data redundancy occur. Data editing encompasses issues such as data reorganization, data file merging, data header or background information updates, repositioning, and inclusion of elevation information with the data. Well-designed GPR systems make collecting and quantifying such steps as easy for the user as possible. Data recorded on equal spatial and temporal time intervals are virtually mandatory for most of the advanced processing methods. As a result, data editing to obtain common sampling intervals is often performed.

7.5.2 BASIC PROCESSING

Basic data processing addresses some of the fundamental manipulations applied to data to make a more acceptable product for initial interpretation and data evaluation. In most instances, this type of processing can also be applied in real-time to vary the real-time display. The advantage of postsurvey processing is that the basic processing can be done more systematically.

7.5.3 TIME GAIN

Radar signals are very rapidly attenuated as they propagate into the ground. Signals from greater depths are very small, such that simultaneous display of this information with signals from a shallower depth requires preconditioning for visual analysis display. Equalizing amplitudes by applying a time-dependent gain function compensates for the rapid falloff in radar signals from deeper depths. Figure 7.14 indicates the general nature of the amplitude of radar signals versus time.

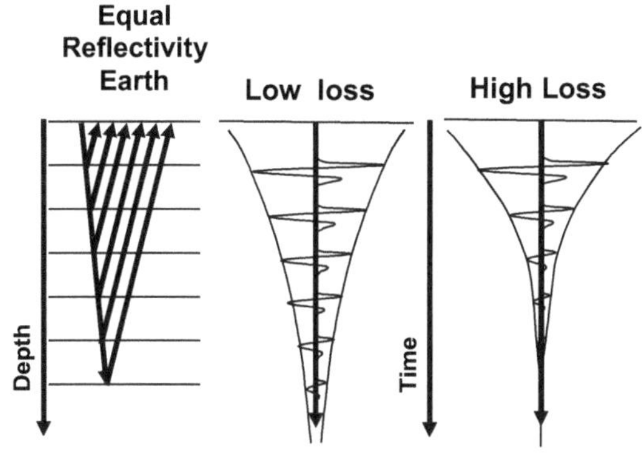

Figure 7.14. Layered earth model of equal reflectivity horizons and impulse response with envelope of reflection amplitude depicted in both low and high attenuation media

Attenuation in the ground can be highly variable. A low attenuation environment may permit exploration to depths of tens of meters. In high-attenuation conditions, depths of less than a meter may be penetrated. Display of GPR data versus time must accommodate the low and high attenuation extremes as depicted in Figure 7.14. The concept of time varying gain is depicted in Figure 7.15.

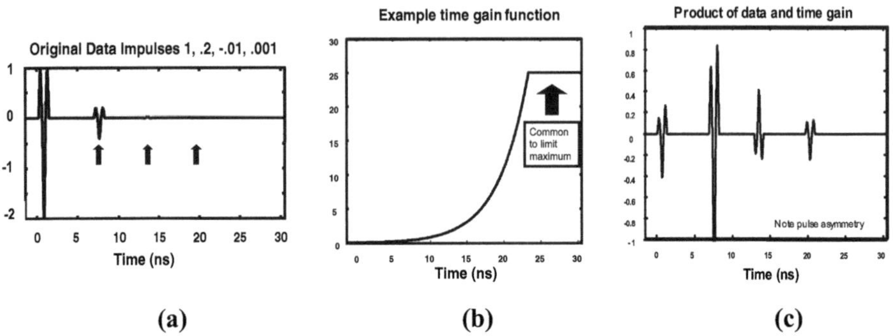

Figure 7.15. Concept of time varying gain where signal amplification varies with time: (a) a radar trace with four signals of decreasing amplitude with time; (b) a time gain function, and (c) the result of multiplying (a) by (b). All four events are visible in (c).

There are a variety of ways of applying time gain to radar data. Gain should be selected based on an *a priori* physical model, not user whim, with the objective of minimizing artifacts created by the process.

Time gain is a nonlinear operation. Filtering operations before and after time gain will not be equivalent. The example in Figure 7.15 clearly shows a change in pulse shape before and after the time gain process.

7.5.4 TEMPORAL AND SPATIAL FILTERING

Temporal and spatial filtering are often the next stage of processing. GPR data sets are most often viewed as two-dimensional arrays, with time (or derived depth) on one axis and space on the other. Temporal filtering means filtering along the time axis of the data set, while spatial filtering means operating on the spatial axis. A whole host of filtering types may be applied. With data being acquired on regular grids, processing is now becoming three-dimensional.

Median and alpha mean trim filters offer powerful data "clean-up" filters for noise spikes. These nonlinear filters can be applied in both the time or space domain. They may be applied before and after time gain, but are usually most effective if applied before time gain and any conventional filtering.

7.5.5 COMPLEX TRACE ATTRIBUTES

Complex trace attribute analysis, common in seismic processing (White, 1991), finds considerable use with GPR. A Hilbert transform is used to create a complementary time series to a data trace, allowing the envelope and the local frequency to be estimated at every point along the trace.

The envelope attribute (Figure 7.16) is a smoothly varying function that reflects the true resolution of the data. For this reason, the envelope is often used to represent reflectivity when generating time/depth slices or creating volume views.

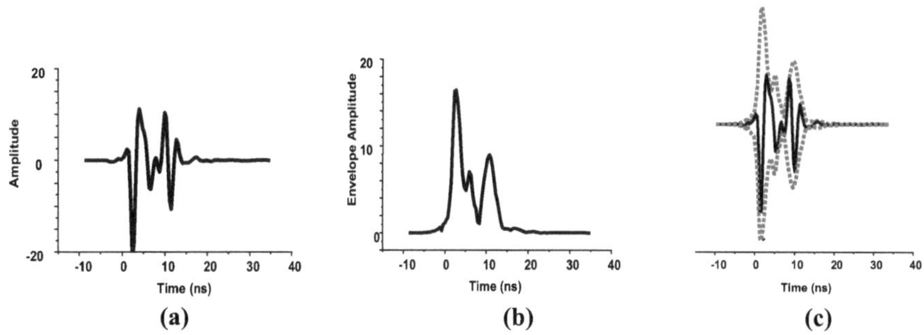

Figure 7.16. Example of the envelope for a GPR trace. The original trace is shown in (a) and the envelope in (b). The envelope encloses the original trace, as illustrated in (c). This data trace is gathered from directly over a pipe in Figure 7-17 (position at 8.6 ft).

7.5.6 DECONVOLUTION

The purpose of deconvolution is normally to maximize bandwidth and reduce pulse dispersion to ultimately maximize resolution. Examples of deconvolution and other types of filtering are given by Todoeschuck (1992) and Turner (1992). Deconvolution ("decon") of GPR data has seldom yielded a great deal of benefit. Part of the reason for this is that the normal GPR pulse is the shortest and most compressed that can be achieved for the given bandwidth and signal-to-noise conditions. Instances where deconvolution has proven beneficial occur when extraneous reverberation or system reverberation are present.

A closely related topic is inverse Q filtering. The higher GPR frequencies tend to be more rapidly attenuated, resulting in lower resolution with increasing depth. Inverse Q filtering attempts to compensate for this effect. Irving and Knight (2003) present an excellent discussion of this topic.

7.5.7 MIGRATION

Migration is spatial deconvolution (Fisher et al., 1992a,b) that attempts to remove source and receiver directionality from reflection data. The goal is to reconstruct the geometrically correct radar reflectivity distribution of the subsurface. Migration requires knowledge of the velocity structure, which often makes it an interactive process because background velocity is iteratively adjusted to optimize the image.

Figure 7.17 shows how migration collapses the hyperbolic response of a pipe. Several types of migration (Kirchoff, Stolt, reverse time, finite difference) are possible. The seismic history is quite applicable to GPR, and Yilmaz (2000) discusses the subject extensively. A unique aspect of GPR is the magnitude of topography compared to depth of exploration. Migration processing that includes topography has been described by Lehman and Green (2000), and Heincke et al. (2002).

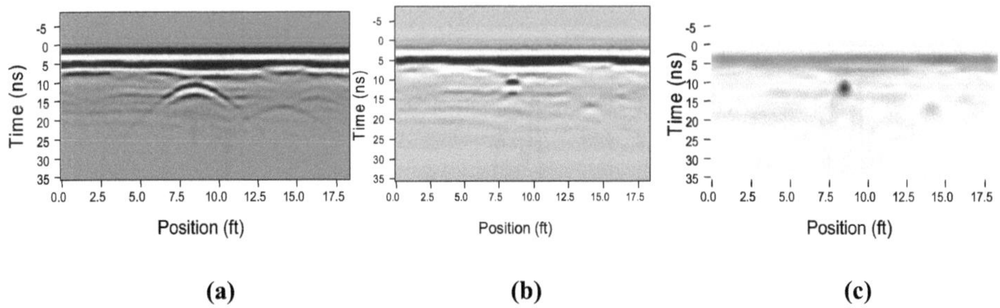

Figure 7.17. Example of migration applied to a buried pipe application: (a) original profile data, (b) the result after a Kirchhoff migration, and (c) the envelope of the migrated data. Enveloped data are often the input to mapping and visualization packages.

7.5.8 ADVANCED DATA PROCESSING

Advanced data processing requires varying degrees of interpreter bias to be applied. The result is data that are significantly different from the raw input information. Such processes include well-known seismic processing operations (Yilmaz, 2000) such as selective muting, dip filtering, deconvolution, and velocity semblance analysis, as well as more GPR-specific operations such as background removal, multiple-frequency antenna mixing, and polarization mixing (Tillard and Dubois, 1992; Roberts and Daniels, 1996).

7.5.9 DATA VISUALIZATION

Initially, GPR data were always displayed as reflection cross sections (essentially vertical slices through the ground along the transect surveyed). As processing and computer processing power have advanced, more and more data presentation is in the form of 3-D volume (voxel) rendering and time/depth slices (plan maps). Excellent examples are given by Sigurdsson and Overgaard 1996), Grasmuek (1996), Lehmann and Green (1999), Green et al. (2002), and McMechan et al. (1997). Examples of 3-D GPR displays are presented in Chapters 13 and 14 of this volume.

7.6 Data Analysis and Interpretation

The ultimate purpose of a GPR survey is to extract useful diagnostic information for a site investigation. The interpretation of GPR is application- and site-specific, and generalization is difficult. The following discussions use field examples for hydrogeologic applications to convey the concepts and procedures.

7.6.1 REFLECTION PROFILING

The common offset reflection profiling example shown in Figure 7.18 is from Canadian Forces Base Borden, in Ontario, Canada. This site is used extensively by the University of Waterloo for hydrogeologic investigation. This particular data set shows stratigraphy and a reflection from the water table. The profile begins in an excavated area and climbs up the side of the excavation pit to the natural terrain surface.

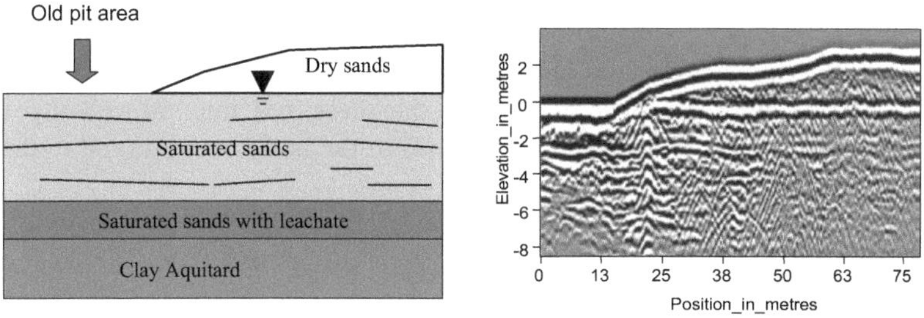

(a) (b)

Figure 7.18. Survey results at the Camp Borden, Ontario, site in Canada, showing the water table reflection: (a) a simplified geologic section and (b) the corresponding GPR elevation-compensated section

The data have been compensated for elevation along the profile line, which renders the water table reflection essentially flat. A spherical attenuation-compensation gain is applied, making the water table and other stratigraphic features visible. A simple elevation scale was created by multiplying one-way travel time by the average velocity estimated for the area. Velocity is usually obtained by examining diffraction tails from localized scatterers, as well as carrying out CMP soundings to determine velocity (as discussed in the following section).

7.6.2 CMP VELOCITY ANALYSIS

Using CMP measurements to obtain velocity for determining depth is common. The example data set shown in Figure 7.19 comes from the 70 m position on the reflection profile, shown in Figure 7.18b. The CMP data were acquired to provide depth calibrations for the reflection profile shown in Figure 7.18.

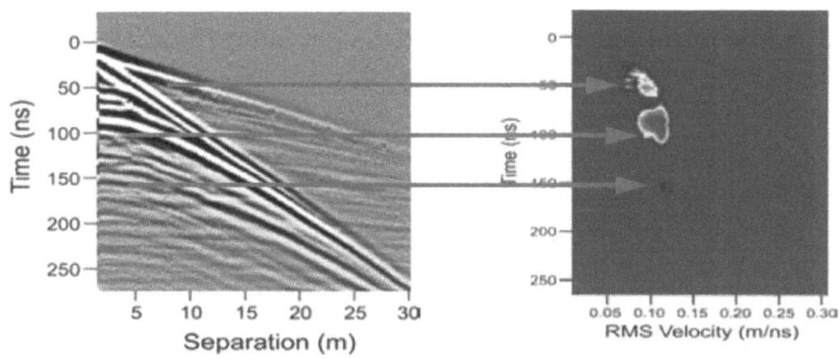

(a) (b)

Figure 7.19. (a) CMP showing reflection from the water table; (b) result of semblance analysis applied to CMP data in (a) to estimate move-out velocities for various horizons

The CMP data, which show amplitude versus offset, are normally analyzed using a semblance analysis process to estimate what velocity correlates with the move-out shape of the reflection responses. The result of applying semblance analysis to the CMP data in Figure 7-19a is presented in Figure 7-19b.

The maximum semblance occurs at what is commonly referred to as the RMS (root mean square) velocity or normal move-out velocity for the particular reflecting horizon. The velocities are not the true material velocities, but rather composites of all of the horizons above the particular reflecting horizon.

The correct interval velocities are obtained using the Dix analysis (Dix, 1956). The Dix process analyzes layer by layer to provide thickness and interval velocity for each horizon. The results from applying Dix analysis to the horizons observed in the data in Figure 7.19 are summarized in Table 7.2.

Table 7.2. Summary reflector intercept times and stacking velocities for semblance analysis in Figure 7.19, using Dix analysis deduced interval velocities and thicknesses

DIX ANALYSIS FOR GPR DATA						
Horizon Number	**Intercept Time (ns)**	**Stacking Velocity (m/ns)**	**Interval Thickness (m)**	**Dix Depth (m)**	**Interval Velocity (m/ns)**	**Topp Water Content**
0	0	0.000		0.00		
1	40	0.095	1.90	1.90	0.095	0.29
2	50	0.098	0.55	2.45	0.109	0.25
3	80	0.105	1.74	4.18	0.116	0.23

In both cases, the stacking velocities and the interval velocities can be used in association with the Topp equation to estimate the average water content throughout the section or for the indicated intervals.

7.6.3 BOREHOLE TRANSILLUMINATION

In a hydrogeological setting very similar to the Camp Borden site, borehole surveys were carried out to examine contamination in the subsurface and to examine water content in the vadose zone. A pair of boreholes were located 7 m apart, as depicted in Figure 7.20. At this particular site, the water table was at a depth of 17 m. Contamination (chloride ion rich leachate) in the groundwater system was coming from a landfill up groundwater flow direction.

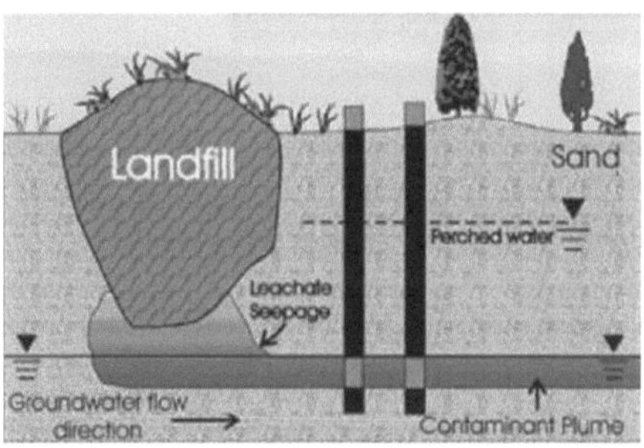

Figure 7.20. Simplified geological section at a landfill site

A quick evaluation of subsurface conditions to estimate water content and variations in subsurface conditions between boreholes was obtained using a zero offset profile. The results of this measurement are shown in Figure 7.21.

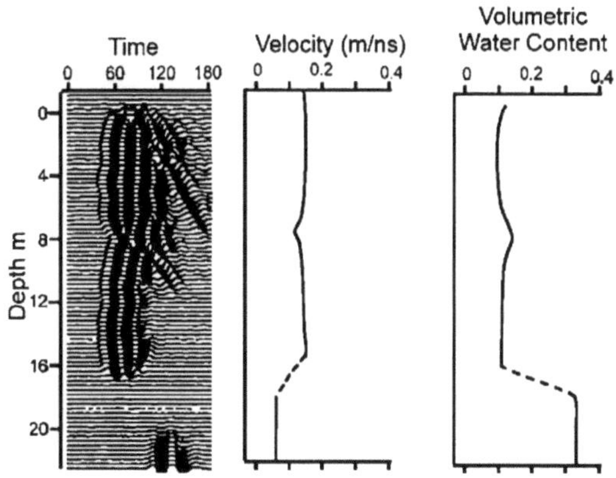

Figure 7.21. Result of zero offset profiling with GPR between two boreholes. The abrupt increase in travel time below 16 m indicates entry into the saturated zone. The loss of signal between 17 and 20 m depth indicates high loss associated with contaminants.

The key aspects of the data are the travel time and signal amplitude. Above the water table, the travel time is approximately half of what it is below the water table. Slight travel-time variations indicate localized water-content changes within the unsaturated zone. The blanking of the signal when the antennas enter the water-saturated zone (between 16 and 20 m) indicates the presence of contaminants in the groundwater, which greatly increase the electrical conductivity and hence the attenuation of the GPR signal. Below the contaminated zone, the signal becomes visible again.

Transformation of this data into water content is achieved by estimating velocity from the signal travel time and hole separation. Velocity is further translated into a first-order estimate of water content using the Topp relationship.

7.6.4 TIME-LAPSE MONITORING

The North Bay, Ontario, landfill site (another University of Waterloo hydrogeology research site) provides an excellent example of using GPR to monitor changes versus time. This glacial outwash area (Figure 7.22) was home to the local landfill. Over the years, leachates had drained out of the landfill and contaminated the groundwater.

The radar section shown in Figure 7.23a are the results of a 50 MHz GPR survey along a transect across the contaminant plume downstream from the landfill. The original data collected in 1985 show a high attenuation zone in the cross section.

Figure 7.22. GPR line at the North Bay landfill area, with 50 MHz antennas in foreground

Subsequent to the initial survey, a considerable amount of remediation work was carried out with pumping-well remediation strategies to minimize the contaminants entering the groundwater system. Several years later, a survey along the same transect showed much less attenuation and indicated the effectiveness of the remediation. The results are shown in Figure 7.23b.

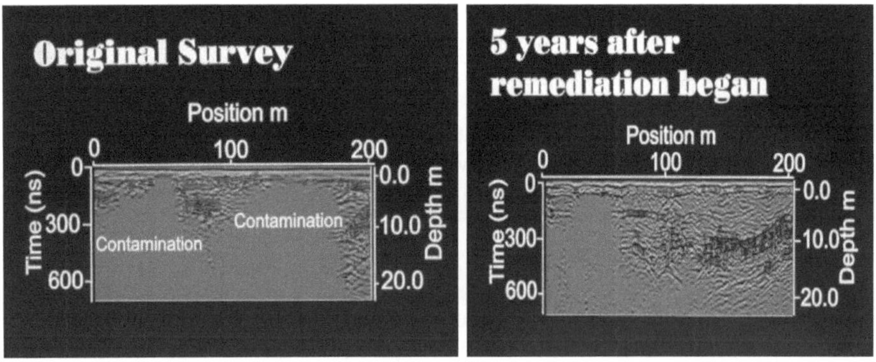

(a) **(b)**

Figure 7.23. (a) Initial 50 MHz GPR cross section along line shown in Figure 7.22, and (b) the same section 5 years later

The difference between the two sections indicates the impact of leachate on groundwater conductivity. Increased TDS (primarily chloride ions in the leachates) raise electrical conductivity, which substantially increases the attenuation of the GPR signals, resulting in the blanked-out GPR response.

7.7 Summary and Conclusions

GPR has many applications in the field of hydrogeology. The GPR wave properties are highly sensitive to the presence of water, as well as those factors that control water movement in the subsurface.

The technique is noninvasive, allows quick measurement of responses, and thus provides a means of remote detection and monitoring of variations in subsurface conditions. GPR effectiveness is limited in areas where signal attenuation is high. As a consequence, the method is highly site specific, and users must use discretion when selecting the method without detailed knowledge of a particular site. In general, environments that have low electrical conductivities are best for using GPR, whereas those with high electrical conductivity will have limited signal penetration (and are not good GPR sites).

Further understanding of the properties of mixtures is needed. Quantitative extraction of desired hydrogeologic information depends on further developments in this area.

While GPR continues to see expanded usage, it has not become a mainstream technique in the hydrogeology community, particularly for environmental site assessment and monitoring of clean up programs. This is because of the site-specific nature of responses and the need for extensive data analysis and interpretation. However, it is expected that the use of GPR will become more routinely used for hydrological investigations as experience with the technique increases.

References

Annan, A.P., Radio interferometery depth sounding: Part I—Theoretical discussion, *Geophysics, 38*, 557–580, 1973.

Annan, A.P., Practical processing of GPR data, *Proceedings of the Second Government Workshop on Ground Penetrating Radar*, Columbus, Ohio, October 1993.

Annan, A.P., Transmission dispersion and GPR, *JEEG, 0*, 125–136, January 1996.

Annan, A.P., The history of ground penetrating radar, *Subsurface Sensing Technologies and Applications, 3*(4), 303–320, October 2002.

Annan, A.P., Ground penetrating radar: Principles, procedures, & applications, *Sensors & Software Inc. Technical Paper*, 2003.

Annan, A.P., and S.W. Cosway, Ground penetrating radar survey design, *Proceedings of the Symposium on the Application of Geophysics to Engineering and Environmental Problems, SAGEEP'92*, Oakbrook, Illinois, pp. 329–351, April 26–29, 1992.

Annan, A.P., and S.W. Cosway, GPR frequency selection, *Proceedings of the Fifth International Conference on Ground Penetrating Radar (GPR'94), 2 of 3*, pp. 747–760, 1994.

Annan, A.P., and J.L. Davis, Methodology for radar transillumination experiments: Report of activities, *Geological Survey of Canada, Paper, 78-1B*, 107–110, 1978.

Archie, The electrical resistivity log as an aid in determining some reservoir characteristics, *Trans. AIME, 146*, 54–62, 1942.

Born, M., and E. Wolf, *Principles of Optics, 6th Edition*, Pergamon Press, 1980.

Brekhovskikh, L.M., *Waves in Layered Media*, New York, Academic Press, 1960.

Brewster, M.L., and A.P. Annan, Ground-penetrating radar monitoring of a controlled DNAPL release: 200 MHz radar, *Geophysics, 59*, 1211–1221, 1994.

Bristow, C.S., and H.M. Jol, *Ground Penetrating Radar in Sediments*, Geological Society, London, *Special Publications, 211*, 2002.

Davis, J.L., and A.P.Annan, Borehole radar sounding in CR-6, CR-7 and CR-8 at Chalk River, Ontario, *Technical Record TR-401*, Atomic Energy of Canada Ltd., 1986.

Davis, J.L., and A.P. Annan, Ground penetrating radar for high-resolution mapping of soil and rock stratigraphy, *Geophysical Prospecting, 37*, 531–551, 1989.

Dix, C.H., *Seismic Prospecting of Oil*, Harper, New York, 1956.

El Said, M.A.H., Geophysical prospection of underground water in the desert by means of electromagnetic interference fringes, *Proc. I.R.E., 44*, pp. 24–30 and 940, 1956.

Endres, A.L. and R. Knight, A theoretical treatment of the effect of microscopic fluid distribution on the dielectric properties of partially saturated rocks, *Geophysical Prospecting, 40*, 307–324, 1992.

Fisher, E., G.A. McMechan, and A.P. Annan, A.P., Acquisition and processing of wide-aperture ground penetrating radar data, *Geophysics, 57*, 495, 1992a.

Fisher, E., G.A. McMechan, A.P. Annan, and S.W. Cosway, Examples of reverse-time migration of single-channel, ground-penetrating radar profiles, *Geophysics, 57*, 577–586, 1992b.

Gilson, E.W., J.D. Redman, J.A. Pilon, and A.P. Annan, Near surface applications of borehole radar, *Proceedings of the Symposium on the Application of Geophysics to Engineering and Environmental Problems*, Keystone, Colorado, pp. 545–553, April 28–May 2, 1996.

Grasmueck, M., 3-D ground-penetrating radar applied to fracture imaging in gneiss, *Geophysics, 61*(4), 1050–1064, 1996.

Greaves, R.J., D.P. Lesmes, J.M. Lee, and M.N. Toksoz, Velocity variation and water content estimated from multi-offset, ground penetrating radar, *Geophysics, 61*, 683–695, 1996.

Green, A., K. Holliger, H. Horstmeyer, H. Maurer, J. Tronicke, and J. van der Kruk, 3-D acquisition, processing and imaging of ground penetrating radar data, Tutorial 2 Notes, *Proceedings of the Ninth International Conference on Ground Penetrating Radar* (GPR 2002), Santa Barbara, California, April 29–May 2, 2002.

Hasted, J.B., Liquid water-dielectric properties, In: *Water: A Comprehensive Treatise, 1, The Physics and Physical Chemistry of Water*, F. Franks, ed., Plenum Press, NY, pp. 255–310, 1972.

Heincke, B., T. Spillman, H. Horstmeyer, and A.G. Green, A.G., 3-D georadar surveying in areas of moderate topographic relief, *Proceedings of the Ninth International Conference on Ground Penetrating Radar*, S.K. Koppenjan and H. Lee, eds., *Proceedings of SPIE Vol. 4758* , 2002.

Hubbard, S., J.E. Peterson, E.L. Majer, P.T. Zawislanski, J. Rovers, K.H. Williams, and F. Wobber, Estimation of permeable pathways and water content using tomographic radar data, *The Leading Edge of Exploration, 16*, 1623–1628, 1997.

Irving, J.D. and R.J. Knight, Removal of wavelet dispersion from ground-penetrating radar data, *Geophysics, 68*(3), 960–970, 2003.

Jackson, J.D., *Classical Electrodynamics*, John Wiley and Sons, 1962.

Lehmann, F. and A.G. Green, Semiautomated georadar data acquisition in three dimensions, *Geophysics, 64*(3), 719–731, 1999.

Lehmann, F., and A.G. Green, Topographic migration of georadar data: Implications for acquisition and processing, *Geophysics, 65*(3), 836–848, 2000.

McMechan, G.A., G.C. Gaynor, and R.B. Szerbiak, Use of ground-penetrating radar for 3-D sedimentological characterization of clastic reservoir analogs, *Geophysics, 62*, 786-796, 1997.

Olhoeft, G.R., Electrical properties of rocks, In: *Physical Properties of Rocks and Minerals, Vol. II*, Y.S. Touloukian, W.R. Judd, and R.F. Roy, eds., McGraw-Hill, 1981.

Olhoeft, G.R., Electrical properties from 10^{-3} to 10^{+9} Hz—Physics and chemistry, *Proceedings of the 2nd International Symposium on the Physics and Chemistry of Porous Media, 154*, American Institute of Physics Conference Proceedings, pp. 281–298, 1987.

Olhoeft, G.R., 1988, Interpretation of hole-to-hole radar measurements, *Proceedings of the Third Technical Symposium on Tunnel Detection*, pp. 616–629, Golden, Colorado, January 12–15, 1988.

Olsson, O., L. Falk, O. Forslund, L. Lundmark, and E. Sandberg, Borehole radar applied to the characterization of hydraulically conductive fracture zones in crystalline rock, *Geophysical Prospecting, 40*, 109–142, 1992.

Owen, T.R., Cavity detection using VHF hole to hole electromagnetic techniques, *Proceedings of the Second Tunnel Detection Symposium*, Colorado School of Mines, Golden, Colorado, U.S. Army MERADOM, Ft. Belvoir, Virginia, pp. 126–141, July 21–23, 1981.

Parkin, G., J.D. Redman, P. von Bertoldi, and Z. Zhang, Measurement of soil water content below a wastewater trench using ground penetrating radar, *Water Resour. Res., 36*(8), 2147–2154, 2000.

Redman, J.D., S.M. DeRyck, and A.P. Annan, Detection of LNAPL pools with GPR: Theoretical modelling and surveys of controlled spill, *Proceedings of the Fifth International Conference n Ground Penetrating Radar*, Kitchener, Ontario, Canada, June 12–14, 1994.

Redman, J.D., G. Parkin, and A.P. Annan, Borehole GPR measurement of soil water content during an infiltration experiment, Proceedings on Ground Penetrating Radar Conference, pp. 501–505, Gold Coast, Australia, May 23–May 26, 2000.

Roberts, R.L. and J.J. Daniels, Analysis of GPR polarization phenomena, *JEEG, 1*(2), 139–157, 1996.

Santamarina, J.C., K. Klein, and A. Fam, *Soils and Waves: Particulate Materials Behavior, Characterization and Process Monitoring*, John Wiley & Sons, 2001.

Sen, P.N., C. Scala, and M.H. Cohen, A self-similar model for sedimentary rocks with application to the dielectric constant of fused glass beads, *Geophysics, 46*, 781–795, 1981.

Sigurdsson, T., and T. Overgaard, Application of GPR for 3-D visualization of geological and structural variation in a limestone formation, *Proceedings of the Sixth International Conference on Ground Penetrating Radar (GPR'96)*, Sendai, Japan, September 30–October 3,1996.

Sihvola, A.H., How strict are theoretical bounds for dielectric properties of mixtures?, *IEEE Trans. On Geoscience & Remote Sensing, 40*, 880–886, 2002.

Smythe, W.R., *Static & Dynamic Electricity*, Taylor & Francis, A SUMMA Book, 1989.

Sommerfeld, *Partial Differential Equations in Physics*, New York, Academic Press, 1949.

Tillard, T., and J.-C. Dubois, Influence and lithology on radar echoes: Analysis with respect to electromagnetic parameters and rock anisotropy, Fourth International Conference on Ground Penetrating Radar Rovaniemi, Finland, June 8–13, 1992; *Geological Survey of Finland, Special Paper 16*, 1992.

Todoeschuck, J.P., P.T. Lafleche, O.G. Jensen, A.S. Judge, and J.A. Pilon, Deconvolution of ground probing radar data, In: *Ground Penetrating Radar*, J. Pilon; ed., Geological Survey of Canada, *Paper 90-4*, pp. 227–230, 1992.

Turner, G., Propagation deconvolution, Fourth International Conference on Ground Penetrating Radar June 8–13, 1992, Rovaniemi, Finland, *Geological Survey of Finland, Special Paper 16*, 1992.

Topp, G.C., J.L. Davis, and A.P. Annan, Electromagnetic determination of soil water content: Measurements in coaxial transmission lines, *Water Resour. Res., 16*(3), 574–582, 1980.

van Overmeeren, R.A., Radar facies of unconsolidated sediments in The Netherlands: A radar stratigraphy interpretation method for hydrogeology, *Journal of Applied Geophysics, 40*, 1–18, 1998.

Wait, J.R., *Electromagnetic Waves in Stratified Media, 2nd edition*: New York, Macmillan, 1970.

Wharton, R. P., G.A. Hazen, R.N. Rau, and D.L. Best, Advancements in electromagnetic propagation logging, *Society Petroleum Engineering, Paper 9041*, 1980.

White, R.E., Properties of instantaneous seismic attributes, *The Leading Edge, 10*(7), 26–32, Society of Exploration Geophysicists, 1991.

Yilmaz, O., Seismic data analysis–Processing, inversion, and interpretation of seismic data, *Society of Exploration Geophysicists J.*, 2027, 2000.

Ground Penetrating Radar Conference References

Proceedings of the International Workshop on the Remote Estimation of Sea Ice Thickness Centre for Cold Ocean Resources Engineering (C-CORE), St. John's Newfoundland, Sept. 25–26, 1979.

Proceedings of the Ground Penetrating Radar Workshop, Geological Survey of Canada, Ottawa, Ontario, Canada, May 24–26, 1988, GSC Paper 90-4.

Abstracts of the Third International Conference on Ground Penetrating Radar, United States Geological Survey, Lakewood, Colorado, USA, May 24–26, 1990.

Proceedings of the Fourth International Conference on Ground Penetrating Radar, Geological Survey of Finland, Rovaniemi, Finland, June 8–13, 1992.

Proceedings of the Fifth International Conference on Ground Penetrating Radar (GPR'94), Kitchener, Ontario, Canada, June 12–16, 1994.

Proceedings of the Sixth International Conference on Ground Penetrating Radar (GPR '96), Sendai, Japan, Sept. 30–Oct. 3, 1996.

Proceedings of the Seventh International Conference on Ground Penetrating Radar (GPR '98), Lawrence, Kansas, USA, May 27–30, 1998.

Proceedings of the Eighth International Conference on Ground Penetrating Radar (GPR 2000), Goldcoast, Australia, May 23–26, 2000, SPIE 4084.

Proceedings of the Ninth International Conference on Ground Penetrating Radar (GPR 2002), Santa Barbara, California, April 29–May 2, 2002.

8 SHALLOW SEISMIC METHODS

DON W. STEEPLES

The University of Kansas, Department of Geology
1475 Jayhawk Blvd.,120 Lindley Hall, Lawrence, Kansas 66045-7613 USA

This chapter covers the fundamentals of near-surface seismic techniques, including refraction, reflection, borehole, and surface-wave methods. Typical seismic applications to hydrogeology include examining sedimentology and stratigraphy, detecting geologic faults, evaluating karst conditions, and mapping the top of bedrock and the base of landslides. Other less common but promising applications include mapping hydrogeological features, assessing hydrological characteristics, measuring the depth to the water table, and assessing the decay of infrastructure (e.g., leaking pipes, storage tanks, and disposal containers).

Links between seismic attributes and hydrogeological parameters are discussed in Chapter 9 of this volume, and case studies using seismic data at various scales are provided in Chapter 12 (regional), Chapter 13 (local), and Chapter 15 (laboratory).

8.1 Foundations of Near-Surface Seismology

Historically, the primary wave (or *P*-wave,) which is a compressional sound wave, has been the type of seismic wave used in near-surface seismic surveying. However, when additional wave types are analyzed as well, supplementary information about the mechanical properties of near-surface geological materials becomes available, although at the cost of invoking more complicated procedures.

To initiate a seismic survey, sources of seismic energy such as sledgehammers, explosives, and vibration devices are used to transmit mechanical waves into the earth. Specialized microphones in the ground known as *geophones* detect the elastic-wave signals generated by the seismic sources. These signals then are transmitted to a *seismograph*, where they are recorded.

In some ways, the modern stereophonic music system is analogous to a seismograph. Like a stereo, a seismograph contains an amplifier, filters that are comparable to treble and bass controls, an analog-to-digital (A/D) converter, and the capacity to store and play back data. Both the fidelity of a stereo and the "quality" of a seismograph are related to *dynamic range*, defined by Sheriff (1991) as "the ratio of the maximum reading to the minimum reading (often noise level) which can be recorded by and read from an instrument or system without change of scale." Thus, the better the dynamic range of a seismograph, the more potential it has to detect subtle wave characteristics,

Y. Rubin and S.S. Hubbard (eds.), Hydrogeophysics, 215–251.

just as a stereo system with a large dynamic range and little distortion replicates musical subtleties with greater fidelity. Also, a seismograph with a large dynamic range is capable of recording small seismic signals in the presence of large amounts of mechanical and electrical *noise* or interference. The dynamic range of a modern seismograph is limited primarily by the noise present in the wiring, geophones, and connectors rather than by any noise inherent in the instrument itself (Owen et al., 1988). High-quality seismographs are readily available and can be used to "listen" to several dozen geophones simultaneously.

8.1.1 SEISMIC WAVE PROPAGATION

Seismic waves include *body waves*, which travel three-dimensionally through solid volumes, and *surface waves,* which travel near the surface of the earth volume in which they propagate. Body waves include *P*-waves and *S*-waves; surface waves are categorized further as *Love waves* and *Rayleigh waves* (Figure 8.1). Because of their greater speed, *P*-waves are the first waves recorded on *seismograms*. Secondary or *S*-waves (shear waves) are distortional, and they typically travel only about 40–60% as fast as *P*-waves. Surface waves are able to travel only within a few seismic wavelengths of the surface of a solid. Often, the term *ground roll* is used to describe Rayleigh waves. Because of our everyday familiarity with sound waves, *P*-waves are the easiest wave type for us to comprehend conceptually.

Since the 1920s, seismic reflection techniques have been used to search for petroleum, and *intercept-time* (ITM) refraction techniques have been used in engineering applications. But since 1980, significant strides have been made in both near-surface *P*-wave seismic reflection surveying (Hunter et al., 1984) and in the development of shallow-seismic refraction methods (Palmer, 1981; Lankston, 1989). Although the use of shallow *S*-wave reflections has not been widespread, examples appear in the refereed literature (e.g., Hasbrouck, 1991; Goforth and Hayward, 1992). The theoretical basis for shallow surface-wave work was developed in the early 1960s (Jones, 1962), but the techniques did not really catch on until the 1980s (e.g., Stokoe et al., 1994). When seismic methods are used for hydrogeology, they should be supplemented by other geophysical and geological practices and supported by verification drilling.

8.1.1.1 *Basic Review of Wave Physics*

In a discussion of seismic waves, the *propagation speed* of a wave, along with the kind, amount, and frequency content of the displacements associated with it, are key measurements. The maximum extent of the peaks and troughs of the waves in seismic data is referred to as *wave amplitude* and usually is designated *A*. Waves are *coherent* when they are arrayed parallel to one another and *incoherent* or *out-of-phase* when the wave train is out of symmetry, as shown in Figure 8.2.

The energy of a seismic wave is proportional to the square of its amplitude. Two important equations relate the frequency of seismic wave data to space and time.

Wavelength (λ) is used to describe the distance in space between successive peaks (or troughs) of waves in the earth or other media. Seismic resolution (i.e., the capacity to discriminate geologic features) is dependent on the wavelength of the data. Widess (1973) showed that the geologic bed-thickness resolution obtainable with seismic reflection methods is no better than λ/4.

Frequency (*f*) is the reciprocal of the period (*T*) between successive wave peaks or troughs. The *propagation velocity* of a wave (more properly termed its *propagation speed*, but seismological custom favors the term *velocity*) is the product of its frequency and wavelength:

$$V = f\lambda \tag{8.1}$$

Because seismic wave velocity is a physical property intrinsic to earth materials, in seismic data, higher frequencies result in shorter wavelengths and better resolution.

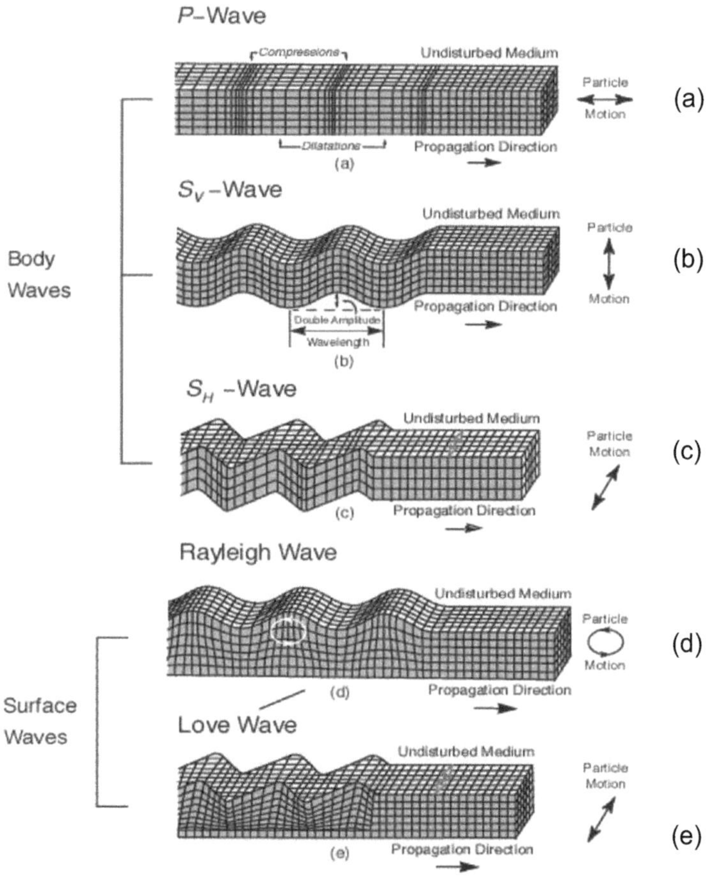

Figure 8.1. Types of waves: (a) Depiction of *P*-waves traveling in a solid medium;(b) the vertical component of *S*-waves traveling in a solid medium; (c) S_H-waves traveling in a solid medium (d) Rayleigh waves traveling along a section of the earth's surface; and (e) Love waves traveling along a section of the earth's surface; (a-e adapted from *Earthquakes*, by Bruce A. Bolt, 1993 [revised], Freeman and Co., with permission).

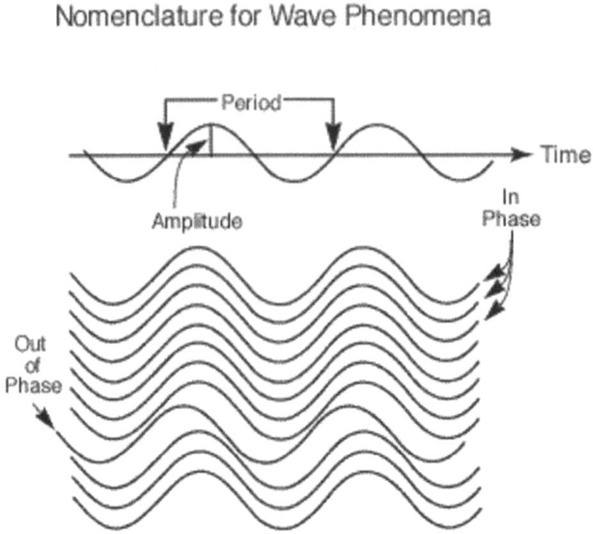

Figure 8.2. The terminology used here refers to a wavefield at a particular location, measured as a function of time. The schematic could equally well represent displace-ment as a function of space, as with wind-driven waves on the surface of a pond. In that case, displacement would be seen as a function of horizontal location at a specific time, *wavelength* would replace *period*, and *distance* would replace *time* in this figure.

Based on the preceding relationships, a seismogram can be used to make several helpful measurements, particularly when the seismogram contains multiple channels of data recorded by sensors whose locations are known relative to the seismic source. Multiple sensors are useful in establishing the directionality of a signal and in separating signals from noise. To illustrate, consider that humans can identify the direction of a sound much more accurately when they use both ears as opposed to when they use only one. Input from two ears allows the brain to process differences in the arrival times (*phase differences*) of sounds as well as differences in loudness (amplitude).

8.1.1.2 *The Wave Equation*

A complete solution of the acoustic and seismic wave problems requires the use of the *wave equation*, shown below in one-dimensional form, where u is the displacement (parallel to the propagation direction for P-waves and perpendicular to the propagation direction for S-waves) caused by the wave as a function of time t and location x:

$$\frac{\partial^2 u}{\partial x^2} = \frac{1}{V^2}\frac{\partial^2 u}{\partial t^2}. \tag{8.2}$$

A useful particular solution of the form $u = A \sin k (Vt - x)$ can be verified by differentiation, where A is amplitude, k is wavenumber, kV is frequency, and $-kx$ represents phase.

8.1.1.3 *Types of Seismic Waves*

Body waves spread three-dimensionally through a volume as either *P*-waves or *S*-waves. The velocity with which *P*-waves propagate is given by

$$V_P = \sqrt{\frac{K + \frac{4}{3}\mu}{\rho}}. \tag{8.3}$$

In this equation, K is the *bulk modulus*, μ is the *shear modulus* (both moduli are expressed as force per unit area), and ρ is the mass density of the material through which the wave is propagating.

In isotropic materials, travel paths for *S*-waves are the same as they are for *P*-waves. Commonly, *S*-waves are used in geotechnical engineering projects, particularly between boreholes, to establish the shear modulus of soils and foundation materials. The equation for *S*-wave velocity is

$$V_S = \sqrt{\frac{\mu}{\rho}}. \tag{8.4}$$

Fluids have no shear strength; therefore, their shear modulus is zero. As a result, *S*-waves cannot propagate through fluids. Also, *S*-wave velocity is relatively insensitive to water saturation in unconsolidated earth materials, whereas *P*-wave velocity is noticeably affected by saturation in the same materials. Shear waves exhibit polarization when reflected or refracted, much as light does.

On a seismogram, surface waves usually produce the largest amplitudes. As noted in Figure 8.1, Rayleigh waves exhibit a retrograde elliptical particle motion, similar to the movement of a cork or fishing bobber that is moved along by gentle waves on the surface of a lake. Love waves are *S*-waves that are reflected multiple times while trapped in a surface layer, and they require a low-velocity layer at the earth's surface in which to propagate. Interference from Love waves can be a major source of difficulties when data from shallow *S*-wave reflection surveys are being analyzed.

8.1.1.4 *Ray Theory*

The wave equation is used when a rigorous analysis of seismic waves is desired, whereas *ray theory* can be used qualitatively in several conceptual areas, including reflection, refraction, interference, diffraction, and scattering.

The idea of seismic reflection can be explained through the familiar analogy of sound echoes. Ray theory is useful in the construction of diagrams that indicate where in the subsurface a reflection will travel when seismic velocity and mass density are known in three dimensions (Figure 8.3). In terms of practical analysis, near-surface seismic techniques measure the elastic deformation that takes place several meters or more from the seismic source. Very near the source, *plastic deformation* or *fracture* sometimes occurs, and in that case, classical seismic analysis does not apply. Under *elastic conditions*, a body can undergo repeated cycles of deformation without sustaining permanent damage. When deformation exceeds the limits of elasticity, however, the high-frequency content of the seismic does not develop properly. Consequently, to produce high frequencies, small seismic sources are required.

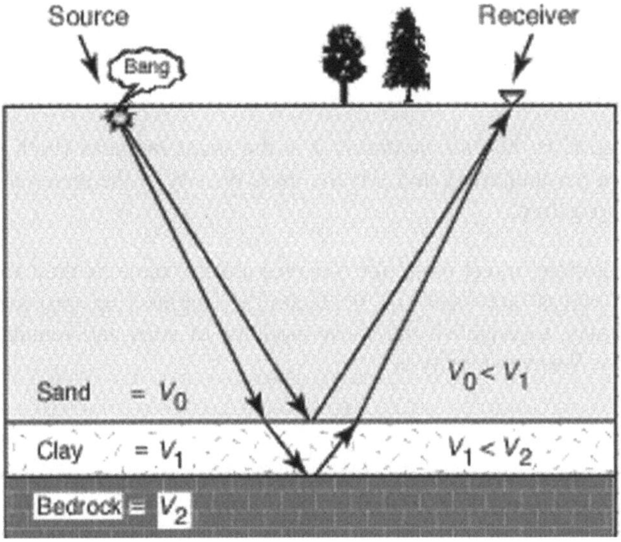

Figure 8.3. A reflection raypath in the case of three layers. In this depiction, velocity (V) increases with each successive layer.

For the purposes of hydrogeological investigation, seismic wave analysis can be reduced to several simple questions:

(1) *How fast do the seismic waves travel, both in the absolute sense and in relation to other waves found on the seismogram?* Absolute velocity provides a clue to the type of material that the waves have traversed and whether the material is water-saturated. Relative velocities help to identify wave types as well as the pathway they have traveled between source and receiver.

(2) *How quickly do the waves attenuate with distance?* The answer to this question can provide information about the hardness or rigidity of the material traversed by the waves. *P*-waves attenuate faster in unsaturated materials than in saturated materials, whereas *S*-wave attenuation usually is not affected greatly by the degree of saturation.

(3) *How much time elapses between the initiation of the source pulse and the arrival of the signal at a specific receiver?* When the velocity of the wave and the shotpoint and receiver locations are known for several of the receivers, the seismologist can determine how deeply the wave has penetrated and whether it has been refracted, reflected, or otherwise changed as it traversed its underground pathway. This information is useful in determining the depth to bedrock, for example.

(4) *In what vector direction and with what type of particle motion does the wave propagate?* These factors can reveal the type of wave under examination, along with whether it is a reflection, refraction, or surface wave. For example, *vector recording* can be used to identify *seismic anisotropy*—a variation of seismic velocity that is dependent on propagation direction—which can be diagnostic of rock-fracture directions and patterns in aquifers.

8.1.1.5 *Mechanisms of Seismic-Energy Loss*

Loss mechanisms that cause decreasing amplitude with increasing distance from the seismic source include *geometrical spreading* and *attenuation*. When we visualize the energy radiating away from a seismic source as a homogeneous, time-dependent sphere that is increasing in radius over time, the energy density of a wavefront will fall off as R^2. Because energy is proportional to amplitude squared, the amplitude decreases as $1/R$ from spherical spreading. This fall-off effect is known as *spherical divergence*, or *geometrical spreading.*

To envision another type of geometrical spreading, consider the case of a rock tossed into a pool of water. The wavefront that forms when the rock strikes the water is ring-shaped rather than spherical, and the energy density of this wavefront falls off as $1/R$. The amplitude decreases as $1\sqrt{R}$ instead of as $1/R$, as it does in the ground. In the case of surface waves, geometrical spreading is analogous to the rock-in-the-water wavefront rather than to the wavefront in a three-dimensional earth volume. Because the loss occurs in two dimensions rather than three, surface-wave energy decreases proportionally to $1/R$ instead of $1/R^2$. The effect in terms of wave amplitude is a decrease as $1\sqrt{R}$. This gives surface waves an advantage over body waves with increasing distance, in terms of relative amplitude.

Additional losses occur through attenuation, which can provide information about the type and condition of earth materials through which seismic waves pass. Attenuation in common earth materials obeys the following equation:

$$A_x = A_0 e^{-\alpha x}, \tag{8.5}$$

where A_0 is the reference amplitude of a wave mode measured at an arbitrary reference distance, α is the attenuation factor, and A_x is the amplitude of the same wave mode at distance x. Attenuation is frequency-dependent; therefore, it is often expressed in terms of λ, so that the "quality factor," i.e., the letter Q, is explicit:

$$Q = \frac{\pi}{\alpha\lambda}. \tag{8.6}$$

The dimensionless number Q is sometimes called the *absorption constant.* The reciprocal of Q represents the fraction of energy dissipated during propagation over a distance of one wavelength. For example, soil or loose sand realistically could have a Q of 10 (Table 8.1). This means that 10% of the energy is lost for each wavelength of propagation. Note that this does not mean that all of the energy has disappeared within ten wavelengths. It means that for each successive wavelength of propagation, 10% of the remainder disappears.

At higher frequencies, wavelengths are shorter; thus, higher frequencies become attenuated more quickly. For example, a lightning bolt striking nearby produces what is perceived as a high-pitched cracking sound, but because of attenuation, the sound of distant thunder is low and rumbling. The rapid loss of high frequencies in near-surface earth materials is probably the largest single technical barrier to the widespread use of seismic methods as effective tools in the search for submeter-scale anomalies in near-surface materials.

Table 8.1. Typical *Q*-Values

Clay and sand	05–25
Sandstone	10–50
Shale	25–75
Granite	55–130
Limestone	50–180

Several interference phenomena can alter the displayed shape of the signal on a seismogram. These include *multiple reflections*, *diffractions*, and *scattering*. Also included are the effects of *direct waves*, *sound waves* that arrive through the air, *interference with refractions*, and *interference with surface waves*.

8.1.2 SEISMIC WAVES IN LAYERED MEDIA: SEISMOGRAM ANALYSIS

Classical petroleum-exploration seismology employs seismic wavelengths of several tens of meters. In contrast, near-surface seismic methods must use wavelengths only a few meters in length to obtain the necessary resolution. Higher-than-normal resolution is possible because high frequencies are still present near the seismic source and because of the low seismic-wave velocities characteristic of near-surface materials.

In a practical sense, when geophones and a seismic source are placed on the surface of the earth, the upper frequency limit appears to be about 1 kHz. In the ground, seismic *P*-wave velocities are commonly a few hundred meters per second, so when *P*-wave methods are used, the wavelength limit is about 1 m. The wavelengths available for *S*-waves and surface waves typically are no shorter than 1/2 meter. To be seismically detectable, at least two dimensions within the feature of interest need to be equivalent to or larger than one seismic wavelength, as required by the *Rayleigh criterion*.

8.1.2.1 *Energy Partitioning and Mode Conversion*

When a source of seismic energy is located beneath the earth's surface, the energy it generates radiates in all directions, eventually striking a geological interface at which refraction or reflection can occur. Two significant phenomena are related to seismic wave refraction at the interface between two layers in the earth. First, at an angle of incidence called the *critical angle*, the refracted wavefront propagates along the interface rather than into the medium itself. The refracted wavefront that results is called a *head wave*. Secondly, when a head wave propagates along an interface, each point on the interface serves as a new source of waves that propagate to the earth's surface according to *Fermat's principle of least time* and *Snell's law*.

When the seismic refraction method is used, the waves emanating from the head wave at the subsurface interface are recorded at the earth's surface. When the wave-propagation velocity on the underside of an interface is greater than it is in the overlying layer, the theoretical basis of the refraction method is satisfied. When the velocity of the propagating seismic wave decreases as the wave enters the lower layer, however, the wavefront is refracted away from the interface instead of toward it. Thus, seismic energy does not return to the earth's surface unless it reaches a higher-velocity layer at some greater depth. In such a case, the low-velocity layer will remain seismically invisible, and the calculated depth to the high-velocity layer below it will be incorrect. Under circumstances such as these, the refraction method will not work, and attempts to use it will yield an inaccurate thickness calculation for the overlying layer.

In addition to the refraction of seismic energy, some of the energy changes from *P*-waves into *S*-waves when it encounters acoustic interfaces. This change from one wave type to another is called *mode conversion*, which is a phenomenon that can become important when source-to-receiver distances are great in comparison to reflector depth. Waves that strike acoustic interfaces at angles of incidence other than approximately 90° produce a predictably complex seismic wavefield. In this situation, at least six things happen, collectively representing a partitioning of seismic wave energy:

(1) Reflected *P*-waves return to the surface with an angle of incidence equal to the angle of reflection.
(2) Refracted *P*-waves propagate as head waves along velocity-contrast interfaces and return to the surface according to Snell's law.
(3) Transmitted *P*-waves pass into the underlying medium.
(4) *P*-waves that have undergone mode conversion can be reflected as *S*-waves.
(5) Mode-converted waves can pass along the interface as head waves and be refracted to the surface as *S*-waves, according to Snell's law.
(6) Transmitted *P*-waves can be mode-converted at the interface and propagate into the lower medium, where they travel as *S*-waves.

The amplitudes of these six types of waves can be calculated using the *Zoeppritz equations,* which contain complex trigonometric relationships. These equations can be found, along with relevant mathematical and technical discussions, in many advanced seismology textbooks, as well as in Sheriff (1991).

8.1.2.2 *Reflections from Near-Vertical Incidence*

In the case of reflection seismology, the *polarity* of the reflected wave (i.e., whether the first motion of the wave is upward or downward) provides information about the geologic interface giving rise to the reflection. If the interface represents a transition from shale to limestone, for example, the polarity of the reflected wave will be the same as that of the wave that strikes the interface. If the transition is from limestone to shale, however, the polarity will be reversed, as discussed below.

Let us assume that seismic waves are reflected vertically from horizontal interfaces at some depth in the earth. The amount of seismic energy that crosses an interface depends upon the acoustic contrast at that interface. The product of velocity and density of a particular i^{th} layer is the acoustic impedance Z_i so that

$$Z_i = \rho_i V_i, \quad (8.7)$$

where ρ_i is density, and V_i is the seismic velocity in the ith layer.

The strength of a reflection from an acoustic-contrast interface depends upon the reflection coefficient R at normal incidence, which is related to the acoustic impedances by

$$R = \frac{Z_2 - Z_1}{Z_2 + Z_1}, \quad (8.8)$$

where Z_1 is the acoustic impedance of the first layer, and Z_2 is the acoustic impedance of the second layer.

Both the polarity and the amplitude of the normal-incidence reflection can be predicted from the reflection coefficient. When the acoustic impedance of the first layer is greater than that of the second layer, R will be negative, and the reflection will return to the surface with its polarity reversed. For example, when the second layer is composed of air, as would be the case with an air-filled cavity (where ρ_2 is approximately zero), reflection would be nearly total, and polarity would be reversed. Notice, too, that when the acoustic impedance of Layer 1 is the same as it is for Layer 2, there will be no reflection.

8.1.3 PHYSICAL PROPERTIES

Among the physical properties that can be measured by seismic methods, the most important are the *seismic velocity* and *attenuation* of the various wave types. The look of a seismogram depends in large part on these two properties, although *density* also affects seismic wave propagation. As can be seen in Equations (8.3), (8.4), and the combination of Equations (8.7) and (8.8), density affects seismic *P*- and *S*-wave velocities as well as the reflectivity that exists across geologic boundaries. However, the seismic method is not a good means of measuring density.

8.1.3.1 *Measuring Velocities in the Vertical and Horizontal Directions*

Vertical-direction velocity can be measured in several ways. The most accurate method is to place a seismic source at the earth's surface and then to position a geophone inside a borehole. The geophone thus can be moved to various depths, and the one-way travel times to the geophone inside the borehole can be measured. An alternative would be to place the seismic source in the borehole and position the receiver at the surface.

The disadvantage of discharging a source within the borehole is that a powerful source can compromise the integrity of the borehole. Either of the borehole techniques yields an average velocity to a particular depth, but in these methods only the first arrival of seismic energy is used, in contrast to the vertical seismic profiling (VSP) technique (see Section 8.3.5 for a discussion of VSP).

Borehole sonic logging is another method of calculating velocity variations in the vicinity of the borehole in the vertical direction. A *sonic log* provides velocity information over distances of about 1 m, using frequencies of about 20 kHz. By integrating velocities over many meter-long segments, an average-velocity estimate can be obtained to various depths in the logged section of the borehole. Sonic-log frequencies produce wavelengths that are about two orders of magnitude smaller than those available in surface-seismic data; therefore, sonic logs may sample different rock volumes and yield somewhat different velocities. A discussion of sonic logging is given in Chapter 10 of this volume.

Seismic velocities in the vertical direction can be calculated from the reflected and refracted waves that appear on seismograms. However, computations are less accurate when sources and receivers are at the surface than when either the source or the receivers are placed within a borehole that extends to the depth of interest.

Velocities in the horizontal direction at depths of more than a few meters are measured most accurately by *crosshole surveys*. The seismic source must be placed in one borehole and the receivers in another, some distance apart. The locations of both the source and the receivers with respect to their three-dimensional locations within the boreholes must be known to an accuracy of a few centimeters.

8.1.3.2 *Typical Values*

Typical seismic body-wave velocities for various earth materials are shown in Table 8.2. One common misconception about seismic velocities is that the velocity of P-waves in earth materials cannot be lower than the velocity of sound waves passing through air (335 m/s). In fact, near-surface materials sometimes have seismic velocities lower than the velocity of sound in air; indeed, they may be as low as 130 m/s (Bachrach and Nur, 1998; Bachrach et al., 1998). As noted in Equation (8.3), the velocity of seismic P-waves depends upon two *elastic moduli*, but it is inversely dependent upon density. Unconsolidated materials can have small moduli relative to their densities. Loose soils are easily compressed and have little shear strength, but they have a mass density several orders of magnitude larger than that of air. The geological and physical constraints upon velocity are discussed in Chapter 9 of this volume.

Table 8.2. Seismic velocities of typical materials

Medium	P-Wave Velocity (m/s)[a]	S-Wave Velocity (V_s) (m/s)[b]	Density (ρ) (g/cm^3)[c]
Air	335	0	0
Soil	100 - 600	--	1.6
Dry sand, gravel	200 - 1000	400	1.6
Wet sand gravel	600 - 1800	500	1.9
Clay	1000 - 2200	--	1.5
Till	1600 - 2200	--	1.7
Water	1500 - 1600	0	1.0
Solid rock	1600 - 6500	--	2.3 - 2.8

a. Modified from Press, Frank, 1966, "Seismic Velocities," in *Handbook of Physical Constants*, Sydney P. Clark, Jr., Editor, GSA Memoir 97, 197-221.
b. Press, Frank, 1966, "Seismic Velocities," in *Handbook of Physical Constants*, Sydney P. Clark, Jr., Editor, GSA Memoir 97, 197-221.
c. Daly, R. A., Manger, G. E., and Clark, S. P., Jr., 1966, "Density of Rocks," in *Handbook of Physical Constants*, Sydney P. Clark, Jr., Editor, GSA Memoir 97, 19-26.

8.2 Seismic Refraction

As seismic body waves penetrate the earth, some of their energy strikes geological layers beneath the surface and then is refracted back to the surface. In near-surface studies, refraction methods often are used to calculate the depth to these layers along with the seismic velocity within the layers. Thus, refraction methods are useful in providing an assessment of gross layering geometry as well as the state of the layer, based on seismic velocities, provided that seismic velocity increases monotonically with depth.

Seismic refraction has a long history of use in engineering geophysics. Refraction surveys devised for civil-engineering projects were reported as early as the 1930s, including bridge-foundation studies (i.e., depth to bedrock) and groundwater exploration. More recently, refraction has been used to characterize shallow subsurface geological conditions. Specifically, refraction methods have been used to map the depth to bedrock at polluted sites that require environmental cleanup. Refraction also can be used to examine soil conditions, which are helpful to know during the foundation-design phase of structures such as dams and large buildings.

8.2.1 FUNDAMENTALS OF THE REFRACTION METHODS

All seismic refraction surveys involve a seismic wave source and several receivers. After a data set has been collected, data can be processed and interpreted in many ways.

One of the techniques used to interpret refracted-wave energy is the *intercept-time method* (ITM; see Section 8.1.1). During most of the 20th century, ITM was the technique of choice for seismic *P*-wave refraction studies. Arguably, except for the simplest two-layer cases, it has since been superseded by more advanced refraction-interpretation methods. To apply ITM, a human interpreter must fit straight lines to groupings of first-arrival *P*-waves (Figure 8.4). The seismic *P*-wave velocities of the layers are the reciprocals of the slopes of the lines (Figure 8.5). The thickness of the layer(s) of interest is determined by the velocities combined with the time-intercept points at which the velocity lines cross the zero-distance axis of the seismograms (see Equation 8.9).

One of the limitations of ITM is that the velocity lines of the first arrivals are established by two arrival times; i.e., an interpreter can draw a straight velocity line on a seismogram on the basis of only two arrival times. More than two arrival times generally are available for each layer, but the additional information often is not used to advantage with ITM. At best, the extra data points are used to perform a least-squares fit of the velocity line to the first-arrival travel times as a function of distance. Sometimes these additional pieces of information are used to determine standard error.

The *generalized reciprocal method* (GRM) of refraction interpretation constitutes an improvement over ITM partly because GRM uses all of the first-arrival data points recorded for each layer. A discussion of GRM can be found in Palmer (1980; 1981) and Lankston (1990). Note that field methods remain the same whether GRM or ITM is selected for use in refraction studies. The main difference among the refraction methods lies in interpretation and data analysis.

8.2.2 THEORETICAL FOUNDATIONS

8.2.2.1 *Basis for the Refraction Method*

Wavefronts can change direction abruptly when they strike an interface at which the propagation velocity of the wave undergoes alteration. We see examples of this phenomenon in everyday life. Consider that a linear object such as a ruler, when partially submerged in a bucket of water, appears to bend abruptly where it meets the water's surface when the viewer is outside the plane of the ruler. The relative wavefront velocities on opposite sides of the interface determine the angle of bending according to *Snell's law*. The same phenomenon occurs when seismic waves cross an interface where media of different seismic wave velocities meet.

8.2.2.2 *Intercept-Time Method Refraction Equations*

Ray theory is used in refraction interpretations, and trigonometry and geometry can be used to derive the equations for depth calculations. The equations are given below without their derivations, which can be found in many basic geophysics texts, including Dobrin and Savit (1988) and Burger (1992).

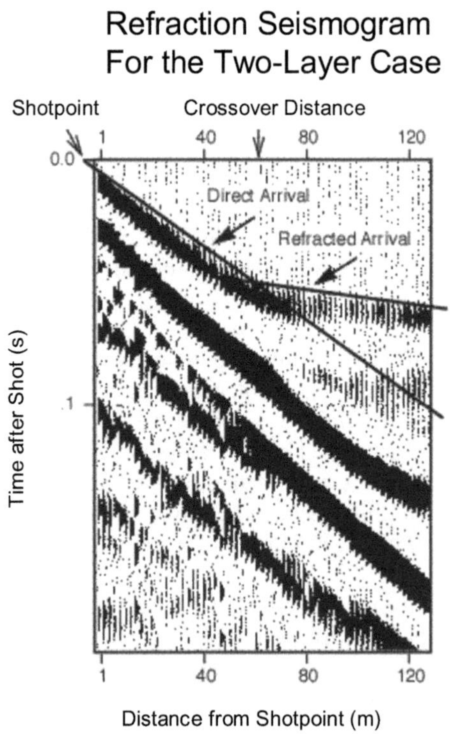

Figure 8.4. Sample seismogram showing a distinct refraction first-arrival crossover at a distance of about 60 m from the shotpoint

We can solve for z because V_0, V_1, and T_i can be measured using the seismograms:

$$z = \frac{T_i}{2} \frac{V_1 V_0}{\sqrt{V_1^2 - V_0^2}}, \tag{8.9}$$

where V_0 is the velocity in the first layer, V_1 is the velocity in the second layer, and T_i is the zero-offset intercept time of the line drawn along the refracted first arrivals. *Offset* refers to the distance between a geophone and the location of a seismic source. When $V_0 > V_1$, the solution becomes imaginary, since the denominator will contain the square root of a negative number. Thus, when seismic waves penetrate a layer whose velocity is lower than that of the overlying layers, the refraction method will not work. At the crossover distance, the direct wave and the refracted wave arrive simultaneously, and algebraic manipulation shows that

$$z = \frac{x_{cros}}{2}\sqrt{\left(\frac{V_1 - V_0}{V_1 + V_0}\right)}. \tag{8.10}$$

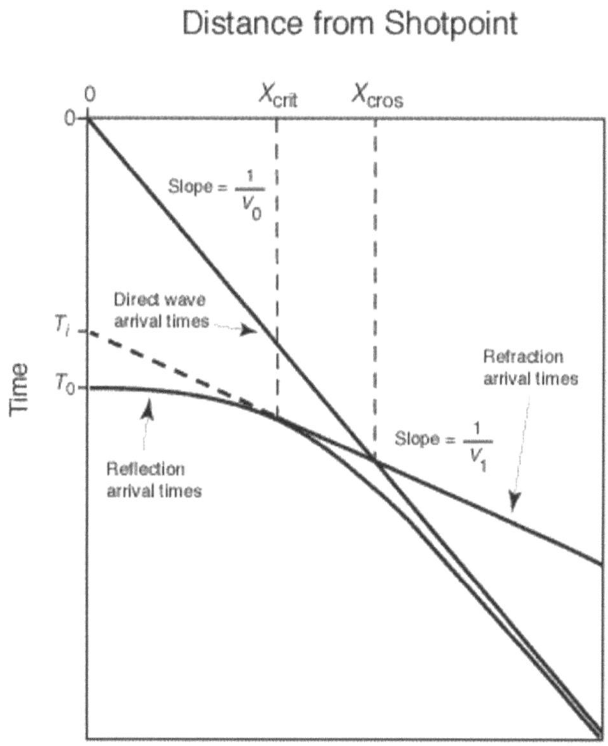

Figure 8.5. Travel times for direct, critically refracted, and reflected rays in the case of a single layer overlying an infinitely thick substrate. T_i is the zero-offset distance intercept time of a line drawn through the first-arriving refracted *P*-waves. The distance at which refracted rays are first physically possible is labeled X_{crit}. The distance at which the direct and refracted rays arrive simultaneously is identified as X_{cros}. The reflection time of a normal-incidence *P*-wave reflection traveling vertically to and from the top of the substrate is T_0.

8.2.3 METHODS OF INTERPRETING REFRACTION

The methods listed here offer varying degrees of reliability when an analyst attempts to compensate for refraction problems or to estimate the size of the errors that those problems can cause:

(1) Intercept-Time Method (ITM), the textbook classic discussed previously;
(2) Generalized Reciprocal Method (GRM), Palmer (1980, 1981);
(3) Reciprocal Method (RM), Hawkins (1961); from Edge and Laby (1931);

(4) Plus-Minus Method of Hagedoorn (1959), similar to the reciprocal method;
(5) Delay-Time Method (DTM), Nettleton, 1940;
(6) Refraction Tomography (RT).

8.2.3.1 *The Generalized Reciprocal Method (GRM)*
The GRM may be thought of as a computationally intensive expansion of Methods (3), (4), and (5) in the preceding list. As tomographic methods become more available, they are likely to replace GRM, most likely because tomography offers robust algorithms requiring less human intervention than does GRM.

Reciprocity is the equivalence in travel time from source *A* to point *B* and its reverse travelpath (Figure 8.6). *Reciprocal time* refers to the difference in travel time between two geophones, i.e., it is the reciprocal of the local apparent velocity.

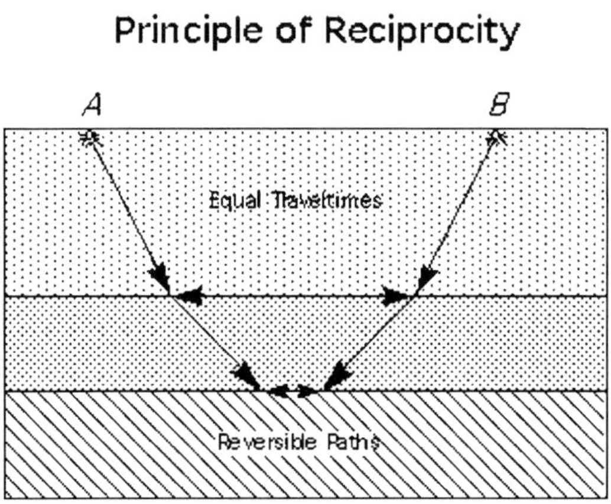

Figure 8.6. Refraction surveying depends in part upon the principle of reciprocity. The travel time from *A* to *B* is the same as the travel time from *B* to *A*.

Compared to ITM, GRM enhances the utility of refraction methods substantially. The GRM is particularly effective when dealing with irregular refractor interfaces and lateral changes in velocity within refractors. However, when layers are flat-lying, homogeneous, and uniform in thickness, the ITM and GRM produce equivalent results, as do the other refraction methods listed in this chapter.

When data are acquired using an adequate number of geophones, and source shots are discharged in multiple locations at each end of the geophone line, some of the interpretational ambiguities commonly present in ITM can be removed (Lankston, 1989). Lankston (1990) provides excellent discussions of GRM and the field procedures necessary to maximize the effectiveness of the near-surface refraction surveys that rely upon this method. Other examples of modern refraction analysis include Hatherly and Neville (1986), Ackerman et al. (1986), and Lankston and Lankston (1986).

8.2.3.2 *Interpreting Data Using the GRM*

The GRM is based on an analysis of rays from forward and reverse seismic shots that exit the refractor at the same point, i.e., at the subsurface interface represented by point *G* in Figure 8.7. The forward and reverse travel times recorded for different geophone separations along a seismic profile allow the calculation of velocity and depth. The important geophone separations used in GRM calculations are referred to as *XY* values (Figure 8.7). The *optimum separation* is one in which the ray traveling from *B* to *X* and the reversed ray traveling from *A* to *Y* leave the refractor at the same point. When data are analyzed using the optimum *XY* distance, the velocity analysis and depth calculations for the refractors afford greater detail. The GRM is an iterative method that involves a set of *XY*-distance test values. An extensive set of calculations is required for each *XY* distance tested in the search for the optimum *XY*. The procedure for recognizing the optimum *XY* after calculations are completed is given in Palmer (1980, 1981) and Lankston (1990). One advantage of GRM over ITM is that GRM provides stable results in the presence of refractor dips at angles of up to about 20°. This is because both forward-profile and reversed-profile data are used in the same calculation (see Palmer, 1980).

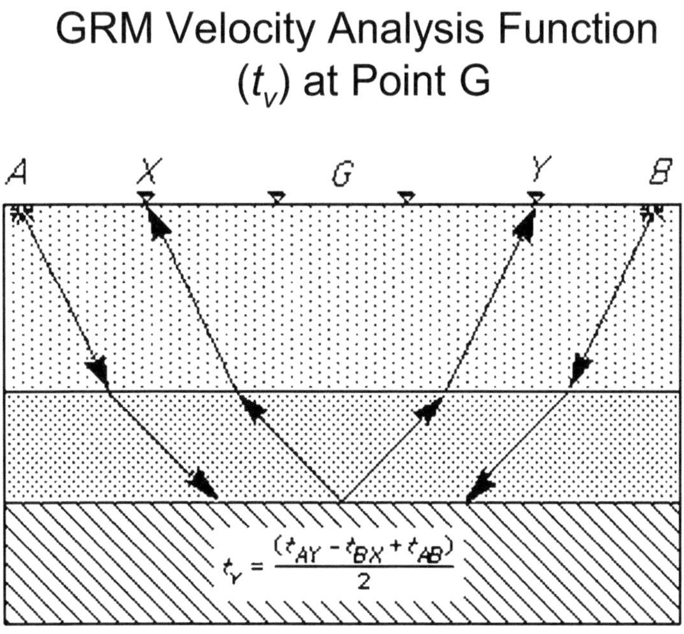

Figure 8.7. Using the equation in this figure, the values of t_v are plotted against distance along the geophone line for different *XY* values. The inverse of the refractor's apparent velocity is the slope of the line fitted to the t_v values for optimum *XY*.

8.3 Seismic Reflection

Seismic reflection data can be useful in hydrogeological investigations. Examples include locating the water table (Birkelo et al., 1987), finding water-filled abandoned coal mines (Branham and Steeples, 1988), mapping a shallow bedrock surface (Miller et al., 1989) and delineating shallow faults (Treadway et al., 1988). Often, the shallow seismic reflection method is used to map geologic structure and stratigraphy because it offers greater detail than the seismic refraction method. Case studies of the seismic reflection method at the regional scale can be found in Chapter 12 of this volume.

8.3.1 ACQUIRING SEISMIC REFLECTION DATA

For shallow seismic reflection methods to be effective, three elements are required. First, *acoustic impedance contrasts* must be present. Next, the *dominant frequency* of the data must be high enough so that direct waves and refractions will not interfere with reflections (Steeples and Miller, 1998). Finally, the *wavefield* must be sampled adequately in space and time, using equipment designed for the task (Steeples and Miller, 1990). The first factor is a function of site geology, the second is determined by a combination of site geology and the type of seismic-energy source used, and the third is a function of experimental design and the kind of equipment available.

Planting geophones closely together generally increases cost. Indeed, the price of reflection surveying at depths ranging from one meter to a few hundred meters is partly an inverse function of depth. That is, shot and geophone intervals must be proportional to reflector depth. For instance, assume that the reflector of principal interest is at a depth z. For a well-designed survey, the geophone whose offset is greatest is typically one to two times z. Then, for a seismograph with n channels, the geophone interval would be approximately z/n, and the smallest offset is often $< z/n$.

Hunter et al. (1984) demonstrated that by selecting the appropriate distance between the geophones and the seismic source using a computation known as the *optimum window offset*, the chances of recording useful shallow reflection data were improved. Several geophones must be located within the area or "window" in which target reflections can be detected most easily. However, as the number of available seismograph recording channels increases, the need to locate Hunter's optimum window becomes less critical. Moreover, smaller geophone group intervals become possible, which helps to minimize the trade-off between the reflection coherency obtained by using short geophone-group intervals and the superior velocity information that can be obtained from the deeper layers by using longer geophone offsets. With a sufficient number of geophones arranged at close intervals on the ground as well as geophones extending to offsets that are at least equivalent to the maximum depth of interest, a seismologist is certain to obtain data from the optimum offsets.

Usually, data are collected so that redundancy is obtained for several points in the subsurface. Figure 8.8a depicts shotpoint and geophone locations as they are moved progressively along the traverse of a two-dimensional (2-D) seismic profile. The data collected from all geophones for an individual shot is called a *common shot gather* or a *field file*. When several seismic shots, at different locations, are fired, more than one seismic trace will contain information about a point or points in the earth's subsurface. These points are known as *common midpoints* or CMPs. (Sometimes the terms *common depth point* [CDP] or *common reflection point* [CRP] are substituted for CMP in the refereed literature. The three terms are equivalent for flat-lying layers and often are used interchangeably.) Most modern seismic surveys are conducted so that reflection data from subsurface points are collected from several different angles of incidence (Figure 8.8b). The result of collecting seismic data traces from a CMP location such as the one depicted in Figure 8.8b) is called a *CMP gather*.

The degree of redundancy at a particular point in the subsurface is known as *fold*. For near-surface data, fold commonly ranges between 6 and 60. Other things being equal, the *signal-to-noise ratio* (S/N) of the reflections increases as the square root of the CMP fold. The minimum level of CMP-fold redundancy needed to obtain interpretable sections should be determined by experimentation with a small amount of CMP data. Modern laptop computers are capable of providing CMP quality-control processing in the field. During experimental processing, the data can be decimated gradually so that the lowest acceptable CMP fold can be found. Using a higher CMP fold than is necessary adds to the cost of fieldwork and data processing. When high-quality data are recorded using many seismograph channels and a low CMP fold, the shotpoint interval used can be several times larger than the geophone interval.

To decrease the cost of shallow reflection surveys, sacrificing CMP fold by increasing the shotpoint interval and fixing the geophones at a small interval is preferable to forfeiting the coherency of the reflector by increasing the geophone interval while keeping the shotpoint interval fixed to maintain high fold. This is because, where shallow reflections are concerned, maintaining coherency is necessary to assure that any coherent events on processed sections are in fact reflections and not refractions or ground roll.

Collecting information from enough but not too many seismic lines is one of the keys to cost containment. Performing near-real-time data processing with powerful, on-site personal computers allows an informed, in-the-field decision to be made about the number of lines necessary.

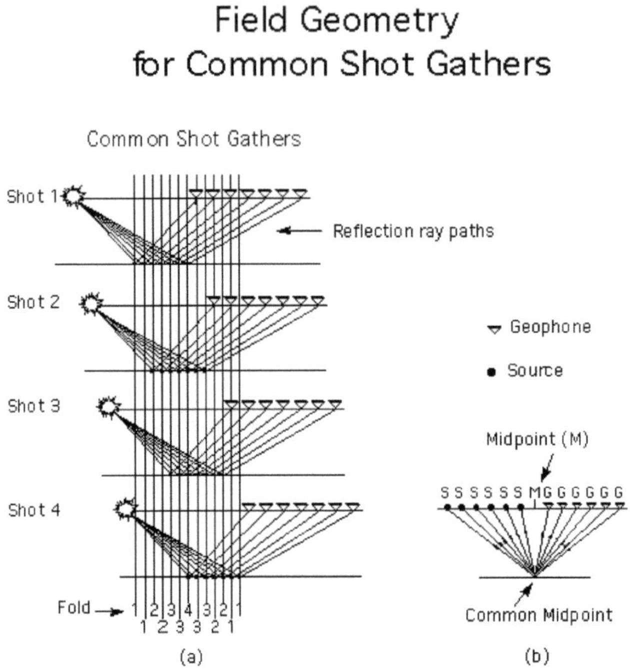

Figure 8.8. Representative common-shot-gather field geometry, CMP acquisition, and resulting fold: (a) In the case of a 12-channel seismograph with shotpoints at all geophone locations, the subsurface reflection points will be sampled six times, which will result in six-fold CMP data after processing; (b) A depiction of the raypaths for a single reflection in which six-fold CMP redundancy was attained.

The geologic target. The first step in planning a near-surface seismic program is to define the hydrogeological target. The next step is to establish whether seismic methods are appropriate for a selected site and, if so, to identify the method likely to be most cost-effective. Seismic surveys are rarely the first hydrogeological investigations to be performed at a site. As a result, some limited but useful drilling information and other geophysical data may be available to the investigator early in the seismic-planning process.

Depending upon local hydrogeological conditions, in theory seismic methods may be hindered by several limitations. For example, in reflection surveying, sufficient acoustic impedance contrast must be present to result in a reflection. In addition, each component of the seismograph system must have enough dynamic range to overcome ground roll and other forms of noise. High frequencies are required to separate shallow reflections from shallow refractions, whereas both high frequencies and wide bandwidths are necessary to produce high resolution. Often, to avoid wide-angle phase distortion, near-vertical angles of incidence are necessary as well (Pullan and Hunter, 1985). At many environmentally sensitive sites, the presence of cultural noise such as that generated by 60 Hz power lines (50 Hz in Europe and some other parts of the world), pavement, or excavated-fill materials that are more than a few meters in thickness may prevent the use of reflection methods. *Ringing* or reverberation in layers in which a shallow, high-velocity layer is present near the surface can also constitute a major hindrance.

Like all geophysical methods, the seismic reflection procedure has limitations. Some of the restrictions to be considered when selecting acquisition parameters are:

(1) The vertical and horizontal limits of resolution;
(2) The wavelength and frequency of the recorded data and the bandpass of the recording components;
(3) The presence of noise from electronic and cultural sources;
(4) Out-of-plane reflections caused by off-line geological structures or three-dimensional features; and
(5) Velocity variations with vertical and horizontal location in the near surface.

Resolution of depth-and-interval geology. After a geologic problem has been defined, the amount and type of knowledge likely to be contributed by each of the various seismic methods can be estimated by assessing several factors:

Resolution requirements. The classic vertical-bed resolution limit is given by the 1/4-wavelength criterion for the dominant frequency (Widess, 1973). The dominant wavelength for seismic reflections normally increases with depth because velocity typically also increases with depth, whereas frequency always decreases with depth (i.e., with increasing time on a seismic section). Hence, vertical resolution for shallow reflections is likely to be better than it is for deeper reflections. For shallow surveys, the recordable dominant frequencies are most commonly in the range of 70–300 Hz. Often, the expected frequency is difficult to estimate during the planning stages of a seismic survey, although gathering information about the near-surface geology and the depth of the water table may be of use. In areas where thick, dry sand is present at the surface, dominant frequencies above a few tens of hertz usually cannot be obtained; conversely, in areas where the water table is near the surface, data with dominant frequencies of several hundred hertz sometimes can be acquired.

Resolution here refers to detecting the top and bottom boundaries of a geologic bed. Note that detecting the presence of a bed may be easier than resolving it. Likewise, the resolution limit does not apply to *fault-offset detection*. When near-surface velocity variations are not a major consideration, fault offset can be detected at a level of 1/10 of one wavelength or less.

Acoustic contrast across lithologic-stratigraphic boundaries. This is another factor that limits the success of reflection seismology. When high-contrast reflection coefficients of more than 0.10 are present, the chance of observing reflections is good. When contrast is low (e.g., in the 0.01 range), the likelihood of reflections is small.

Drilling information and geophysical logs. Information derived from boreholes can provide useful data regarding both acoustic-contrast estimates and uphole travel times, which provide average seismic velocities.

Timing errors. When seismic data are tied to borehole data, and determining the absolute depth to the reflector is important, both variable and systematic timing errors must be tracked. When information about relative depth alone is needed, however, systematic timing errors can be tolerated. Absolute timing errors include *time breaks* plus topographic elevation, along with the horizontal survey location errors for the geophones and the shotpoints.

Energy-source selection. Shallow reflection seismology admits a wide range of potentially useful seismic energy sources, partly because the amount of energy needed is only a fraction of that required for traditional petroleum-exploration surveys, which analyze objectives at greater depths. The factors involved in energy-source selection include frequency content, the amount of energy required, initial price, cost per shotpoint, shotpoint-preparation requirements, speed of operation, and safety, which is of paramount importance.

With respect to many available seismic sources, the results of tests conducted in the United States (i.e., in New Jersey, California, and Texas) have been published in *Geophysics* (Miller et al., 1986, 1992a, 1994). Additional field tests have been conducted by others, including Pullan and MacAulay (1987) and Doll et al. (1998).

8.3.2 PROCESSING SEISMIC REFLECTION DATA

Processing can be performed during acquisition as well as subsequent to acquisition. In this section, several common on-field processing steps are discussed.

8.3.2.1 *Digital Sampling*

To transform data into a format suitable for digital signal processing (DSP), the data must be "digitized" or converted into a discrete form. The procedure is detailed in Yilmaz (1987). In the following sections, the most basic digital geophysical signal-processing tools are introduced.

8.3.2.2 *Field Processing*

In a practical sense, data were "processed" in the field even before digital computers were available. By adjusting the gain control on a seismograph and regulating the analog filters (consisting of resistors and capacitors), the amplitudes of the data as well as their displayed frequency content could be modified. By means of an analog tape-recording device, the data could be played back using various gain and filter settings. Then, the settings producing the best S/N could be selected. Modern digital data can be filtered either in the field or during the main processing phase.

Processing data in the field offers several advantages, including maintaining quality control over the data, allowing the remainder of a project to be planned dynamically, and decreasing the amount of time needed to prepare a final report. Moreover, when data are processed onsite, the quality of the final output can be previewed. Sometimes a preview will induce a field crew to increase the rate of sampling, decrease the spacing between observations, or even modify the size of the geographic area to be surveyed.

Enhancing data by stacking. Strengthening existing data can be as simple as adding color to a display or as complex as implementing algorithms based on information theory. Only one of the many methods commonly used to improve S/N is (briefly) described here. *Stacking* is a method used widely to enhance the quality of the data in studies that include *ground-penetrating radar* (GPR), *direct-current* (DC) *resistivity*, and *near-surface seismology*. Each of these geophysical techniques introduces some form of electromagnetic, electrical, or elastic energy into the earth. When an energy source is activated repeatedly at a single point on the earth's surface and the resulting outputs are summed, the resulting data are known as *stacked data.*

The improvement in S/N conferred by stacking can be striking, especially in environments in which a high level of random noise is present. Generally, when a source pulse is identical from one input to the next at a particular point, and noise is random, the S/N increases as the square root of the number of pulses stacked. When these assumptions hold and, for example, four source-pulse outputs are stacked, the S/N increases by a factor of two (6 dB); when 16 outputs are stacked, S/N increases by a factor of four (12 dB); and when 64 outputs are stacked, S/N increases by a factor of eight (18 dB). Clearly, stacking large numbers of pulse outputs (i.e., hundreds or thousands) reaches the point of diminishing returns quite rapidly. When a system is capable of resetting itself quickly, i.e., cycling from one input pulse to another in a matter of seconds, stacking the output from tens of source pulses becomes feasible.

8.3.3 PROCESSING AND INTERPRETING SEISMIC REFLECTIONS

Interpreting shallow seismic reflection data differs slightly from interpreting the more traditional types of seismic reflection data acquired at greater depths. In shallow reflection work, the first objective is to make sure that *reflections* are being interpreted as opposed to any of the other coherent wave trains that may appear on near-surface seismic sections. For example, refractions, ground roll, and *air waves* can all stack coherently when data are not processed properly (Steeples and Miller, 1998). Thus, shallow seismic data cannot be interpreted adequately without access to a representative selection of the unprocessed *field files.* So that their origins can be identified, any coherent wave trains found on final, processed seismic sections should be traced back through the intermediate steps in the processing flow to the original field files.

Shallow reflection work presents three main areas of concern. In the case of very shallow reflections, *interference of refractions* with reflections is a major problem. Refractions must be removed (*muted*) during processing so that they will not appear on the stacked sections. When refractions stack on seismic sections, they usually appear as wavelets whose frequencies are lower than those of reflections. This occurs because the *normal moveout* (NMO) process time-adjusts the seismic traces to fit a hyperbolic curve. The refractions occur along straight lines on a seismogram, so time-adjusting them by hyperbolic transformation and then stacking causes a decrease in their apparent frequency. Thus, when the apparent frequency increases with depth on a seismic section, it may be an indication that refractions have stacked together in the shallow part of the section, whereas reflections may have stacked in the deeper part. Decreased apparent frequency in the shallowest wavelets also could be caused by *NMO stretch* (Miller, 1992).

A second cause for concern is the presence of *air-wave vestiges*. These remnants may appear even when seismic sources are relatively small, e.g., when a sledgehammer is used as a source. The bandwidth of air waves tends to be very broad, so frequency filtering cannot be used to isolate them. Because of their broad bandwidth, air waves also generate some very short wavelengths, commonly less than 1/2 m long. Hence, unless the geophone interval is less than 1/4 m, attempting to remove air waves by using *frequency-wavenumber* (*f-k*) filtering is futile, because doing so will cause spatial aliasing. However, one of the positive elements of air waves is that they have a velocity of ~335 m/s, which makes them easy to identify in field files. This attribute also makes removing air waves by *surgical muting* fairly simple. A cautionary note: when the arrival of air waves has been established on a seismogram, the analyst should be aware that the data later in time may be contaminated, because air waves can produce echoes from objects such as trees, fences, buildings, vehicles, or people.

The third area of concern is the emergence of *stacked surface waves* on shallow seismic sections. Stacking of this type is usually less serious than are refractions or the presence of air waves because, in most shallow surveys, surface waves can be attenuated by *frequency filtering*. Frequency filtering works here because the dominant frequency of the surface waves usually is lower than the frequency of the desired shallow-reflection information. A final issue is the potential inclusion of *instrument noise* and *processing artifacts* in the processed data. Instrument noise often takes the form of impulses or spikes that have the potential to replicate any filter operators applied to the data. Likewise, a data set may be sensitive to the parameters applied during automatic gain control (AGC) and *deconvolution*, among other processes (Steeples and Miller, 1998). When the interpreter is satisfied that the data in a stacked section are reflections, a serious interpretation effort can begin.

A gross, first-pass interpretation generally involves identifying the coherent waves on a seismic section and then following them laterally across the section. Faults can be located by looking for vertical or near-vertical offsets in a coherent reflector.

Diffraction patterns also can be of use diagnostically with respect to faulting, particularly when data have not undergone migration processing.

An example of an interpreted shallow seismic reflection section is shown in Figure 8.9, from Miller et al. (1989). In this particular case, a bedrock reflection at a depth between 5 and 15 m was interpreted easily on the seismic section (Figure 8.9a). The objective of the seismic survey was to find low areas in the bedrock where pollutants might collect in the event of a leak at a chemical storage pond. Initial test drilling had suggested the presence of a valley or channel on the surface of the bedrock. The seismic reflection data confirmed the presence, location, and depth of a channel in the bedrock.

The lower half of the figure (Figure 8.9b) is an interpretation of the seismic cross section. To interpret this and other seismic data, it is sometimes best to look at the seismic section at a slight angle relative to the plane of the page, with the line of sight parallel to the horizontal axis. This allows the eye of the interpreter to benefit from the coherency of the reflection; in this case, the eye is drawn to the signal that echoed from the surface of the bedrock.

Advanced seismic data-processing techniques sometimes make use of *trace attributes*, including *instantaneous amplitude*, *instantaneous phase*, and *instantaneous frequency*. These attributes are shown to best advantage by color plotting. Color displays allow greater detail because their dynamic range is greater than that of black-and-white or gray-scale plots. Yilmaz (1987) and Taner et al. (1979) discuss trace-attribute processing and provide examples.

Trace-attribute analysis has not been used extensively with shallow data. For these techniques to function properly, the true amplitude and frequency content of the data must be preserved, and variations in the reflection wavelet that are not related to the geology must be eliminated. In deeper reflection data, these variations in the reflection wavelet are commonly removed by *deconvolution*. However, deconvolution often does not work well on shallow reflection data, partly because of low S/N (Steeples and Miller, 1998; Stephenson et al., 1993; House et al., 1996).

8.3.4 BOREHOLE SEISMOLOGY

Seismic methods attain higher resolutions when the source and/or the receiver can be placed in boreholes, where signal attenuation is smaller. When sources as well as receivers are placed in boreholes, recorded seismic-data frequencies and bandwidth may increase by an order of magnitude. However, few noteworthy instances of high-resolution borehole seismic data of the sort used in hydrogeological characterization and monitoring are available in the refereed literature (see Chapter 13 of this book for existing examples).

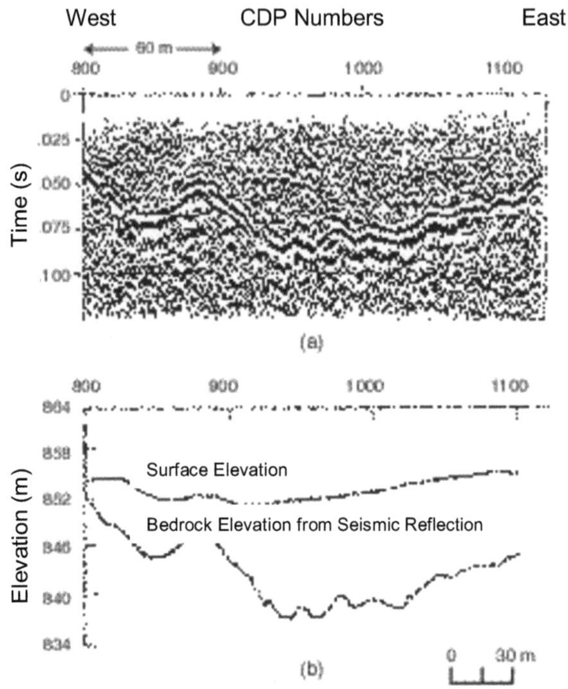

Figure 8.9. (a) The 12-fold CDP seismic reflection section for the seismic line; (b) An elevation cross section of the topography and bedrock surface. Seismic data are static-corrected to an elevation of 860 m (from Miller et al., 1989).

8.3.4.1 *Seismic Tomography*

Tomography has been used to study the interior of the earth on scales ranging from thousands of kilometers to tens of meters (e.g., Humphreys et al., 1984; Clayton and Stolt, 1981). The term is derived from *tomo* (to slice) and *graph* (picture). The word "slice" conjures a likeness in two dimensions; however, a three-dimensional picture can be obtained from a series of two-dimensional slices. To obtain a slice-picture of part of the earth, a collection of seismic raypaths is analyzed to take advantage of the information contributed by each. Commonly, seismic tomography involves the estimation of velocity or attenuation. Whereas the mathematical formulation for tomography is a rigorous inversion process, the technique can be thought of as the logical analysis of physical observations.

Conceptually, *crosshole tomography* can be diagrammed as in Figure 8.10. Assume that velocity tomography is being used and that each ray either has a "normal" travel time or it does not. For this exercise, suppose that the abnormal rays are delayed. A slice of the earth is plotted, with the raypaths shown in the plane of the slice. If the suite of raypaths is sufficiently complete and a low-velocity body is present, the low-velocity body will affect some of the raypaths but not others. Abnormal rays that have penetrated the low-velocity body are plotted as thicker lines. Note that some abnormal rays intersect other abnormal rays and that the intersecting points of the two fall either within or near the abnormal body. When the area of the slice defined by the normal rays has been eliminated, only those rays whose travel times are abnormal remain, as illustrated in Figure 8.10c. At this point, the area within the slice has been constrained within the boundaries defined by the abnormal rays.

In practice, seismic tomographic surveys rely on the same rigorous mathematical approach used successfully by the medical profession to develop three-dimensional x-ray imaging of the human body such as computed axial tomography (the CAT scan), such as described by Peterson et al. (1985). Tomographic surveying depends upon measuring the travel times of large numbers of seismic raypaths through a body of earth material. Although the technique as illustrated here involves timing the raypaths between boreholes, the timing of surface-to-borehole and/or borehole-to-surface raypaths is common as well. Unless preexisting boreholes are available, however, seismic tomography is both computationally intensive and costly because of the drilling required. Tomography often produces a very detailed velocity model of the area between the boreholes. Moreover, to be correct theoretically, it does not require that any initial assumptions be made, provided that the locations of all points in both boreholes are known accurately in three dimensions. Examples of the use of seismic tomography for hydrogeological investigations are given in Chapter 13 of this volume.

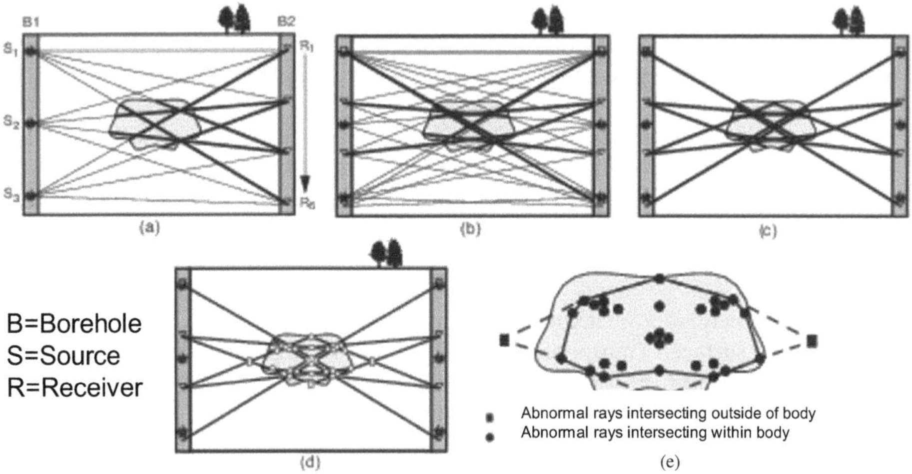

Figure 8.10. Schematic of crosshole raypaths in which (a) delayed (abnormal) rays are represented as heavy lines; (b) sources and receivers have been activated and their positions exchanged (heavy lines indicate raypaths whose arrival times have been delayed anomalously); (c) only raypaths whose arrival times are anomalous raypaths intersect (black dots) begin to convey a rough "image" of the anomalous body (enlarged for visibility). Generally, the greater the number of raypaths, the more accurate the image.

Diving-wave tomography may be of some value to near-surface geophysics. This type of tomography is similar in concept to crosshole tomography, except that the sources and receivers are located at the earth's surface. Raypaths that penetrate downward into the earth and then return to receivers at the surface are analyzed either by *inversion processes* or by *iterative ray-trace forward modeling*. The latter technique is a sophisticated means of analyzing seismic refraction data. It involves using much more of the available information than was used in classical intercept-time refraction methods. And when used in conjunction with surface-wave methods, refraction tomography has the potential to assist in the examination of the upper few tens of meters of the earth, thus decreasing the number of boreholes needed to delineate near-surface features of geologic interest (Ivanov, 2002).

8.3.4.2 *Vertical Seismic Profiling* (VSP)

This technique is seldom used alone; rather, it is used to provide better interpretation of seismic reflection data. It is commonly performed using a string of *hydrophones*, *three-component geophones*, or *three-component accelerometers* that are placed in a borehole. The seismic source is located at the surface, within a few seismic wavelengths of the borehole.

Vertical seismic profiling allows the accurate determination of the one-way travel time of seismic waves to various geologic units. It also allows the analysis of attenuation and *acoustic impedances*, both of which are needed to construct *synthetic seismograms*. Synthetic seismograms are compared to seismic reflection data to identify specific geologic formations and to refine depth estimates. Often, reflections that are deeper than the bottom of the borehole are available for analysis in VSP data.

Synthetic-seismogram results are typically better with VSP data than with sonic logs, because the seismic wavelengths are of the same order of magnitude for VSP as they are for reflection, and because VSP data are less vulnerable to error from borehole washouts than are sonic log data. References concerning VSP include Balch and Lee (1984) and Hardage (1983).

Transposed or *reverse VSP* involves a source located in a borehole, with the receivers at the earth's surface. Transposed VSP has the advantage of being faster and less expensive to perform, provided that a suitable seismic source can be placed in the borehole and used repeatedly at different depths without damaging the borehole.

To convert seismic reflection data into accurate depth measurements, uphole, downhole, or *VSP travel times* are needed. Absolute-velocity data also are used to produce better models and synthetic seismograms, and they can provide an initial stacking velocity during CMP processing. Without velocity constraints, the selection of an accurate stacking velocity can be very difficult, especially when the reflection data are of marginal quality.

8.3.5 *S*-WAVE AND THREE-COMPONENT (3C) SEISMOLOGY

The use of shallow *S*-wave reflection methods has not been widespread, but a few examples can be found in the scientific literature (Hasbrouck, 1991; Goforth and Hayward, 1992). Separating *S*-wave reflections from interfering Love waves that appear at the same time on the seismograms has proved difficult. In fact, distinguishing Love waves from the shallowest *S*-wave reflections is especially troublesome because Love waves are actually multiple *S*-wave reflections that have been trapped in a low-velocity surface layer.

Another problem associated with near-surface *S*-wave reflection surveying is the narrow bandwidth of the signal sources. The vast majority of the *S*-wave reflection work described in the technical literature has been performed at dominant frequencies of <100 Hz. Experiments with several *S*-wave sources (Miller et al., 1992b) have shown that the source spectrum tends to be less than 50 Hz wide, which is narrower by a factor of perhaps 3 to 5 than it is for *P*-wave sources. However, near-surface *S*-waves have lower velocities, so the wavelengths at specific sites are often shorter for *S*-waves than for *P*-waves. The shallow *S*-wave reflection technique has great potential for helping to solve engineering and environmental problems. However, until the source-bandwidth and Love-wave interference problems are solved, the technique is not likely to be used widely. Unpublished work using a horizontal vibratory source with a mass of less than 100 kg has shown promise.

Three-component recording methods. Most seismic reflection and refraction work employs vertical geophones intended to measure only a single component of the three-component vector motion of the ground. A method of observation that is limited in this way is a major handicap to extracting the desirable information contained in the vector wave field. For example, information about the horizontal motion of *S*-waves, along with all of the information concerning Love waves, is lost when only vertical geophones are used. *S*-waves, because they are polarized, carry more information than *P*-waves, provided that the wavelengths are comparable.

To perform 3C work, three orthogonally mounted geophones are placed at a single location. Classically, this has involved two geophones mounted orthogonally to each other in a horizontal plane, with a third geophone mounted vertically, thus forming an orthogonal, three-vector frame of reference as data are recorded. From this configuration, the true vector motion of the ground can be extracted. A horizontal geophone is designed to operate optimally when its axis of motion is exactly horizontal; thus, its response to ground motion decreases when its orientation is not fully horizontal. For this reason, a leveling bubble is attached to horizontal geophones.

8.4 Surface-Wave Analysis

Surface waves in layered solids include Love and Rayleigh waves. A concise mathematical discussion of surface waves can be found in Grant and West (1965), with extensive treatment provided by Ewing et al. (1957).

8.4.1 SURFACE WAVES

Surface waves have multiple modes in much the same way that pipes of different lengths produce the multiple tones characteristic of a pipe organ. Generally, however, the *fundamental mode* is the most important mode with respect to surface waves. Rix et al. (1990) concluded that neglecting the effects of the higher modes does not introduce significant errors when using the two-station spectral analysis of surface waves. Alternatively, Park et al. (1999d) and Xia et al. (2000) discussed some of the advantages of examining the higher modes of surface waves. Two promising higher-mode properties were observed by Xia et al. (2000). First, for fundamental and higher-mode Rayleigh-wave data of the same wavelength, the higher modes were able to "see" more deeply than the fundamental mode. That is, the higher modes can sample depths that are greater than one wavelength, whereas the fundamental mode normally samples depths shallower than that. Moreover, higher-mode data can increase the resolution of the inverted *S*-wave velocities. Consequently, a better *S*-wave velocity profile can be produced from the inversion of surface-wave data when higher-mode data are included.

8.4.2 PHASE VELOCITY AND GROUP VELOCITY

Group velocity is the speed with which seismic energy travels from one point to another. For each of the wave types as they propagate through homogeneous, isotropic materials, group velocity is an intrinsic physical property of the material. *Phase velocity*, however, is a form of *apparent velocity*; therefore, it may be much higher than group velocity.

As an example of high phase-velocity, imagine a line of geophones attached to the ground at a point on the earth's surface that is antipodal, i.e., 180 Earth-circumference degrees away from a site at which an underground nuclear bomb is detonated. If the bomb is large enough to dispatch detectable *P*-waves completely through the earth, the first *P*-waves will arrive at all of the geophones approximately 22 minutes after the blast. The apparent velocity of the wave front will be very high—approaching infinity. Meanwhile, at most, the group velocity beneath the line of geophones will be a few thousand meters per second.

Seismic waves may disperse as they travel from one place to another, much as a large group of runners tends to disperse in the latter stages of a long-distance foot race. Because different wavelengths of surface waves sample different depths in the earth, their degree of dispersion is a measure of seismic wave velocity as a function of depth.

Dispersive seismic wave trains can exhibit characteristics that permit the measurement of both group- and phase-velocity, as seen in Figure 8.11, which is also an example of how the two types of velocity differ. The phase velocity is the apparent velocity obtained as the result of projecting a line, in-phase, from one seismic trace to another along a particular peak or trough of the seismic signal. In the example, three lines have been drawn to indicate the location at which phase velocity can be measured. Inspection of the middle phase-velocity line reveals that the wavelet appears to travel across ten traces, which is a distance of about 10 m, in 50 ms. This represents a phase velocity of about 200 m/s. In contrast, the wave train arriving just after the group-velocity line covers about 10 m in 100 ms, representing an energy propagation speed of only about 100 m/s. Whereas group velocity represents the rate of propagation of the wave train just below (in time) the group-velocity line, the phase-velocity lines identify the higher apparent velocity of the individual waveforms.

Inspecting the surface waves on a seismogram and then performing some basic calculations can yield useful near-surface geologic information. Nevertheless, most exploration seismologists consider surface waves to be nothing more than useless noise. Civil engineers, however, have had some success in applying surface waves, particularly Rayleigh waves, to the study of the engineering properties of the shallow subsurface. Using the *spectral analysis of surface waves* (SASW)(Stokoe et al., 1994), the *stiffness profile* or *shear modulus* of near-surface materials can be obtained by one of two methods. *Forward modeling* or *data inversion*, in which observed data are inverted to obtain the velocity of surface-wave propagation as a function of wavelength, can be used. Velocity is then exhibited as a function of depth and horizontal position. Moreover, by using Rayleigh waves and a broad band of frequencies, different depths can be sampled.

Often, surface-wave velocity is referred to in the scientific literature as being the equivalent of 92% of the shear-wave velocity of a material. This is true when Poisson's ratio is 0.25, which is typical for hard rocks such as granite, basalt, and limestone, and no layering is present. However, for a Poisson's ratio of 0, velocity is 87.4% of the *S*-wave; for a Poisson's ratio closely approaching 0.5, velocity is 95.5%. For wet (unlayered), unconsolidated materials, *Poisson's ratio* is often in the range of 0.40 to 0.45. For dry materials, the range commonly is 0.15 to 0.2. If one were to assume that the Rayleigh-wave velocity was 94% of the *S*-wave velocity for wet, unconsolidated materials, the typical margin of error would be less than 1%.

Rayleigh-wave velocity is normally thought to be independent of *P*-wave velocity. However, one of the determinants of Poisson's ratio is *P*-wave velocity; therefore, because Rayleigh-wave velocity is weakly dependent on Poisson's ratio, it is also weakly dependent on *P*-wave velocity.

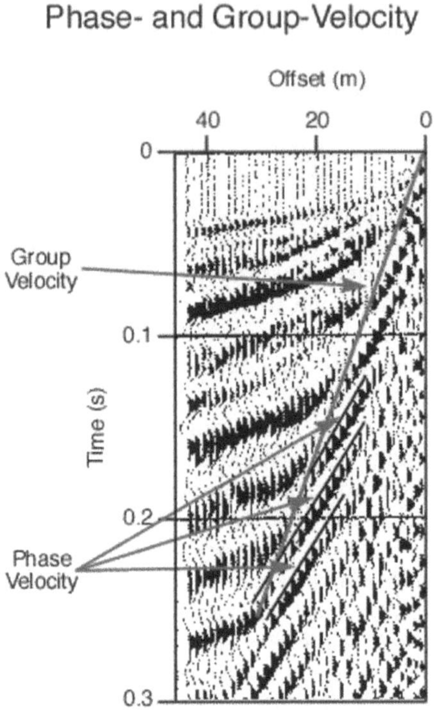

Figure 8.11. This seismogram illustrates the difference between group velocity and phase velocity. Whereas group velocity represents the rate of propagation of the wave train just below (in time) the group-velocity line, the phase-velocity lines identify the higher apparent velocity of the individual waveforms as well as the dispersive nature of the surface waves.

8.4.3 THE SPECTRAL ANALYSIS OF SURFACE WAVES (SASW)

One of the most promising developments in the use of Rayleigh waves for the evaluation of geotechnical sites has been the spectral analysis of surface waves, also called the SASW method (e.g., Stokoe et al., 1994). This method has been used to evaluate the shear strength of pavement and to profile, to a depth of several meters, the stiffness of materials of interest to civil engineers. By using a broad band of wavelengths, the SASW method allows different parts of the profile of a material under consideration to be sampled. The SASW method evolved from the *steady-state Rayleigh-wave method,* which is based on a vibrating source with a known, fixed frequency. In this method, a single vertical geophone is moved radially and progressively away from a fixed geophone located a few seismic wavelengths from the source until successive in-phase positions are located, at which point the distance between the geophone positions will be one wavelength.

Because different wavelengths sample different depth combinations, a velocity profile can be constructed by varying the frequency and measuring the wavelength again. However, this technique is time-consuming. Data collection speed can be increased greatly by simultaneously measuring the velocities associated with several frequencies.

By 1994, SASW involved a *swept-frequency Rayleigh-wave source* and two or more geophones. All of the equipment was located at the surface. The *Fast-Fourier transform* was used to convert signals into the frequency domain, and then the phase difference was calculated for each frequency.

The travel-time difference is given by

$$t(f) = \left[\frac{\phi(f)}{(2\pi f)} \right], \tag{8.11}$$

for each frequency, where ϕ is the phase difference in radians, and f is the frequency in hertz. The distance (d) between receivers is known, so the Rayleigh-wave velocity for each frequency can be calculated by

$$V(f) = \left\lfloor \frac{d_2 - d_1}{t(f)} \right\rfloor. \tag{8.12}$$

Then, the wavelength of the Rayleigh waves can be extracted by

$$\lambda = \frac{V}{f}.$$

The calculations for various frequencies can be plotted as a *dispersion curve* of Rayleigh-wave velocity versus wavelength, which also suggests S-wave velocity as a function of depth along the seismic survey line. Subsequently, the dispersion curves either are matched to a set of theoretical curves derived from forward analytical modeling, or they are subjected to an inversion procedure to extract a stiffness model. Most of the SASW work reported in the scientific literature employs only two geophones to develop a dispersion curve.

Recent developments in shallow surface-wave analysis include the *multichannel analysis of surface waves* (Figure 8.12), a technique known as MASW developed and discussed by Park et al. (1999a,b) and Xia et al. (1999). The principal difference between SASW and MASW is in the number of geophones used simultaneously. In contrast to the SASW technique, MASW allows the analyst to differentiate and separate coherent sources of noise (i.e., direct, refracted, and reflected P- and S-waves). One of the goals of MASW is to identify trace-to-trace coherent arrivals that are not fundamental-mode Rayleigh waves and to remove them from the data. Acquiring MASW data along linear transects has shown promise as a tool for detecting shallow voids and tunnels, mapping the surface of bedrock, locating remnants of underground mines, and delineating fracture systems (Miller et al., 1999a,b; Park et al., 1999c; Xia et al., 1999).

Several features characteristic of surface-wave imaging make it applicable to sites at which other geophysical tools produce inadequate results or fail altogether. Foremost among these is the ease with which surface waves can be generated. The relatively large amplitude of surface waves (compared to body waves) makes them usable in areas where mechanical noise levels are high. Rayleigh-wave dispersion methods require neither increasing velocity with increasing depth nor a seismic-impedance boundary contrast. Thus, the acquisition flexibility of MASW, and its relative insensitivity to environmental and electrical noise, allows it to be used for *S*-wave velocity profiling in areas where other geophysical methods may not be practicable.

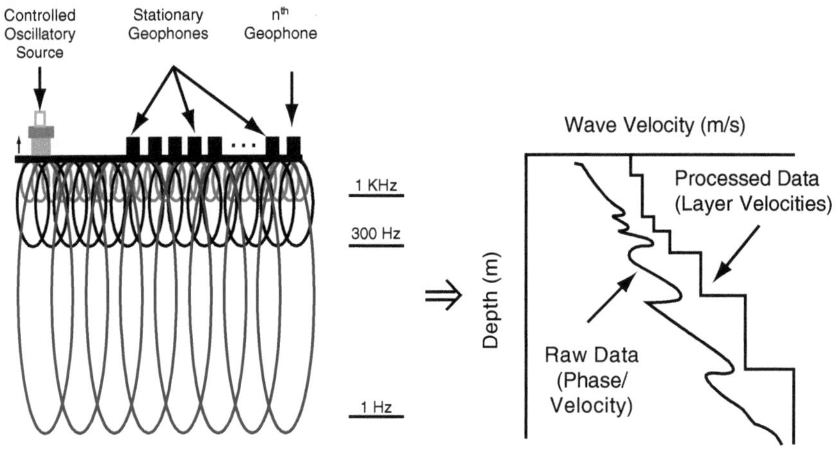

Figure 8.12. MASW Method. The MASW method commonly employs 48 or more geophones to establish redundancy in a data set and to allow the higher wave modes to be analyzed (figure adapted from *Shear Wave Velocity Profiling*, ENSOL Corp., with permission).

8.5 Summary Discussion

Seismic methods provide information about the mechanical properties of earth materials. Some of those properties, such as bulk compressibility, are sensitive to the presence of water. Consequently, *P*-wave seismic methods can be useful in examining saturation in unconsolidated materials. Seismic methods may be thought of as providing information about the framework of earth materials, whereas electrical and electromagnetic methods offer information about fluid found in the pore spaces of materials.

The resolving power of seismic methods is directly related to the wavelength of seismic energy. In a practical sense, surface seismic methods are limited to frequencies below 1 kHz (usually below 300 Hz), although crosshole seismic methods can employ frequencies of more than 1 kHz. With shallow seismic methods, resolution is generally limited to features of about 1 m or more in the least dimension.

The historical roots of the shallow seismic methods lie in the 1930s, when the intercept-time method was used to perform limited shallow refraction work. Because of high costs and the lack of appropriate equipment, particularly data-processing equipment and software, the shallow-reflection and surface-wave techniques were not accepted as quickly as other methods. Since 1980, however, substantial progress has been made in the development of all of the shallow seismic techniques. The seismic reflection method has been used increasingly in applications at depths shallower than 30 m, especially in such hydrologically important areas as mapping bedrock elevation, locating stratigraphic pinch outs, and detecting shallow faults.

In refraction studies, the generalized reciprocal method largely has replaced the classical intercept-time method, and tomographic approaches are rapidly gaining popularity. Also, civil engineers have developed the spectral analysis of surface waves, and surface-wave analysis involving many seismograph channels (MASW) recently has shown promise.

In conclusion, it should be emphasized that selecting suitable seismic recording equipment, energy sources, data acquisition parameters, data processing regimens, and interpretation strategies is essential to the success of a seismic survey.

Acknowledgments

The preparation of this manuscript was supported in part under Grant No. DE-FG07-97ER14826, Environmental Management Science Program, Office of Science and Technology, Office of Environmental Management, United States Department of Energy (DOE). However, any opinions, findings, conclusions, or recommendations expressed herein are those of the author and do not necessarily reflect the views of DOE.

References

Ackermann, H.D., L.W. Pankratz, and D. Dansereau, Resolution of ambiguities of seismic refraction travel time curves, *Geophysics, 51*, 223–235, 1986.

Bachrach, R., and A. Nur, High-resolution shallow-seismic experiments in sand, Part I: Water table, fluid-flow and saturation, *Geophysics, 63,* 1225–1233, 1998.

Bachrach, R., J. Dvorkin, and A. Nur, High-resolution shallow-seismic experiments in sand, Part II: Velocities in shallow, unconsolidated sand, *Geophysics, 63,* 1234–1240, 1998.

Balch, A.H., and M.W. Lee, Vertical seismic profiling: Technique, applications and case histories, International Human Resources Development Corp., Boston, MA., 1984.

Birkelo, B.A., D.W. Steeples, R.D. Miller, and M.A. Sophocleous, Seismic reflection study of a shallow aquifer during a pumping test, *Ground Water, 25,* 703–709, 1987.

Bolt, Bruce A., *Earthquakes*, Freeman and Co., 1993.

Branham, K.L., and D.W. Steeples, Cavity detection using high-resolution seismic reflection methods, *Mining Engineering, 40,* 115–119, 1988.

Burger, H.R., *Exploration Geophysics of the Shallow Subsurface*, Prentice-Hall PTR, Englewood Cliffs, NJ, 1992.

Clark, S.P., ed., *Handbook of Physical Constants*: GSA Memoir 97, 1966.

Clayton, R.W., and R.H. Stolt, A Born-WKBJ inversion method for acoustic reflection data, *Geophysics, 46,* 1559–1567, 1981.

Daly, R.A., G.E. Manger, and S.P. Clark, Jr., Density of rocks, In: *Handbook of Physical Constants*, S.P. Clark, Jr., Ed., GSA Memoir 97, 19–26, 1966.
Dobrin, M.B., and C.H. Savit, *Introduction to Geophysical Prospecting*, John Wiley and Sons, Inc., 1988.
Doll, W.E., R.D. Miller, and J. Xia, A noninvasive shallow seismic source comparison on the Oak Ridge Reservation, Tennessee, *Geophysics, 63,* 1318–1331, 1998.
Edge, A.B., and T.H. Laby, *The Principles and Practice of Geophysical Prospecting*, Cambridge University Press, 1931.
Ewing, W. M., W.S. Jardetzky, and F. Press, Elastic waves in layered media, *Lamont Geological Observatory Contribution No. 189,* McGraw-Hill Book Company, 1957.
Goforth, T. and C. Hayward, Seismic reflection investigations of a bedrock surface buried under alluvium, *Geophysics, 57,* 1217–1227, 1992.
Grant, F.S., and G.F. West, *Interpretation Theory in Applied Geophysics,* McGraw-Hill Book Co., New York, 1965.
Hagedoorn, J. G., The plus-minus method of interpreting seismic refraction sections, *Geophys. Prosp., 7,* 158–182, 1959.
Hardage, B.A., Vertical seismic profiling, Part A—Principles, *Geophysical Press*, 1983.
Hasbrouck, W.P., Four shallow-depth, shear-wave feasibility studies, *Geophysics, 56*, 1875–1885, 1991.
Hatherly, P.J., and M.J. Neville, Experience with the generalized reciprocal method of seismic refraction interpretation for shallow engineering site investigation, *Geophysics, 51,* 255–265, 1986.
Hawkins, L.V., The reciprocal method of routine shallow seismic refraction investigations, *Geophysics, 26,* 806–819, 1961.
House, J.R., T.M. Boyd, and F.P. Haeni, Haddam Meadows, Connecticut: A case study for the acquisition, processing, and relevance of 3-D seismic data as applied to the remediation of DNAPL contamination, In: *Applications of 3-D Seismic Data to Exploration and Production,* P. Weimer and T.L. Davis, eds., AAPG Studies in Geology No. 42, SEG Geophysical Developments Series No. 5, 257–265, 1996.
Humphreys, G., R.W. Clayton, and B.H. Hager, A tomographic image of mantle structure beneath Southern California, *Geophys. Res. Lett., 11,* 625–627, 1984.
Hunter, J. A., S.E. Pullan R.A. Burns, R.M. Gagne, and R.L. Good, Shallow seismic reflection mapping of the overburden-bedrock interface with the engineering seismograph—Some simple techniques, *Geophysics, 49,* 1381–1385, 1984.
Ivanov, J.M., JASR—Joint Analysis of Surface Waves and Refractions, Ph.D. dissertation, The University of Kansas, Lawrence, Kansas, May 2002.
Jones, R., Surface wave technique for measuring the elastic properties and thickness of roads: Theoretical development, *British Journal of Applied Physics, 13,* 21–29, 1962.
Lankston, R. W., The seismic refraction method: A viable tool for mapping shallow targets into the 1990s, *Geophysics, 54,* 1535–1542, 1989.
Lankston, R.W., High-resolution refraction seismic data acquisition and interpretation, In: *Geotechnical and Environmental Geophysics*, *1, Review and Tutorial,* S. Ward, ed., *Soc. Expl. Geophys.*, 45–74, 1990.
Lankston, R.W., and M.M. Lankston, Obtaining multilayer reciprocal times through phantoming: *Geophysics, 51,* 45–49, 1986.
Miller, R.D., Normal moveout stretch mute on shallow-reflection data, *Geophysics, 57,* 1502–1507, 1992.
Miller, R.D., D.W. Steeples, and M. Brannan, Mapping a bedrock surface under dry alluvium with shallow seismic reflections, *Geophysics, 54,* 1528–1534, 1989.
Miller, R.D., S.E. Pullan, J.S. Waldner, and F.P. Haeni, Field comparison of shallow seismic sources, *Geophysics, 51*, 2067–2092, 1986.
Miller, R.D., S.E. Pullan, D.W. Steeples, and J.A. Hunter, Field comparison of shallow seismic sources near Chino, California, *Geophysics, 57,* 693–709, 1992a.
Miller, R.D., S.E. Pullan, D.A. Keiswetter, D.W. Steeples, and J.A. Hunter, Field comparison of shallow *S*-wave seismic sources near Houston, Texas, *Kansas Geological Survey Open-File Report 92-33*, 1992b.
Miller, R.D., S.E. Pullan, D.W. Steeples, and J.A. Hunter, Field comparison of shallow *P*-wave seismic sources near Houston, Texas, *Geophysics, 59,* 1713–1728, 1994.
Miller, R.D., J. Xia, C.B. Park, and J. Ivanov, Using MASW to map bedrock in Olathe, Kansas [Exp. Abs.], *Soc. Explor. Geophys.*, 433–436, 1999a.
Miller, R.D., J. Xia, C.B.Park, J.C. Davis, W.T. Shefchik, and L. Moore, Seismic techniques to delineate dissolution features in the upper 1,000 ft at a power plant site [Exp. Abs.], *Soc. Explor. Geophys.*, 492–495, 1999b.
Nettleton, L.L., *Geophysical Prospecting for Oil,* McGraw-Hill, New York, 1940.
Owen, T.E., J.O. Parra, and J.T. O'Brien, Shear wave seismic reflection system for detection of cavities and tunnels (Exp. Abstr), *Tech. Program of SEG 58th Annual Meeting,* 294–297, 1988.

Palmer, D., The generalized reciprocal method of seismic refraction interpretation, *Soc. Expl. Geophys.*, Tulsa, Oklahoma, 1980.

Palmer, D., An introduction to the generalized reciprocal method of seismic refraction interpretation, *Geophysics, 46,* 1508–1518, 1981.

Park, C.B., R.D. Miller, and J. Xia, Detection of near-surface voids using surface waves, *SAGEEP99*, Oakland, California, pp. 281–286, March 14–18, 1999a.

Park, C.B., R.D. Miller, J. Xia, J.A. Hunter, and J.B. Harris, Higher mode observation by the MASW method [Exp. Abs.], *Soc. Explor. Geophys.,* 524–527, 1999b.

Park, C.B., R.D. Miller, and J. Xia, Multi-channel analysis of surface waves, *Geophysics, 64,* 800–808, 1999c.

Park, C.B., R.D. Miller, and J. Xia, Multimodal analysis of high frequency surface waves, *SAGEEP 99*, Oakland, California, pp. 115–121, March 14–18, 1999d.

Peterson, Jr., J.E., B.N.P. Paulsson and T.V. McEvilly, Applications of algebraic reconstruction techniques to crosshold seismic data, Geophysics, *vol. 50*, no. 10, pp.1566-1580, 1985.

Press, F., Seismic velocities, In: *Handbook of Physical Constants,* S.P. Clark, Jr., ed., *GSA Memoir 97*, 197–221, 1966.

Pullan, S.E., and J.A. Hunter, Seismic model studies of the overburden-bedrock problem, *Geophysics, 50,* 1684–1688, 1985.

Pullan, S.E., and H.A. MacAulay, An in-hole shotgun source for engineering seismic surveys, *Geophysics, 52,* 985–996, 1987.

Rix, G.J., K.H. Stokoe, II, and J.M. Roesset, Experimental determination of surface wave mode contributions, In: *Expanded Abstracts, Soc. Expl. Geophys.,* 447–450, 1990.

Sheriff, R.E., *Encyclopedic Dictionary of Exploration Geophysics, 3rd ed.,* Society of Exploration Geophys., Tulsa, Oklahoma, 1991.

Steeples, D.W., and R.D. Miller, Seismic-reflection methods applied to engineering, environmental, and ground-water problems, In: *Soc. Explor. Geophys. Volumes on Geotechnical and Environmental Geophysics,* Stan Ward, ed., Volume 1: Review and Tutorial, 1–30, 1990.

Steeples, D.W., and R.D. Miller, Avoiding pitfalls in shallow seismic reflection surveys, *Geophysics, 63,* 1213–1224, 1998.

Stephenson, W.J., R.B. Smith, and J.R. Pelton, A high-resolution seismic reflection and gravity survey of Quaternary deformation across the Wasatch Fault, Utah, *J. Geophys. Res, 98,* 8211–8223, 1993.

Stokoe, II, K.H., G.W. Wright, A.B. James, and M.R. Jose, Characterization of geotechnical sites by SASW method, In: *Geophysical Characterization of Sites,* R.D. Woods, ed., ISSMFE Technical Committee #10, Oxford Publishers, New Delhi, 1994.

Taner, M.T., F. Koehler, and R.E. Sheriff, Complex seismic trace analysis, *Soc. of Expl. Geophys., 44,* 1041–1063, 1979.

Treadway, J.A., D.W. Steeples, and R.D. Miller, Shallow seismic study of a fault scarp near Borah Peak, Idaho, *Journal of Geophysical Research, 93,* n. B6, 6325–6337, 1988.

Widess, M.B., How thin is a thin bed? *Geophysics, 38,* 1176–1180, 1973.

Xia, J., R.D. Miller, and C.B. Park, Estimation of near-surface shear-wave velocity by inversion of Rayleigh waves, *Geophysics, 64,* 691–700, 1999.

Xia, J., R.D. Miller, and C.B. Park, Advantage of calculating shear-wave velocity from surface waves with higher modes, *Technical Program with Biographies, Society of Exploration Geophysicists 70th Annual Meeting.,* Calgary, Canada, pp. 1295–1298, 2000.

Yilmaz, O., Seismic data processing: In: *Investigations in Geophysics, No. 2,* S.M. Doherty, ed., E.B. Neitzel, series ed., Soc. Expl. Geophys., Tulsa, Oklahoma, 1987.

9 RELATIONSHIPS BETWEEN SEISMIC AND HYDROLOGICAL PROPERTIES

STEVEN R. PRIDE
Lawrence Berkeley National Laboratory, Berkeley, CA 94720

9.1 Introduction

Reflection seismology is capable of producing detailed three-dimensional images of the earth's interior at the resolution of a seismic wavelength. Such images are obtained by filtering and migrating the seismic data and give geometrical information about where in the earth the elastic moduli and mass densities change. However, information about which specific property has changed and by how much is not contained in the images. Hydrologists can use such migrated images to place geometrical constraints on their possible flow models, but must rely on well data to place constraints on the actual values of the hydrological properties.

With more effort, the seismic data may be inverted to yield the seismic velocity and attenuation structure within a region probed by the waves at a resolution of one or two wavelengths. However, the problem of further relating the seismic velocities and attenuation to hydrological properties such as permeability, porosity, and fluid type is still an ongoing research problem despite more than 50 years of effort. A computer program does not presently exist that can read in seismic data and output reliable information about permeability, porosity, or saturation.

Nonetheless, such hydrological properties can influence seismic properties, and much about the relationship is known. The theoretical framework for studying the connection is poroelasticity, which simultaneously provides the laws governing the hydrological response of a porous material caused by pumping and the seismic response caused by seismic sources. This chapter provides an up-to-date review of porous-media acoustics. It shows how seismic properties can change due to hydrological pumping and acts as a user's guide to the forward problem of predicting seismic wave speeds and attenuation from knowledge of porosity, saturation, and permeability. Similar recommendations for how to optimally perform the inverse problem and obtain hydrological properties from the seismic data are not yet available.

In Section 9.2, general expressions for the seismic wave speeds and attenuation in a porous material are obtained. The pedagogic exercise of obtaining the hydrological and seismic limiting cases of poroelasticity is also demonstrated. In Section 9.3, the definition and model of the dynamic permeability is obtained. In Section 9.4, definitions and models of the various poroelastic moduli are provided in the quasi-static limit where fluid-pressure is effectively uniform throughout each sample of the earth. In Section 9.5, these same moduli are modeled over the entire range of frequencies where the fluid-pressure equilibration on different spatial scales must be allowed for. Such wave-induced flow is responsible for attenuation and dispersion in porous materials and provides the possible link for obtaining permeability from seismic properties. Finally, in Section 9.6, the results of the chapter are discussed and interpreted with an eye toward understanding the possible connection between permeability and seismic properties.

Y. Rubin and S.S. Hubbard (eds.), Hydrogeophysics, 253–291.

9.2 Acoustics of Isotropic Porous Materials

The governing equations for porous media acoustics are generally credited to Biot (1956a,b, 1962), although Frenkel (1944) produced a similar set of equations.

Implicit in the standard form of Biot theory is that the porous material is uniform at "mesoscopic scales," which are those length scales lying between the "microscopic" grain scale and the "macroscopic" wavelength scale. As a compressional wave propagates through such a porous material, a fluid-pressure gradient is established between the peaks and troughs of the wave. The fluid tries to equilibrate by flowing from the peaks to the troughs. Such "macroscopic" flow is the only source of compressional-wave attenuation in standard Biot theory; however, this flow simply does not produce enough loss to explain the attenuation measured in both field work and most laboratory experiments involving geological materials.

To remedy this situation, a microscopic flow mechanism called "squirt" was proposed (Mavko and Nur, 1975, 1979; O'Connell and Budiansky, 1977; Dvorkin et al., 1995). Squirt flow is based on the fact that any broken grain contacts or microcracks in the grains are necessarily more compliant than the main part of the pore space. When a compressional wave squeezes the material, there is a larger fluid-pressure response in the microcracks than in the main pores, which results in a fluid flow from the microcracks to the main pore space. Such microscopic squirt flow can effectively explain ultrasonic attenuation data (frequencies near 1 MHz) obtained in the laboratory under ambient conditions. However, as will be explained in Section 9.5, squirt flow does not seem capable of explaining the measured attenuation in the seismic band of exploration frequencies (say, 10 Hz to several kHz).

One way to account for the attenuation in exploration work is if mesoscopic-scale heterogeneity is present. This heterogeneity may be due to patches of different immiscible fluids or to lithological variations (such as sand/shale mixtures, or pockets/fingers where the grains are less well cemented together, or the presence of joints/fractures within a sedimentary host material). When a compressional wave squeezes a material having such mesoscopic heterogeneity, the effect is similar to squirt: the more compliant parts of the material respond with a greater fluid pressure than the stiffer portions. There results a mesoscopic flow of fluid that does seem capable of explaining the measured levels of attenuation in the seismic band as shown in Section 9.5. Early models of such mesoscopic flow were developed by White (1975) and White et al. (1975), while more recent models are those of Johnson (2001), Pride and Berryman (2003a,b), and Pride et al. (2003, 2004). An interesting aspect of the mesoscopic-loss mechanism is that it depends on the permeabilities of the materials present. Squirt loss, however, depends on h/R where h is the effective aperture of any microcracks in the grains of a rock and R is a characteristic grain radius. Squirt is only indirectly related to the permeability through the effective grain size R.

As shown in Section 9.5, both the micro and meso mechanisms result in the poroelastic moduli being complex and frequency dependent. The macro mechanism credited to Biot/Frenkel (wavelength scale equilibration) is normally modeled using real frequency-independent elastic moduli that are the focus of Section 9.4.

9.2.1 GOVERNING EQUATIONS OF POROUS MEDIA ACOUSTICS

Many authors, including Levy (1979), Burridge and Keller (1981), Pride et al. (1992) and Pride and Berryman (1998), have demonstrated that Biot's (1962) theory is the correct general model for porous-media acoustics. Assuming an $e^{-i\omega t}$ time dependence so that cumbersome time-convolution integrals (associated with the loss processes) are not explicitly present, Biot's (1962) equations controlling isotropic poroelastic response are

$$\nabla \cdot \boldsymbol{\tau}^D - \nabla P_c = -\omega^2(\rho \mathbf{u} + \rho_f \mathbf{w}) \quad (9.1)$$

$$-\nabla p_f = -\omega^2 \rho_f \mathbf{u} - i\omega \frac{\eta}{k(\omega)} \mathbf{w} \quad (9.2)$$

$$\boldsymbol{\tau}^D = G\left[\nabla \mathbf{u} + (\nabla \mathbf{u})^T - \frac{2}{3}\nabla \cdot \mathbf{u}\mathbf{I}\right] \quad (9.3)$$

$$-P_c = K_U \nabla \cdot \mathbf{u} + C\nabla \cdot \mathbf{w} \quad (9.4)$$

$$-p_f = C\nabla \cdot \mathbf{u} + M\nabla \cdot \mathbf{w}. \quad (9.5)$$

The various fields $\mathbf{u}$, $\mathbf{w}$, $\boldsymbol{\tau}^D$, P_c and p_f represent the average response in volumes that are much larger than the grains of the material but much smaller than the wavelengths. If $\overline{\mathbf{u}}_s$ is the average displacement of the solid grains throughout an averaging volume and $\overline{\mathbf{u}}_f$ the average displacement of the fluid in the pore space, then the displacement fields of the theory are $\mathbf{u} = \overline{\mathbf{u}}_s$ and $\mathbf{w} = \phi(\overline{\mathbf{u}}_f - \overline{\mathbf{u}}_s)$ where ϕ is the porosity of the averaging volume. When the area porosity defined on a slice of the porous material is equivalent to the volume porosity (for isotropic media, this can be considered exact in the limit of large averaging volumes), $-i\omega\mathbf{w}$ corresponds to the Darcy filtration velocity. The dilatation $\nabla \cdot \mathbf{u}$ can be shown (e.g., Pride and Berryman, 1998) to accurately represent the total volume dilatation of a sample of material (exactly so when the geometrical center of the grain space is coincident with the geometrical center of the pore space). The dilatation $\nabla \cdot \mathbf{w}$ corresponds to the accumulation or depletion of fluid in the sample and is often called the "increment of fluid content" (denoted $-\zeta$ by Biot). Let the tensor $\boldsymbol{\tau}$ represent the average stress tensor throughout the entire averaging volume (both solid and fluid). Upon separating this average total-stress tensor into isotropic and deviatoric portions $\boldsymbol{\tau} = -P_c\mathbf{I} + \boldsymbol{\tau}^D$ (where $\mathbf{I}$ is the identity tensor), the scalar $P_c = -\mathrm{tr}\{\boldsymbol{\tau}\}/3$ represents the average total pressure acting on a sample (the so-called "confining pressure"), while the tensor $\boldsymbol{\tau}^D$ represents the average shear stress. Last, p_f is the average fluid pressure throughout the pores of a sample.

Generally, the fields and moduli change with the size of the averaging volume. This size is ideally chosen so that the averaged fields correspond to those that a measuring device such as a geophone or geophone group records. In this manner, modeled fields and recorded data may be directly compared in the inverse problem. Alternatively, the size of the averaging volume implicitly corresponds to the size of the discretization element when these equations are solved numerically by finite difference, and is necessarily much smaller than the seismic wavelength. If laboratory experiments are performed to determine the moduli of the theory, the sample size corresponds to the averaging volume (and, therefore, discretization element) employed in the forward model (Pride and Berryman, 1998). Since averaging volumes on the order of cubic meters (and even much larger) are commonly used in seismic forward modeling, one can properly say that no laboratory experiment has ever been performed that measures the moduli/coefficients at the scale required for the seismic forward modeling. Measurement of the moduli at the seismic length scale can only come indirectly from inversion of seismic data.

Equation (9.1) is the total balance of forces acting on each sample, while Equation (9.2) is a generalized Darcy law and is itself a force balance on the fluid from a frame of reference fixed on the skeletal framework of grains. The apparent force $\omega^2\rho_f\mathbf{u}$ that acts along with $-\nabla p_f$ to drive the fluid flux $-i\omega\mathbf{w}$ is caused by wave-induced acceleration of this reference frame. The bulk density ρ of the rock is $\rho = (1-\phi)\rho_s + \phi\rho_f$ where ρ_f is the average density of the pore fluid and ρ_s the average density of the solid grains in a sample. The so-called "dynamic permeability" $k(\omega)$ is a complex frequency-dependent quantity such that at low-frequencies (to be precisely defined in Section 9.3.1) it is exactly the hydrological permeability k_o of a sample ($\lim_{\omega\to 0} k(\omega) = k_o$) while at high-enough frequencies it includes inertial effects associated with the relative fluid-solid movement. The viscosity of the pore fluid is η.

Equations (9.3)–(9.5) are the constitutive equations. Since linear wave propagation is being assumed, the displacements can always be considered small. As such, we have foregone placing a differential "d" in front of the various stresses and strains appearing in Equations (9.3)–(9.5). However, these constitutive equations are obtained by differentiating a strain-energy function and, as such, should always be thought of as differential equations (changes in strain related to changes in stress). Models for the three poroelastic incompressibilities K_U, C, M and for the shear modulus G are the focus of Sections 9.4 (quasi-static) and 9.5 (frequency dependent).

9.2.2 SEISMIC WAVE PROPERTIES

The seismic wave speeds, attenuation, and fluid-pressure diffusivity implicitly contained in Equations (9.1)–(9.5) are now obtained. These results are independent of whether the coefficients k, K_U, C, M, and G are complex and frequency dependent. The low-frequency limit of Equations (9.1)–(9.5) is also shown to yield both the laws of hydrology and seismology.

To obtain these results, one need only consider a homogeneous porous continuum and insert the stress/strain relations into the force balances to obtain

$$\left[(K_U + G/3)\nabla\nabla + (G\nabla^2 + \omega^2\rho)\mathbf{I}\right]\cdot\mathbf{u} + \left[C\nabla\nabla + \omega^2\rho_f\mathbf{I}\right]\cdot\mathbf{w} = 0 \quad (9.6)$$

$$\left[C\nabla\nabla + +\omega^2\rho_f\mathbf{I}\right]\cdot\mathbf{u} + \left[M\nabla\nabla + \omega^2\tilde{\rho}\mathbf{I}\right]\cdot\mathbf{w} = 0 \quad (9.7)$$

where $\mathbf{I}$ is again the identity tensor and where the relative-flow resistance in the Darcy law has been written as if it were an inertial property

$$\tilde{\rho}(\omega) = \frac{i}{\omega}\frac{\eta}{k(\omega)}. \quad (9.8)$$

We consider a plane wave that, by definition, has a material response of the form

$$\mathbf{u} = U \exp(i\mathbf{k}\cdot\mathbf{r})\hat{\mathbf{u}} \qquad \text{and} \qquad \mathbf{w} = W \exp(i\mathbf{k}\cdot\mathbf{r})\hat{\mathbf{w}}. \tag{9.9}$$

Here, $\hat{\mathbf{u}}$ and $\hat{\mathbf{w}}$ are unit vectors defining the polarization of the response, while $\mathbf{k}$ is the wave vector that we write in the form

$$\mathbf{k} = \omega s(\omega)\hat{\mathbf{k}}, \tag{9.10}$$

where $s(\omega)$ is the complex slowness (to be determined) and $\hat{\mathbf{k}}$ is a unit vector in the direction of propagation.

The quantities of interest are the phase velocity v and attenuation coefficient a (units of inverse length) for the various wave types that are related to the complex slowness as

$$v(\omega) = 1/\mathrm{Re}\{s(\omega)\} \quad \text{and} \quad a(\omega) = \omega \mathrm{Im}\{s(\omega)\}. \tag{9.11}$$

Taking $\hat{\mathbf{k}}$ to be in the x direction, the plane-wave spatial response is then of the form $e^{-a(\omega)x}e^{i\omega x/v(\omega)}$ and a is seen to control the exponential decay of the wave amplitude with distance x propagated. It is sometimes convenient to use the inverse of the quality factor Q (dimensionless) given by

$$Q^{-1}(\omega) = \frac{2a(\omega)v(\omega)}{\omega} = 2\frac{\mathrm{Im}\{s(\omega)\}}{\mathrm{Re}\{s(\omega)\}} \tag{9.12}$$

as the measure of attenuation. By definition, Q^{-1} represents the energy lost in a wave period divided by the average stored strain energy (and divided by 4π).

Putting the plane-wave response into Equations (9.6) and (9.7) gives

$$\begin{bmatrix} (K_U + G/3)s^2(\hat{\mathbf{k}}\cdot\hat{\mathbf{u}})\hat{\mathbf{k}} - (\rho - Gs^2)\hat{\mathbf{u}} & Cs^2(\hat{\mathbf{k}}\cdot\hat{\mathbf{w}})\hat{\mathbf{k}} - \rho_f\hat{\mathbf{w}} \\ Cs^2(\hat{\mathbf{k}}\cdot\hat{\mathbf{u}})\hat{\mathbf{k}} - \rho_f\hat{\mathbf{u}} & Ms^2(\hat{\mathbf{k}}\cdot\hat{\mathbf{w}})\hat{\mathbf{k}} - \tilde{\rho}\hat{\mathbf{w}} \end{bmatrix} \begin{bmatrix} U \\ W \end{bmatrix} = 0. \tag{9.13}$$

By definition, a transverse wave has a material response perpendicular to the wave direction ($\hat{\mathbf{k}}\cdot\hat{\mathbf{u}} = \hat{\mathbf{k}}\cdot\hat{\mathbf{w}} = 0$) while a longitudinal wave has a material response parallel with the wave direction ($\hat{\mathbf{k}}\cdot\hat{\mathbf{u}} = \hat{\mathbf{k}}\cdot\hat{\mathbf{w}} = 1$). A plane wave having "mixed response" (i.e., $\hat{\mathbf{u}} \neq \hat{\mathbf{w}}$) leads to the trivial solution $U = W = 0$ and thus does not exist. We may therefore take $\hat{\mathbf{u}} = \hat{\mathbf{w}}$ for plane waves in a homogeneous material.

9.2.2.1 *Transverse Waves*

Upon placing $\hat{\mathbf{k}}\cdot\hat{\mathbf{u}} = \hat{\mathbf{k}}\cdot\hat{\mathbf{w}} = 0$ in Equation (9.13) we obtain

$$\begin{bmatrix} \rho - Gs^2 & \rho_f \\ \rho_f & \tilde{\rho} \end{bmatrix} \begin{bmatrix} U \\ W \end{bmatrix} = 0, \tag{9.14}$$

which has nontrivial solutions only when the determinant vanishes. This occurs at the complex wave slowness s_s given by

$$s_s^2 = \frac{\rho - \rho_f^2/\tilde{\rho}}{G}. \tag{9.15}$$

When G can be considered real (no mesoscopic or microscopic flow), the imaginary part of s_s is due entirely to relative fluid-solid motion induced by the acceleration of the framework of grains.

The amplitude W of this relative displacement (as normalized by the amplitude U of the displacement of the solid grains) can then be read from Equation (9.14)

$$\frac{W}{U} = i\omega \frac{\rho_f k(\omega)}{\eta} \tag{9.16}$$

where the definition of $\tilde{\rho}$ was used. If it were possible to measure the relative displacement $\mathbf{w}$ of a passing shear wave in addition to the average grain velocity $-i\omega\mathbf{u}$ recorded by a geophone, Equation (9.16) would provide a direct measure of permeability. Unfortunately, a device for measuring $\mathbf{w}$ does not presently exist.

9.2.2.2 *Longtitudinal Waves*

Upon placing $\hat{\mathbf{k}} \cdot \hat{\mathbf{u}} = \hat{\mathbf{k}} \cdot \hat{\mathbf{w}} = 1$ in Equation (9.13) we obtain

$$\begin{bmatrix} Hs^2 - \rho & Cs^2 - \rho_f \\ Cs^2 - \rho_f & Ms^2 - \tilde{\rho} \end{bmatrix} \begin{bmatrix} U \\ W \end{bmatrix} = 0 \tag{9.17}$$

where we have introduced the "undrained" P-wave modulus H as

$$H = K_U + 4G/3. \tag{9.18}$$

An undrained elastic response is one in which fluid exchanges between each sample of the earth and its surroundings do not take place. Again, nontrivial solutions of Equation (9.17) are obtained when the determinant vanishes, and this occurs at two values of the slowness that are called here s_{pf} and s_{ps}. Upon solving the quadratic characteristic equations for s^2, one finds that

$$2s^2_{pf,ps} = \gamma \mp \sqrt{\gamma^2 - \frac{4(\tilde{\rho}\rho - \rho_f^2)}{MH - C^2}} \tag{9.19}$$

where taking the "$-$" gives the so-called "fast" P-wave slowness, and taking the "$+$" gives the "slow" P-wave slowness and where γ is the auxiliary parameter

$$\gamma = \frac{\rho M + \tilde{\rho} H - 2\rho_f C}{HM - C^2}. \tag{9.20}$$

Equation (9.19) is what allows P-wave attenuation and dispersion to be determined once models for the complex frequency dependence of H, C, and M are determined. When these three moduli are taken to be real and frequency independent, the associated fast-wave attenuation is called "Biot loss" and results entirely from the fluid flow occuring between the peaks and troughs of the wave.

The relative-displacement amplitude W is then read from Equation (9.17)

$$\frac{W}{U} = \beta_{pf,ps} = -\frac{Hs^2_{pf,ps} - \rho}{Cs^2_{pf,ps} - \rho_f}. \tag{9.21}$$

The nature of $\beta_{pf,ps}$ for fast and slow waves is key to understanding the entirely different nature of these two disturbances. It is informative to investigate the $\beta_{pf,ps}$ in the low-frequency limit defined by the dimensionless condition $\omega\rho_f k_o/\eta \ll 1$. For water in the pores ($\rho_f = 10^3$ kg/m^3 and $\eta = 10^{-3}$ Pa s) and for permeabilities on the order of 10 Darcy or smaller, this limit holds for frequencies satisfying $f < 10^4$ Hz, which corresponds to the entire band of interest in exploration seismology. From Equations (9.19)–(9.21), it is straightforward to obtain

$$\beta_{pf} = \frac{i\omega\rho_f k_o}{\eta}\left(1 - \frac{\rho}{\rho_f}\frac{C}{H}\right)[1 + O(\omega\rho_f k_o/\eta)] \tag{9.22}$$

$$\beta_{ps} = -\frac{H}{C}[1 + O(\omega\rho_f k_o/\eta)]. \tag{9.23}$$

Equation (9.22) shows that in exploration work, the macroscopic relative fluid-solid displacement vanishes for the fast wave, and the response becomes effectively undrained. For the slow wave, the relative fluid-solid displacement is much larger than the displacement of the solid grains (because normally $H \gg C$) and is exactly out of phase with the solid displacement. We will now show that the slow wave at low frequencies is a pure fluid-pressure diffusion and thus not a wave at all.

9.2.2.3 *Understanding the Fast and Slow Response*

As just seen, in either a fast or a slow disturbance, the fluid accumulations are related to the volume changes of a sample by the relation

$$\nabla \cdot \mathbf{w} = \beta_{pf,ps}\nabla \cdot \mathbf{u}. \tag{9.24}$$

This relation is also trivially satisfied by the equivoluminal shear waves. If $\nabla \cdot \mathbf{w} = \beta_{pf}\nabla \cdot \mathbf{u}$ is introduced into the governing equations along with Equation (9.22) for β_{pf}, then to leading order in $\omega\rho_f k_o/\eta$, Biot's equations reduce to

$$(K_U + G/3)\nabla\nabla \cdot \mathbf{u} + G\nabla^2\mathbf{u} + \omega^2\rho\mathbf{u} = 0. \tag{9.25}$$

This is exactly the usual elastodynamic wave equation, having compressional and shear wave slowness given by the standard expressions

$$s_{pf} = \sqrt{\frac{\rho}{K_U + 4G/3}} \quad \text{and} \quad s_s = \sqrt{\frac{\rho}{G}}. \tag{9.26}$$

If K_U and G are complex and frequency dependent in the seismic band of frequencies due to mesoscopic-scale flow, both of these fast and shear waves will be attenuative despite the response being undrained. Again, the term *undrained* means only that fluid does not enter or leave an averaging volume and is independent of any fluid redistributions that occur within the averaging volume.

If we introduce $\nabla \cdot \mathbf{w} = \beta_{ps}\nabla \cdot \mathbf{u}$ into the governing equations and use Equation (9.23) for β_{ps}, then to leading order in $\omega\rho_f k_o/\eta$, Biot's equations reduce to

$$\frac{k_o}{\eta}M\left(1 - \frac{C^2}{MH}\right)\nabla^2 p_f + i\omega p_f = 0 \tag{9.27}$$

which is exactly a diffusion equation for the fluid pressure having a fluid-pressure diffusivity given by

$$D = \frac{k_o}{\eta}M\left(1 - \frac{C^2}{MH}\right). \tag{9.28}$$

So in the seismic band of frequencies ($f < 10$ kHz), the slow wave is in fact a pure fluid-pressure diffusion. Because $C^2 \ll MH$, the small correction in the parentheses that is due to fluid-pressure-induced frame expansion is often neglected in hydrology. The undrained P-wave modulus $H = K_U + 4G/3$ depends on the shear modulus. The shear modulus is present in D because fluid-pressure diffusion, like P-wave propagation, is a uniaxial response (a sum of both compression and shear) with no displacements occuring perpendicular to the diffusion front; i.e., all displacement of the solid is in the direction of diffusion. When seismologists or rock physicists talk about a "slow wave," all that is being referred to is hydrological fluid-pressure diffusion.

So the two low-frequency longtitudinal modes of Biot theory correspond to seismology (fast-wave or "P-wave" response) and hydrology (slow-wave response). In the presense of heterogeneity in the porous continuum, seismology and hydrology are coupled (fast waves generate slow waves), and a possible connection between seismic attenuation and permeability can be envisioned. Details of such a possible relation are explored in Section 9.5.

9.3 Relaxation Processes in Standard Biot Theory

The term "standard Biot theory" simply means that the poroelastic moduli K_U, C, M, and G are all real frequency-independent constants as in Biot's (1956a,b, 1962) work. There are two P-wave relaxation processes in this case. One relaxation occurs at the frequency where viscous boundary layers first develop in the pores of a rock and is allowed for within the so-called dynamic permeability $k(\omega)$. The other relaxation occurs when the fluid pressure between the peaks and troughs of a compressional wave just has time to equilibrate during a wave period.

9.3.1 DYNAMIC PERMEABILITY

At low-enough wave frequencies, the relative fluid flow in the pores has a locally "parabolic" flow profile, and the resistance η/k_o to the average fluid-solid movement is entirely a result of the viscous shearing associated with this flow. However, as frequency increases, so does the importance of inertial forces in the local force balance on the fluid (the Navier-Stokes equations that hold throughout the pore space). Since every fluid in earth materials (water, air, natural gas, oil, LNAPL, DNAPL, CO_2) remains attached to the grain surfaces, there develop viscous boundary layers in the pores that connect the purely inertial "plug-profile" flow in the center of the pores to the no-slip condition on the grains surfaces.

Johnson et al. (1987) exactly determine the nature of the flow in the high-frequency limit where viscous-boundary layers become so thin as to be considered locally planar relative to the curved grain surfaces. They connect this exact high-frequency limit to the low-frequency limit (where η/k_o controls the flow resistance) using a simple frequency function that respects causality constraints. Their final model is

$$\frac{k(\omega)}{k_o} = \left[\sqrt{1 - i\frac{4}{n_J}\frac{\omega}{\omega_J}} - i\frac{\omega}{\omega_J}\right]^{-1} \tag{9.29}$$

where the relaxation frequency ω_J, which controls the frequency at which viscous-boundary layers first develop, is given by

$$\omega_J = \frac{\eta}{\rho_f F k_o}. \tag{9.30}$$

Here, F is exactly the electrical formation factor when grain-surface electrical conduction is not important and is conveniently (though crudely) modeled using Archie's (1942) law $F = \phi^{-m}$. The cementation exponent m is related to the distribution of grain shapes (or pore topology) in the sample and is generally close to 3/2 in clean sands, close to 2 in shaly sands, and close to 1 in rocks having fractured porosity. For an extremely clean sandstone, one might have $\phi = 0.3$ and $k_o = 10^{-12}$ m^2 (1 Darcy), in which case the relaxation frequency is (assuming water in the pores) $f_J = \omega_J/(2\pi) = 10$ kHz, which can almost be considered a lower limit for sedimentary rock in the earth. A more common shaly sandstone might have $\phi = 0.15$ and $k_o = 10^{-14}$ m^2 (10 mD), in which case $f_J = 1$ MHz, a value much more typical of consolidated sandstones. Generally, the onset of viscous boundary layers occurs at a frequency well above the seismic band of exploration frequencies (10 Hz to a few kHz) and, as such, it is safe to simply take $k(\omega) = k_o$ for most seismic applications. Note as well that the condition for the neglect of viscous boundary layers $\omega \ll \omega_J$, is equivalent to the low-frequency limit of the Biot theory, where the fast-wave becomes an undrained seismic response and the slow-wave becomes a pure fluid-pressure diffusion.

Proper modeling of the remaining parameter n_J in Equation (9.29) is, therefore, not essential in seismic exploration. Nonetheless, this parameter is defined

$$n_J = \frac{\Lambda^2}{k_o F} \tag{9.31}$$

where Λ is a weighted pore-volume to pore-surface ratio with the weight favoring constricted regions of the pore space (the pore throats). See Johnson et al. (1987) for the precise mathematical definition of Λ. The modeling choice of convenience is simply to take $n_J = 8$ for all materials, which is experimentally observed to do a fine job in unconsolidated materials and clean sandstones and is the theoretically expected result for cylindrical tube models of the pore space. In shaly sands, the local pore-volume to pore-surface ratio in the throats becomes smaller, owing to the presence of clays on the grains, and in this case the value of n_J becomes smaller than 8 (potentially much smaller). However, since the precise value of n_J is not of great importance in the seismic band of frequencies, we will not propose here how Λ and k_o might depend on clay content.

The frequency dependence of the attenuation in Biot theory is strongly affected by the nature of $k(\omega)$. Recall that Q^{-1} represents the energy lost to heat in a wave period (Darcy flux multiplied by the fluid-pressure gradient and divided by frequency) as normalized by the strain energy (stress multiplied by strain). In symbols, this may be approximated as

$$Q^{-1} \approx \frac{\mathrm{Re}\{k(\omega)\}}{\omega\eta} \frac{|\nabla p_f \cdot \nabla p_f|}{H|\nabla \cdot \mathbf{u}|^2} \tag{9.32}$$

$$\approx \frac{\omega\rho\mathrm{Re}\{\mathrm{k}(\omega)\}}{\eta} \left(\frac{C}{H}\right)^2 \tag{9.33}$$

where a P wavelength of $\lambda = 2\pi\sqrt{(H/\rho)}/\omega$ was used in estimating the pressure gradient, and it is assumed that fluid-pressure equilibration at the scale of the wavelength does not have time to occur at each frequency.

Now the exact high-frequency asymptotic behavior of $k(\omega)$ has real and imaginary parts given by

$$k(\omega) \sim \sqrt{\frac{2}{n_J}}\left(\frac{\omega_J}{\omega}\right)^{3/2} + i\frac{\omega_J}{\omega} \quad \text{as} \quad \frac{\omega_J}{\omega} \to 0. \tag{9.34}$$

Equation (9.33) then shows that when $\omega \ll \omega_J$, Q^{-1} increases linearly with frequency because $\mathrm{Re}\{k(\omega)\} \to k_o$. However, when $\omega \gg \omega_J$, Q^{-1} decreases as the square root of frequency because $\mathrm{Re}\{k(\omega)\}$ decreases as $\omega^{-3/2}$. This behavior is also seen in the exact results based on Equation (9.19) displayed in Figure 9.1 for the curve $\omega_B/\omega_J > 1$.

9.3.2 WAVELENGTH-SCALE FLUID-PRESSURE EQUILIBRATION

The other relaxation in standard Biot theory occurs when (and if) the fluid pressure between the peaks and troughs of the wave just has time to equilibrate in a wave period.

To approximate the frequency $\omega_B = 2\pi f_B$ at which this occurs, we can use the standard result of any diffusion process $\tau = L^2/D$ for the time τ required to diffuse a distance L into a material having a diffusivity D. In the case of a P wave, the distance to be equilibrated L is roughly half of one wavelength λ, which is stated (for algebraic convenience) as $L \approx \lambda/\sqrt{2\pi}$. Now, $\lambda \approx \sqrt{H/\rho}/f$, where f is the wave frequency (Hertz) and the diffusivity is roughly $D \approx Mk_o/\eta$. Thus, the critical frequency is defined when $1/f_B = \tau = \lambda^2/(2\pi D) = H/(2\pi D\rho f_B^2)$ or

$$\omega_B = \frac{H}{M}\frac{\eta}{\rho k_o}. \tag{9.35}$$

Comparing the expressions for ω_J and ω_B, one obtains

$$\frac{\omega_B}{\omega_J} = \frac{H}{M}\frac{\rho_f}{\rho}F. \tag{9.36}$$

When $\omega_B < \omega_J$, the wavelength-scale equilibration occurs at a lower frequency than the onset of viscous boundary layers. In this case, as the wave frequency becomes greater than ω_B, the fluid-pressure equilibrates rapidly in each wave period and the seismic response becomes increasingly drained (i.e., as if the fluid in the pores were no longer present). Accordingly, the velocity decreases when $\omega > \omega_B$ as is seen in Figure 9.1 for the curve $\omega_B/\omega_J = 0.1$. When $\omega_B > \omega_J$, the viscous boundary layers develop and the attenuation peaks and begins to fall as the square root of frequency before wavelength-scale equilibration ever has a chance to occur. In this case, the wavelength-scale equilibration never has time to occur at any frequency. Such behavior is also demonstrated in Figure 9.1 using Equation (9.19) for the complex slowness. For most sedimentary rocks, ω_B/ω_J as given by Equation (9.36) is close to, but greater than, unity.

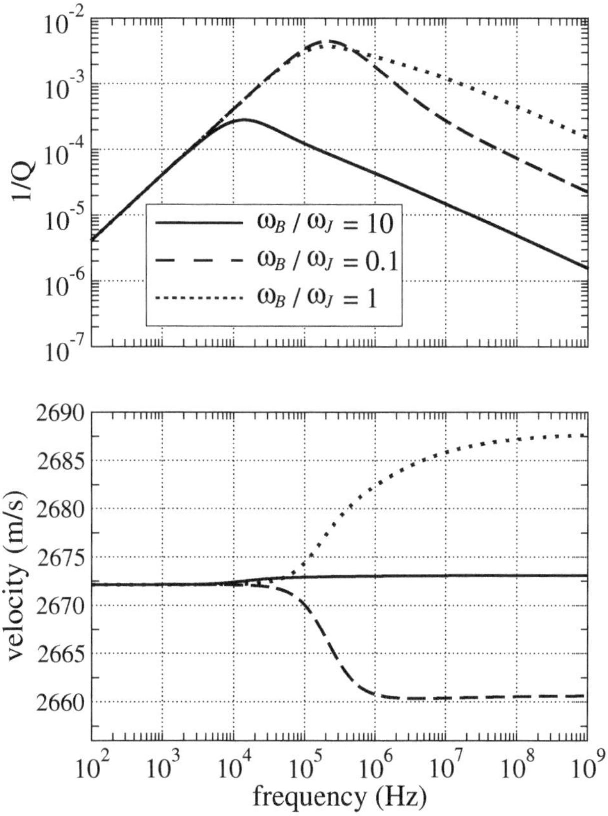

Figure 9.1: Various types of dispersion curves in standard Biot theory. The most common situation in porous rocks is that $\omega_B/\omega_J > 1$ (i.e., viscous boundary layers develop in the pores at a lower frequency compared to when fluid-pressure equilibration at the scale of the wavelength has time to occur). The material properties here correspond to a weakly consolidated sandstone; however, to achieve the unusual condition that $\omega_B/\omega_J = 0.1$, we took the shear modulus to be several times smaller than and the fluid-pressure diffusion modulus M to be several times larger than in the other two curves.

However, the level of dispersion and attenuation associated with these effects is almost totally negligible in the seismic frequency band of interest. Attenuation-dispersion mechanisms of more pertinence to seismic exploration are the focus of Section 9.5.

9.4 The Quasi-Static Poroelastic Moduli

In naturally occuring rocks, the three poroelastic incompressibilities K_U, C, and M, as well as the shear modulus G of the material, will be complex and frequency dependent due to both mesoscopic and squirt flow. Only in the special case of porous materials having uniform frame properties and being free of grain-scale cracks/damage (such as in some artificially created "rocks") will these elastic moduli be real constants at all frequencies. Alternatively, for real rocks in the limit of very low frequencies (possibly below the seismic band of frequencies), the imaginary parts of these moduli will tend to zero and the real parts will become frequency-independent constants. The present section concerns this low-frequency (quasi-static) limit, where fluid pressure may be taken as uniform throughout a porous rock sample.

9.4.1 THE THREE POROELASTIC INCOMPRESSIBILITIES

A multitude of names have been given to the three incompressibilities of poroelasticity (here denoted as K_U, C, and M), which sometimes makes the theory seem more difficult than it really is. From both a laboratory and pedagogic perspective, the three moduli that have the clearest definition are

$$K_U = -\left(\frac{\delta P_c}{\delta V/V_o}\right)_{\nabla\cdot\mathbf{w}=0} \tag{9.37}$$

$$K_D = -\left(\frac{\delta P_c}{\delta V/V_o}\right)_{\delta p_f=0} \tag{9.38}$$

$$B = \left(\frac{\delta p_f}{\delta P_c}\right)_{\nabla\cdot\mathbf{w}=0}. \tag{9.39}$$

The modulus K_U is called the "undrained bulk modulus" because it is defined under the condition $\nabla\cdot\mathbf{w} = 0$, where fluid is not allowed to either enter or leave the sample during deformation. Recall that $\nabla\cdot\mathbf{u} = \delta V/V_o$ where δV is the volume change of a sample that initially occupied a volume V_o. The modulus K_D is called the "drained bulk modulus" and is defined under the condition that the fluid pressure in a sample does not change. The modulus K_D is thus not affected by the presence of fluid in the rock. A drained experiment with any of oil, water or air in the pores should all produce the same value for K_D. Last, the undrained fluid-pressure to confining-pressure ratio B is called "Skempton's coefficient," after the work of Skempton (1954).

These three moduli can be conveniently measured in the lab (c.f., Wang, 2000). An impermeable jacket is usually fitted around a lab sample and a tube is inserted through the jacket, allowing the fluid in the pores to be connected to a reservoir whose pressure p_f may be controlled and measured. The jacketed sample is then immersed in a second fluid reservoir characterized by a confining pressure P_c. First, an increment δP_c is applied in such a manner that no fluid flows through the tube (this is done by controlling the pressure p_f). The change in volume of the jacketed sample is recorded to give K_U and the fluid pressure increment δp_f is read off to give B. Next, δp_f is returned to zero, allowing fluid to flow through the tube. The subsequent change in sample volume then yields K_D.

Using these definitions, it is easy to demonstrate that the moduli C and M in Equations (9.4)–(9.5) are given by

$$C = BK_U \quad \text{and} \quad M = BK_U/\alpha \tag{9.40}$$

where the new constant α is called the "Biot-Willis constant," after Biot and Willis (1957), and is exactly defined as

$$\alpha = \left(\frac{\delta P_c}{\delta p_f}\right)_{\nabla\cdot\mathbf{u}} = \frac{1 - K_D/K_U}{B}. \tag{9.41}$$

Berge et al. (1993) have demonstrated the not-so-obvious and important fact that although K_U and B are both dependent on the type of fluid in the pores, α is not. All of these definitions and relations are independent of the possible presence of anisotropy at either the sample or grain scale, and are also independent of whether the grains making up the rock have different mineralogies. The modulus M is called the "fluid-storage coefficient," since it represents how much fluid can accumulate in a sample when the fluid-pressure changes at constant sample size. It is the elastic modulus principally involved in fluid-pressure diffusion, as Equation (9.27) demonstrates. There is no standard name for the modulus C other than Biot's (1962) "coupling modulus."

It is sometimes convenient to rewrite the incompressibility laws [Equations (9.4) and (9.5)] as compressibility laws

$$\begin{pmatrix} \nabla\cdot\mathbf{u} \\ \nabla\cdot\mathbf{w} \end{pmatrix} = \frac{1}{K_D}\begin{pmatrix} 1 & -\alpha \\ -\alpha & \alpha/B \end{pmatrix}\begin{pmatrix} -\delta P_c \\ -\delta p_f \end{pmatrix}. \tag{9.42}$$

We often use K_D, B, and α as the fundamental suite of three poroelastic constants.

9.4.2 THE BIOT-GASSMANN FLUID SUBSTITUTION RELATIONS

From the perspective of connecting seismic velocities and hydrological properties, the pioneering contribution of Gassmann (1951) is without rival. Gassmann showed that if the solid material making up the grains is both isotropic and uniform throughout a sample, then in the limit of low frequencies, the undrained modulus K_U and Skempton's coefficient B are frequency-independent constants that can be exactly expressed in terms of the drained modulus K_D, the porosity ϕ, the mineral modulus of the grains K_s, and the fluid modulus K_f

$$B = \frac{1/K_D - 1/K_s}{1/K_D - 1/K_s + \phi(1/K_f - 1/K_s)}, \tag{9.43}$$

$$K_U = \frac{K_D}{1 - B(1 - K_D/K_s)}. \tag{9.44}$$

From these results, one also has $\alpha = 1 - K_D/K_s$ which is indeed seen to be independent of K_f. The importance of Equations (9.43) and (9.44) is that all dependence on the fluid type is entirely confined to the fluid modulus K_f. These relations tell us how the poroelastic incompressibilities change when one fluid is substituted for another. Gassmann also made the reasonable assumption that at sufficiently low frequencies, the fluid has no influence on the shear modulus G.

After some algebra, the following forms for the Biot-Gassmann incompressibilities are instructive

$$K_U = \frac{K_D + [1 - (1+\phi)K_D/K_s]\, K_f/\phi}{1+\Delta}, \tag{9.45}$$

$$C = \frac{(1 - K_D/K_s)K_f/\phi}{1+\Delta}, \tag{9.46}$$

$$M = \frac{K_f/\phi}{1+\Delta}, \tag{9.47}$$

where Δ is a dimensionless parameter defined as

$$\Delta = \frac{1-\phi}{\phi}\frac{K_f}{K_s}\left(1 - \frac{K_D}{(1-\phi)K_s}\right). \tag{9.48}$$

These are called the "Biot-Gassmann" relations, because although Gassmann (1951) treated K_U and C, it was Biot and Willis (1957) who first treated M (Gassmann only considered the undrained response). These particular algebraic forms are useful because Δ is always a small number. In an extreme stiff-frame limit defined by $K_D \to (1-\phi)K_s$ (which actually lies above the Hashin and Shtrikman, 1963, upper bound), $\Delta \to 0$. The opposite limit of an infinitely compliant frame $K_D \to 0$ occurs when the grains no longer form connected paths across the sample. In sediments, this percolation threshhold occurs when $\phi \approx 0.5$, with the precise value depending on details of the grain-size distribution and packing configuration. Thus, Δ takes its largest value of K_f/K_s when there is an infinitely compliant frame and is never outside the range $0 < \Delta < K_f/K_s$ for any material type. For a liquid, one generally has $K_f/K_s \approx 10^{-1}$, while for a gas, $K_f/K_s \approx 10^{-5}$.

Equations (9.45)–(9.47) demonstrate that $K_U \geq M \geq C$ in a Gassmann material. Only for very soft unconsolidated materials in which $K_D \ll K_f/\phi$, does one obtain $K_U \approx M \approx C$.

9.4.3 THE DRAINED BULK MODULUS K_D AND SHEAR MODULUS G

For the fluid-substitution relations to be useful in interpreting and/or inverting seismic data, it will usually be necessary to have a theoretical model for K_D. In the low-frequency limit, both K_D and G depend only on the microgeometry of how the framework of grains is put together and on the minerals making up the grains. There exists a vast literature surrounding the various theoretical models for such "frame moduli" (c.f., Berryman, 1995, and Mavko et al., 1998). No one universal model is valid for all porous materials. The goal here is restricted to recommending a few simple models that are consistent with data.

The single most important factor in choosing a theoretical model for K_D and G is whether or not the grains are cemented together; i.e., is the material consolidated or unconsolidated?

9.4.3.1 *Unconsolidated Materials*

If a material is unconsolidated (e.g., a sandpack or soil), the stress concentration and deformation in the micron-scale regions surrounding the individual grain-to-grain contact points is what controls the overall deformation. Such contact points are most often modeled as idealized sphere-to-sphere "Hertzian" contacts from which theoretical results for the overall packing can be approximately derived. A key concept is that the area of the individual grain contacts increases with effective stress $P_e = P_c - p_f$ (see Section 9.4.5 for a discussion of effective stress). As the grain-contact area increases, both the contact and the material become stiffer and more rigid. Data on natural sands (e.g., Hardin and Richart, 1963; Domenico, 1977) indicate that both G and K_D increase with effective stress as $P_e^{1/2}$ for $0 < P_e < P_o$, where P_o is observed to be on the order of 10 MPa (corresponding to a depth of roughly 1 km). For $P_e > P_o$, both G and K_D roll over to a more gradual $P_e^{1/3}$ increase.

For natural soils having a spectrum of grain sizes, there need not be a clear relation between the porosity and the frame moduli. Imagine a packing of large 500 μm sand grains. If smaller 10 μm grains are loosely present in the voids between the larger grains, but do not act as stress bridges, they can markedly change the porosity of the soil while leaving the frame moduli effectively unchanged. Most theoretical models are based on packings of single-radius spheres, in which case known relations exist between the porosity and the type of packing.

Walton (1987) has produced perhaps the simplest model for a random packing of identical spheres:

$$K_D = \frac{1}{6}\left[\frac{3(1-\phi_o)^2 n^2 P_e}{\pi^4 C_s^2}\right]^{1/3} \quad \text{and} \quad G = RK_D \tag{9.49}$$

where n is the average number of contacts per grain, known as the "coordination number," ϕ_o is the porosity of the random packing at $P_e = 0$, and C_s is a compliance parameter defined as

$$C_s = \frac{1}{4\pi}\left(\frac{1}{G_s} + \frac{1}{K_s + G_s/3}\right) \tag{9.50}$$

where G_s and K_s are the single-mineral moduli of the grains. The parameter R takes on a value somewhere in the range

$$\frac{3}{5} \le R \le \frac{18}{5}\left(\frac{K_s + G_s}{3K_s + 2G_s}\right) \tag{9.51}$$

where the lower limit corresponds to grains so smooth that tangential slip always occurs, which prevents shear force from being transmitted at the contact, and the upper limit corresponds to grains so rough that no slip occurs which results in a maximum transmitted shear at a contact. Our experience is that the lower limit of $R = 3/5$ does a somewhat better job matching data.

Walton (1987) derived these results assuming that n was a constant (all contacts in place at $P_e = 0$). His result is at odds with the experimentally observed $P_e^{1/2}$ dependence when $P_e < P_o$, which corresponds to the entire depth range of interest in hydrogeophysics. One way to account for the observed pressure dependence is to assume that the average coordination number increases with the negative dilatation $\epsilon = -\nabla \cdot \mathbf{u}$ of the material. At $P_e = 0$, the material is assumed to have, effectively, no contacts in place, but as the material is compressed, the contacts between the grains are created. Goddard (1990) has proposed that stress-free grain rotations create grain-to-grain contacts that result in the law

$$n(\epsilon) = \begin{cases} n_o \left(\epsilon/\epsilon_o\right)^{1/2} & \text{when } P_e \ll P_o, \\ n_o & \text{when } P_e \gg P_o, \end{cases} \tag{9.52}$$

where ϵ_o is the negative dilatation when $P_e = P_o$, and where n_o is the maximum number of contacts per grain that can arrive. Using this law within the Walton (1987) theory gives the relation between dilatation and pressure to be $\epsilon(P_e) = \{6\pi^2 C_s \epsilon_o^{1/2} P_e/[(1 - \phi_o)n_o]\}^{1/2}$ whenever $P_e \ll P_o$. It is then straightforward to re-express the Walton result as

$$K_D = \frac{1}{6}\left[\frac{4(1-\phi_o)^2 n_o^2 P_o}{\pi^4 C_s^2}\right]^{1/3} \frac{(P_e/P_o)^{1/2}}{\{1 + [16P_e/(9P_o)]^4\}^{1/24}} \tag{9.53}$$

which is a new result, though anticipated by Goddard (1990). The new parameter compared to Walton's (1987) theory is P_o, which is the pressure beyond which $n = n_o$ and no new contacts are created. It is taken here as an empirical constant on the order of 10 MPa in sands. If we put $P_o = 0$, the Walton result [Equation (9.49)] is exactly recovered. The exponents of 4 and 1/24 in the denominator of Equation (9.53) were chosen somewhat arbitrarily, the only requirement being that their product is 1/6. (If we use 8 and 1/48, for example, the results are imperceptibly different.)

Equation (9.53), along with $G = 3K_D/5$, is our recommendation for the drained moduli of unconsolidated materials which are the key properties affecting wave speeds in hydrogeophysics. If the same values for K_s and G_s are always employed (a reasonable modeling choice), there are two free parameters; namely, P_o and $(1-\phi_o)n_o$. For random sand packs, we have $0.32 < \phi_o < 0.38$ and $8 < n_o < 11$.

In Figure 9.2, this model with $P_o = 18$ MPa, $R = 3/5$, $\phi_o = 0.36$, $n_o = 9$, $K_s = 37$ GPa, and $G_s = 44$ GPa is compared to data collected by Murphy (1982) on a random glass-bead pack in which the bead-diameters ranged from 74 to 105 μm (common soil grain sizes) and in which the pore space was empty. Murphy (1982) chose this pressure range because it exhibits the $P_e^{1/2}$ to $P_e^{1/3}$ transition in the modulus dependence. The theory does an adequate job of modeling this transition. Particularly interesting is that the model parameters have not been adjusted to give an optimal fit; they are the parameters one *a priori* expects to hold for a dense random pack of identical quartz spheres. Although the difference between Equation (9.53) and Equation (9.49) appears to be minimal in this example, if the velocities are determined over the lower-pressure ranges of interest in hydrogeophysics, there is a pure $P_e^{1/2}$ modulus dependence, as has been documented in many data sets including Hardin and Richart (1963) and Domenico (1977).

In conclusion, we emphasize that Equation (9.53) is entirely independent of grain size and, therefore, permeability (permeability depends on the size of the pores and, therefore, grains). Thus, the predicted low-frequency seismic velocities for any unconsolidated material having a unimodal grain-size distribution is expected to be independent of grain

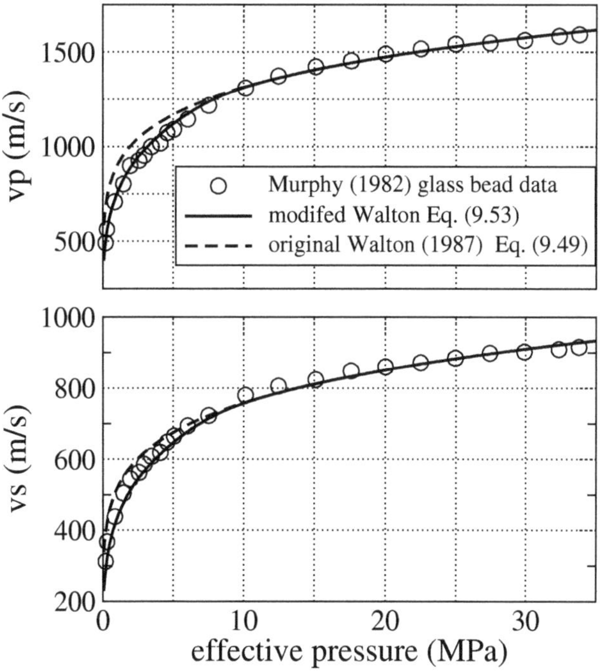

Figure 9.2: Comparison of the Murphy (1982) data set on a dry glass-bead pack to the theoretical model discussed in the text. The $P_e^{1/2} \rightarrow P_e^{1/3}$ transition occurs at an effective pressure of 10 MPa corresponding to P_o = (16/9)10 MPa $\approx$ 18 MPa.

size and permeability, as has been experimentally demonstrated by Murphy (1982) using glass-bead packs.

But, as is shown in Chapter 13 of this volume, field measurements of permeability and seismic velocities indicate the two are sometimes at least weakly correlated in unconsolidated sediments (although in some cases the observed correlation is positive, while in others it is negative). Presently existing models for the seismic velocities in unconsolidated sediments are all based on single grain-size distributions. Any link between permeability and velocity must necessarily come from having a spectrum of grain sizes present. Unfortunately, statistical relations between grain-size distribution and quantities like average number of contacts per grain are generally not available since they depend strongly on how the grain packing was prepared. Possible connections between permeability and seismic wave speeds will be more thoroughly discussed in Section 9.6.

9.4.3.2 *Consolidated Materials*

In a consolidated material, diagenetic clay or quartz has been deposited around the grain contacts, and the Hertzian nature of such contacts is largely if not entirely removed. The material in this case behaves more like a pure solid with cavities. Numerous effective-medium theories (e.g., Berryman, 1980a,b) have been proposed that treat how various shaped cavities (mostly ellipsoidal) at various volume fractions influence the drained moduli. Although such embedded cavities do not form a connected porosity as in real rocks, they are often used in estimating the drained frame moduli so long as each cavity is empty. Upon using sufficient numbers of relatively stiff spherical cavities and relatively compliant penny-shaped cavities, effective-medium models can usually predict fairly well the measured frame moduli of consolidated rock, including the pressure dependence (which is controlled largely by crack closure).

Models based on randomly placed ellipsoidal inclusions often have the implicit form $K_D = f_1(K_D, G, K_s, G_s, a_r, \phi)$ and $G = f_2(K_D, G, K_s, G_s, a_r, \phi)$ where a_r is the aspect ratio of the embedded cavities and ϕ the porosity. The functions f_1 and f_2 are different depending on the theory, but are highly nonlinear in the variables. Uncoupled analytical expressions for K_D and G are generally not available. If a spectrum of ellipsoidal shapes is to be included (e.g., Toksoz et al., 1976), the model becomes even more complicated.

Our recommendation is to work with the following simple forms

$$K_D = K_s \frac{1-\phi}{1+c\phi} \quad \text{and} \quad G = G_s \frac{1-\phi}{1+3c\phi/2} \tag{9.54}$$

where the parameter c is called the "consolidation parameter," since it characterizes the degree of consolidation between the grains. Again, it seems a reasonable modeling choice to fix K_s and G_s and to only allow the consolidation parameter c and porosity ϕ to change from one rock type to the next (though one may want to distinguish between carbonates for which $K_s \approx 2G_s$ and silicate grains for which $K_s \approx G_s$). Effective-medium theories (e.g., Korringa et al., 1979; Berryman, 1980a,b) can be approximately manipulated into expressions of this form and predict that c depends both on the shape of the cavities and the ratio G_s/K_s. The factor of $3/2$ in the expression for G is somewhat arbitrary (working with $5/3$ or 2 is also reasonable) but yields plots of P-wave velocity versus S-wave velocity that are consistent with data on sandstones (e.g., Castagna et al., 1993). Depending on the degree of cementation, one can expect the approximate range $2 < c < 20$, for consolidated sandstones (2 being extremely consolidated and 20 poorly consolidated). An unconsolidated sand in this model can require $c \gg 20$ in which case it may be more appropriate to use the above modified Walton theory [Equation (9.53)]. The frame moduli for various ϕ and c in this model are given in Figure 9.3. With K_s and G_s fixed, both ϕ and c become the targets of seismic inversion.

9.4.4 POROSITY CHANGE

A general expression exists for how porosity changes $\delta\phi$ are related to changes in confining pressure δP_c and fluid pressure δp_f. This is a central result of poroelasticity that is useful in hydrological (and other) applications of poroelasticity, even if it is not directly needed in modeling linear wave propagation.

Let V_ϕ define the pore volume within a sample of total volume V. By definition, the porosity is $\phi = V_\phi/V$. Taking the derivative yields $\delta\phi = \delta V_\phi/V - \phi\delta V/V$ where

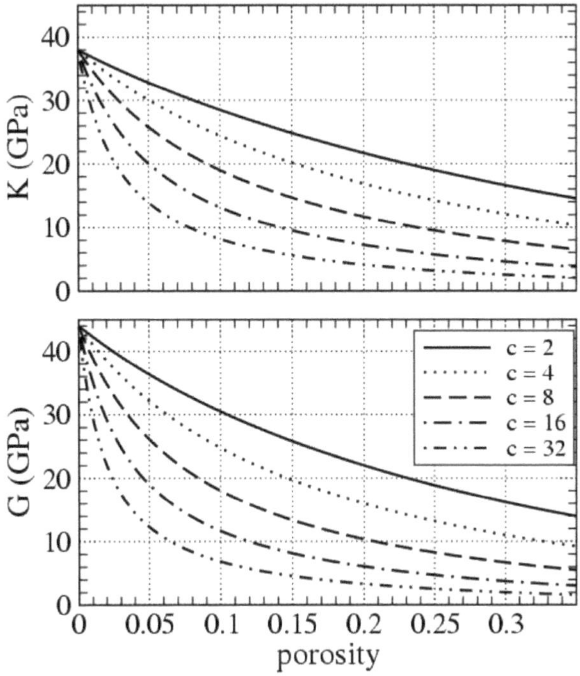

Figure 9.3: The frame moduli of Equation (9.54) for various porosities and consolidation factors c. The solid was taken to be quartz ($K_s = 38$ GPa and $G_s = 44$ GPa).

$\delta V/V = \nabla \cdot \mathbf{u}$. The only way the pore volume changes is if the increment of fluid content $-\nabla \cdot \mathbf{w}$ differs from the fluid volume change due to compression $\phi \delta p_f / K_f$. Thus,

$$\delta\phi = -\phi \nabla \cdot \mathbf{u} - \nabla \cdot \mathbf{w} - \phi \delta p_f / K_f \tag{9.55}$$

which, when combined with the compressibility laws of Equation (9.42), gives exactly

$$\delta\phi = -C_\phi \left(\delta P_c - \alpha_\phi \delta p_f\right) \tag{9.56}$$

where the porosity compliance C_ϕ is defined as

$$C_\phi = \frac{1}{B}\left(\frac{1}{K_D} - \frac{1}{K_U}\right) - \frac{\phi}{K_D} \tag{9.57}$$

and where the porosity effective-stress coefficient α_ϕ is

$$\alpha_\phi = \frac{\phi/K_f + (\phi - 1/B)(1/K_D - 1/K_U)/B}{\phi/K_D - (1/K_D - 1/K_U)/B}. \tag{9.58}$$

In the special case of a Gassmann material [mono-mineral isotropic grains for which Equations (9.43) and (9.44) apply], one then has the convenient result

$$C_\phi = \frac{1-\phi}{K_D} - \frac{1}{K_s} \quad \text{and} \quad \alpha_\phi = 1. \tag{9.59}$$

Note that the porosity changes being modeled here are the purely elastic reversible changes. The actual porosity variation with depth in the crust is controlled almost exclusively by diagenetic changes (deposition and dissolution of minerals). The usefulness of Equation (9.56) is for modeling porosity change when fluids are being pumped into or out of sand formations, such as in CO_2 sequestration.

9.4.5 EFFECTIVE STRESS

Effective stress refers to what linear combination of confining pressure change δP_c and fluid pressure change δp_f are required to affect change of a particular material property. So far, we have seen three effective-stress laws; namely, the law for how the volume V of a sample changes,

$$-\frac{\delta V}{V} = -\nabla \cdot \mathbf{u} = \frac{1}{K_D} \left(\delta P_c - \alpha \delta p_f\right), \tag{9.60}$$

the law for how much fluid volume δV_ς either enters or leaves a porous sample,

$$\frac{\delta V_\varsigma}{V} = \nabla \cdot \mathbf{w} = \frac{\alpha}{K_D} \left(\delta P_c - \frac{1}{B}\delta p_f\right), \tag{9.61}$$

and the just-mentioned law for how the porosity changes $\delta\phi = -C_\phi\left(\delta P_c - \alpha_\phi \delta p_f\right)$ where C_ϕ and α_ϕ are defined in Equations (9.57) and (9.58).

All porous material properties change to some degree when either δP_c or δp_f change and there is an effective-stress variable $\delta P_c - \alpha_\pi \delta p_f$ for each such property π (c.f., Berryman, 1992). Alternatively stated, if $\delta P_c = \alpha_\pi \delta p_f$, property π will not change. The coefficient α_π is known as the effective-stress coefficient of the property π (note that the effective-stress coefficient for volume change is $\alpha_V = \alpha$, which is what we also call the Biot-Willis constant).

In near-surface hydrological problems, the relatively soft materials are quite sensitive to the fluid-pressure changes, and it is of interest from a monitoring perspective to know how the geophysical properties of the subsurface are changing due to fluid-pressure changes caused, for example, by pumping.

Any physical property that is scale invariant (i.e., a property that does not change if the grain space is uniformly expanded or contracted) should have an effective-stress variable $\delta P_c - \alpha_\phi \delta p_f$ that is identical to the porosity variable, since porosity is scale invariant (Berryman, 1992). Such properties include soil density, electrical conductivity, and the low-frequency and high-frequency limits of the poroelastic moduli. Since $\alpha_\phi = 1$ in a Gassmann material (isotropic mono-mineral grains), the effective-stress combination $\delta P_c - \delta p_f$ is often employed (such as in Section 9.4.3.1 for the drained moduli of unconsolidated sands). So the general rule for a scale-invariant property π that, accordingly, depends only on the porosity ϕ is that $\delta\pi = [d\pi(\phi)/d\phi]\,\delta\phi$, where the porosity change $\delta\phi$ is given by Equation (9.56) and where the derivative $d\pi/d\phi$ requires a model for the property $\pi(\phi)$.

Permeability is an important example of a material property that is not scale invariant. The general form of the permeability (e.g., Thompson et al., 1987) is $k = \ell^2/F$, where F is the electrical formation factor and ℓ is some appropriately defined pore-throat dimension. This gives

$$\frac{\delta k}{k} = 2\frac{\delta \ell}{\ell} - \frac{\delta F}{F}. \tag{9.62}$$

Upon stressing the material, the length ℓ will change only if the pore volume V_ϕ changes. Thus, it is reasonable to assume that $\ell = \text{const}\, V_\phi^{1/3}$ which results in

$$\frac{\delta \ell}{\ell} = \frac{1}{3}\frac{\delta V_\phi}{V_\phi} = -\frac{1}{3}\left(\frac{\nabla \cdot \mathbf{w}}{\phi} + \frac{\delta p_f}{K_f}\right). \tag{9.63}$$

If Archie's (1942) law $F = \phi^{-m}$ for the electrical formation factor is employed, one obtains

$$\frac{\delta F}{F} = -m\frac{\delta \phi}{\phi} = m\left(\frac{\nabla \cdot \mathbf{w}}{\phi} + \frac{\delta p_f}{K_f} + \nabla \cdot \mathbf{u}\right). \tag{9.64}$$

Putting this together then gives an effective-stress model for permeability

$$\frac{\delta k}{k} = -\frac{[2\alpha/3 + m(\alpha - \phi)]}{\phi K_D}\left(\delta P_c - \alpha_k \delta p_f\right) \tag{9.65}$$

where

$$\alpha_k = \frac{\phi \alpha m - (2/3 + m)(\alpha/B - \phi)}{\phi m - (2/3 + m)\alpha} \tag{9.66}$$

is the effective-stress coefficient for permeability in the model.

9.5 Attenuation and Dispersion in the Seismic Band of Frequencies

The intrinsic attenuation Q^{-1} so far allowed for in standard Biot theory is inadequate to explain the levels of attenuation observed on seismograms in the field.

For transmission experiments (VSP, crosswell tomography, sonic logs), the total attenuation inferred from the seismograms can be decomposed as $Q_{\text{total}}^{-1} = Q_{\text{scat}}^{-1} + Q^{-1}$ where both the scattering and intrinsic contributions are necessarily positive. In transmission experiments, multiple scattering transfers energy from the coherent first-arrival pulse into the coda and into directions that will not be recorded on the seismogram, and is thus responsible for an effective "scattering attenuation" Q_{scat}^{-1}. Techniques have been developed that attempt to separate the intrinsic loss from the scattering loss in transmission experiments (e.g., Wu and Aki, 1988, and Sato and Fehler, 1998). In seismic reflection experiments, back-scattered energy from the random heterogeneity can sometimes act to enhance the amplitude of the primary reflections. At the present time, techniques that can reliably separate the total loss inferred from reflection experiments into scattering and intrinsic portions are not available.

Inverting seismic data for attenuation is not yet standard practice, and there are only a limited number of published examples. Quan and Harris (1997) use tomography to invert the amplitudes of crosswell P-wave first arrivals to obtain the Q^{-1} for the layers of a stratified sequence of shaly sandstones and limestones (depths ranging from 500–900 m). Crosswell experiments in such horizontally stratified sediments produce negligible amounts of scattering loss, so that essentially all apparent loss (except for easily corrected spherical spreading) is attributable to intrinsic attenuation. The center frequency in the Quan and Harris (1997) measurements is roughly 1750 Hz, and they find that $10^{-2} < Q^{-1} < 10^{-1}$ for all the layers in the sequence. Sams et al. (1997) have also measured the intrinsic loss in a stratified sequence of water-saturated sandstones, siltstones, and limestones (depths ranging from 50–250 m) using VSP (30–280 Hz), crosswell (200–2300 Hz), sonic logs (8–24 kHz), and ultrasonic laboratory (500–900 kHz) measurements. They calculate that in the VSP experiments, $Q^{-1}/Q^{-1}_{\text{scat}} \approx 4$, while in the sonic experiments, $Q^{-1}/Q^{-1}_{\text{scat}} \approx 19$; i.e., for their sequence of sediments, the intrinsic loss dominates the scattering loss at all frequencies. Sams et al. (1997) also find that $10^{-2} < Q^{-1} < 10^{-1}$ across the seismic band.

One way to explain these measured levels of intrinsic attenuation in the seismic band is if mesoscopic heterogeneity is present within each averaging volume. A P wave creates a larger fluid-pressure change where the sediments are relatively more compliant than where they are relatively more stiff. A fluid pressure equilibration ensues that attenuates wave energy. This mechanism has a Q^{-1} that can peak anywhere within the seismic band, depending on the length scale of the heterogeneity and the permeability of the material. White et al. (1975), Norris (1993), Gurevich and Lopatnikov (1995), and Gelinsky and Shapiro (1997) have all modeled such mesoscopic loss in the case of waves normally incident to a sequence of thin porous layers.

The approach taken here is to model the mesoscopic-scale heterogeneity as an arbitrary mixture of two porous phases. Each porous phase is assumed to locally obey Biot's Equations (9.1)–(9.5) and both porous phases are present in an averaging volume of the earth. In this case, Pride and Berryman (2003a,b), following up on work by Berryman and Wang (1995), have shown that the macroscopic-scale compressibility laws (the laws controlling the response of each sample of a two-porous phase composite) are

$$\begin{bmatrix} \nabla\cdot\mathbf{v} \\ \nabla\cdot\mathbf{q}_1 \\ \nabla\cdot\mathbf{q}_2 \end{bmatrix} = i\omega \begin{bmatrix} a_{11} & a_{12} & a_{13} \\ a_{12} & a_{22} & a_{23} \\ a_{13} & a_{23} & a_{33} \end{bmatrix} \cdot \begin{bmatrix} P_c \\ \overline{p}_{f1} \\ \overline{p}_{f2} \end{bmatrix} + i\omega \begin{bmatrix} 0 \\ \zeta_{\text{int}} \\ -\zeta_{\text{int}} \end{bmatrix}, \tag{9.67}$$

$$-i\omega\zeta_{\text{int}} = \gamma(\omega)\,(\overline{p}_{f1} - \overline{p}_{f2}). \tag{9.68}$$

Here, $\mathbf{v}$ is the average particle velocity of the solid grains throughout an averaging volume, $\mathbf{q}_i$ is the average Darcy flux across phase i, P_c is the average total pressure acting on the averaging volume, $\overline{p}_{fi}$ is the average fluid pressure within phase i, and $-i\omega\zeta_{\text{int}}$ is the average rate at which fluid volume is being transferred from phase 1 into phase 2 as normalized by the total volume of the averaging region. The dimensionless increment ζ_{int} represents the "mesoscopic flow."

The compressibilities a_{ij} are real and control the elastic response at high frequencies before there is time for fluid-pressure equilibration between the two porous phases to occur. The internal transport coefficient $\gamma(\omega)$ controls the rate at which fluid is being transferred from one porous phase to the other within a porous sample. It is complex and frequency dependent, and has been shown by Pride and Berryman (2003b) and Pride et al. (2004) to take the form

$$\gamma(\omega) = \gamma_o \sqrt{1 - i\frac{\omega}{\omega_o}}. \tag{9.69}$$

At sufficiently high frequencies defined by $\omega \gg \omega_o$, the fluid-pressure diffusion penetrates only a small distance from one phase into the other during a wave cycle. At sufficiently low frequencies defined by $\omega \ll \omega_o$, the fluid-pressure front has time to advance through the entire sample during a wave cycle, and the final stages of equilibration are better characterized as nearly uniform fluid-pressure gradients that are quasi-statically decreasing in amplitude until they ultimately vanish in the dc limit. Models for the a_{ij}, γ_o, and the transition frequency ω_o will be presented in the sections that follow in various geometrical circumstances.

We now reduce these "double-porosity" compressibility laws to an effective Biot theory having complex frequency-dependent coefficients. The easiest way to do this is to assume that phase 2 is entirely embedded in phase 1, so that the flux $\mathbf{q}_2$ into and out of the averaging volume is zero. By placing $\nabla \cdot \mathbf{q}_2 = 0$ into the compressibility laws of Equation (9.67), the fluid pressure $\overline{p}_{f2}$ can be entirely eliminated from the equations. If we then introduce the average solid displacement $\mathbf{u}$ using $-i\omega\mathbf{u} = \mathbf{v}$ and the relative fluid-solid displacement $\mathbf{w}$ using $-i\omega\mathbf{w} = \mathbf{q}_1$, Biot's compressibility laws [Equations (9.4) and (9.5)] are exactly recovered, but now with complex effective moduli given by

$$\frac{1}{K_D(\omega)} = a_{11} - \frac{a_{13}^2}{a_{33} - \gamma(\omega)/i\omega}, \tag{9.70}$$

$$B(\omega) = \frac{-a_{12}(a_{33} - \gamma(\omega)/i\omega) + a_{13}(a_{23} + \gamma(\omega)/i\omega)}{(a_{22} - \gamma(\omega)/i\omega)(a_{33} - \gamma(\omega)/i\omega) - (a_{23} + \gamma(\omega)/i\omega)^2}, \tag{9.71}$$

$$\frac{1}{K_U(\omega)} = \frac{1}{K_D(\omega)} + B(\omega)\left(a_{12} - \frac{a_{13}(a_{23} + \gamma(\omega)/i\omega)}{a_{33} - \gamma(\omega)/i\omega}\right). \tag{9.72}$$

The effective complex C and M moduli are obtained from these results using Equations (9.40) and (9.41).

In the three sections that follow, examples of attenuation and dispersion will be presented corresponding to: (1) a double-porosity model of the mesoscopic heterogeneity; (2) a patchy-saturation model of the mesoscopic heterogeneity; and (3) a squirt-flow model of the microcracks in the grains.

9.5.1 DOUBLE POROSITY OR PATCHY SKELETAL PROPERTIES

The mesoscopic heterogeneity in this case is modeled as a mixture of two porous skeletons uniformly saturated by a single fluid.

Various scenarios can be envisioned for how two distinct porosity types might come to reside within a single geological sample. For example, even within an apparently uniform sandstone formation, there can remain a small volume fraction of less-consolidated (even non-cemented) sand grains. This is because diagenesis is a transport process sensitive to even subtle heterogeneity in the initial grain pack, resulting in spatially variable mineral deposition (e.g., Thompson et al., 1987) and spatially variable skeletal properties. Alternatively, the two porosities might correspond to interwoven lenses of detrital sands and clays; however, any associated anisotropy in the deviatoric seismic response will not be modeled in this chapter. Jointed rock is also reasonably modeled as a double-porosity material. The joints or macroscopic fractures are typically more compressible and have a higher intrinsic permeability than the background host rock they reside within.

The constants a_{ij} in this case are given exactly as (c.f., Berryman and Pride, 2002; Pride and Berryman, 2003a; and Pride et al., 2004)

$$a_{11} = 1/K \tag{9.73}$$

$$a_{22} = \frac{v_1\alpha_1}{K_1}\left(\frac{1}{B_1} - \frac{\alpha_1(1-Q_1)}{1-K_1/K_2}\right) \tag{9.74}$$

$$a_{33} = \frac{v_2\alpha_2}{K_2}\left(\frac{1}{B_2} - \frac{\alpha_2(1-Q_2)}{1-K_2/K_1}\right) \tag{9.75}$$

$$a_{12} = -v_1 Q_1 \alpha_1/K_1 \tag{9.76}$$

$$a_{13} = -v_2 Q_2 \alpha_2/K_2 \tag{9.77}$$

$$a_{23} = -\frac{\alpha_1\alpha_2 K_1/K_2}{(1-K_1/K_2)^2}\left(\frac{1}{K} - \frac{v_1}{K_1} - \frac{v_2}{K_2}\right) \tag{9.78}$$

where

$$v_1 Q_1 = \frac{1-K_2/K}{1-K_2/K_1} \quad \text{and} \quad v_2 Q_2 = \frac{1-K_1/K}{1-K_1/K_2}. \tag{9.79}$$

Here, v_i is the volume-fraction of phase i in each averaging volume ($v_1 + v_2 = 1$), K_i is the drained frame modulus of phase i, B_i is the Skempton's coefficient of phase i, and α_i is the Biot-Willis constant of phase i. Models for all of these parameters have been given earlier in Section 9.4.

The one parameter in these a_{ij} that has not yet been modeled is the overall drained modulus $K = 1/a_{11}$ of the two-phase composite. It is through K that all dependence on the mesoscopic geometry of the two phases occurs. As more thoroughly discussed by Pride et al. (2004), a reasonable modeling choice for most geological scenarios in which a more compressible phase 2 is embedded within a stiffer phase 1 is the Hashin and Shtrikman (1963) lower bound given by

$$\frac{1}{K+4G_2/3} = \frac{v_1}{K_1+4G_2/3} + \frac{v_2}{K_2+4G_2/3} \tag{9.80}$$

$$\frac{1}{G+\zeta_2} = \frac{v_1}{G_1+\zeta_2} + \frac{v_2}{G_2+\zeta_2} \tag{9.81}$$

where ζ_2 is defined as

$$\zeta_2 = \frac{G_2}{6}\frac{(9K_2+8G_2)}{(K_2+2G_2)}. \tag{9.82}$$

Roscoe (1973) has shown that this lower bound is exactly realized when the Bruggeman (1935) differential-effective-medium theory is used to model phase 2 as a collection of arbitrarily oriented penny-shaped oblate (squashed) spheroids or disks embedded within a stiffer host phase 1.

When phase 2 is much more permeable than phase 1, the low-frequency limit of the internal transport coefficient γ_o is given by

$$\gamma_o = -\frac{k_1 K_1^d}{\eta L_1^2}\left(\frac{a_{12} + B_o(a_{22} + a_{33})}{R_1 - B_o/B_1}\right)[1 + O(k_1/k_2)]\,. \tag{9.83}$$

where the parameters B_o, R_1, and L_1 are now defined. The dimensionless number B_o is the static Skempton's coefficient for the composite and is exactly

$$B_o = -\frac{(a_{12} + a_{13})}{a_{22} + 2a_{23} + a_{33}}. \tag{9.84}$$

The dimensionless number R_1 is the ratio of the average static confining pressure in the hose phase 1 of a sealed sample divided by the confining pressure applied to the sample and is exactly

$$R_1 = Q_1 + \frac{\alpha_1(1 - Q_1)B_o}{1 - K_1^d/K_2^d} - \frac{v_2}{v_1}\frac{\alpha_2(1 - Q_2)B_o}{1 - K_2^d/K_1^d} \tag{9.85}$$

where the Q_i are given by Equation (9.79). Last, the length L_1 is the distance over which the fluid-pressure gradient still exists in phase 1 in the final approach to fluid-pressure equilibrium and is formally defined as

$$L_1^2 = \frac{1}{V_1}\int_{\Omega_1} \Phi_1\, dV \tag{9.86}$$

where Ω_1 is the region of an averaging volume occupied by phase 1 and having a volume measure V_1. The potential Φ_1 has units of length squared and is a solution of an elliptic boundary-value problem that, under conditions where the permeability ratio k_1/k_2 can be considered small, reduces to

$$\nabla^2\Phi_1 = -1 \ \text{in}\ \Omega_1, \tag{9.87}$$

$$\mathbf{n}\cdot\nabla\Phi_1 = 0 \ \text{on}\ \partial E_1, \tag{9.88}$$

$$\Phi_1 = 0 \ \text{on}\ \partial\Omega_{12} \tag{9.89}$$

where $\partial\Omega_{12}$ is the surface separating the two phases within a sample of composite and ∂E_1 is the external surface of the sample that is coincident with phase 1. Pride et al. (2004) suggest values to use for L_1 in various circumstances. For example, if phase 2 is modeled as penny-shaped inclusions of radius a, then $L_1^2 \approx a^2/12$. If it is more appropriate to consider phase 2 as being less permeable than phase 1 (e.g., embedded lenses of clay), then one need only exchange the indices 1 and 2 throughout Equations (9.83)– (9.89) with the exception of B_o, which is independent of permeability.

The transition frequency ω_o corresponds to the onset of a high-frequency regime in which the fluid-pressure-diffusion penetration distance becomes small relative to the scale of the mesoscopic heterogeneity, and is given by

$$\omega_o = \frac{\eta B_1 K_1}{k_1 \alpha_1}\left(\gamma_o \frac{V}{S}\right)^2\left(1 + \sqrt{\frac{k_1 B_2 K_2 \alpha_1}{k_2 B_1 K_1 \alpha_2}}\right)^2 \tag{9.90}$$

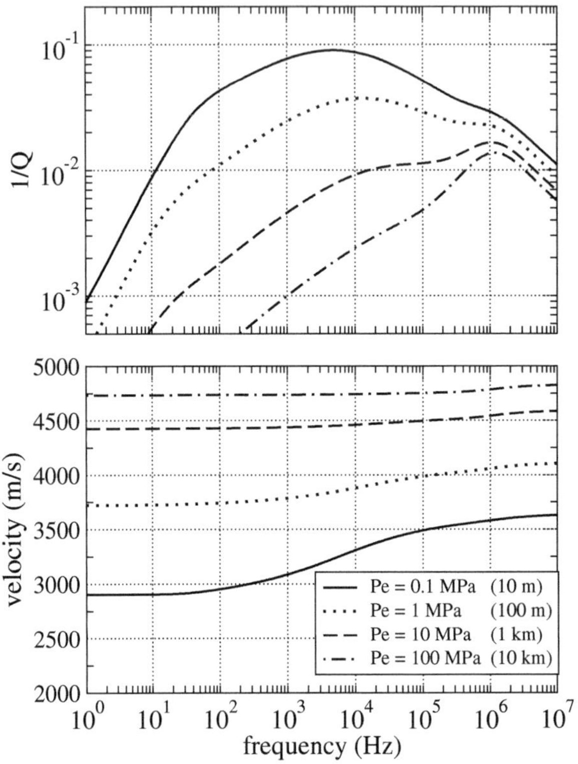

Figure 9.4: The attenuation and phase velocity of compressional waves in the double-porosity model. The thin lenses of phase 2 have frame moduli (K_2 and G_2) modeled using the modified Walton theory given in Section 9.4.3.1 in which both K_2 and G_2 vary strongly with the background effective pressure P_e (or overburden thickness). These lenses of porous continuum 2 are embedded into a phase 1 continuum modeled as a consolidated sandstone.

where S is the surface area of the interface between the two phases in each volume V of composite. For penny-shaped inclusions of phase 2 having radius a, an aspect ratio ϵ and a volume concentration v_2, this volume-to-surface ratio is $V/S = a\epsilon/(2v_2)$.

In Figure 9.4, we give the attenuation and phase determined using the complex slowness of Equation (9.19) and the above complex K_U, C and M moduli. The embedded phase 2 was modeled as 3 cm radius penny-shaped discs having an aspect ratio $\epsilon = 10^{-2}$ and volume concentration of $v_2 = 0.03$. The frame moduli of the embedded spheres were determined using the modified Walton theory of Equation (9.53). These compressible disks of phase 2 are embedded in a consolidated sandstone (phase 1) modeled using Equation (9.54) with $\phi_1 = 0.15$ and $c = 4$. The permeabilities of the two phases are taken as $k_1 = 10^{-14}$ m^2 and $k_2 = 10^{-12}$ m^2. The invariant peak near 10^6 Hz is that caused by the Biot loss (fluid equilibration at the scale of the seismic wavelength), while the principal peak that changes with P_e is that caused by mesoscopic-scale equilibration.

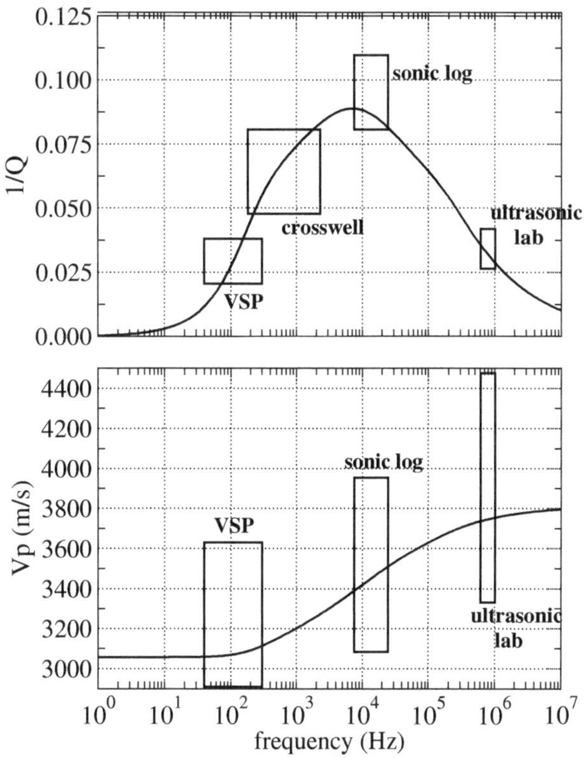

Figure 9.5: Attenuation and dispersion predicted by the double-porosity model (the solid curves) as compared to the data of Sams et al. (1997) [rectangular boxes]. The number of Q^{-1} estimates determined by Sams et al. (1997) falling within each rectangular box are: 40 VSP, 69 crosswell, 854 sonic log and 46 ultrasonic core measurements. A similar number of velocity measurements were made.

This example demonstrates that the degree of attenuation in the model is controlled principally by the contrast of compressibilities between the two porous phases; the greater the contrast, the greater the mesoscopic fluid-pressure gradient and the greater the mesoscopic flow and attenuation. In Figure 9.5, we compare the double-porosity model to the data of Sams et al. (1997), who used different seismic measurements (VSP, crosswell, sonic log, and ultrasonic lab) to determine Q^{-1} and P-wave velocity over a wide band of frequencies at their test site in England. The variance of the measurements falling within each rectangular box results from the various rock layers present at this site. Data collection was between four wells that are a few hundred meters deep. The geology at the site is a sequence of layered limestones, sandstones, siltstones, and mudstones. In this example, phase 2 is modeled as unconsolidated penny-shaped inclusions in which $a = 5$ cm (inclusion radius), $\varepsilon = 6 \times 10^{-3}$, $v_2 = 1.2$ %, $k_1 = 80$ mD, $V/S = 1.25$ cm, and $L_1 = 1.45$ cm. The phase 1 host is taken to be a well-consolidated sandstone ($\phi_1 = 0.20$ and $c = 1$).

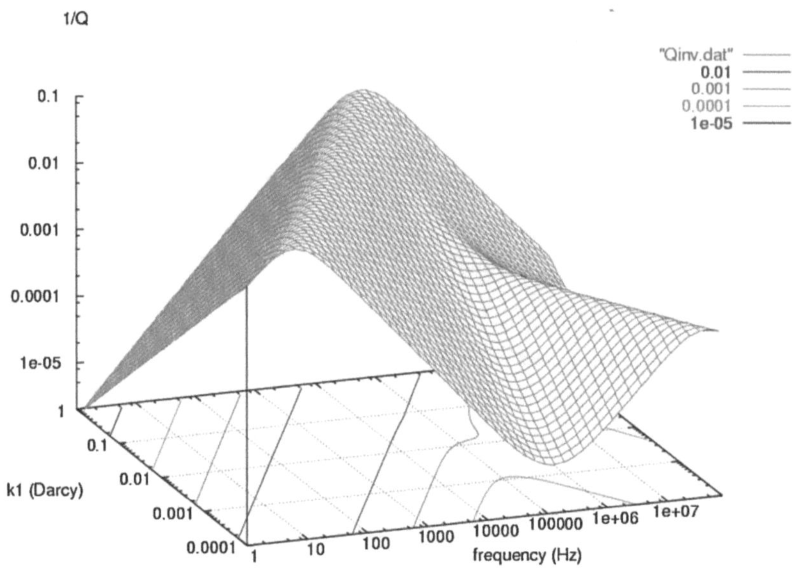

Figure 9.6: The permeability and frequency dependence of the P-wave attenuation in the double-porosity model. The smaller ridge in the attenuation surface between 0.1 and 1 MHz corresponds to the Biot losses. The dominant ridge that has the opposite permeability dependence is the result of mesoscopic flow.

In Figure 9.6, we allow the permeability k_1 of the host to vary while keeping all other properties fixed (with $P_e = 1$ MPa). The peak value of Q^{-1} is independent of k_1, while the critical frequency at which Q^{-1} is maximum is directly proportional to k_1. Thus, the slope $\partial Q^{-1}/\partial\omega$ in the approach to peak attenuation is inversely proportional to k_1. Field measurements of $Q^{-1}(\omega)$ over a range of frequencies could potentially yield information about k_1.

9.5.2 PATCHY SATURATION

All natural hydrological processes by which one fluid nonmiscibly invades a region initially occupied by another result in a patchy distribution of the two fluids. The patch sizes are distributed across the entire range of mesoscopic length scales and for many invasion scenarios are expected to be fractal. As a compressional wave squeezes such a material, the patches occupied by the less-compressible fluid will respond with a greater fluid-pressure change than the patches occupied by the more-compressible fluid. The two fluids will then equilibrate by the same type of mesoscopic flow already modeled in the double-porosity model.

Johnson (2001) provides a theory for the complex frequency-dependent undrained bulk modulus in the patchy-saturation model. Alternatively (and equivalently), Pride et al. (2004) provide a patchy-saturation analysis similar to that in the double-porosity model. This analysis leads to the same effective poroelastic moduli given by Equations (9.70)–(9.72) but with different definitions of the a_{ij} constants and internal transport coefficient $\gamma(\omega)$. In the model, a single uniform porous frame is saturated by patches of fluid 1 and fluid 2. We define porous phase 1 as those regions (patches) occupied by the less mobile fluid and phase 2 as the patches saturated by the more mobile fluid; i.e., by definition $\eta_1 > \eta_2$. This most often (but not necessarily) corresponds to $K_{f1} > K_{f2}$ and to $B_1 > B_2$.

A possible concern in the analysis is whether capillary effects at the local interface separating the two phases need to be allowed for. Tserkovnyak and Johnson (2003) have recently addressed that question in detail. Pride et al. (2004) have demonstrated that the condition for the neglect of surface tension σ on the meniscii separating the two fluids is

$$\frac{\sigma}{k_o K}\frac{V}{S} < 1 \tag{9.91}$$

where S is again the area of the surface separating the two phases in each volume V of composite, k_o is the permeability, and K is the drained bulk modulus taken to be a constant everywhere throughout each sample. As this dimensionless group of terms becomes much larger than one, the meniscii become stiff, and little or no fluid equilibration occurs between the two phases. Accordingly, the attenuation and dispersion vanishes in the limit of very large surface tension. When this inequality is satisfied, the fluids equilibrate under the usual condition that the fluid-pressure changes are continuous across the interface of separation, and there is a correspondingly large amount of attenuation and dispersion. Using the common sandstone values of $k = 100$ mD, $K = 10$ GPa, and $\sigma \approx 10^{-2}$ Pa m, one finds that V/S should be smaller than roughly 10^{-1} m in order to neglect surface tension and capillary effects. This can be considered the more normal situtation in the earth and will be the only situation treated here.

The a_{ij} constants in the patchy saturation model are given by (Pride et al., 2004)

$$a_{11} = 1/K \tag{9.92}$$

$$a_{22} = (-\beta + v_1/B_1)\,\alpha/K \tag{9.93}$$

$$a_{33} = (-\beta + v_2/B_2)\,\alpha/K \tag{9.94}$$

$$a_{12} = -v_1\alpha/K \tag{9.95}$$

$$a_{13} = -v_2\alpha/K \tag{9.96}$$

$$a_{23} = \beta\alpha/K \tag{9.97}$$

where α is the Biot-Willis constant for the material (always independent of fluid type), K is again the drained modulus for the material (also independent of fluid type), and the B_i are the Skempton's coefficients of each patch (all fluid dependence arrives through these terms). Under the assumption that the shear modulus of the material is independent of the fluid type and, accordingly, uniform throughout the patchy-saturation composite, the parameter β is given by

$$\beta = v_1 v_2\left(\frac{v_1}{B_2} + \frac{v_2}{B_1}\right)\left[\frac{\alpha - (1 - K/K_H)/(v_1 B_1 + v_2 B_2)}{\alpha - (1 - K/K_H)(v_1/B_1 + v_2/B_2)}\right]. \tag{9.98}$$

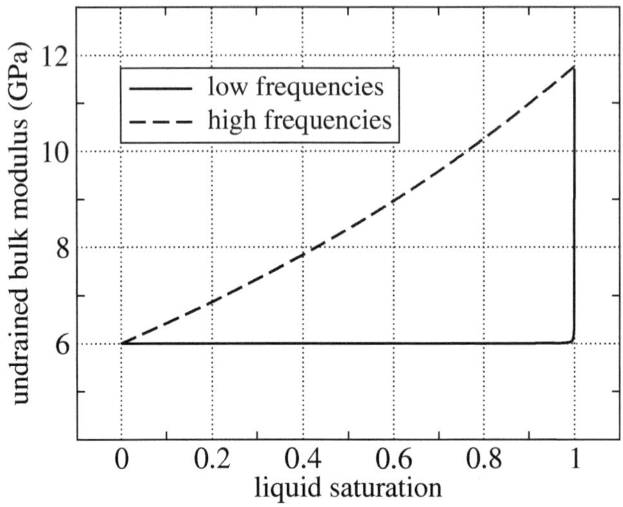

Figure 9.7: The undrained bulk modulus K_U in the limits of low and high frequencies as a function of liquid saturation. The liquid is taken to be water and the gas is air. The properties of the rock correspond to a lightly consolidated sandstone.

The modulus K_H is called the Hill modulus and corresponds to the undrained bulk modulus in the high-frequency limit where no fluid-pressure equilibration has a chance to occur between the two phases [it would also be the undrained bulk modulus at any frequency in the limit of infinite surface tension as defined by the dimensionless group in Equation (9.91)]. It is given by (c.f., Hill, 1963)

$$\frac{1}{K_H + 4G/3} = \frac{v_1}{K/(1-\alpha B_1) + 4G/3} + \frac{v_2}{K/(1-\alpha B_2) + 4G/3} \tag{9.99}$$

where the v_i are again the volume fractions of each phase.

The internal transfer coefficients are given by

$$\gamma_o = \frac{v_1 k_o}{\eta_1 L_1^2}\left[1 + O(\eta_2/\eta_1)\right] \tag{9.100}$$

$$\omega_o = \frac{K B_1}{\eta_1 \alpha}\frac{k_o (v_1 V/S)^2}{L_1^4}\left(1 + \sqrt{\frac{\eta_2 B_2}{\eta_1 B_1}}\right)^2 . \tag{9.101}$$

When the mobility ratio η_2/η_1 can be considered small, the definition of L_1 is again given by Equations (9.86) and (9.89). If phase 2 (the more mobile fluid) is modeled as a sphere of radius a embedded within each larger sphere R of the patchy-saturation composite, then $v_2 = (a/R)^3$, $V/S = av_2/3$, and $L_1^2 = (9v_2^{-2/3}a^2/14)(1 - 7v_2^{1/3}/6)$. These expressions are particularly appropriate when $v_2 \ll v_1$. For other scenarios of the patchy fluid distributions, see Pride et al. (2004) for modeling suggestions.

In the quasi-static limit $\omega/\omega_o \to 0$, the effective Skempton's coefficient B reduces exactly to $1/B = v_1/B_1 + v_2/B_2$. For a Gassmann material (uniform isotropic grains) this is equivalent to using an effective fluid modulus given by $1/K_f = v_1/K_{f1} + v_2/K_{f2}$ in the Gassmann fluid-substitution relations of Equations (9.43) and (9.44). So in the quasi-static limit, the effective moduli have absolutely no dependence on the geometrical nature of the fluid patches; they depend only on the volume fractions (saturation) of the patches. As already stated, in the opposite limit of very high frequencies, the undrained bulk modulus becomes K_H, which is also independent of the specific shape of the fluid patches. All dependence on the geometry of the patches is restricted to the value of the relaxation frequency separating the high-frequency regime from the low-frequency regime. The relaxation frequency can reside anywhere within the seismic band of frequencies depending on the permeability of the material and the effective patch size. The liquid saturation dependence of the undrained bulk modulus K_U in both the low-frequency and high-frequency limits is shown in Figure 9.7.

9.5.3 SQUIRT FLOW

Laboratory samples of consolidated rock often have broken grain contacts and/or microcracks in the grains. Much of this damage occurs as the rock is brought from depth to the surface. Since diagenetic processes in a sedimentary basin tend to cement microcracks and grain contacts, it is uncertain whether *in situ* rocks have significant numbers of open microcracks. Nonetheless, if such grain-scale damage is present, as it always is in laboratory rock samples at ambient pressures, the fluid-pressure response in the microcracks will be greater than in the principal pore space when the rock is compressed by a P wave. The resulting flow from crack to pore is called "squirt flow," and Dvorkin et al. (1995) have obtained a quantitative model for fully-saturated rocks.

In the squirt model of Dvorkin et al. (1995), the grains of a porous material are themselves allowed to have porosity in the form of microcracks. The effect of each broken grain contact is implicitly taken to be equivalent to a microcrack in a grain. The number of such microcracks per grain is thus limited by the coordination number of the packing, and so the total porosity contribution coming from the grains is negligible compared to the porosity of the main pore space.

Pride et al. (2004) have shown that the Dvorkin et al. (1995) squirt model can be analyzed using the double-porosity framework of the previous sections. Phase 1 is now defined to be the pure fluid within the main pore space of a sample and is characterized elastically by the single modulus K_f (fluid bulk modulus). Phase 2 is taken to be the porous (i.e., cracked) grains and characterized by the poroelastic constants K_2 (the drained modulus of an isolated porous grain), α_2 (the Biot-Willis constant of an isolated grain), and B_2 (Skempton's coefficient of an isolated grain) as well as by a permeability k_2. The overall composite of porous grains (phase 2) packed together within the fluid (phase 1) has two distinct properties of its own that must be specified: an overall drained modulus K and an overall permeability k associated with flow through the main pore space. The fraction of each averaging volume occupied by phase 1 is defined to be the "porosity" of the sample $v_1 = \phi$, whereas the volume fraction occupied by the cracked grains is $v_2 = 1 - \phi$ (i.e., the porosity within the grains does not contribute to this definition of ϕ).

The a_{ij} constants are given exactly by (Pride et al., 2004)

$$a_{11} = 1/K \tag{9.102}$$
$$a_{22} = 1/K - (1+\phi)/K_2 + \phi/K_f \tag{9.103}$$
$$a_{33} = \frac{(1-\phi)\alpha_2}{B_2 K_2} \tag{9.104}$$
$$a_{12} = -1/K + 1/K_2 \tag{9.105}$$
$$a_{13} = -\alpha_2/K_2 \tag{9.106}$$
$$a_{23} = \phi\alpha_2/K_2 \tag{9.107}$$

while the internal transfer coefficients are given by

$$\gamma_o = \frac{(1-\phi)k_2}{\eta L^2} \tag{9.108}$$
$$\omega_o = \frac{B_2 K_2}{\eta \alpha_2}\frac{k_2}{L^2}\left(\frac{(1-\phi)V/S}{L}\right)^2. \tag{9.109}$$

To make predictions, one must propose models for the phase 2 (porous grain) parameters.

If the grains are modeled as spheres of radius R, the fluid-pressure gradient length within the grains can be estimated as $L = R/\sqrt{15}$ and the volume to surface ratio as $V/S = R/[3(1-\phi)]$. The grain porosity is assumed to be in the form of microcracks, so it is natural to define an effective aperture h for these cracks. If the cracks have an average effective radius of R/N_R where N_R is roughly 2 or 3 and if there are on average N_c cracks per grain where N_c is also roughly 2 or 3, then the permeability and porsity of the grains is reasonably modeled as

$$\phi_2 = \frac{3N_c}{4N_R^2}\frac{h}{R} \quad \text{and} \quad k_2 = \phi_2 h^2/12. \tag{9.110}$$

The dimensionless parameters k_2/L^2 and $(1-\phi)(V/S)/L$ required in the expressions for γ_o and ω_o are given by

$$\frac{k_2}{L^2} = \frac{15N_c}{16N_R^2}\left(\frac{h}{R}\right)^3 \quad \text{and} \quad \left(\frac{(1-\phi)V/S}{L}\right)^2 = \frac{5}{3}. \tag{9.111}$$

The normalized fracture aperture h/R is the key parameter in the squirt model.

Pride et al. (2004) have argued that the elastic moduli of the porous grains are reasonably modeled as

$$K_2 = K_s(1-b\phi_2) \tag{9.112}$$
$$\alpha_2 = 1 - K_2/K_s \tag{9.113}$$
$$\frac{1}{B_2} = 1 + \phi_2\frac{K_2}{K_f}\left(\frac{1-K_f/K_s}{1-K_2/K_s}\right) \tag{9.114}$$

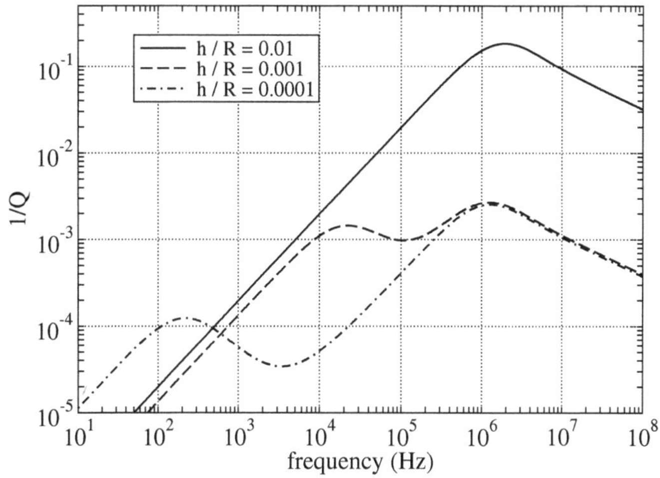

Figure 9.8: The squirt-flow model of P-wave attenuation when the grains are modeled as being spherical of radius R and containing microcracks having effective apertures h. Only the ratio h/R enters the squirt model. The overall drained modulus of the rock corresponds to a consolidated sandstone.

where the parameter b is taken to be independent of h/R and needs to be on the order of 10^2 to explain lab ultrasonic attenuation data. It is true that effective medium theories predict that b should be inversely proportional to the aspect ratio of true open cracks. However, as a crack closes and asperities are brought into contact, there is naturally a decrease in ϕ_2, but there should also be a decrease in b due to the fact that the remaining crack porosity becomes more spherical as new asperities come into contact. Taking b to be constant as crack porosity h/R decreases thus yields a conservative estimate for how the drained modulus increases. The Gassmann fluid-substitution relations were used for α_2 and B_2.

In Figure 9.8, we plot the P-wave attenuation predicted using the above model when the overall drained modulus corresponds to a sandstone [$\phi = 0.2$ and $c = 15$ in Equation (9.54)] having a permeability of 10 mD. For the grain properties, one takes $b = 300$, $N_R = 3$, $N_c = 3$, and $K_s = 37$ GPa (quartz) as fixed constants. The peak in Q^{-1} near 1 MHz that is invariant to h/R is that caused by the macroscopic Biot loss (fluid-pressure equilibration at the scale of the wavelength). The peak that shifts with the square of h/R is that caused by the squirt flow. This figure indicates that although the squirt mechanism is probably operative and perhaps even dominant at ultrasonic frequencies, it does not seem to be involved in explaining the observed levels of intrinsic attenuation in exploration work.

9.6 Discussion and Summary

Both equations and guidance have been given for modeling seismic wave speeds and attenuation in isotropic porous materials. A central question that has only been peripherally answered so far is: Can the permeability of a region be determined if the seismic wave speeds and/or attenuation in that region are known?

The permeability of a rock sample depends on the cross-sectional area ℓ^2 of the smallest pore throat along the flow path that is dominating the fluid transport (the so-called "percolation backbone"). It also depends on the electrical formation factor F, which is the appropriate measure of how tortuous the dominant flow paths are compared to straightline trajectories. Indeed, Thompson et al. (1987) use percolation theory to obtain the law $k = \ell^2/(226F)$ and experimentally demonstrate that this law does an excellent job of predicting rock permeabilities spanning some seven orders of magnitude when ℓ is measured as the breakthrough radius in a mercury porosimetry (invasion) experiment, and when $F = \sigma_{\text{fluid}}/\sigma_{\text{rock}}$ is obtained by measuring the electrical conductivity σ of the rock and pore fluid. Conceptually, one can change ℓ significantly by placing small amounts of solid (e.g., secondary mineralization) at key points along the percolation backbone without significantly affecting the elastic moduli of the rock. This is sufficient to demonstrate that permeability and seismic velocities are not necessarily linked to each other.

If a connection exists between permeability and seismic velocities, it is related to the case-by-case details of how a rock is built. For example, if secondary mineralization is uniform over the grain surfaces, one could develop a model in which the rock stiffness goes up as the permeability goes down. However, no such model for this presently exists in the literature, and it is known from x-ray tomography and SEM micrographs that secondary mineralization is not uniform, which greatly complicates any such model. As another example, consider unconsolidated sediments. It was shown earlier (Section 9.4.3.1) that grain packs having unimodal grain-size distributions have elastic moduli that are independent of grain size and, therefore, permeability. If smaller radius grains are introduced into the grain packing, one expects a reduction in permeability; however, whether the elastic moduli increase depends on whether these smaller grains form stress-bridging grain-to-grain contacts, which in turn depends sensitively to where the smaller grains are positioned in the packing. Statistics for, and even the very definition of (c.f., Torquato, 2000), random grain packs is an open field of research, especially when there is a range of grain sizes present.

This discussion is meant to emphasize that although one can imagine scenarios in which permeability and elastic stiffness might both vary as some control knob (like the degree of secondary mineralization or the grain-size distribution) is turned, any such relation depends on the details of how the rock was intially formed and how it has evolved. Perhaps the only thing that can be said with certainty is that no universal relation valid for all rock types can exist between permeability and seismic velocity.

A more promising connection exists between permeability and intrinsic seismic attenuation. It was shown in Sections 9.5.1 and 9.5.2 that mesoscopic heterogeneity is capable of producing the amount of seismic attenuation required to explain the admittedly sparse amount of field data that is available (see Figure 9.5). When a compressional wave squeezes a sample having mesoscopic heterogeneity, the different types of porosity respond with different changes in their fluid pressure. In the type of double-porosity models discussed in Section 9.5, fluid pressure diffuses from an embedded phase into a host phase, and the permeability of the host phase directly controls the time necessary for equilibration to take place. By observing how Q varies with frequency, one can, in principle, obtain information about the permeability of the host phase in these models. However, inverting seismic data to obtain $Q(\omega)$ over a broad range of frequencies using a single seismic experiment (such as crosswell tomography) is an active point of ongoing research, with no published results presently available. Note that it is only the permeability of the host phase that affects the frequency dependence of Q in such double-porosity models. If the embedded phase represents disconnected inclusions at small volume fractions, then the host-phase permeability will be controlling both $\partial Q^{-1}/\partial\omega$ and the sample's permeability. However, if the embedded phase represents through-going connected joints or faults, then the host-phase permeability will again be controlling $\partial Q^{-1}/\partial\omega$, but may be unrelated to the sample's permeability, which is being controlled by the through-going joints.

The squirt mechanism presented in Section 9.5.3 has an indirect connection to a sample's permeability. The connection is through the effective size of the grains, which simultaneously affects both the time necessary for the crack porosity to equilibrate with the main pores (the grain size is thus influencing the frequency dependence of Q in the squirt model) and the permeability of the sample. However, it was shown in Section 9.5.3 that squirt is not likely to be operative over the range of frequencies used in seismic exploration.

Arguably the greatest opportunity for connecting permeability to seismic properties is through time-dependent pumping of the fluids in the rock. As the pore pressure changes, Sections 9.4.4 and 9.4.5 show how both the porosity and permeability are changed. The change in seismic velocity is proportional to the change in porosity. Although the percent change in permeability will not, in general, be the same as the percent change in porosity (they have different effective-stress variables), one nonetheless can expect that where seismic velocities change the most, permeability will change the most.

On account of space limitations, anisotropic porous materials were not treated in this chapter. Most earth materials have some degree of anisotropy at the macroscale because of fine layering and/or fracture networks. It is a reasonable postulate that any anisotropic symmetry class determined from seismic measurements must also be satisfied by the permeability tensor. However, no rigorous work along these lines has apparently been published. In his landmark paper, Gassmann (1951) gave the proper fluid-substitution relations that determine how the fluid affects the elastic moduli of an anisotropic porous material. Putting anisotropy into the double-porosity model of mesoscopic and microscopic flow greatly complicates the analysis, since it couples together the response of different tensorial orders within the constitutive laws. So the scalar increment $\zeta_{\rm int}$ characterizing fluid transfer between the mesoscopic "patches" becomes coupled to the vectorial Darcy flow, and the average fluid pressure in each patch becomes coupled to the deviatoric (shear) tensor. These complications have not yet been worked through in detail.

On a related note, the possible dispersive nature of the shear modulus was not discussed here. In the presence of local anisotropy (e.g., an embedded fracture or other elongated local inclusion), an applied shear can result in lobes of compression and dilation surrounding the elongated inclusion. If the material is macroscopically isotropic (e.g., the inclusions or fractures are oriented in all directions), the average fluid pressure will remain zero throughout a sample; however, there will be local regions of enhanced and decreased fluid pressure that result in local fluid flow and an associated shear dispersion and attenuation. These effects have been discussed by Berryman and Wang (2001), and an approximate model has been proposed by Mavko and Jizba (1991); however, no rigorous theory presently exists that accounts for the complex frequency-dependent nature of a rock's shear modulus.

References

Archie, G. E., The electrical resistivity log as an aid in determining some reservoir characteristics, *Trans. AIME*, *146*, 54–62, 1942.

Berge, P. A., H. F. Wang, and B. P. Bonner, Pore-pressure buildup coefficient in synthetic and natural sandstones, *Int. J. Rock Mech. Min. Sci. Geomech. Abstr.*, *30*, 1135–1141, 1993.

Berryman, J. G., Long-wavelength propagation in composite elastic media I. Spherical inclusions, *J. Acoust. Soc. Am.*, *68*, 1809–1819, 1980a.

Berryman, J. G., Long-wavelength propagation in composite elastic media II. Ellipsoidal inclusions, *J. Acoust. Soc. Am.*, *68*, 1820–1831, 1980b.

Berryman, J. G., Effective stress for transport properties of inhomogeneous porous rock, *J. Geophys. Res.*, *97*, 17,409–17,424, 1992.

Berryman, J. G., Mixture theory for rock properties, in *Rock Physics and Phase Relations – A Handbook of Physical Constants*, edited by T. J. Ahrens, Am. Geophys. Union, Washington, DC, 1995.

Berryman, J. G., and S. R. Pride, Models for computing geomechanical constants of double-porosity materials from the constituents' properties, *J. Geophys. Res.*, *107*, 1000–1015, 2002.

Berryman, J. G., and H. F. Wang, The elastic coefficients of double-porosity models for fluid transport in jointed rock, *J. Geophys. Res.*, *100*, 24,611–24,627, 1995.

Berryman, J. G., and H. F. Wang, Dispersion in poroelastic systems, *Phys. Rev. E*, *64*, 011303, 2001.

Biot, M. A., Theory of propagation of elastic waves in a fluid-saturated porous solid. I. Low-frequency range, *J. Acoust. Soc. Am.*, *28*, 168–178, 1956a.

Biot, M. A., Theory of propagation of elastic waves in a fluid-saturated porous solid. II. Higher frequency range, *J. Acoust. Soc. Am.*, *28*, 179–191, 1956b.

Biot, M. A., Mechanics of deformation and acoustic propagation in porous media, *J. Appl. Phys.*, *33*, 1482–1498, 1962.

Biot, M. A., and D. G. Willis, The elastic coefficients of the theory of consolidation, *J. Appl. Mech.*, *24*, 594–601, 1957.

Bruggeman, D. A. G., Berechnung verschiedener physikalischer Konstanten von heterogenen Substanzen, *Ann. Physik. (Leipzig)*, *24*, 636–679, 1935.

Burridge, R., and J. B. Keller, Poroelasticity equations derived from microstructure, *J. Acoust. Soc. Am.*, *70*, 1140–1146, 1981.

Castagna, J. P., M. L. Batzle, and T. K. Kan, Rock physics–the link between rock properties and avo response, in *Offset-Dependent Reflectivity – Theory and Practice of AVO Analysis*, edited by J. P. Castagna and M. M. Backus, Soc. Explor. Geophys., Tulsa, 1993.

Domenico, S. N., Elastic properties of unconsolidated porous reservoirs, *Geophysics*, *42*, 1339–1368, 1977.

Dvorkin, J., G. Mavko, and A. Nur, Squirt flow in fully saturated rocks, *Geophysics*, *60*, 97–107, 1995.

Frenkel, J., On the theory of seismic and seismoelectric phenomena in a moist soil, *J. Physics (Soviet)*, *8*, 230–241, 1944.

Gassmann, F., Über die Elastizität poröser Medien, *Vierteljahrsschrift der Naturforschenden Gesellschaft in Zürich*, *96*, 1–23, 1951.

Gelinsky, S., and S. A. Shapiro, Dynamic-equivalent medium approach for thinly layered saturated sediments, *Geophys. J. Internat.*, *128*, F1–F4, 1997.

Goddard, J. D., Nonlinear elasticity and pressure-dependent wave speeds in granular media, *Proc. R. Soc. Lond. A*, *430*, 105–131, 1990.

Gurevich, B., and S. L. Lopatnikov, Velocity and attenuation of elastic waves in finely layered porous rocks, *Geophys. J. Internat.*, *121*, 933–947, 1995.

Hardin, B. O., and F. E. Richart, Elastic wave velocities in granular soils, *J. Soil Mech. Found. Div. ASCE*, *89*, 33–65, 1963.

Hashin, Z., and S. Shtrikman, A variational approach to the theory of the elastic behavior of multiphase materials, *J. Mech. Phys. Solids*, *11*, 127–140, 1963.

Hill, R., Elastic properties of reinforced solids: Some theoretical principles, *J. Mech. Phys. Solids*, *11*, 357–372, 1963.

Johnson, D. L., Theory of frequency dependent acoustics in patchy-saturated porous media, *J. Acoust. Soc. Am.*, *110*, 682–694, 2001.

Johnson, D. L., J. Koplik, and R. Dashen, Theory of dynamic permeability and tortuosity in fluid-saturated porous media, *J. Fluid Mech.*, *176*, 379–402, 1987.

Korringa, J., R. J. S. Brown, D. D. Thompson, and R. J. Runge, Self-consistent imbedding and the ellipsoidal model for porous rocks, *J. Geophys. Res.*, *84*, 5591–5598, 1979.

Levy, T., Propagation of waves in a fluid-saturated porous elastic solid, *Int. J. Engng. Sci.*, *17*, 1005–1014, 1979.

Mavko, G., and D. Jizba, Estimating grain-scale fluid effects on velocity disperion in rocks, *Geophysics*, *56*, 1940–1949, 1991.

Mavko, G., and A. Nur, Melt squirt in the asthenosphere, *J. Geophys. Res.*, *80*, 1444–1448, 1975.

Mavko, G., and A. Nur, Wave attenuation in partially saturated rocks, *Geophysics*, *44*, 161–178, 1979.

Mavko, G., T. Mukerji, and J. Dvorkin, *The Rock Physics Handbook: Tools for Seismic Analysis in Porous Media*, Cambridge University Press, Cambridge, 1998.

Murphy III, W. F., Effects of microstructure and pore fluids on the acoustic properties of granular sedimentary materials, Ph.D. thesis, Stanford University, Stanford, California, 1982.

Norris, A. N., Low-frequency dispersion and attenuation in partially saturated rocks, *J. Acoust. Soc. Am.*, *94*, 359–370, 1993.

O'Connell, R. J., and B. Budiansky, Viscoelastic properties of fluid-saturated cracked solids, *J. Geophys. Res.*, *82*, 5719–5735, 1977.

Pride, S. R., and J. G. Berryman, Connecting theory to experiment in poroelasticity, *J. Mech. Phys. Solids*, *46*, 719–747, 1998.

Pride, S. R., and J. G. Berryman, Linear dynamics of double-porosity and dual-permeability materials. I. Governing equations and acoustic attenuation, *Phys. Rev. E*, *68*, 036603, 2003a.

Pride, S. R., and J. G. Berryman, Linear dynamics of double-porosity and dual-permeability materials. II. Fluid transport equations, *Phys. Rev. E*, *68*, 036604, 2003b.

Pride, S. R., A. F. Gangi, and F. D. Morgan, Deriving the equations of motion for porous isotropic media, *J. Acoust. Soc. Am.*, *92*, 3278–3290, 1992.

Pride, S. R., J. M. Harris, D. L. Johnson, A. Mateeva, K. T. Nihei, R. L. Nowack, J. W. Rector, H. Spetzler, R. Wu, T. Yamomoto, J. G. Berryman, and M. Fehler, Permeability dependence of seismic amplitudes, *The Leading Edge*, *22*, 518–525, 2003.

Pride, S. R., J. G. Berryman, and J. M. Harris, Seismic attenuation due to wave-induced flow, *J. Geophys. Res.*, *109*, B01201,doi:10.1029/2003JB002639, 2004.

Quan, Y., and J. M. Harris, Seismic attenuation tomography using the frequency shift method, *Geophysics*, *62*, 895–905, 1997.

Roscoe, R., Isotropic composites with elastic or viscoelastic phases: General bounds for the moduli and solutions for special geometries, *Rheologica Acta*, *12*, 404–411, 1973.

Sams, M. S., J. P. Neep, M. H. Worthington, and M. S. King, The measurement of velocity dispersion and frequency-dependent intrinsic attenuation in sedimentary rocks, *Geophysics*, *62*, 1456–1464, 1997.

Sato, H., and M. Fehler, *Seismic wave propagation and scattering in the heterogeneous earth*, Springer-Verlag, New York, 1998.

Skempton, A. W., The pore-pressure coefficients A and B, *Geotechnique*, *4*, 143–147, 1954.

Thompson, A. H., A. J. Katz, and C. E. Krohn, The microgeometry and transport properties of sedimentary rock, *Advances in Physics*, *36*, 625–694, 1987.

Toksoz, M. N., C. H. Cheng, and A. Timur, Velocities of seismic waves in porous rocks, *Geophysics*, *41*, 621–645, 1976.

Torquato, S., T. M. Truskett, and P. G. Debendetti, Is random close packing of spheres well defined?, *Phys. Rev. Lett.*, *84*, 2064–2067, 2000.

Tserkovnyak, Y., and D. L. Johnson, Capillary forces in the acoustics of patchy-saturated porous media, *J. Acoust. Soc. Am.*, *114*, 2596–2606, 2003.

Walton, K., The effective elastic moduli of a random packing of spheres, *J. Mech. Phys. Solids*, *35*, 213–226, 1987.

Wang, H. F., *Theory of Linear Poroelasticity with Applications to Geomechanics and Hydrogeology*, Princeton University Press, Princeton, NJ, 2000.

White, J. E., Computed seismic speeds and attenuation in rocks with partial gas saturation, *Geophysics*, *40*, 224–232, 1975.

White, J. E., N. G. Mikhaylova, and F. M. Lyakhovitsky, Low-frequency seismic waves in fluid-saturated layered rocks, *Izvestija Academy of Sciences USSR, Phys. Solid Earth*, *11*, 654–659, 1975.

Wu, R. S., and K. Aki, Multiple scattering and energy transfer of seismic waves: separation of scattering effect from intrinsic attenuation. II. Application of the theory to Hindu Kush region, *Pure and Applied Geophys.*, *128*, 49–80, 1988.

10 GEOPHYSICAL WELL LOGGING
BOREHOLE GEOPHYSICS FOR HYDROGEOLOGICAL STUDIES: PRINCIPLES AND APPLICATIONS

MIROSLAV KOBR[1], STANISLAV MAREŠ[1], and FREDERICK PAILLET[2]

[1]*Charles University in Prague, Faculty of Natural Sciences, Institute of Hydrogeology, Engineering Geology and Applied Geophysics, Albertov 6, 128 43 Prague, Czech Republic*

[2]*Department of Earth Sciences, University of Maine, Orono, Maine, USA*

10.1 Introduction

Borehole geophysics includes all methods for making continuous profiles or point measurements at discrete depth stations in a borehole. These measurements are made by lowering different types of probes into a borehole and electrically transmitting data in the form of either analog or digital signals to the surface, where they are recorded as a function of depth or distance along the borehole. The measurements are related to the physical and chemical properties of the rocks surrounding the borehole, the properties of the fluid saturating the pore spaces in the formation, the properties of fluid in the borehole, the construction of the well, or some combination of these factors.

Many geophysical logging techniques were developed in the petroleum industry. Most of the textbooks and published reports are written about petroleum applications, which differ markedly from groundwater and environmental applications. In addition, the characteristics (probe weight and diameter, pressure rating, and sensor configuration) of the equipment used for nonpetroleum logging is generally different from that used to log oil and gas wells. This chapter addresses the application of borehole geophysics to the discipline of hydrogeology, including such topics as groundwater resource development, aquifer characterization, water-quality sampling and assessment, and aquifer remediation. Additional background on specific borehole geophysical techniques and geophysical log data processing can be obtained by consulting Hearst et al. (2000), Hurst et al. (1992; 1993), and Doveton (1994).

10.1.1 LOGGING METHODS

One of the most important attributes of geophysical well logs is the ability to make several different physical (nuclear, acoustic, electrical, etc.) measurements in a borehole. This is the so-called synergistic property of borehole geophysics described by Keys (1997). Logging methods are summarized in Table 10.1 for the most widely used logs.

Y. Rubin and S.S. Hubbard (eds.), Hydrogeophysics, 291–331.

Table 10.1. The most widely used logs in water wells (adapted after Keys, 1997)

Type of Log	**Potential Uses**	**Borehole Conditions Required**	**Limitations**
Spontaneous potential SP	Lithology, correlation, water quality	Uncased, mud filled only	Needs contrast in solutes in borehole versus formation
Normal and/or lateral resistivity logs Ra	Lithology, correlation, water quality	Uncased, fluid filled only	Influenced by fluid resistivity and bed boundaries
Conductively focused-current logs, micro-focused logs LL, MLL	Lithology, correlation, water quality, monitoring formation fluid conductivity	Uncased, fluid filled only	Removes some but not all fluid and bed boundary effects
Inductively focused-current logs IL	Lithology, correlation, water quality, monitoring formation fluid conductivity	Uncased or cased with PVC or fiberglass, air or fluid filled	Poor response in massive resistive formations
Gamma-ray log GR	Lithology, correlation	Most conditions	Unusual lithology interpretations in arkose and phosphate sands
Gamma-gamma log GGL-D	Bulk density, porosity, moisture, lithology, correlation, well construction	Wide range	Mineralogy may influence calibration through Z effect
Neutron log NNL	Saturated porosity, moisture, lithology, correlation	Wide range, preferably in small holes, fresh or brackish fluid filled	Ineffective porosity of clay minerals included in total porosity
Elastic wave propagation log AL	Saturated porosity, lithology, correlation	Uncased, fluid filled only	Measures primary or matrix porosity rather than total porosity
Cement bond log CBL	Bonding to casing and formations	Cased, fluid filled only	Response gives a circumferential average of bonding; may not detect channels
Acoustic televiewer AT	Character and orientation of fractures, fractures openings, and bedding	Uncased, fluid filled, water or mud	Fracture interpretation may be affected by drilling damage to borehole wall
Borehole television BTV	Well construction, secondary porosity, lost objects	Uncased or cased, air or clear fluid	No information in opaque fluid; dark mineral bands mimic fractures

Caliper log CL	Borehole diameter, log correction, well construction, fractures, lithology	Most conditions	Probe performance often affected by borehole deviation
Temperature log TM	Fluid movement in the borehole and/or behind casing	Water filled	Affected by disturbance to fluid column by probe motion
Fluid conductivity FRL	Fluid movement in the borehole	Water filled	Affected by disturbance to fluid column by probe motion
Flowmeter FM	Vertical fluid movement	Fluid filled	Spurious results may be caused by convection, or by movement of fluid in annulus behind screens
Fluid transparency or turbidity log PHL	Water movement in the borehole, water transparency	Water filled	Degree of borehole fluid transparency affects operation and calibration
Well completion methods	Well construction, numerous techniques for checking casing and completion materials, borehole deviation, etc.	All conditions	Labor and equipment intensive; completions may be irreversible and often depend on the specific interpretation

10.1.2 FIELD OPERATIONS

A schematic block diagram of the various components in a combination analog and digital logging system is shown in Figure 10.1, where geophysical data files are digitized and stored by up-hole electronics, and the analog recorder (strip chart) provides a preliminary field record of the log. The logging system components shown in this figure can be interfaced with a computer system installed in the logging vehicle, or imported into a portable (laptop) computer provided by the user.

The most recent logging systems are fully digital, with control of the logging equipment implemented from a computer keyboard. Logs are displayed by the computer, and a field record is obtained from a printer. Winches controlled by the computer usually hold a few hundred meters of single-conductor cable, and may be powered by a hand-crank or by an electric motor run using an inverter (running off of a vehicle battery) or a portable generator. It is usually not possible to run complex logging probes using a small portable logger because of its size and limited additional surface instrumentation, but many of the smaller probes are essentially scaled-down versions of those operated on truck-mounted units.

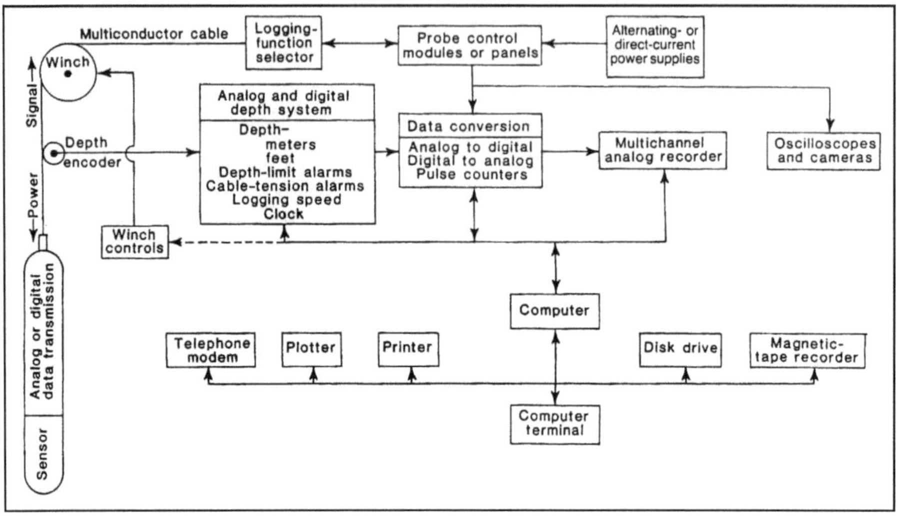

Figure 10.1. Block diagram of a logging system (after Keys, 1997)

Logging probes, also called sondes or tools, enclose the sensors, electronics for transmitting and receiving signals, and power supplies. The probes are connected to a cable by a cable head screwed onto the top of the probe. Some electric- and acoustic-logging devices are flexible; most are rigid. Probes vary in diameter from 25 mm to 100 mm. The probes are either axially symmetric (centralized) or use bowsprings or spring-loaded arms to force the probe sensor pad or pads against the borehole wall (side collimation). Measurements are usually made with the probe pulled upward to insure steady tension on the logging cable when data are recorded. Many probes are run at logging speeds determined only by the limits on rate of data transmission and the need to prevent "road noise" produced by jolting of the probe. Fluid-column probes are often run at speeds of less than 2 meters (m) per minute to minimize disturbance to the borehole fluid. Borehole nuclear measurements are subject to nuclear statistics, so that effective sampling of properties depends on logging speed, where probe retrieval rate is limited to satisfy criteria imposed by nuclear statistical requirements.

10.1.3 BENEFITS AND LIMITATIONS OF LOGGING

The main objective of borehole geophysics is to obtain more information than can be obtained from conventional drilling, sampling, and testing. Drilling a borehole or well is an expensive procedure, but the borehole provides access to the subsurface for geophysical probes. Logs may be interpreted in terms of lithology, thickness, and continuity of aquifers; porosity and bulk density; resistivity; moisture content; specific capacity; groundwater chemical and physical characteristics and parameters of water movement; and integrity of well construction.

Log data are readily reproduced over long periods. The repeatability and comparability provide the basis for measuring changes in yields of water wells with time. Changes in quality of casing, such as extent of corrosion or clogging of screens, or changes in water quality may also be recorded. Thus, logs may be used to establish baseline aquifer characteristics to determine the extent of changes from that baseline or to identify the movement of contaminant plumes through networks of observation wells.

Borehole logs provide continuous records that can be analyzed in real time at the well site. Thus, they can be used to guide completion or testing procedures. Logs also aid the lateral and vertical extrapolation of geologic and water sampling data or hydraulic test data obtained from wells. In contrast, samples of rock or fluid from a borehole provide data only from sampled depth intervals and only after laboratory analysis. Data from some geophysical logs, such as acoustic velocity and resistivity, are also useful in interpreting surface geophysical surveys.

Laboratory analysis of core is essential either for direct calibration of logs or for verifying calibration carried out by other means. Calibration of logs for one rock type may not be valid in other rock types, because of how the rock-matrix chemical composition may affect log measurements. For this reason, geophysical logging cannot completely replace sampling. A log analyst cannot evaluate a set of logs properly without information on the local geological and hydrological conditions. To maximize results from logs, at least one core hole should be drilled or core recovered from selected intervals in one borehole at each study site. Correct interpretation of logs should be based on a thorough understanding of the operating principles of each logging technique. Geophysical logs can be analyzed in the field to guide the location and frequency of sampling, and thus may reduce the number of samples needed, along with the cost of sample processing and equipment decontamination. Log data may also be used to identify situations in which potential cross-contamination between aquifer units may occur unless aquifers are immediately isolated from each other by well completion.

10.1.4 BOREHOLE PARAMETERS INFLUENCING THE QUALITY OF LOGS

The manner in which a well is drilled affects the logging environment, through such factors as: the resistivity of the fluid in the well bore, the diameter of the well, the roughness (rugosity) of the borehole wall, and the mud-filtrate invasion of porous formations surrounding the well. Boreholes may be drilled by percussion or rotary techniques, and by mechanical auger in unconsolidated materials. In rotary drilling, the mud is circulated down through the drill pipe and returns to the surface through the well bore. The drilling fluid serves as a coupling medium for electrical and acoustic logs. The formations are therefore subjected to the hydrostatic pressure of a column of mud, which alters the conditions in pores of the formation within the immediate vicinity of the borehole walls. When this happens, a mud cake is developed on the borehole wall, and the mud filtrate can penetrate the aquifer (a process known as invasion) for a distance of several borehole diameters from the borehole wall. If the mud filtrate has a solute content different from that of the fluids in the pore spaces, the resistivity of the drilling mud should be measured and recorded at the time of logging, and interpretation of resistivity and SP logs should take the measured mud properties into account

Drilling can also induce changes in elastic properties along the borehole wall in some sedimentary formations. Fluids in the drilling mud cause the swelling of clay minerals, reducing the cohesion of the surrounding formation. This produces a low-velocity zone adjacent to the borehole wall, where the width of the low-velocity annulus and the magnitude of the velocity contrast may increase with time after drilling.

The thickness and composition of material between a logging probe and the borehole wall can have an effect on the data set. In general, the larger the borehole diameter, the poorer the quality of the geophysical logs obtained. Spacing between source and detector, between transmitter and receiver, or between electrodes affects the radius of investigation of logging tool. Although some tools are called "borehole compensated" on the basis of tool design, almost all logs are adversely affected by abrupt changes in borehole diameter. Thus, a caliper log is essential to the interpretation of most logs. All drill holes, except those drilled in very hard rocks, have intervals (often denoted as washouts) where the borehole diameter exceeds bit size by an amount that can cause anomalous log response. In unconsolidated formations, boreholes may be lined with casing or screens, but voids may exist in the annulus outside of the casing or screen. The best procedure for quantitative log analysis is to eliminate those depth intervals demonstrating significant diameter changes (with respect to the sampled volume for a particular probe) from the analysis.

10.1.5 CORRELATION OF LOGS WITH ROCK CORE OR SEDIMENTS

Borehole geophysics consists of measurements that are obtained with a carefully controlled depth scale and applied to sediments around a borehole that have not been disturbed by drilling, disaggregation, or desiccation. However, these measurements are almost always given in quantities (such as gamma activity or electrical conductivity) that are not of direct interest to hydrogeologists. In contrast, samples recovered from boreholes provide direct and unambiguous information about the geology of the subsurface, but these samples are disturbed by drilling and are derived from a poorly constrained or undetermined depth during the drilling process. Most effective use of geophysical logs and geologic descriptions is made when the two different data sets are combined (Figure 10.2). In this example, a gamma log is compared to a geologist's description of cuttings brought to the surface during the auger drilling of a borehole for a nested set of observation wells. The gamma log has a precise depth scale and applies to the formation *in situ*, but the data are given in units that are not of direct interest to the hydrogeologist. Moreover, the entire formation is described on a scale of relative gamma activity, whereas the hydrogeology of the formation can probably be described using several different scales, such as the percent of fine particles, pore size distribution, or sediment particle lithology. With no other information, the gamma log alone can be used to express sediment properties according to only one of these scales. The sediment description can be made more precisely in terms of quantities of interest to the geologist, but cannot be assigned to well-defined depth intervals because of the mixing that occurred during the drilling process. A logical comparison of the two data sets allows the definition of aquifer units, as indicated in Figure 10.2. The geologist's descriptive log shows the locations of aquifers. The gamma log indicates the precise depth of the tops and bottom of those units, and indicates where contacts are abrupt and where they are gradational. That there can be more than one explanation for features on a gamma log is indicated by the fact that one thin zone of high gamma activity (24 m in depth) probably corresponds to a clay layer at the bottom of a coarsening upward sedimentary unit, whereas another (8 m in depth) probably corresponds to an unweathered granitic cobble in the overlying glacial sediments. Otherwise, the effective design of a set of nested piezometers results when both geophysical log and sediment description data sets are combined in the identification of aquifers and confining units at a given study site.

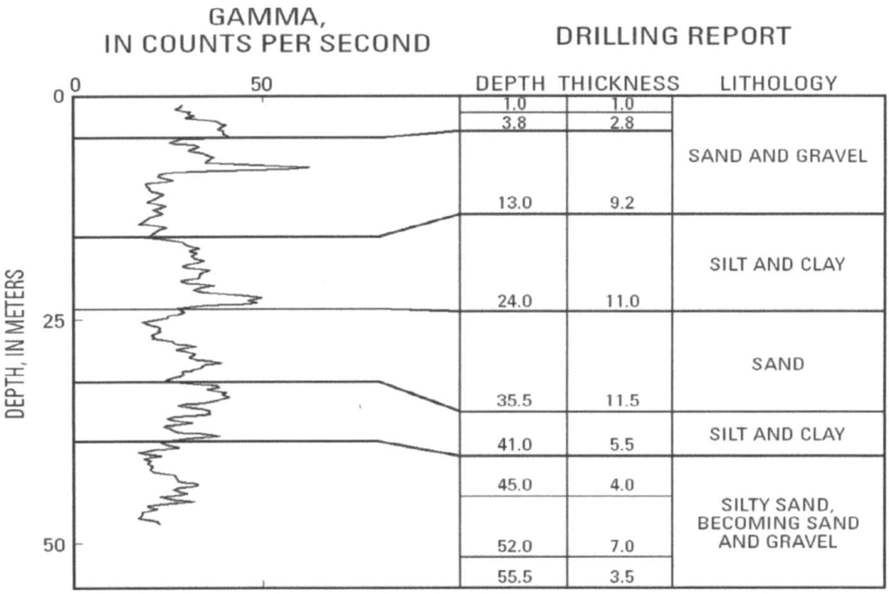

Figure 10.2. Gamma log compared to a qualitative description of cuttings sampled from an augur-drilled borehole in an alluvial aquifer showing correlation of lithologic contacts in the two data sets (Paillet, unpublished data)

10.1.6 CALIBRATION AND STANDARDIZATION OF LOGS

Standardization is the process of checking the response of logging probes in the field, using some type of portable field standard, usually before and after logging a well, to provide the basis for system drift correction in output with time. Calibration is the process of converting measuring units of log response to units that represent rock characteristics. Calibration facilities should closely simulate the lithology to be logged. Boreholes that have been carefully cored, with the cores analyzed quantitatively in a laboratory, also may be used to calibrate logging probes. This point is discussed more thoroughly below.

In many applications, it is impossible to define a single laboratory calibration for geophysical logs, because the local lithologic environment where the logging is conducted may differ from standard laboratory calibration facilities, such as the API porosity calibration blocks located at the University of Houston. A direct and site-specific calibration can sometimes be made using core sample measurements from the formation of interest. The calibration is performed by developing a regression between the measured geophysical property (the log) and the property of interest determined from laboratory tests using recovered core. In principle, the quality of the calibration is indicated by the correlation coefficient given from the regression. The amount of variability in the geophysical log represented by changes in the property of interest is given as a fraction of total variability by the r-squared coefficient. In practice, core calibration involves other important statistical issues (Figure 10.3). The nominal depth given on the recovered core almost never corresponds exactly with the depth scale on the geophysical log. In Figure 10.3A, a gamma log is compared to a list of core values for a borehole in a sandstone and shale aquifer. Because the lithology in this example is assumed to be some combination of sand and shale, and because the gamma log is often assumed to be directly proportional to clay mineral content, the gamma log might be related to the percent of drainable porosity given by the core sample measurements. However, core porosity and gamma log cannot be directly compared at each nominal depth station because there may be a difference in depth reference point. We also recognize that the gamma log measurement is averaged over a sample volume of ~15 cm, whereas the core porosity is given for a much smaller plug taken from the core.

Without accounting for depth and scale mismatch, simple regression shows a very weak correlation between the two data sets (Figure 10.3B). To address these issues, we perform a spatial smoothing of the core sample data by fitting those data points to a polynomial, and then adjusting the depth scale on the core data set so that the shape of the polynomial aligns with the log (Figure 10.3C). This requires a 4 ft (1.2 m) depth shift to make the shape of the polynomial fit to the core data and the gamma log coincide. Note that the porosity scale is reversed to allow for the expected negative correlation between gamma activity and porosity. These steps insure that the depth scales are aligned, and that both sets of data correspond to approximately the same sample volume, greatly improving the quality of the regression (Figure 10.3D). With this regression, the gamma log can be replotted to change the measurement scale, from gamma activity in counts per second to effective porosity in percent. Because gamma activity may reflect the percentage of fine particles in the sample volume, it might also be possible to use the same techniques to generate a direct calibration of the gamma log in units of hydraulic conductivity, as described by Keys (1986).

CALIBRATION USING CORE SAMPLES

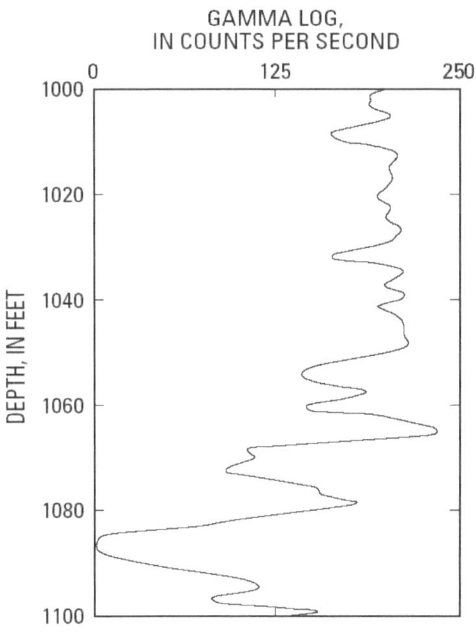

CORE DATA

POROSITY	DEPTH
3.3	1067.2
3.3	1068.0
5.7	1069.0
5.1	1070.0
7.8	1071.0
9.1	1072.0
6.2	1073.0
6.0	1074.0
9.5	1075.5
7.5	1076.5
7.6	1077.5
6.4	1078.5
7.8	1079.2
7.4	1080.0
4.7	1081.6
7.7	1082.5
8.9	1083.5
7.3	1084.5
6.6	1085.5
10.4	1086.5
9.7	1087.3
9.4	1088.3
8.6	1089.3
9.8	1090.3
10.6	1091.2
9.7	1092.2
8.6	1093.1
9.4	1094.1
7.4	1095.9
6.4	1096.8

A

Figure 10.3A. Example of core sample data used to calibrate a gamma log in porosity units (from Paillet and Crowder, 1996): Gamma log and core porosity data

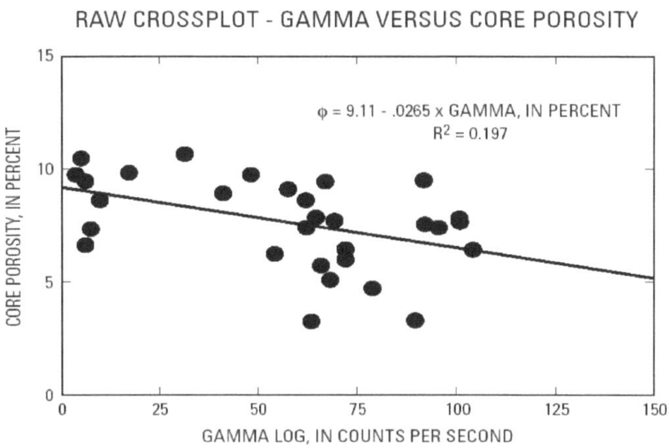

B

Figure 10.3B. Example of core sample data used to calibrate a gamma log in porosity units (from Paillet and Crowder, 1996): Regression of gamma log data and core porosity data using the nominal depth values given with the data

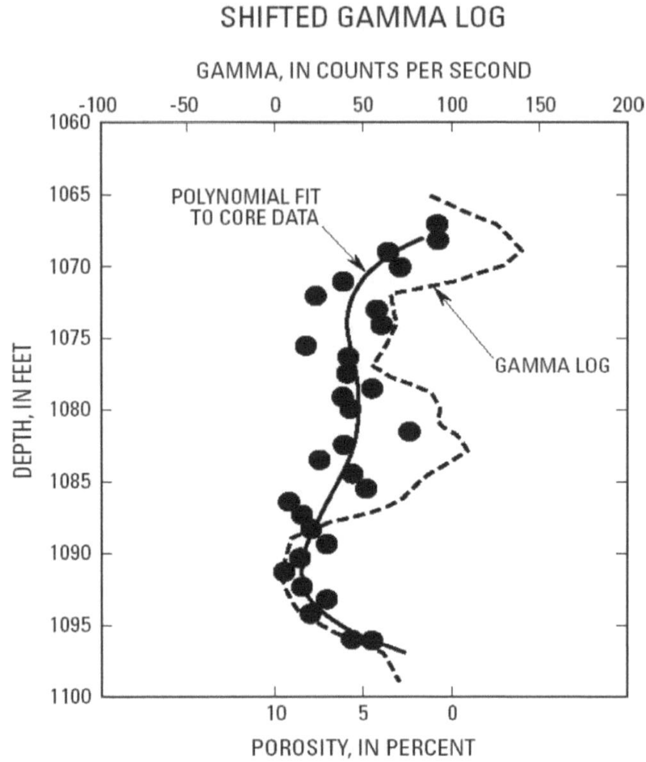

C

Figure 10.3C. Example of core sample data used to calibrate a gamma log in porosity units (from Paillet and Crowder, 1996): Polynomial fit to core data points after depth adjustment to maximize the depth correlation of the logs and core values

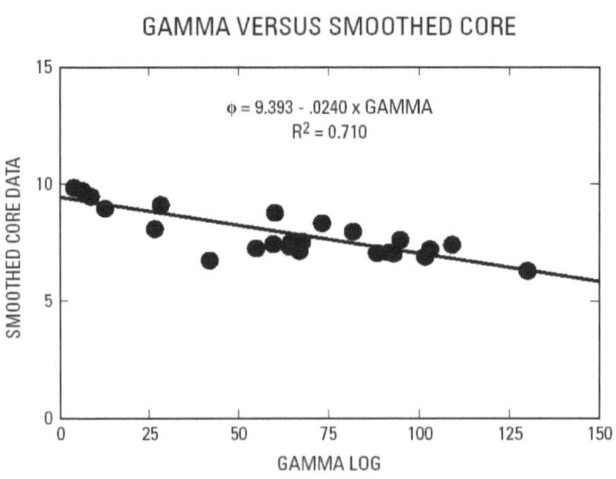

D

Figure 10.3D. Example of core sample data used to calibrate a gamma log in porosity units (from Paillet and Crowder, 1996): Regression of depth shifted data sets.

10.2 Fundamentals of Log Analysis

The process of log analysis for an environmental or hydrogeological project can be simplified into several basic steps:

- Data processing, including depth matching and merging all logs and other data from a single well and editing and smoothing the data set
- Correcting borehole effects and other errors
- Converting log measurements to hydrogeologic and engineering parameters of interest; this is best done with calibration data for each probe
- Combining logs and other data from all wells on a project, so that the data can be extrapolated laterally in cross sections or on maps
- Writing a report describing the analytical procedures and results from the logs

The analysis of log data by computer offers a number of time-saving advantages: a large mass of data can be collated and displayed; logs can be corrected and replotted; scales can be changed; smoothing and filtering operations can be carried out; cross plots can be made between different kinds of logs and core data; and calibration curves, correlation functions, and ratios can be plotted, as well as cross sections and isopach maps.

10.2.1 DEMONSTRATION OF THE USE OF LOGS TO CHARACTERIZE POROUS AQUIFERS

Geophysical well logs can be used to define hydrogeologic parameters such as hydraulic conductivity (permeability) and drainable (effective) porosity. One approach is to use the known relationship between porosity given by neutron, density, or acoustic logs calibrated in porosity units and hydraulic conductivity, where log-derived porosity values are substituted into semi-empirical relations between porosity and permeability (Nelson, 1994; Jorgensen, 1988, 1991). Although simple in theory, this direct approach is difficult to carry out in practice, because other parameters that appear in such equations are difficult to know, and geophysical porosity logs are not often tightly calibrated for the wide range of possible geologic environments encountered in hydrogeologic applications. An alternative approach to estimating hydraulic conductivity *in situ* is to use the dependence of electrical conductivity in porous material on the mobility of ions in pore spaces (Alger and Harrison, 1990; Kwader, 1985). If all other variables besides hydraulic conductivity are held constant, the electrical conductivity of a porous formation increases with increasing permeability. However, other factors, such as clay mineral fraction, influence formation electrical conductivity, and these factors may obscure the dependence of electrical conductivity on permeability (Biella et al., 1983).

The simplest relation between electrical conductivity and formation properties is given by an empirical relation between formation factor F (the ratio of the resistivity of a porous medium R_t to the resistivity of the water R_w with which the medium is saturated) and porosity n (Archie's law; Lynch, 1962):

$$F = R_t / R_w = a / n^m \quad , \qquad (10.1)$$

where the coefficient a and the exponent m (the cementation factor) can be determined independently from laboratory measurements on core samples and are usually consistent within a lithologic unit.

Equation (10.1) applies only in the limit that the electrical conductivity through interconnected pore spaces is much greater than all other forms of electrical conductivity in the formation. The limit cannot be based on a simple magnitude of mineral grain conductivity for a specific formation, because the ratio of conductivities of alternate pathways depends on the electrical conductivity of the water saturating pores. Various studies show that pore-surface conduction mechanisms become important, even when mineral grains are perfect insulators, whenever the salinity of pore water falls below about 1,000 μS/cm (Alger and Harrison, 1989; Worthington and Johnson, 1991; Hearst et al., 2000). If there are any electrically conductive clay mineral particles mixed with mineral grains, or if there are clay films coating quartz grains, Archie's law effectively fails over the 1,000–5,000 μS/cm range of pore-water conductivity. There are various equations designed to account for mineral and pore-surface conduction mechanisms, but these theories generally require additional information not normally available in groundwater studies, and that cannot be readily obtained from other geophysical logs (Worthington and Johnson, 1991). In an elaborate study of formation electrical properties conducted by Biella et al. (1983), those authors conclude that there is a weak dependence of formation electrical conductivity on permeability, which is generally masked by a much stronger dependence on grain-size distribution in typical fresh-water aquifers. Table 10.2 presents specific features of porosity logs.

Table 10.2. Response of logs to porosity (adapted from Keys, 1997)

Log	Property Measured	Response to the Type of Porosity	Extraneous Effects	Borehole Requirements
Gamma-gamma	Electron density	Total porosity, best for high porosity	Matrix, salinity, water saturation, well construction	Wet or dry, cased or open
Neutron	Hydrogen content	Total porosity, best for low porosity	Bound water, matrix chemistry, salinity, water saturation	Wet or dry, cased or open
Acoustic-velocity	Average P-wave velocity	Total porosity, only primary and intergranular	Cycle skipping, poor in high-porosity sediments	Fluid filled and open
Resistivity	Pore fluid and inter-connected porosity	Effective porosity both primary and secondary	Pore shape, fluid chemistry, water saturation	Open or with perforated PVC casing

The presence of shale (i.e., mixture of clay minerals and silt) in a sedimentary formation can cause erroneous values for porosity derived from logs. Whenever shale is present in a formation, all the porosity tools (sonic, density, and neutron) will tend to record porosity values (representing total formation water content) that exceed the amount of drainable (effective) porosity. The resistivity logs record values that are significantly lower than those that would be predicted by Archie's law under the assumption of non-conducting mineral grains.

Quantitative interpretation of logs from sedimentary formations typically involves estimation of clay content or shaliness (clay and silt content), the total void volume (porosity), and the matrix volume. Clay content percentage is determined from other geophysical log measurements that serve as clay-content indicators with respect to a baseline established for clay-free sediments. Measurements of resistivity, SP, gamma-ray, neutron-hydrogen index and/or clay content deduced from crossplots may all serve as such clay indicators. If clay content (the clay-mineral fraction of the formation) is known, log values may be shifted to correct for clay content. This amounts to the subtraction of one geophysical log (for example, the gamma log properly scaled as the clay indicator) from a "porosity" log (Asquith and Gibson, 1982). Alternatively, a more sophisticated numerical inversion scheme can be applied in which coupled inversion equations are solved simultaneously for both porosity and shale fraction (Serra, 1986; Paillet and Crowder, 1996). Determination of permeability is often based on porosity measurements because of the correlation (petrophysical relations) between permeability and intergranular porosity in clastic reservoir rocks (Jorgensen 1988; 1991). Because some logs such as gamma and resistivity can be related to mineral grain-size distributions and degree of sorting, it may be possible to generate an effective calibration of these logs in units of permeability, by regressing log measurements against the logarithm of permeability from aquifer tests (e.g., Keys, 1986).

10.2.2 STRATIGRAPHIC CORRELATION

One of the simplest applications of borehole geophysics in hydrogeology is the use of well logs to identify the depth and thickness of aquifers within the stratigraphic column. Geological maps or surface geophysical soundings can provide a general idea of the approximate depth and thickness of an aquifer, but more precise information is almost always needed to determine the specific depth interval for well completion or water sampling. The characteristic profile of such logs as natural gamma and induction resistivity indicate the stratigraphic sequence with a precise depth scale, so that a target aquifer can be located, or a water-producing unit in one borehole can be extrapolated to adjacent boreholes. A typical example of this application is indicated in Figure 10.4. In this example, solution openings along bedding planes in a calcareous aquifer were identified as possible conduits for the migration of contaminant from an industrial spill site. Several boreholes in the study area were logged with a suite of geophysical probes, and each data set indicated strong ambient flow entering at one or more bedding planes near the top of the borehole and exiting at other bedding planes near the bottom of the borehole. All inflow and outflow zones were located along bedding planes indicated on caliper and image (acoustic televiewer) logs, but it was not clear whether individual water-producing solution openings were continuous over the region. The identity of each bedding plane on each televiewer log was determined by correlating gamma logs across the data set (Figure 10.4A). The cross-correlation of log traces showed an unambiguous maximum when the logs were aligned at the elevations indicated in the figure, indicating a stratigraphic dip of about 5° to the east. When this stratigraphic dip is plotted on the televiewer logs, the lateral projection of individual bedding planes becomes evident (Figure 10.4B). Spinner (impeller) flowmeter logs showed inflow or outflow (hydraulically active fractures) within specific sets of dissolution-enlarged bedding planes, but inflow or outflow was associated with different planes within the sets. This information was of obvious use in assessing the potential for contaminant dispersal along laterally continuous subsurface conduits in this geologic environment.

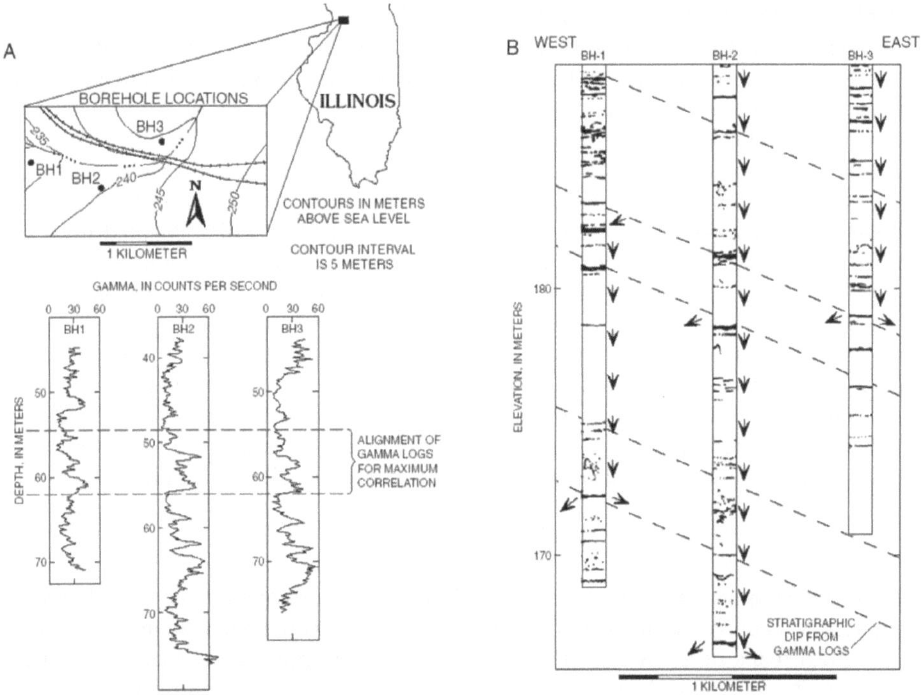

Figure 10.4. Example of gamma log correlation to identify laterally continuous bedding-plane conduits in a carbonate aquifer: (A) correlation of gamma log signatures to identify stratigraphic sequence in three boreholes; and (B) stratigraphic dip used to identify individual bedding plane aquifers on the televiewer logs obtained in the same three boreholes (from Paillet and Crowder, 1996).

10.2.3 FORMATION FLUID CONTENT

Borehole geophysical measurements provide the unique ability to measure the properties of pore water *in situ* in situations where the water in the borehole is not representative of the water in the formation, or where the borehole is filled with drilling mud. Figure 10.5 shows a typical example for a large-diameter water production well in sandstone. The well had been producing water with an unacceptably high level of dissolved minerals, but fluid-column logs showed the borehole was filled with fresh water after the well was taken off line and the pump removed. On that basis, the borehole was filled with water having a resistivity of about 35 ohm-m, corresponding to fresh water with a dissolved solids content of ~225 mg/L.

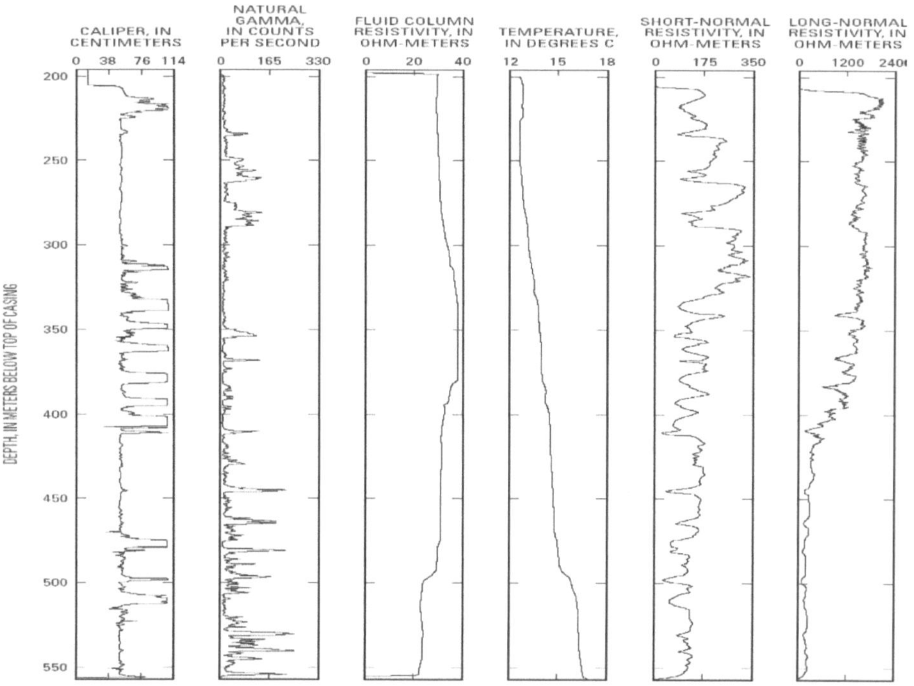

Figure 10.5. Example of formation resistivity logs used to indicate decreasing formation resistivity with depth attributed to increasing pore-water dissolved solids in a situation where fluid-column logs indicate fresh water throughout the borehole and where the logs indicate the decreasing resistivity trend cannot be attributed to lithology (from Paillet et al., 2002)

However, the fluid column logs also exhibit a step-like shape indicative of strong borehole flow. Water was probably entering the abandoned well in the upper part of the formation, and exiting the abandoned well in the lower 100 m at the time of logging. Therefore, the fluid column properties are those of the pore water at the inflow point, and not necessarily that of the pore water at the same nominal depth in the formation. The short normal resistivity log shows very little variation because the borehole diameter exceeds the sample volume of that log. In contrast, the long-normal resistivity provides information over a sample volume extending well beyond the 50 cm diameter of the borehole. The long-normal log shows an abrupt decrease of resistivity below ~400 m in depth, suggesting a significant increase in pore-fluid salinity unless the change can be accounted for by lithology. Analysis of this data (Paillet et al., 2002) yields an estimate of 1,800 mg/L of dissolved solids for the lower aquifer, consistent with water quality produced from this well when it was taken out of production.

10.2.4 USING LOGS TO CHARACTERIZE FRACTURED AQUIFERS

Fractured zones are of great interest in consolidated rocks, because fractures often provide the only drainable water and permeable flow paths in otherwise impermeable rocks. Open fractures may affect the response of some logging tools. The effect, however, is usually rather subtle, because the water-filled volume provided by the fracture is a small fraction of the total geophysical sample volume. Extensive reviews of the use of geophysical logs to characterize fractured igneous and sedimentary rocks are given by Keys (1979), Kobr et al. (1996), and Long et al. (1996).

A reliable determination of lithological types of carbonates (limestones, dolomites with some quartz or silicate admixture) can be made on the basis of gamma-gamma (density), neutron (neutron porosity), acoustic-velocity (sonic), and caliper logs using cross-plots of two porosity measurements (Schlumberger, 1986). The complete set of well logging methods used for identification of rock types within igneous and metamorphic rocks will include the natural gamma-ray, or (preferably) the spectrometric gamma-ray log, enabling determination of U, Th, and ^{40}K concentrations, the magnetic susceptibility, neutron, gamma-gamma density, and photoelectric logging.

Fractured aquifers in carbonate, igneous, and metamorphic rocks are indicated on logs in the following way (Figure 10.6):

- On spontaneous potential curves by a narrow and comparatively sharp negative anomaly produced mainly by electrokinetic potentials
- On caliper logs by major or minor cavities
- On formation density logs as a zone with a lower bulk density, owing to the presence of fracture and cracks filled with water
- On neutron logs as a zone of a higher neutron porosity
- On formation resistivity logs as conspicuous conductive zones, in contrast to solid rock blocks showing relatively high resisitivities, ranging from one to tens of kΩm

A number of specialized logging tools were developed in the past ten years for detecting fractures in the borehole wall. Schlumberger (1987) and the SPWLA image logging review monograph (Paillet et al., 1990) give thorough reviews of all classes of image logs and their interpretations. Image-logging technology has been advancing very rapidly in recent years; the latest developments are described by Williams and Johnson (2004) and Deltombe and Schepers (2004).

The permeability of fractured aquifers is most often characterized in units of relative production (proportion of total borehole specific capacity or discharge per unit drawdown obtained from the individual fracture):

$$q_i = Q / S \qquad [m^2\, s^{-1}], \qquad (10.2)$$

where q_i is the inflow from fracture i when the borehole is pumped at a steady rate of Q and the drawdown has stabilized at a value S. A representative value of hydraulic fracture (as opposed to the total depth interval within which the fracture is embedded) cannot be measured. In those situations where flowmeter logs are analyzed or individual fractures are subjected to hydraulic testing after isolation with straddle packers, the hydraulic properties of the fracture or set of fractures is given in units of transmissivity (T) representing the product of permeability and thickness. Hydrologists sometimes define the hydraulic aperture of a fracture as the aperture of a perfectly planar opening providing the same specific capacity. In that case, the transmissivity of the fracture is related to the hydraulic aperture by the cubic law:

$$b = 0.24T^{1/3} \tag{10.3}$$

where b is the aperture in millimeters and T is the fracture transmissivity in square meters per day. This equation applies to fresh water at approximately 25°C.

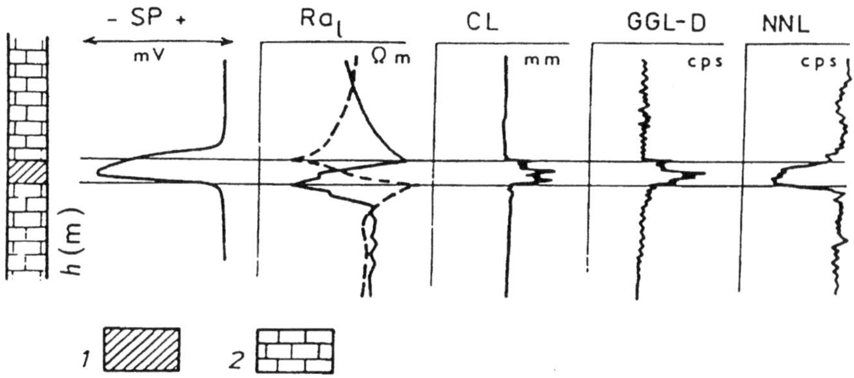

Figure 10.6. Characteristic features of a fractured zone on an SP log, lateral resistivity logs (Ra_l) with potential electrodes above (dashed line) or below (solid line) the current electrode, caliper log (CL), gamma-gamma density log (GGL-D) and neutron-neutron log (NNL). 1—fractured aquifer, 2—solid limestone (from Kobr et al., 1996)

10.2.5 EXAMPLE OF MULTIVARIATE PATTERN RECOGNITION

This section presents an attempt to distinguish among individual rock types in the Devonian metamorphic rocks, using only the data from geophysical measurements in boreholes. Based on a comparison of logging data with the chemical composition of rock samples and multivariate statistical analysis, suitable parameters were identified for an *in situ* classification of rock types in the geologically complex conditions of epimetamorphites of the Hrubý Jeseník Mountains.

A 150 m deep borehole (R-150) was logged in the village of Rejvíz (Jeseníky Mountains, Czech Republic) (Figure 10.7), and the logged section was divided into quasi-homogeneous intervals. Statistical analysis was applied to individual depth intervals. Frequency diagrams were plotted to evaluate the range and magnitude of individual physical parameters. Some parameters are apparently controlled by fractured zones. For example NN–neutron, density, Zef (photoelectric log), acoustic, and three electrode laterolog (LL-3) traces all show sharp deflections in fractured intervals (a process denoted in the logging literature as zoning). The responses in the fractured intervals were assumed to be unrelated to lithologic properties and were not included in further analysis.

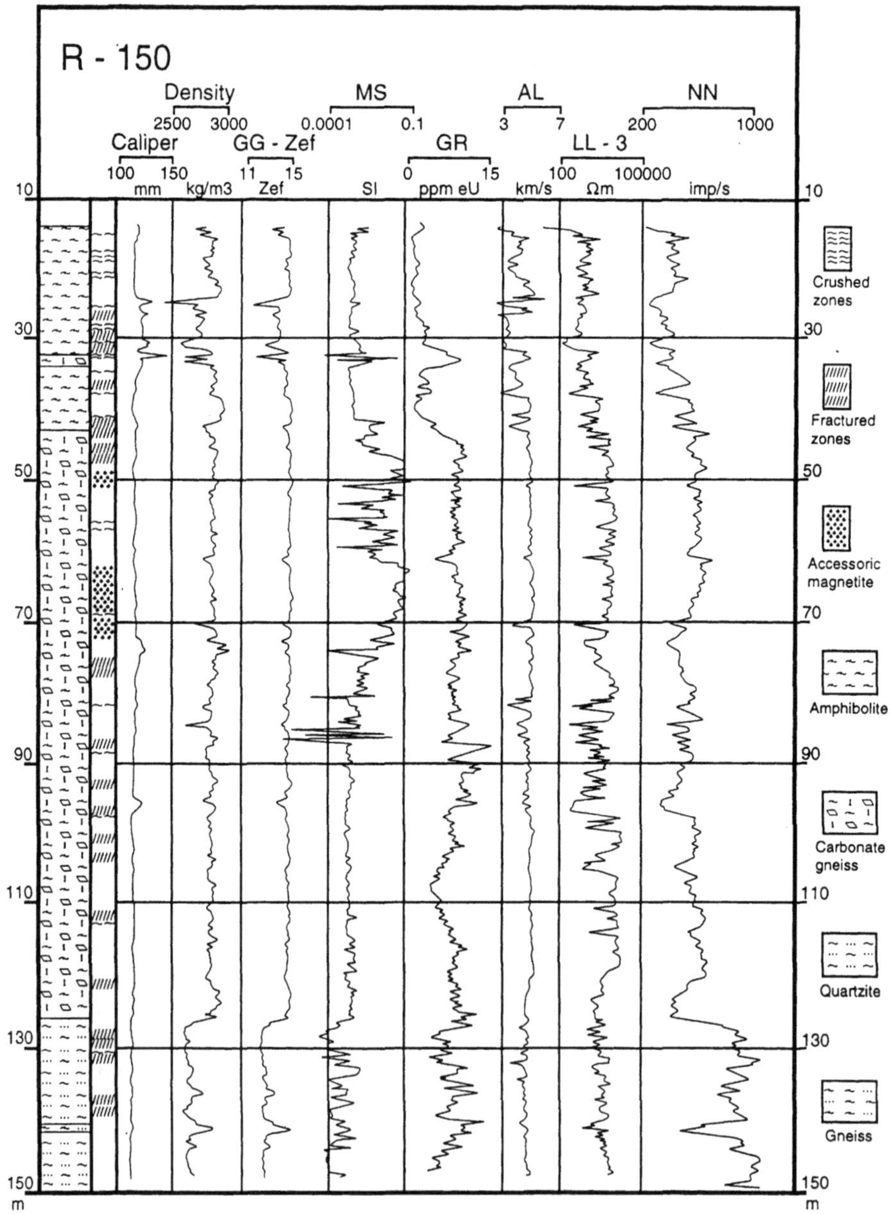

Figure 10.7. Suite of well logs from the borehole R–150 (from Kobr et al., 1997)

Some histograms for the nonfractured intervals show more than one maximum. These cases are suitable for the determination of rock types. Descriptive statistics, including the Shapiro-Wilk test of normality, show a much smaller dispersion within each of the three groups A (amphibolite), G (gneiss), and Q (quartzite)—if subgroups with caverns (N=8 cases) and thin samples (6 cases) are eliminated. The coincidence of means for all logging parameters was tested for individual rock types to quantitatively evaluate the significance of the variation divergence of the means. The nonparametric Mann-Whitney test (Table 10.3) gives three groups: A, G, and Q. Within this table, three asterisks signify that the delimited rock groups differ from each other on a level of significance greater than 0.1%; two asterisks signify a level greater than 1%; and one asterisk a level greater than 5%. The minus sign stands for no significant difference between investigated means, i.e., less than 5%.

Table 10.3. The results of testing of accordance of means of individual log parameters between studied rock types: A (amphibolite), G (gneiss), and Q (quartzite) (from Kobr et al., 1997)

Parameter	**Rock type**		
	A vs. G	A vs. Q	G vs. Q
Gamma ray	***	***	-
Magnetic susceptibility	-	**	***
GG – Zef.	-	***	***
Density	-	***	***
Neutron neutron	-	***	***
LL-3	*	***	-
Acoustic	*	-	-

Cross plots of selected physical parameters, obtained by logging, can also be used to separate individual rock types between them (Figure 10.8). These approaches are used for identification of lithology based on physical properties of the borehole section. On the other hand, those logging methods that are not fitted for lithology reasons (acoustic, laterolog) can be used for determination of fractures.

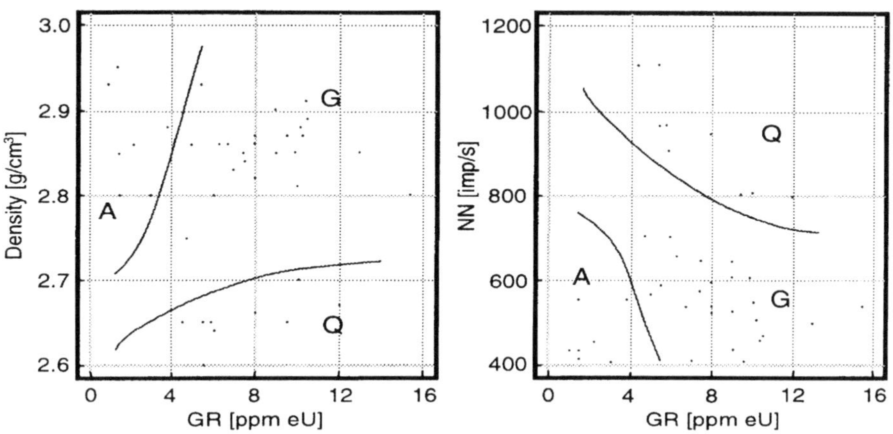

Figure 10.8. Cross plots of selected physical parameters from the R–150 borehole section (from Kobr et al., 1997): A (amphibolite), G (gneiss), and Q (quartzite)

The reliability of the sample differentiation into three groups A, G, and Q was verified by consecutive discriminate analysis. Using the 1st and 2nd canonical variable, the result can be depicted as a projection onto the plain with the best differentiation of individual rock types (Figure 10.9). The larger distance between the center of gravity of samples of Q from A and G documents a better contrast of physical parameters between quartzites and amphibolites (and also between quartzites and gneisses), than between gneisses and amphibolites. This exactly corresponds to geological conditions where transitional rock types such as amphibolite gneisses (amphibolite or gneiss-carbonate stromatites) are common.

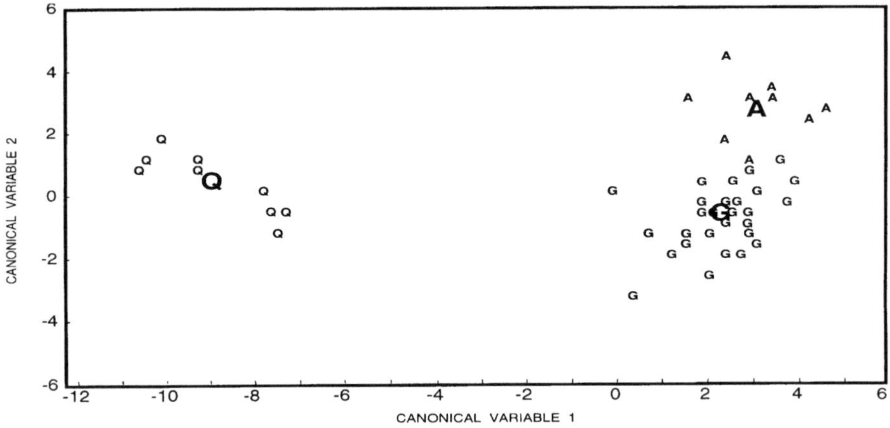

Figure 10.9. Application of discriminant function analysis to lithofacies distinction between groups of samples of logging parameters by the representation of canonic variables (from Kobr et al., 1997): A (amphibolite), G (gneiss), and Q (quartzite)

10.2.6 AQUIFER SAMPLING AND TESTING

One advantage of borehole geophysics in hydrogeology is that geophysical measurements can be correlated with hydraulic measurements or water sample information over specific depth intervals. This is a real improvement over surface geophysical soundings, in which there is no direct way to correlate the noninvasive sounding made at the surface with hydraulic properties *in situ*. In borehole data sets, formation electrical conductivity can be directly related to the electrical conductivity of water samples recovered from discrete zones in screened wells, or from zones in open borehole isolated with straddle packers. In one example (Figure 10.10), induction logs obtained through plastic well screens or casing provide estimates of formation electrical conductivity adjacent to the borehole. Surface electromagnetic soundings in the borehole area provide similar estimates of subsurface electrical conductivity, but cannot be compared to water sample conductivity. In a series of water-sample wells in alluvial sediments such as that in Figure 10.10A, the electrical conductivity of water samples was compared to the interval-average formation electrical conductivity over short screened intervals. Before performing the regression, both formation and water sample conductivities were corrected to a standard temperature of 25°C. After correction, the regression gives a linear relationship between water sample conductivity and formation conductivity (Figure 10.10B). This data set allows the estimation of water conductivity *in situ* from any electrical measurements as long as that measurement applies to this same aquifer, and can be used to interpret surface electromagnetic soundings as well as borehole induction logs.

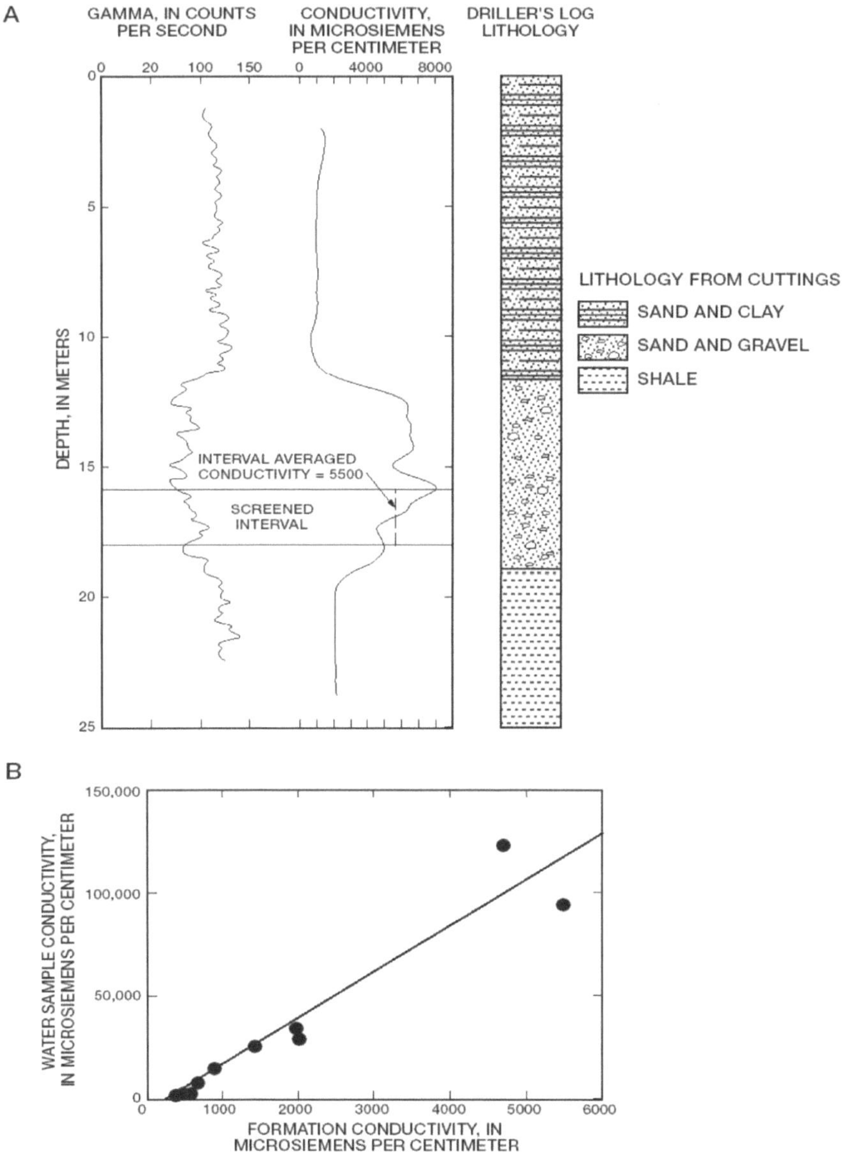

Figure 10.10. Example of water-sample data used to generate a relation between formation electrical conductivity and the electrical conductivity of pore water within the formation: (A) induction and gamma logs compared to well construction and lithology and illustrating the estimation of formation electrical conductivity over the screened interval in the sampling well; and (B) regression between water-sample electrical conductivity and the electrical conductivity of the formation from which that water was derived (from Paillet, 2002)

10.2.7 DATA INTEGRATION

An important advantage of borehole geophysics is that the same measurement can be made in the borehole and on the surface at two very different scales of investigation (Figure 10.11). Such measurements provide direct information about the effects of aquifer scale on physical properties. Another important advantage is that the borehole allows several different physical properties to be measured, whereas noninvasive surface geophysical measurements are almost always made using a single method, such as electrical resistivity or seismic propagation. Any one physical property such as acoustic velocity or electrical resistivity can be determined by several different properties of the subsurface, such as lithology, pore-size distribution, permeability, and groundwater salinity. The full suite of geophysical logs from a borehole can often be used to develop a site-specific interpretation model for the interpretation of surface geophysics. A typical example is described by Paillet et al. (1999). Surface time-domain electromagnetic soundings were used to map the subsurface distribution of salinity in an aquifer in south Florida. Induction logs from representative boreholes were used in the analysis in two important ways: (1) Induction logs demonstrated that although the addition of several subsurface layers to the interpretation model nominally reduced the residual error in modeling the data, the model interpretation could only estimate the thickness and resistivity of the surficial aquifer and the average resistivity of all underlying aquifers; (2) Combined interpretation of gamma, neutron, and induction logs showed that subsurface electrical conductivity depended almost exclusively on pore-water salinity.

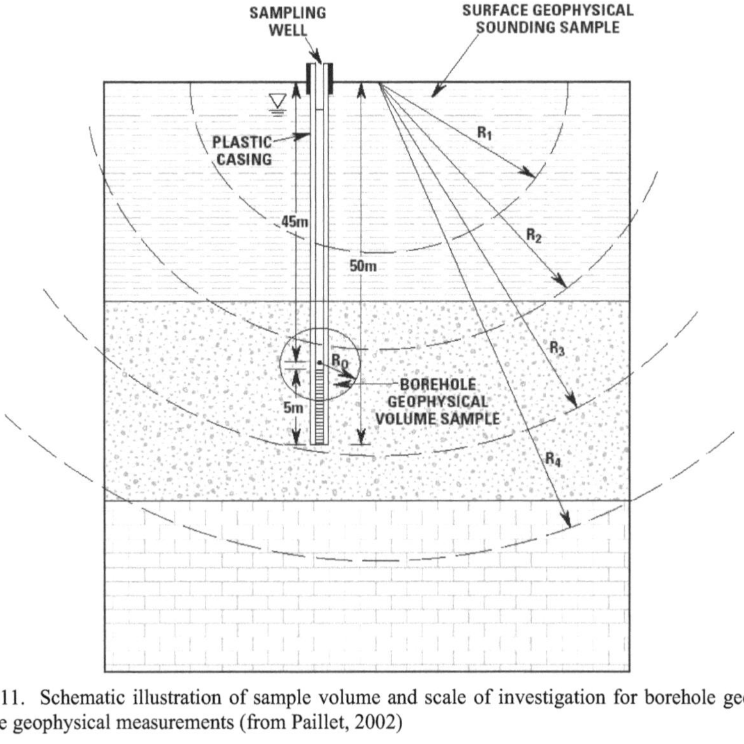

Figure 10.11. Schematic illustration of sample volume and scale of investigation for borehole geophysical and surface geophysical measurements (from Paillet, 2002)

Subsurface cross sections constructed from the electromagnetic soundings showed a partially confined surficial aquifer of uniform water quality, except for a wedge of sea-water intrusion on the south and a variable pattern of electrical conductivity (and presumably variable water salinity) in the underlying strata (Figure 10.12). This irregular distribution of deeper groundwater salinity was used to infer that groundwater quality in the study area was influenced by variable rates of brine seepage from deeper aquifers, and not by systematic intrusion of sea water from the adjacent coast. Borehole geophysics further contributed to this interpretation in two important ways. First, borehole flow logs conclusively demonstrated the presence of a large upward hydraulic head gradient driving water upwards in open or long-screened boreholes throughout the study area. Second, test boreholes drilled at strategic locations along the electromagnetic sounding profiles verified that the thickness of the top layer and the salinity of the underlying layer given by model predictions agreed with induction logs in the verification boreholes shown in Figure 10.12. This one example illustrates the several important ways in which the integration of surface and borehole geophysics contributes to the interpretation of the properties of aquifers *in situ*.

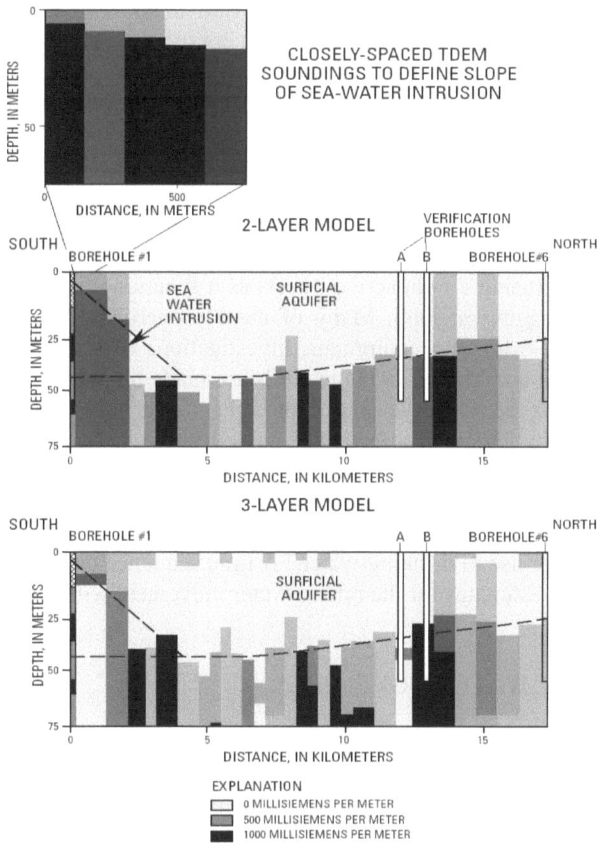

Figure 10.12. Examples of 2-layer and 3-layer models fit to surface electromagnetic soundings where model parameters are conditioned by comparison with induction log data (from Paillet et al., 1999)

10.3 Borehole Flow Logging

Flowmeters (Schlumberger, 1973) are probes designed to measure groundwater flow rates along boreholes under natural or ambient and forced gradient (pump-test) conditions. Impellor flowmeters (Hill, 1990; Keys and Sullivan, 1979) require a significant velocity to rotate impellor blades, so that this probe cannot measure velocities less than about 10^{-1} m/s. Heat-pulse (Hess, 1986) and electromagnetic (Molz et al., 1994) flowmeters can detect velocities as low as 10^{-3} m/s. Taking into account the sensitivity limit of flowmeters, vertical flow rates are often determined indirectly from the vertical velocity W [$m.s^{-1}$] estimated from fluid-column logging, where the movement of the interface of water of different properties can be traced by time sequences by repeated logging over a specified time interval (Tsang et al., 1990). The use of repeat fluid-column logging to identify water movement in boreholes extends the effective measurement range of vertical velocity in boreholes as low as 10^{-6} m/s. Situations in which a well is connected to a single aquifer can be treated with conventional aquifer test methods described elsewhere (Chapter 2 of this volume). Geophysical logging and especially borehole flow logging are useful in situations where a well is completed in multiple zones, and there is a need to quantify the hydraulic and/or water quality parameters of each contributing zone independently.

Techniques based on the time sequence of fluid logs are applied only in clean wells, where mud has been replaced by water and the filter cake does not affect aquifer permeability in the vicinity of the borehole. Under these conditions, changes in the physical properties of the borehole fluids under ambient or stressed conditions reflect the properties of the formation and the water contained in the formation. Various properties of water in boreholes, such as the temperature, conductivity, transparency and/or radioactivity (using a radioactive tracer) as a function of depth, may be used in this fluid-column logging method. Many of these properties can be measured using conventional borehole logging equipment. Investigations of turbidity by photometric measurement (Zboril and Mares, 1971) in wells are made by measuring the attenuation of the luminous flux between the source of light and a photovoltaic cell. The flux impinges on a photovoltaic cell after it has passed through the fluid to be measured. Cell illumination depends on the absorption properties of the fluid, i.e., on the intensity of coloring, or on the amount of the dispersed particles. The substance used for the dye tracer (for example, Brilliant Blue FCF, Type E 133) must be an approved nontoxic substance suitable for use in drinking water. A time sequence of such profiles can then be analyzed to yield estimates for the rate of water movement within the fluid column.

10.3.1 VERTICAL WATER MOVEMENT

Inflow sites, loss sites, and sections with vertical movement of water are usually indicated on high-resolution temperature logs or temperature-gradient logs (differential temperature logs). The depth intervals where natural movement of the water occurs along the borehole axis are indicated on temperature logs by a constant or nearly constant temperature (Conaway, 1987, Prensky, 1992). If the solute content (and thus also the resistivity) of individual inflows is not the same, the inflow sites appear on the fluid resistivity logs, and often on the photometry logs, because of insufficient cleaning of the borehole from the original mud. The sites of inflows and losses can be emphasized by introducing sodium chloride into the borehole to decrease the resistivity of the water in the borehole. The inflow sites then appear on the fluid-resistivity-log curves as sudden increases of fluid resistivity, because fresh groundwater usually exhibits resistivity values in the range of 20–100 ohm-m. However, fluid-column logs can sometimes be ambiguous indicators of flow in boreholes. For example, a fluid-column temperature log obtained in a borehole with downflow might be characterized by a long isothermal interval. The bottom of the isothermal interval will coincide with the depth where the downflow regime ends, even though significant outflow may occur at other depths. Thus, several different interpretations can often be derived from a given fluid-column log.

A more direct detection of the vertical velocity W is based on temporal plots of fluid properties following the injection of a tracer into the well bore. The water entering the well or injected into the well creates an interface whose vertical displacement can be traced on time-sequence plots of fluid resistivity, photometry, or temperature log. Figure 10.13 shows a schematic illustration of the situation in which a borehole penetrates two aquifers with different piezometric levels (hydraulic heads). Thus, natural flow occurs along the well bore, from the aquifer with higher hydraulic head to the aquifer with lower hydraulic head. Examples of vertical flow measurement using the borehole dilution method are given by Vernon et al. (1993) and Paillet and Pedler (1996).

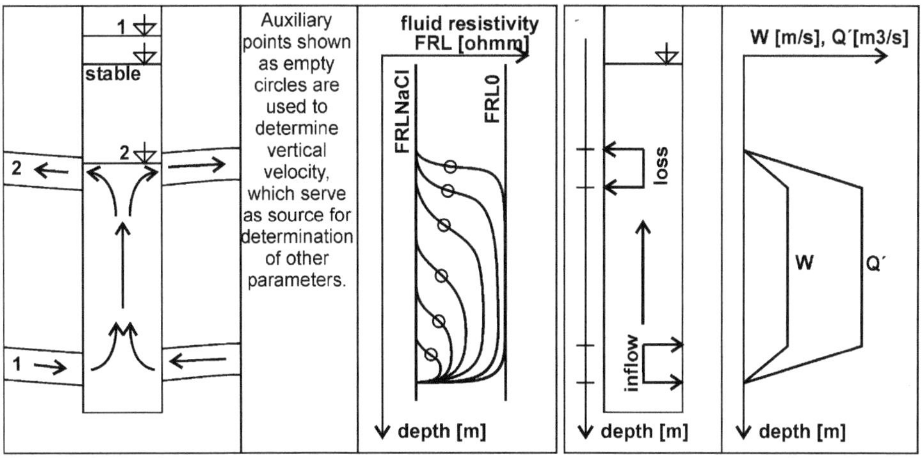

Figure 10.13. Schematic illustration of the characterization of vertical water flow (velocity W and yield Q′) in a borehole using time-sequence fluid-resistivity logging (Kobr, unpublished data)

Dilution logging techniques are regularly applied by commercial contractors in North America under the generic description of hydrophysical logging. This service provides estimates of the distribution of transmissivity along the borehole, along with estimates of the electrical conductivity of water derived from each producing interval delineated in the analysis.

10.3.2 LOG-DERIVED HYDRAULIC PROPERTIES OF AQUIFERS

10.3.2.1 *Hydraulic Properties from Flowmeter Analysis*

High-resolution flow logging with recently available heat-pulse (Hess, 1986) and electromagnetic (Molz et al., 1994) flowmeters provides the ability to include direct hydraulic analysis of borehole conditions in the geophysical data interpretation. For example, flow profiles during pumping can be used to identify the fractures, bedding planes, or geological strata where water enters the borehole. Although this analysis appears simple in principle, there are a number of complications in practice. First of all, the flowmeter data sets need to be corrected to account for various conditions, such as variation in drawdown during the experiment and changes in borehole diameter (Paillet, 2004). Once these corrections are applied, the data consist of estimates for the amount of inflow to or outflow from the borehole under a specified hydraulic condition at a number of depth intervals. However, inflow or outflow cannot be taken as directly proportional to hydraulic conductivity, because the amount of flow depends upon the product of hydraulic conductivity and the pressure gradient driving the flow. If there is a vertical gradient in pressure and the pressure difference varies with distance along the well bore, then the pressure driving the flow will vary along with the hydraulic conductivity. In such a situation, there are effectively two sets of unknowns for the interpretation of hydraulic properties. Molz et al. (1989) show that one approach is to measure flow under two hydraulic conditions (ambient and steady pumping for example), and then subtract the inflows under the two conditions. This approach effectively subtracts out the pressure effect, so that the inflow difference is directly proportional to hydraulic conductivity expressed as a percent of total borehole hydraulic conductivity.

Note that several practical drawbacks exist with the Molz et al. (1989) subtraction-of-inflows approach to borehole flowmeter log analysis. First of all, the method provides only relative estimates of zone transmissivity and gives no estimate at all for the hydraulic head difference driving the flow. In practice, the subtraction-of-flows interpretation involves fitting the flow profiles to straight-line segments (Figure 10.14) so that the size of the "steps" gives the amount of inflow or outflow. There is no guarantee that any one set of lines fit to the data actually satisfies the equations of flow. It is possible to difference the data such that one or more inflow zones are assigned a negative permeability value. A more effective method of analysis is to find a set of lines that satisfy the equations of flow and fit the measured flow data in a least-squares sense. This has been done in Figure 10.14, using the Paillet (1998) borehole flow model. The results are given in Table 10.4. Paillet (2000) shows that a unique solution can be found by matching the model to two different flow profiles (usually ambient and pumping) for a measured change in water level (drawdown) between the two states. This has the advantage of yielding direct estimates of transmissivity for each zone, and the far-field water level (hydraulic head) for each zone. Although the transmissivity values apply to the immediate vicinity of the borehole and may not relate to large-scale aquifer properties, the hydraulic head estimates can be useful in identifying the large-scale structure of aquifers (Paillet et al., 2000; Paillet, 2001).

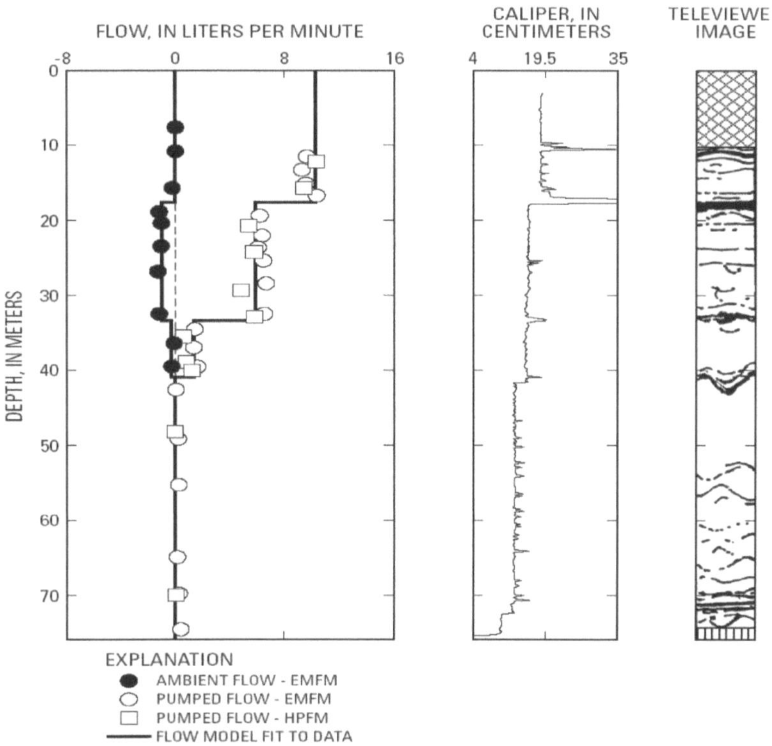

Figure 10.14. Example of high-resolution flowmeter log data obtained with both the electromagnetic (EM) and heat pulse (HP) flowmeters in a fractured bedrock aquifer, where the hydraulic properties of water-producing fractures (transmissivity and hydraulic head; Table 10.1) are estimated by fitting the Paillet (1998, 2000) flow model to the flowmeter data set (Paillet, unpublished data)

Table 10.4. Flow model inversion applied to the flowmeter data in Figure 10.14

FLOW ZONE (depth in meters)	TRANSMISSIVITY (meters squared per second)	ZONE HYDRAULIC HEAD (meters below top of casing)
16.8	2.0×10^{-5}	5.95
32.0	4.0×10^{-5}	6.87
39.8	1.3×10^{-5}	6.87

The method can be taken one step further and used in the cross-borehole configuration. When observation boreholes are available, flow measurements can be made in observation boreholes over a period of time after pumping starts in an adjacent borehole (Lapcevic et al., 1993; Paillet, 1998). Measurements are made at depth stations between each water-producing zone in the observation borehole and at a location above all inflow zones. In these experiments, there are usually estimates of zone transmissivity values, T, from preliminary single-borehole experiments. Then the analysis consists of adjusting the aquifer storage coefficient, S, in model equations until the shape and magnitude of predicted response curves matches the flowmeter data sets. Field studies show that this method can be used to identify discrete fracture connections involving fracture segments that connect fracture zones without intersecting any of the boreholes (Williams and Paillet, 2002).

10.3.2.2 *Hydraulic Properties Derived from the Time Sequence of Fluid Logs*

Log-derived hydraulic conductivities K_s or transmissivities T are based on the theory of steady-state flow to the ideal water well. For a homogeneous confined aquifer penetrated by a water well, and for steady-state radial flow to the well during constant pumping or from the well during constant injection, Dupuit's assumption applies:

$$K_s = [Q \,.\, \ln (R/r)] \,/\, [2\pi.\ H.\ \Delta S\,] \qquad [m.s^{-1}] \qquad (10.4)$$

where ΔS [m] is the difference in the water levels in the well between that at the start of the hydrodynamic test and that during the test (drawdown), r [m] is the well radius and R [m] is the radius of the cone of water level depression. Grinbaum (1965) shows that when the water-level difference ΔS is small in either confined or unconfined aquifers:

$$K_s = Q \,/\, (H \,.\, \Delta S\,) \qquad [m.s^{-1}] \qquad (10.5)$$

To estimate the hydraulic conductivity of an aquifer, the parameters Q, H, and ΔS occurring in Equation (10.5) have to be determined. The ΔS is given directly by the measurement of water levels before and during the hydraulic test. The thickness H_i and the yield or water loss Q_i of individual aquifers are evaluated from the diagram representing the depth changes of vertical flow rates Q' during the hydrodynamic test. The vertical flow rate is constant over impermeable intervals, and there is a linear change in flow rate within permeable intervals. Over permeable intervals $\Delta Q' = Q'_i - Q'_{i-1}$ corresponding to the yield or water loss Q_i of the aquifer.

The water-level drawdown ΔD caused by the pumping test corresponds to ΔS only if one confined aquifer appears in the borehole section. If more confined aquifers with different piezometric levels are penetrated by the water well, two hydrodynamic tests under different yields of pumped or injected water must be carried out to determine ΔS_i. The piezometric level of each aquifer is taken as a limit of ΔD_j for $Q_{ij} \rightarrow 0$, supposing there is a linear the relation between the yield of an aquifer Q_{ij} and steady-state water level D_j during the pumping test (Mares et al., 1994).

Transmissivity T of an aquifer is related to hydraulic conductivity K_s and thickness H through the relation

$$T = K_s \,.\, H \qquad [m^2.s^{-1}] \qquad (10.6)$$

A schematic illustration of the application of time-sequenced fluid-column resistivity logging during pumping to the determination of aquifer hydraulic parameters is illustrated in Figure 10.15. An example of the application of this technique is given in Figure 10.16 (constant pumping column).

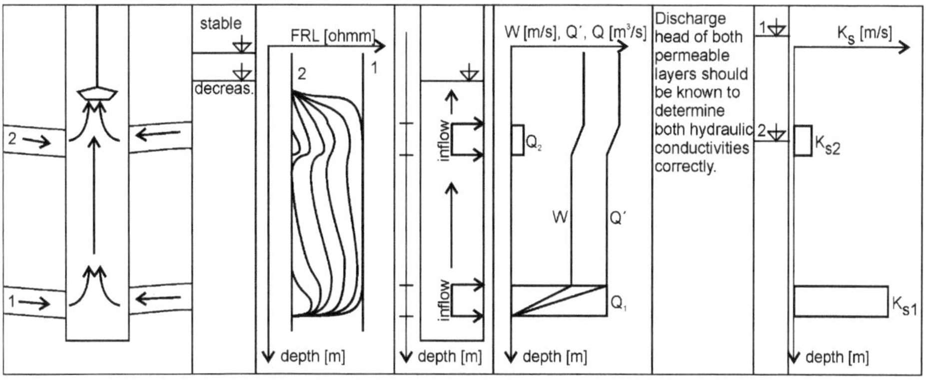

Figure 10.15. Schematic illustration of the application of the fluid resistivity log during a pumping test to identify water-producing zones in a borehole (Kobr, unpublished data)

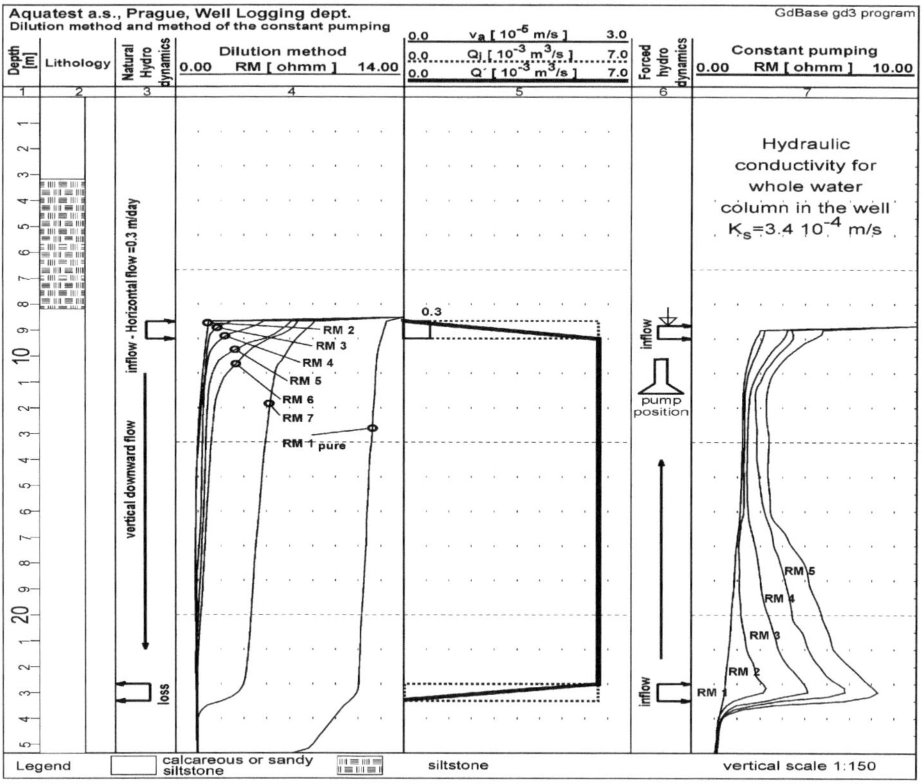

Figure 10.16. Logging of a water well using fluid-resistivity logging under steady-state conditions (dilution method) and during pumping test (constant pumping). Courtesy of Aquatest Praha

10.4 Logging to Study Water Flow in the Rock Medium

Movement of the groundwater flow in sandy aquifers is oriented in the direction of the hydraulic gradient. The subhorizontal groundwater flow in the aquifer is characterized by filtration velocity v_f, by the hydraulic conductivity K_s and the azimuth (direction) of the flow. Logging techniques for estimating hydraulic conductivity are described above (see Section 10.3.2).

10.4.1 FILTRATION VELOCITY

The filtration velocity v_f of an aquifer is defined as volumetric flow rate Q through a unit area perpendicular to the flow direction. There is no way to measure the filtration velocity directly. The logging procedure measures the dilution velocity (the apparent filtration velocity) v_a in the following way. If the water column in the well is marked by a tracer, the dilution velocity v_a can be related to the decrease in concentration of the tracer over time. Relationships between the filtration velocity v_f and the measured dilution velocity v_a are given by potential theory (Drost et al., 1968). The contrast between the finite permeability of the aquifer and the effectively infinite permeability of the borehole causes the local flow field to converge and accelerate in passing across the borehole. Thus, the measured rate of flow in the borehole is greater than the far-field filtration velocity in the aquifer. Although the distribution of flow varies across the borehole, flow measured near the center of the borehole v_a is assumed to be related to the far-field flow v_f by the parameter α (e.g., Schneider, 1973):

$$v_a = \alpha \, . \, v_f. \qquad (10.7)$$

Detailed investigations of the parameter α (Halevy et al., 1967; Bergmann, 1970, 1972; Klotz and Chand, 1979) confirmed the influence of the hydraulic properties of the well completion (screen, gravel, etc.). In uncased boreholes with ideally clean walls, a value $\alpha \approx 2$ provides a reasonable fit in relating the average horizontal flow in the borehole to the far-field filtration velocity in the aquifer.

Dilution velocity v_a can be determined using different tracer substances, e.g. NaCl (Ogilvi and Fedorovich, 1964; Grinbaum, 1965; Lukes, 1973), soluble colours (Schön, 1993; Mares and Zboril, 1995), and radioactive substances (Moser et al., 1957; Halevy et al., 1967; Drost et al., 1972). The dilution process results in the decrease of concentration C with time t (starting with an initial value C_1 at time t_1). This decrease is directly connected with v_a and the borehole diameter d:

$$v_a = (\pi/4).d. \ln (C_1 / C)/ (t - t_1). \qquad (10.8)$$

Thus, v_a can directly be determined from the slope in a diagram log C = f (t). The technique is illustrated schematically in Figure 10.17 and by a practical example from one monitoring well in the Cretaceous of Bohemia in Figure 10.18.

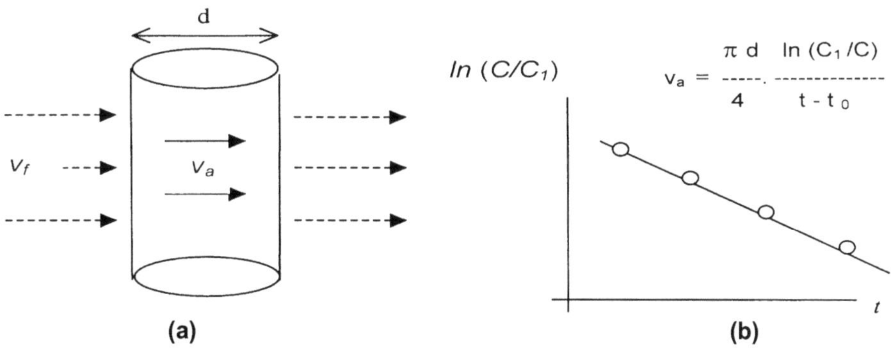

Figure 10.17. Principle of the borehole dilution technique: (a) definition of filtration and dilution velocities; (b) decrease of tracer concentration C with time t (from Mares and Schön, 1999).

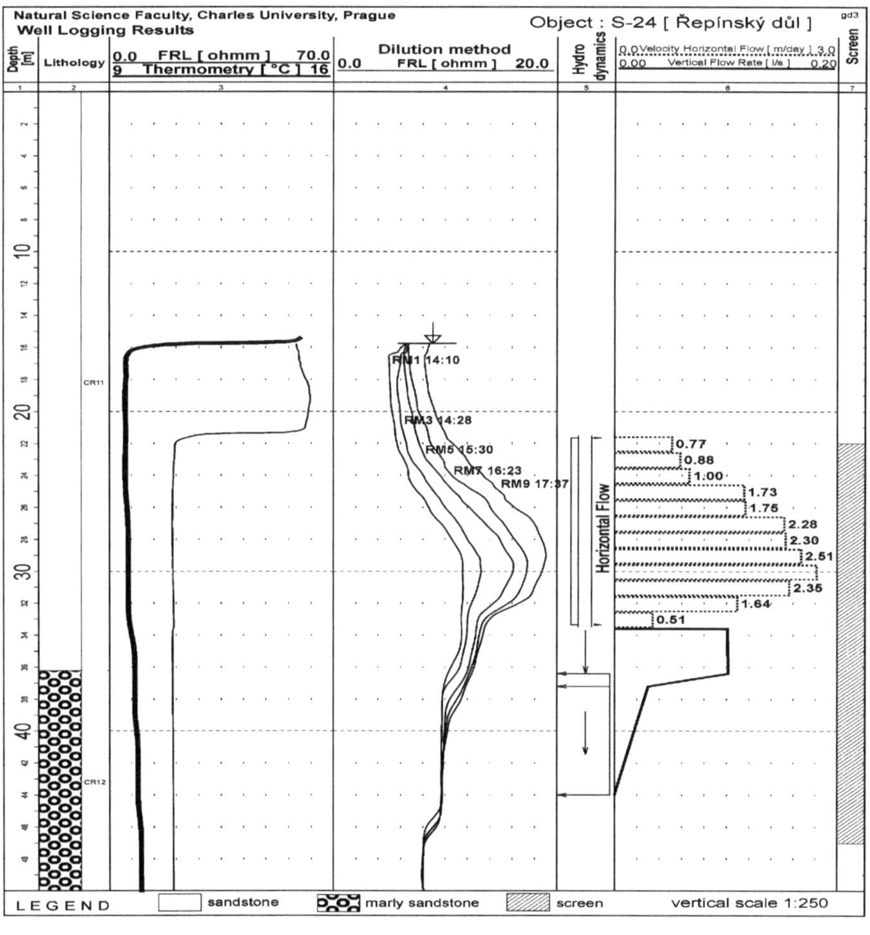

Figure 10.18. Time sequence of fluid-resistivity logs in the Well S-24. Determination of horizontal flow (dilution) velocity at a depth of 22–33 m (Kobr, unpublished data).

10.4.2 AZIMUTH OF THE GROUNDWATER FLOW

Three geophysical techniques can be used to estimate the azimuth of groundwater flow: (1) analyzing the field of streaming potentials, (2) application of the mise-a-la masse method, and (3) using a specially designed tracer technique. As described below, the acquisition geometry varies depending on what method is employed.

10.4.2.1 *Streaming Potential and Mise-a-la Masse Methods Using Surface Geophysical or a Combination of Surface and Downhole Techniques*

Streaming potentials are characterized by a positive electrical potential gradient in the direction of groundwater flow (Kelly and Mares, 1993). This technique can deliver useful information in relatively shallow clastic aquifers, but only in regions not affected by industrial electrical noise. It thus has the advantage of being a simple and inexpensive field technique that does not require drilling, and is a common surface geophysical measurement. The mise-a-la-masse method requires the presence of a well wellbore within the target aquifer. If casing is necessary, then casing should be made of insulating material (e.g., PVC) and should be screened over the aquifer interval. The principle of the mise-a-la masse method is based on the comparison of time-lapse images of the potential field created by an electrically charged, highly conductive water body surrounding the well after adding an electrolyte to well (Mazac et al., 1980). The charged body changes its shape with time, becoming elongated in the direction of groundwater flow. The changes in the shape of the charged body are reflected in changes in the electrical field measured on the surface. The applicability of these methods is limited to shallow aquifers with depth not exceeding 50 m.

10.4.2.2 *Tracer Techniques Using Single-Well Logging Approaches*

The single-well logging method for determining the azimuth of horizontal flow is based on the tracer distribution in a cross section perpendicular to the borehole axis during some period after the tracer addition. Construction of the logging tool depends on the type of tracer used. Sensors must be able to detect the spatial distribution of the tracer in different parts of the fluid or borehole wall surrounding the tool or filling the horizontal measuring slit of the tool. Tracers can include: temperature (Pitschel, 2000), radioactive elements bound to colloidal particles (Drost et al., 1972), or dye (Mares et al., 2003). All tools must be equipped to measure the orientation of the sensor to magnetic or geographic north. One such probe (Figure 10.19) is equipped with a special circular photometric sensor able to detect changes in the distribution of the Brillant Blue FCF food color (E133) on the margin of the horizontal measuring slit. The voltage distribution on the set of 32 phototransistors indicates (with a sensitivity of $\pm 5^{\circ}$) the direction of the groundwater flow corresponding to the position of the voltage minimum on the screen. Other examples of applications for the directional flow logging tool include the registration of very small illuminated particles on the dark-field background (Niesner and Reindl, 2000), the tracking of temperature changes across the borehole, and the monitoring of suspended particles (Wilson et al., 2001).

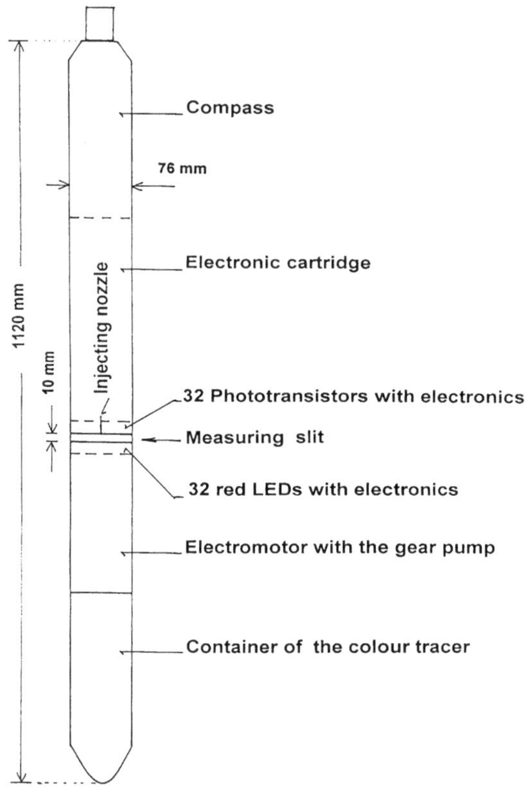

Figure 10.19. Schematic illustration of probes used to determine direction of horizontal flow in a borehole (from Mares et al., 2003—with EAGE permission)

10.5 Logging to Check the Quality of Well Completion

Wells are often protected against collapse in soft formations by casing tubes of different lengths and diameters, made of such materials as steel, PVC, and occasionally wood. Individual sections of casing are screwed or welded in a string and put into the well. Some pieces of casing are perforated to allow hydraulic communication with aquifer fluids. The space between the casing and the borehole wall (the annulus) can be filled with different materials (sand, gravel in screened intervals; clay and cement as a sealing material to avoid fluid circulation behind casing), or formation materials may simply collapse around the casing or screen after installation. There is sometimes a need to separate individual aquifers or to protect an aquifer from surface pollution. In such cases, it is necessary to have some method to check for horizontal leakage through the casing or for vertical leakage through the annulus around the casing. Table 10.5 summarizes logging methods used in casing and cement bond inspection. Typical geophysical log responses used to determine the location of screens and casing leaks in wells are illustrated in Figure 10.20.

Table 10.5. Logging methods used for checking the quality of the well completion

Parameters Measured	Optimum Set of Logging Methods
True borehole diameter	Caliper log CL
Trend of borehole in space	Inclinometer IM
Bottom of casing	Resistivity log using lateral array Ra_l
Perforated casing (Figure 10.20)	Ra_l (PVC casing), cement-bond log (steel casing)
Casing collars	Casing collars locator (steel casings), resistivity log and sensitive CL (PVC and wood casings)
Lost casing, tools	Resistivity log Ra, magnetic susceptibility log MSL
Deflection of the bit from the original hole	IM, CL, in flowing wells also high-resolution temperature log TL
Top of cement behind casing	TL until 24 hrs after cementation, gamma-gamma density GGL-D, gamma-ray GR using tracers
Quality of cementation	Cement-bond log
Tightness of casing	TL, fluid resistivity log FRL, photometry PHL, GR after treatment of water with appropriate tracer
Effect of casing on the yield of aquifers	Repeated yield assessment of water inflows before and after the borehole has been cased
Communication between two adjacent boreholes	Tracer tests using FRL, PHL, GR

Casing physical properties in relation to casing materials:

Steel: high electric conductivity, magnetic susceptibility and acoustic velocity

PVC: electrical insulator, very low magnetic susceptibility and low acoustic velocity

Wood: medium electric conductivity, low magnetic susceptibility and sound velocity

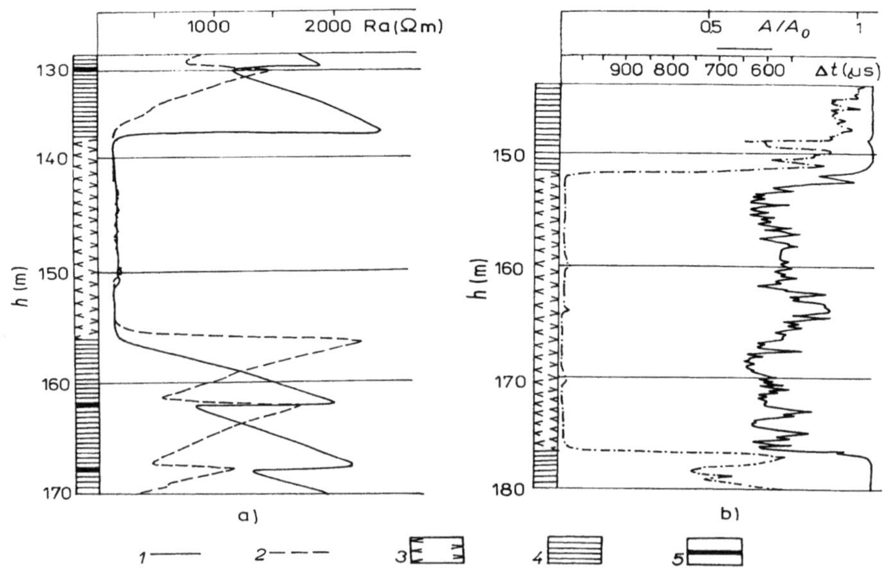

Figure 10.20. Character of logs in screened sections of casings (Havelka and Tezky, 1979): (a) lateral resistivity logs (AO = 0.55 m) in PVC casing, (b) cement bond log in steel casing. A/A 0—relative amplitude, Δt—interval transit time, 1 and 2—measured curves with potential electrodes below or above current electrode, 3—screened interval, 4—massive casing, 5—leaky weld of PVC casing.

10.6 Logging in Contamination Studies

Industrial fluids—polluting or contaminating the soil and rock medium and consequently the groundwater as well—can be divided into two principal groups according to their chemical composition, physical properties, and detection possibilities by geophysical techniques.

1. *Inorganic compounds* (acids, lyes, salts) are usually easily soluble in water and dissociate to positive and negative ions. The conductivity γ_{wc} of contaminated waters is proportional to the total dissolved chemicals, usually exceeding the conductivity of fresh water by one to two orders of magnitude, which has a similar influence on the formation conductivity γ_t. Electric permittivity ε_r of inorganic solutions, however, is comparable to the permittivity of fresh water ($\varepsilon_r = 80$).

2. *Organic compounds* (typically hydrocarbons) are usually characterized by very high resistivity (practically insulators) but very low electric permittivity ($\varepsilon_r < 5$), approaching the permittivity of air ($\varepsilon_r = 1$). They usually have very low solubility in water and can be considered practically immiscible with water. Their transport and accumulation in formation media is controlled by their density. The light nonaqueous phase liquids (LNAPL)—crude oil, engine oil, gasoline, kerosene, heating oil, aromatic hydrocarbons—have a density lower than water and, when passing through the vadose zone, they accumulate in a thin layer above the water table. The dense nonaqueous phase liquids (DNAPL)—e.g., aliphatic halo-hydrocarbons—have densities higher than water and pass through the vadose and saturated zones to accumulate at the bottom of an aquifer.

We will concentrate on the first group of contaminants, because their detectability by geophysical methods is much higher than that of the second group.

10.6.1 MONITORING WELLS IN THE VICINITY OF WASTE DEPOSITS

Waste deposits represent accumulations of either industrial wastes or unsorted municipal wastes. Industrial waste deposits usually contain high concentrations of inorganic chemical compounds. These wastes are often exposed to the environment after their deposition, when various physico-chemical processes result in the release of electrically conductive fluids (leachate) that contain both inorganic chemical compounds and organic acids. The pollutants present in both types of deposits are mobilized by the infiltration of precipitation and constitute a great hazard to groundwater. Geoelectrical methods are usually capable of detecting the contaminated parts of the rock medium, revealing the spatial extent and depth range of the contamination. If monitoring wells are situated around the waste deposit, fluid logging can check for the presence of pollutants in the groundwater. Figure 10.21 documents the characteristic features of fluid logs in a monitoring well situated near an old industrial depository. This depository contains the wastes from a facility extracting uranium from radioactive raw materials. The extracted fluid (sulfuric acid) was apparently not fully removed from the treated raw materials before they were deposited at the site. This is indicated by the concentration of sulfate ions (3 to 4 g/L) in groundwater samples taken from wells. The spatial extent of the deposited material (no longer visible in the field) was successfully mapped by surface geoelectrical methods (resistivity profiling and electromagnetic dipole profiling—see Mares et al., 2000).

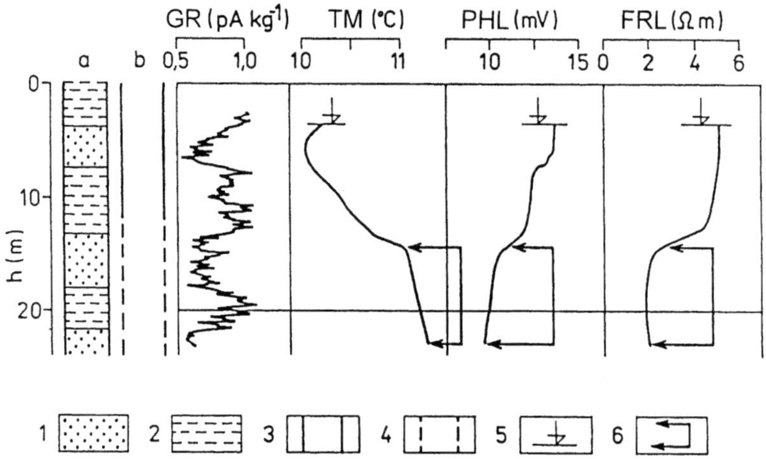

Figure 10.21. Test site Mydlovary-Zahaji, logging of the shallow monitoring well PM3 (Mares et al., 2000). Legend: GK—gamma-ray log in the well equipped with metal casing, TM—temperature log, PHL—photometry, FRL—fluid resistivity log, a— lithological section documented on the basis of gamma-ray log, b—well completion according to technical documentation, 1—sand, 2—clay, 3—metal casing, 4—perforated parts of the metal casing, 5—water level in the well, 6—water crossflow

10.6.2 LOGGING OF WELLS RELATED TO MINING ACTIVITIES

Some mining technologies use chemical solutions for *in situ* leaching of some raw materials. One example is uranium leaching from Cenomanian sandstones in Northern Bohemia, near the village of Straz pod Ralskem. Within the Cretaceous sediments, there is an overlying Turonian sandstone aquifer. Both aquifers are separated by a 60–80 m thick sequence of siltstones and marlstones, which acts effectively as an aquiclude. A dense network of PVC injection and recovery wells screened in the ore zone was used to recover the uranium. The leaching solution was sulfuric acid in concentrations from 30 to 50 g/L. Application of this technology requires control of leaching solutions in the Cenomanian sandstones to prevent their spreading to adjacent areas beyond the leached fields. A time series of induction logs (Figure 10.22) clearly documents the continuing movement of leach solution beyond the boundary of the *in situ* mining area. A baseline formation lateral log is shown to indicate aquifer structure. The measured formation conductivities can be converted to estimates of pollutant concentrations Q_{acp} in the pore water, using the mean formation factor for the Cenomanian sandstones (Karous et al., 1993).

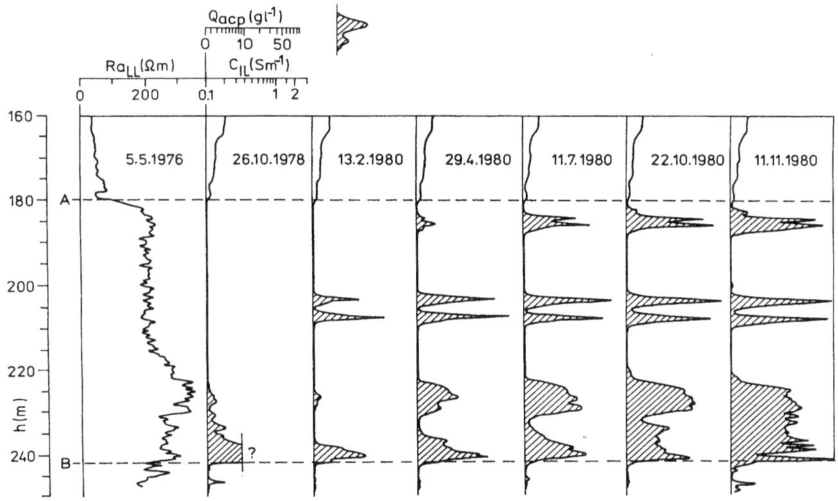

Figure 10.22. Repeated induction logs in the Cenomanian monitoring well outside the mined area (Karous et al., 1993). Ra_{LL}—formation resistivity for the precontamination period, C_{IL}—formation conductivities, Q_{acp}—pollutant concentration in the pore water

Mining activity in this area stopped in the early 1990s. Time sequences of induction logs are now used to control the efficiency of the remediation procedure in this area. It has been recognized that sulfuric acid in certain parts of the mining district has also polluted the Turonian sandstones. The pollution was probably caused by leakage from pipes distributing leach solutions to injection wells during the leaching process. In areas with massive contamination, a recovery well was installed and pumped for long time periods to clean up the Turonian aquifer. The progress of remediation is clearly documented by the decreasing conductivity of contaminated zone on the time series of induction logs (Figure 10.23) in a monitoring well near the recovery well.

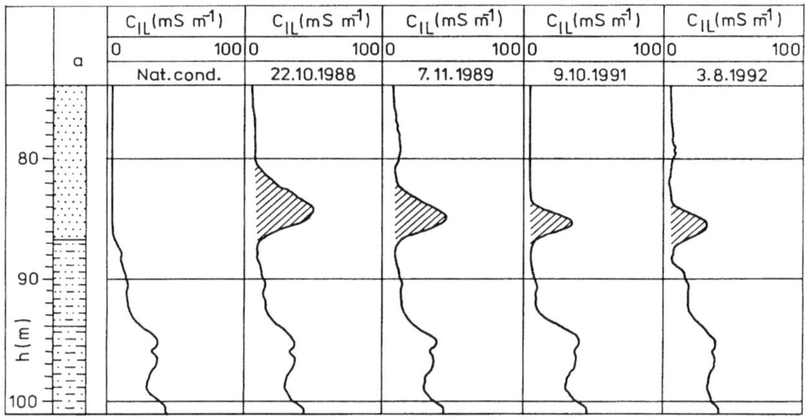

Figure 10.23. Repeated induction logs in a Turonian monitoring well from the remediated area (Karous et al., 1993). Nat. cond.—under natural conditions, 1988–1992, with pollution in the lower part of the Turonian sandstone.

10.7 Conclusions

Drilling and sampling are labor- and equipment-intensive techniques required to sample the hydraulic and water quality parameters of aquifers *in situ*. Geophysical well logs provide a cost-effective way to obtain profiles of the properties of formations surrounding such boreholes. Geophysical log data are provided as continuous profiles and can be related to the undisturbed properties of the formation. This data may be collected repeatedly over time to characterize any changes in formation properties, well completion conditions, or groundwater quality that may occur. Several different physical properties of the formation and formation fluids can be obtained with a typical suite of well logs. These different measurements can be used to identify the way in which different physical properties such as porosity, permeability, pore size distribution and mineralogy contribute to the composite geophysical response. Hydraulic test results and water-quality data from boreholes can be related to geophysical log measurements using various physical models, or by statistical regression. Flow measurements in boreholes made directly with flowmeters or inferred from dilution methods provide additional ways to estimate the hydraulic and water quality properties of aquifers *in situ*. The various examples illustrated in this chapter illustrate these contributions, and show the many ways in which geophysical well logging complement the results from other aquifer characterization techniques.

Acknowledgments

The authors deeply appreciate the professional approach of Michal Pitrák to the final technical arrangement of this chapter. The chapter has been prepared in the frame of the research project J 13/98: 113100006 funded by the Ministry of Education, Youth and Sports of the Czech Republic.

References

Alger, R.P., and C.W. Harrison, Improved fresh-water assessment in sand aquifers utilizing geophysical logs, *Log Analyst, 30*, 31–44, 1989.

Asquith, G., and C. Gibson, Basic well log analysis for geologists, *AAPG*, 1982.

Bergmann, H., Über die Grundwasserbewegung am Filterrohr, GSF-Bericht R24, Institut für Radiohydrometrie, Gesellschaft für Strahlenforschung m.b.H. München, 1970.

Bergmann, H., Hydromechanische Fragen zur Interpretation von Tracermessungen, GSF-Bericht R36, Institut für Radiohydrometrie, Gesellschaft für Strahlenforschung m.b.H. München, 1972.

Biella G., A. Lozeij, and I. Tobacco, Experimental study of some hydrogeological properties of unconsolidated porous sediments, *Ground Water, 21*, 741–749, 1983.

Conaway, J., Temperature logging as an aid to understanding groundwater flow in boreholes, *Proceedings of the 2nd International MGLS Symposium, Paper F*, pp. 51–59, 1987.

Deltombe, J.-L., and R. Schepers, New developments in real time processing of full waveform acoustic televiewer data, *Journal of Applied Geophysics, 55*, 161–168, 2004.

Doveton, J.H., Geologic log analysis using computer methods, *AAPG*, 1994.

Drost, W., D. Klost, A. Koch, H. Moser, F. Neumaier, and W. Rauert, Point dilution methods of investigating groundwater flow by means of radioisotopes, *Water Resour. Res., 4*, 125–146, 1968.

Drost, W., H. Moser, F. Neumaier, and W. Rauert, Isotopenmethoden in der Grundwasserkunde. *Informationsheft des Bureau Eurisitop 61*, Brussels, 1972.

Grinbaum, I.I., Geophysical methods investigating filtration properties of rocks, *Nedra* (in Russian), 1965.
Halevy, E., H. Moser, and A. Zuber, Borehole dilution techniques: A critical review, In: *Isotopes in Hydrology*, I.A.E.A., pp. 531–564, 1967.
Havelka, J., and A. Tezky, Determination of perforated casing position by logging, *Geologicky pruzkum, 8*, 244–245 (in Czech), 1979.
Hearst, J.R., P.H. Nelson, and F.L. Paillet, *Well Logging for Physical Properties (2nd ed.)*, New York, John Wiley and Sons, Ltd., 2000.
Hess, A.E., Identifying hydraulically conductive fractures with a slow velocity borehole flowmeter, *Canadian Journal of Earth Sciences, 23*, 69–78, 1986.
Hill, A.D., Production logging—Theoretical and interpretive elements, *Society of Petroleum Engineers Monograph 14*, Richardson, Texas, 1990.
Hurst, M.A., C.M. Griffiths, and P.F. Worthington, eds., *Geological Applications of Wireline Logs II*, The Geological Society Publishing House, 1992.
Hurst, M.A., M.A. Lovell, and A.C. Morton, eds., *Geological Applications of Wireline Logs*, The Geological Society Publishing House, 1993.
Jorgensen, D.G., Estimating permeability in water-saturated sediments, *Log Analyst, 29*, 401–409, 1988.
Jorgensen, D.G., Estimating geohydrologic properties from corehole geophysical logs, *Ground Water Monitoring and Remediation, 10*, 123–129, 1991.
Karous, M., S. Mares, W.E. Kelly, J. Anton, J. Havelka, and V. Stoje, Resistivity method for monitoring spatial and temporal variations in groundwater contamination, *Proceedings of the GQM 93 Conference*, IAHS Publ. No. 220, 1993.
Kelly, W.E. and S. Mares, *Applied Geophysics in Hydrogeological and Engineering Practice*, Elsevier, 1993.
Keys, W.S., Analysis of geophysical logs of water wells with a microcomputer, *Ground Water, 24*, 750–760, 1986.
Keys, W.S., Borehole geophysics in igneous and metamorphic rocks, *Transactions of the SPWLA 20th Annual Logging Symposium*, Houston, Texas, pp. P1–P26, 1979.
Keys, W.S., *A Practical Guide to Borehole Geophysics in Environmental Investigations*, CRC Press Inc, Lewis Publishers, 1997.
Keys, W.S., and J.K. Sullivan, Role of borehole geophysics in defining the physical characteristics of the Raft River geothermal reservoir, *Geophysics, 44*, 1116–1141, 1979.
Klotz, D., and R. Chand, Laboruntersuchungen mit dem Tracerverfahren, GSF-Bericht R188, Institut für Radiohydrometrie, Gesellschaft für Strahlenforschung m.b.H. München, 1979.
Kobr, M., S. Mares, and J. Lukes, Contribution of geophysical well-logging techniques to evaluation of fractured rocks, *Acta Universitatis Carolinae, 40*, 257–268, 1996.
Kobr, M., M. Vanecek, and H. Maslowska, Relationship between petrography and well-logging parameters in the borehole R-150, Rejviz. Hruby jesenik Mts. (in the Bohemian Massif, Central Europe), *Bulletin of the Czech Geological Survey, 72*, 181–188, 1997
Kwader, T., Estimating aquifer permeability from formation resistivity logs, *Ground Water, 23*, 762-766, 1985.
Lapcevik, P.A., K.C. Novakowski, and F.L. Paillet, Analysis of flow in an observation well intersecting a single fracture, *Journal of Hydrology, 151*, 227-239, 1993.
Long, J.C.S., A. Aydin, S.R. Brown, H.H. Einstein, K. Hestir, P.A. Hsieh, L.R. Myer, K.G. Nolte, D.L. Norton, O.L. Olsson, F.L. Paillet, J.L. SMITH, and L. Thompson, *Rock Fractures and Fluid Flow: Contemporary Understanding and Applications*, Washington, D.C., National Academy Press, 1996.
Lukes, J., Filtration properties of rocks determined from digital well-logging measurements, *Acta Universitatis Carolinae, Geologica, 4*, 289–304, 1973.
Lynch, E.J., *Formation Evaluation*, New York, Harper and Row, 1962.
Mares, S., J. Dohnal, Z. Jane, J. Knez, L. Zima, L. Alexajeva , and V. Iliceto, Geophysikalische Prospektion der Deponie bei Mydlovary: Zahaji in Sued, Boehmen, Freiberger Forschungshefte, Geoingenieurwesen C482, 335–344. 2000.
Mares, S., T. Henzel, L. Woltaer, K.-N. Lux, and U. Stump, New logging tool for measuring azimuth of groundwater flow, *Near Surface Geophysics, 1*, 31–37, 2003.
Mares, S., and H.J. Schön, Logging studying water movement in hydrogeological boreholes and wells: Estimating hydraulic parameters of aquifers, *Proceedings of the Workshop on Hydrogeological Logging*, in the Frame of the 9th Meeting of the EEGS-ES, Budapest, Hungary, September 6–9, 1999.
Mares, S., and A. Zboril, Dilution technique in the logging variant: State of the art, Proceedings of the EEGS-ES First Meeting, Torino, 115–118, September 25–27, 1995.
Mares, S., A. Zboril, and W.E. Kelly, Logging for determination of aquifer hydraulic properties, *The Log Analyst, 35*(6), 28–36, 1994.

Mazac, O., I. Landa, and A. Kolinger, The feasibility of a hydrogeological variant of the mise-a-la masse method: An analysis, *Journal of Geological Sciences, Hydrogeology and Engineering Geology, 14,* 147–177, 1980.
Molz, F.J., R.H. Morin, A.E. Hess, J.G. Melville, and O. Guven, The impeller meter for measuring aquifer permeability variations--evaluation and comparison with other tests. *Water Resour. Res., 25(7),* 1677–1683, 1989.
Molz, F. J., G.K. Bowman, S.C. Young, and W.R Waldrop, Borehole flowmeters—Field application and data analysis, *J. Hydrology, 163,* 347–371, 1994.
Moser, H. and F. Neumaier, Die Anwendung radioaktiver Isotopen in der Hydrologie, I., *Atomkernenergie, I,* 26–34, 1957.
Moser, H., F. Neumaier, and W. Rauert, Die Anwendung radioaktiver Isotopen in der Hydrologie, II, *Atomkernenergie, 2,* 225–231, 1957.
Nelson, P. H., Permeability-porosity relationships in sedimentary rocks, *Log Analyst, 35,* 38–62, 1994.
Niesner, E., and I. Reindl, Optical darkfield flowmeter well logging tool, *Proceedings of the 6th Meeting of EEGS-ES, Paper WL03*, Bochum, September 3–7, 2000.
Ogilvi, N.A., and D.I. Fedorovic, Electrolytic method determining the velocity of ground water filtration and conditions of its application (in Russian), *Nedra,* Moscow, 1964.
Paillet, F.L., Flow modeling and permeability estimation using borehole flow logs in heterogeneous fractured formations, *Water Resour. Res.,, 34*(5), 997–1010, 1998.
Paillet, F.L., A field technique for estimating aquifer parameters using flow log data, *Ground Water, 38*(4), 510–521, 2000.
Paillet, F. L., Hydraulic head applications of flow logs in the study of heterogeneous aquifers, *Ground Water, 39*(5), 667–675, 2001.
Paillet, F. L., Spatial scale analysis in geophysics—Integrating surface and borehole geophysics in ground water studies, In: *Spatial Methods for Solution of Environmental and Hydrologic Problems*, V.F. Singhroy, D.T. Hansen, R.R. Pierce, and I.A. Johnson, eds., ASTM STP 1420, pp. 77–91, 2002.
Paillet, F.L., Borehole flowmeter applications in irregular and large-diameter boreholes, *Journal of Applied Geophysics, 55,* 39–60, 2004.
Paillet, F.L., C. Barton, S. Luthi, F. Rambow, and J. Zemanek, *Borehole Imaging*, SPWLA Reprint Series, Houston, Texas, Society of Professional Well Log Analysts, 1990.
Paillet, F.L., and R.E. Crowder, A generalized approach for the interpretation of geophysical well logs in ground water studies—Theory and application, *Ground Water, 34*(5), 883–898, 1996.
Paillet, F.L., L. Hite, and M. Carlson, Integrating surface and borehole geophysics in ground water studies—An example using electromagnetic soundings in south Florida, *Journal of Environmental and Engineering Geophysics, 4*(1), 45–55, 1999.
Paillet, F.L., and W.H. Pedler, Integrated borehole logging methods for wellhead protection applications, *Engineering Geology, 42,* 155–165, 1996.
Paillet, F.L., Y. Senay, A. Mukhopadhyay, and F. Szekely, Flowmetering of drainage wells in Kuwait City, Kuwait, *J. of Hydrology, 234,* 208–227, 2000.
Paillet, F.L., J.H. Williams, D.S. Oki, and K.D. Knutson, Comparison of formation and fluid column logs in a heterogeneous basalt aquifer, *Ground Water, 40,* 577–585, 2002.
Pitschel, B., Bohrlochsonde zur Messung von Grundwasserstroemungen auf der Basis der kontinuirlich thermischen Anregung, Ph.D. Thesis, Technical University of Freiberg, 2000.
Prensky, S., Temperature measurements in boreholes: An overview of engineering and scientific applications, *The Log Analyst,* 313–333, 1992.
Schlumberger, *Production Log Interpretation*, Schlumberger Educational Services, 1973.
Schlumberger, *Log Interpretation: Principles/Applications*, Schlumberger Educational Services, 1986.
Schneider, H., *Die Wassererschließung*, Vulkan-Verlag, 1973.
Serra, O., *Fundamentals of Well-Log Interpretation*, Elsevier, 1986.
Schön, J.H., Hydrogeological information for water exploration and environmental investigation from well log measurements, *Proceedings of the 5th International MGLS Symposium, Paper F,* 1–8, 1993.
Tsang, C-F., P. Hufschmied, and F.V. Hale, Determination of fracture inflow parameters with a borehole fluid conductivity logging method, *Water Resour. Res., 26,* 561–578, 1990.
Vernon, J.H., F.L. Paillet, W.H. Pedler, and W.J. Griswold, Application of borehole geophysics in defining the wellhead protection area for a fractured crystalline bedrock aquifer, *Log Analyst, 34*(1), 41–57, 1993.
Williams, J.H., and C.D. Johnson, Acoustic and optical borehole-wall imaging for fracture rock aquifers, *Journal of Applied Geophysics, 55,* 150–160, 2004.

Williams, J.H., and F.L. Paillet, Using flowmeter pulse tests to define hydraulic connections in the subsurface: A fractured shale example, *Journal of Hydrology, 265*, 100–117, 2002.

Wilson, J.T., W.A. Mandell, F.L. Paillet, E.R. Bayless, R.T. Hanson, P.M. Kearl, W.B. Keerfoot, M.W. Newhouse, and W.H. Pedler, An evaluation of borehole flowmeters used to measure horizontal ground-water flow in limestones of Indiana, Kentucky, and Tennessee, 1999, *U.S. Geological Survey Water-Resources Investigations Report 01-4139*, 2001.

Worthington, P.F., and P.W. Johnson, Quantitative evaluation of hydrocarbon saturation hydrocarbon saturation in shaly freshwater reservoirs, *The Log Analyst 32,* 358–370, 1991.

Zboril, A., and S. Mares, Photometry in the solution of complicated conditions in hydrologic wells, *Journal for Mineralogy and Geology, 16*(2), 113–131, 1971.

11 AIRBORNE HYDROGEOPHYSICS

JEFFREY G. PAINE[1] and BRIAN R. S. MINTY[2]

[1]*Bureau of Economic Geology, Jackson School of Geosciences, The University of Texas at Austin, University Station, Box X, Austin, Texas 78713 U.S.A.*
[2]*Geoscience Australia, GPO Box 378, Canberra, ACT 2601 Australia*

11.1 Introduction

The usefulness of geophysics in hydrogeological studies depends on the existence of a contrast in one or more physical properties that can be measured by appropriately applying a geophysical method. Available properties where measurable contrasts might exist include electrical conductivity, radioactivity, magnetic susceptibility, bulk density, seismo-acoustic impedance, and heat capacity and conductance. The magnitudes of physical property contrasts vary widely among methods, as does the achievable measurement accuracy and precision. Electrical conductivity and magnetic susceptibility, for example, may vary over several orders of magnitude, whereas density may vary over only a few percent. Some properties may be more directly related to hydrogeologic properties of interest than others, thereby reducing the analytical and interpretational uncertainty inherent in translating proxy geophysical measurements into hydrogeologic properties. Depending on the goal, one or more geophysical methods may be suitable. As discussed in other chapters of this volume, measurements of these physical attributes commonly can be used to estimate important hydrogeologic parameters such as water content, salinity, groundwater and bedrock depth, and host material properties.

Available platforms also influence the choice of physical properties to exploit and methods to employ. Airborne measurements can be made for most but not all potentially applicable properties. The most common airborne hydrogeophysical studies are electromagnetic induction (both frequency and time domain), gamma-ray spectrometry (radiometrics), and magnetics. Less commonly used are airborne gravity, thermal, and radar surveys. Surveys requiring ground contact, such as seismic imaging (Chapter 9 of this volume), and electrical methods (Chapter 5 of this volume), such as resistivity, cannot be carried out from an aircraft.

Just as each method has advantages and disadvantages for a range of problems, the chosen platform also has advantages and disadvantages. In general, the sampling footprint (the surface area or subsurface volume from which the measured signal arises) of airborne instruments is larger than that for ground surface and borehole instruments measuring the same physical property. Vertical resolution decreases from borehole to airborne instruments, owing to the increased distance from the instrument to the material being investigated. The ability to resolve small targets is also diminished when using airborne as opposed to ground-based and borehole instruments.

Y. Rubin and S.S. Hubbard (eds.), Hydrogeophysics, 333–357.

The decrease in lateral and vertical resolution accompanied by the larger footprint and greater distance to the target typical of airborne instruments is offset by the extensive areal coverage and high spatial density of measurements obtained in an airborne survey. Airborne instruments aboard low-flying helicopters and fixed-wing aircraft commonly acquire multisensor data along a flight path at rates of a few to a few hundred samples per second, translating into an on-the-ground spacing of tens of centimeters to tens of meters. Depending on the flight-line spacing chosen for the survey, large areas can be surveyed over a short time at a spatial density that would be impractical or impossible using ground or borehole instruments. The greatest advantage common to all airborne geophysical surveys is the ability to capture dense measurements of physical properties over large spatial areas. These measurements allow interpolation or extrapolation from sparse but more precise ground or borehole measurements of the same physical properties, provided that they are scalable to the airborne platform. Pre-survey modeling and ground measurements can help assess whether airborne surveys are likely to render usable results.

This chapter surveys the airborne geophysical methods that are most commonly used (electromagnetic [EM], gamma-ray spectrometry, and magnetics). We discuss the measured physical property, survey design, data acquisition and processing, typical survey products, and how the method applies. We briefly mention other airborne methods—such as gravity, multispectral imaging, and laser terrain mapping—that are either less commonly used, have less obvious application, or provide strictly surficial information. We include airborne laser terrain mapping (lidar), for example, because of its ability to produce highly accurate digital elevation models that can either directly identify significant hydrogeological features or enable better processing, analysis, and interpretation of other types of airborne geophysical data.

11.2 Electromagnetic Induction

Electromagnetic induction (or simply EM) is the most versatile of the airborne geophysical methods and is perhaps the most widely applied in hydrogeological studies. This family of methods, consisting of both frequency- and time-domain approaches (Chapter 6 of this volume), measures the apparent electrical conductivity of the ground to depths ranging from a few to a few hundred meters, depending on the instrument chosen and the ground conductivity.

Electrical conductivity is an important parameter in most groundwater studies (see Chapter 4 of this volume). Bulk conductivity of the ground is a function of water content, water chemistry, pore volume and structure, and the electrical properties of the host mineral grains (McNeill, 1980a). In most aquifers, host sediment or rock is relatively nonconductive. Typical conductivities of common unsaturated sediment and rock types are a few to a few tens of millisiemens per meter (mS/m). Ground conductivity arises as a result of ohmic (flow through and along mineral grains), dielectric (flow as a result of net shifts of electrically polarized molecules), and electrolytic (flow as a result of ions migrating in the pore fluid) currents. Electrolytic flow is the dominant contributor to aquifer conductivity, particularly where water has high ionic concentrations. Because of their high cation exchange capacity and large surface area per unit volume, clay-bearing materials typically have higher conductivities than do sediments or rocks dominated by non-clay minerals.

Water quality is classified according to total dissolved solids (TDS) concentration (Figure 11.1). Minerals dissociate into anions and cations when dissolved in water, increasing the TDS concentration. Water is considered fresh at TDS concentrations below 1,000 mg/L, slightly saline at 1,000 to 3,000 mg/L, moderately saline at 3,000 to 10,000 mg/L, very saline at 10,000 to 35,000 mg/L, and briny above 35,000 mg/L (Robinove et al., 1958). With the increasing TDS concentrations come elevated ionic concentrations that enhance the ability of the water to carry electrical current. Over water qualities ranging from fresh to briny, there is a nearly linear relationship between TDS concentration and water conductivity. Conductivities are about 180 mS/m for fresh water, increasing to about 180 to 540 mS/m for slightly saline, 540 to 1,800 mS/m for moderately saline, 1,800 to 6,350 mS/m for very saline, and more than 6,350 mS/m for brine (Figure 11.1). Considering the typically low conductivities of relatively dry sediments and rocks, bulk conductivities of water-saturated strata with porosities of 25–50% are dominated by electrolytic conduction in pore fluid. As TDS increases, the measured conductivities are increasingly influenced by TDS in pore water. The relative dominance of TDS in ground conductivity measurements makes EM a useful tool in groundwater resource, water quality, and salinity investigations.

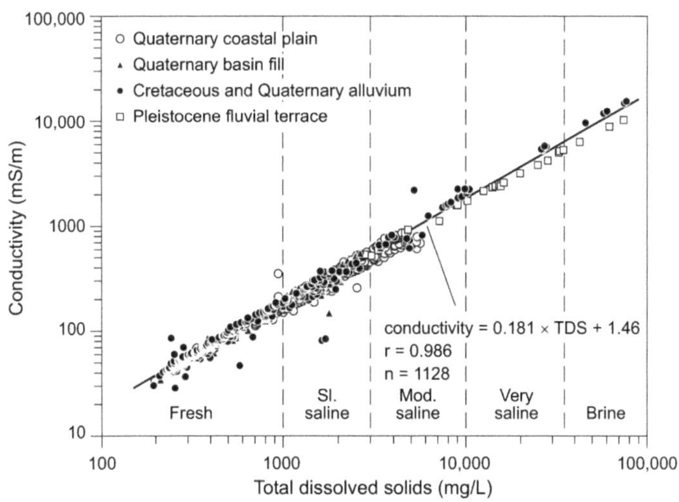

Figure 11.1. Relationship between total dissolved solids (TDS) concentration and measured electrical conductivity in groundwater samples from several clastic aquifers. Water-quality data from the Texas Water Development Board. Salinity classification from Robinove et al. (1958).

11.2.1 METHODS, INSTRUMENTS, AND PLATFORMS

Both frequency- and time-domain EM methods employ a changing primary magnetic field created around a transmitter loop or coil to induce current to flow in the ground, which in turn creates a secondary magnetic field sensed by the receiver coil (Parasnis,

1986; Frischknecht and others, 1991; West and Macnae, 1991). These methods are described in Chapter 6 of this volume. Here we describe relevant aspects of these methods as they are employed in airborne surveys (Palacky, 1986; Becker, 1988; Palacky and West, 1991; Morrison and others, 1998).

Frequency-Domain EM. In frequency-domain EM (FDEM), the transmitter coil operates continuously at a fixed frequency. The strength of the secondary field is a complex function of EM frequency and ground conductivity (McNeill, 1980b), but generally increases with ground conductivity at constant frequency. Airborne FDEM surveys are typically completed using several coil pairs mounted in a single bird and towed by a low-flying helicopter (Figure 11.2; Fraser, 1978). Available systems acquire data at primary (transmitter) frequencies ranging from a few hundred Hz to about 100 kHz. Because exploration depth decreases as either frequency or ground conductivity increases, different coils will measure different apparent conductivities over ground where conductivity varies with depth. Exploration depth is approximated by skin depth, which is defined as the depth at which the field strength generated by the transmitter is reduced to 1/e times its original value. It is calculated using the equation

$$\delta = 503\sqrt{\frac{\rho}{f}} = 15{,}681\sqrt{\frac{1}{\sigma f}} \qquad (11.1)$$

where δ = skin depth (in meters), ρ = resistivity (in ohm-m), f = EM frequency (in cycles/s), and σ = conductivity (in mS/m) (Telford and others, 1990).

Figure 11.2. Helicopter acquiring frequency-domain electromagnetic induction data (lower bird) and magnetic field data (upper bird) over cultivated salinized ground adjacent to an oil field. Photograph by Alan R. Dutton.

Measured apparent conductivities (equivalent horizontal halfspace conductivities that would produce the observed signal at a given frequency) in typical groundwater and salinity investigations range from a few tens to a few hundred millisiemens per meter, and approximate bulk conductivities at the scale of the measurement. Whereas exploration depth increases with decreases in transmitter frequency (Figure 11.3), it decreases with increasing ground conductivity. Using 25 kHz coils as an example, theoretical exploration depth decreases from about 14 m over relatively nonconductive ground (50 mS/m) to about 5 m over relatively conductive ground (500 mS/m). Similarly, lower frequency coils (1,500 Hz) explore to deeper depths of about 58 m over relatively nonconductive ground (50 mS/m) and about 18 m over relatively conductive ground (500 mS/m). In addition to frequency and ground conductivity, actual achieved exploration depth depends on signal strength, coil orientation, ambient noise levels, and the three-dimensional conductivity structure.

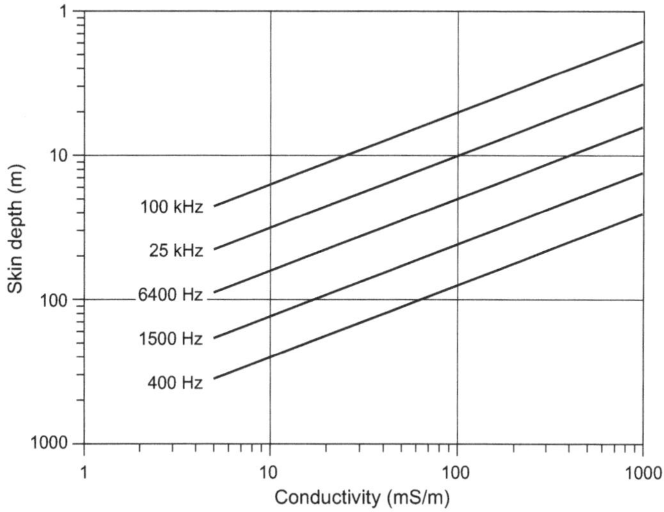

Figure 11.3. Changes in estimated exploration depth (skin depth) with ground conductivity for airborne coil frequencies commonly used in helicopter-based FDEM surveys and a representative half-space conductivity range

Achievable ground resolution is determined by subsurface conditions and instrument frequency, instrument height, sample rate, flight speed, and line spacing. The volume of ground contributing to the recorded EM signal decreases as the transmitter frequency increases, reflecting a decrease in exploration depth as frequency increases. The ground footprint increases with instrument height because of geometric spreading of the field generated by the transmitter coil. Common sampling frequencies of 10 Hz, when combined with typical helicopter survey speeds of 30 m/s, yield along-line sample spacings of 3 m, which is much smaller than the lateral extent of the subsurface volume being sampled. Survey resolution also depends on flight-line spacing. Depending on the resolution required for the application, most surveys are flown at line spacings of 100 m or more, although closer spacings have been flown. Flight-line spacings often are a compromise among competing parameters that include area to be flown, the smallest feature to be resolved, and cost. Flight-line orientation should also be chosen with the recognition that measurement density is much higher along the flight lines than it is across the flight lines.

Data from airborne FDEM surveys are processed to produce maps of apparent conductivity as measured by the coil pairs at differing frequencies and exploration depths. In addition, the dependence of exploration depth on frequency and conductivity allows data acquired at multiple frequencies to be converted to profiles of conductivity variations with depth (Sengpiel, 1983; Newman and Alumbaugh, 1995; Avdeev et al., 1998; Sengpiel and Siemon, 2000; Siemon, 2001; Beamish, 2002). The limited number of frequencies in available systems mandates simplified vertical conductivity models of the subsurface and limits the theoretically achievable vertical resolution and layer conductivity accuracy.

Time-Domain EM. Time-domain EM (TDEM) devices measure the decay of a transient, secondary magnetic field produced by currents induced to flow in the ground by the termination of a primary electric current (Chapter 6 of this volume; Kaufman and Keller, 1983; Spies and Frischknecht, 1991) flowing in a transmitter loop. The secondary field strength is measured by the receiving coil at discrete time intervals following transmitter current termination. In horizontally layered media, secondary field strength at early times gives information on conductivity in the shallow subsurface; field strength at later times is influenced by conductivity at depth.

Airborne TDEM surveys are typically deployed on fixed-wing aircraft (see Fountain, 1998, for historical development) capable of supporting a large transmitter loop draped around the aircraft (Figure 11.4), but are also deployed in smaller, helicopter-towed configurations (e.g., Sorensen et al., 2004). A discontinuous current of alternating polarity is passed around the loop, creating a primary magnetic field that collapses when the current is shut off. A receiver towed by the aircraft records the decay of the transient field generated by eddy currents flowing in the ground in response to the collapsing primary field. The strength and shape of the transient signal are analyzed to determine the apparent conductivity of the ground and its variation with depth.

Figure 11.4. MEGATEM II system (Fugro Airborne Surveys) acquiring TDEM and magnetic-field data during a groundwater-resource evaluation. The transmitter loop is attached to the wingtips and booms at the front and rear of the aircraft. Trailing cables tow a triaxial EM receiver and magnetometer (not shown).

Critical instrument parameters affecting exploration depth and data quality in airborne TDEM systems include (1) the dipole moment, a measure of transmitter strength that is calculated by multiplying the transmitter loop current by its effective area, (2) the transmitter frequency, calculated as the number of times per second that the transmitter is pulsed in both polarities, (3) the transmitter pulse duration and shape, and (4) the duration of the recording window, which may begin during the transmit pulse and extend to the latest possible time before onset of the next transmit pulse. Common instrument configurations include dipole moments of tens of thousands to more than 1 million ampere-m^2, transmitter frequencies of 12.5 to 150 Hz, transmit pulses of 1 to 8 ms, and off-time recording durations of a few to a few tens of milliseconds during which the ground response is recorded.

TDEM exploration depth is greater than that of FDEM systems, owing to greater transmitter field strength and lower frequency content. Potential exploration depth can be estimated using the skin-depth equation, by replacing the frequency term with the reciprocal of the off-time duration, or decay period. Exploration depth thus increases with decreasing conductivity and with a longer recording window, assuming the decaying ground signal (transient) is at least as long as the recording window (Figure 11.5). For example, using a high transmitter frequency with a short recording window (2 ms) results in a potential exploration depth that increases from about 32 m over conductive ground (500 mS/m) to about 100 m over relatively nonconductive ground (50 mS/m). At lower transmitter frequencies and a longer recording window (11 milliseconds), potential exploration depths increase from about 75 m over relatively conductive ground (500 mS/m) to about 240 m over relatively nonconductive ground (50 mS/m), again assuming the signal from the ground remains above ambient noise for the duration of the recording window. Achievable exploration depth depends on signal strength, ambient noise, and the three-dimensional conductivity structure; and may be deeper or shallower than simple estimates, based on field strength attenuation in a homogeneous half-space.

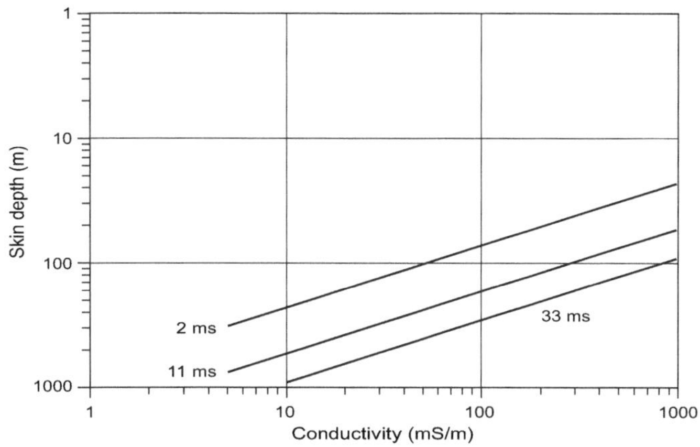

Figure 11.5. Changes in estimated exploration depth (skin depth) with conductivity for common airborne TDEM transient-decay periods (in milliseconds) and apparent conductivity ranges encountered in most hydrogeologic studies

Airborne TDEM surveys are flown on fixed-wing aircraft at higher altitudes (100 to 150 m) and at higher speeds than helicopter-based FDEM surveys, resulting in larger ground footprints and lower measurement density. At typical airspeeds of about 70 m/s, transients are acquired at intervals of 3 m for a 12.5 Hz transmitter and 0.2 m for a 150 Hz transmitter, but these individual transients are stacked to increase the signal-to-noise ratio at a cost of reduced lateral resolution (Christensen et al., 2000). At a sample frequency of 4 Hz, stacked transients are spaced about every 17 m along the flight line, a distance that is still about an order of magnitude smaller than the diameter of the ground volume contributing to the signal.

The larger aircraft, higher speeds, and larger TDEM subsurface sampling volume resulting from a larger transmitter loop and higher flight heights—all dictate longer flight-line spacing for fixed-wing TDEM surveys than is typical for helicopter-based FDEM or TDEM surveys. Although surveys have been completed at 100 m line spacing, most are flown at 400 m or more, enabling large areas to be covered quickly. Flight-line spacing is a compromise among competing parameters, including the area to be covered, the resolution needed, and the available time and budget. Depending on the spacing chosen, measurement spacing is significantly shorter along the flight lines than it is across them, making line orientation an important survey design decision.

Unprocessed TDEM data consist of the vertical (z) and two horizontal (x and y) components of secondary field strength (B) or decay rates (dB/dt) over time (Smith and Keating, 1996; Smith and Annan, 1998). During processing, the magnitude and slope of the decaying signals are used to produce maps of apparent conductance at selected times that correspond to shallow, intermediate, and deep exploration depths. Recent work has focused on transforming or inverting recorded TDEM data into models that accurately depict subsurface conductivity (Chapter 6 of this volume; Macnae et al., 1991; Wolfgram and Karlik, 1995; Ellis, 1998; Macnae et al., 1999; Hunter and Macnae, 2001; Beamish, 2002). Although the vertical resolution of the resulting models does not approach that obtainable with ground or borehole instruments (Christiansen and Christensen, 2003), depth inversions do allow reasonably consistent conductivity cross sections to be constructed. Further, apparent conductivity values can be extracted at similar depths across a survey area, producing apparent conductivity "slices" at high lateral resolution. The lateral and vertical slices facilitate mapping of stratigraphic, structural, and water-quality trends that may not be apparent from sparse wells or surface geophysical measurements.

11.2.2 APPLICATIONS AND EXAMPLES

Airborne EM surveys are widely applied in hydrogeological investigations because the measurements respond to both lithologic and water-chemistry variations. Proven and potential applications include geologic mapping and aquifer structure (Palacky, 1981; Street and Anderson, 1993; Beard et al., 1994; Bromley et al., 1994; Barongo, 1998; Gamey et al., 1998; Puranen et al., 1999; Bishop et al., 2001; Smith et al., 2003), delineation of soil and groundwater salinization arising from natural, agricultural, and oil-field causes (Smith et al., 1992; Street and Roberts, 1995; Paine et al., 1997; Smith et al., 1997; Coppa et al., 1998; Brodie, 1999; Paine, 2003), salt-water intrusion into coastal aquifers (Fitterman and Deszcz-Pan, 1998), wetlands mapping, groundwater assessments (Becker, 1988; Wynn and Gettings, 1998; Christensen et al., 2000; Wynn, 2002; Christiansen and Christensen, 2003; Paine and Collins, 2003; Sattel and Kgolthang, 2003), mining and mine-tailing impact on groundwater quality (Smith et al., 2000), aquifer recharge feature (soil and karst) identification (Cook and Kilty, 1992; Beard et al., 1994; Doll et al., 2000), and bathymetry (Vrbancich et al., 2001).

Airborne EM has been applied extensively to salinity problems in Australia and the United States. In parts of Australia, increased recharge related to agricultural development has led to rising groundwater, which in turn has carried stored salt to the evapotranspiration zone, increasing soil and water salinity, altering the natural environment, and impairing agricultural productivity (e.g. Coppa et al., 1998). In studies of natural and oil-field salinization in the United States, apparent conductivity maps produced from multifrequency airborne EM data (Figure 11.6) have been used to delineate salinity-plume boundaries, assess the intensity of salinization, identify salinity sources, estimate plume volumes and contaminant masses, and evaluate remediation approaches (Paine, 2003).

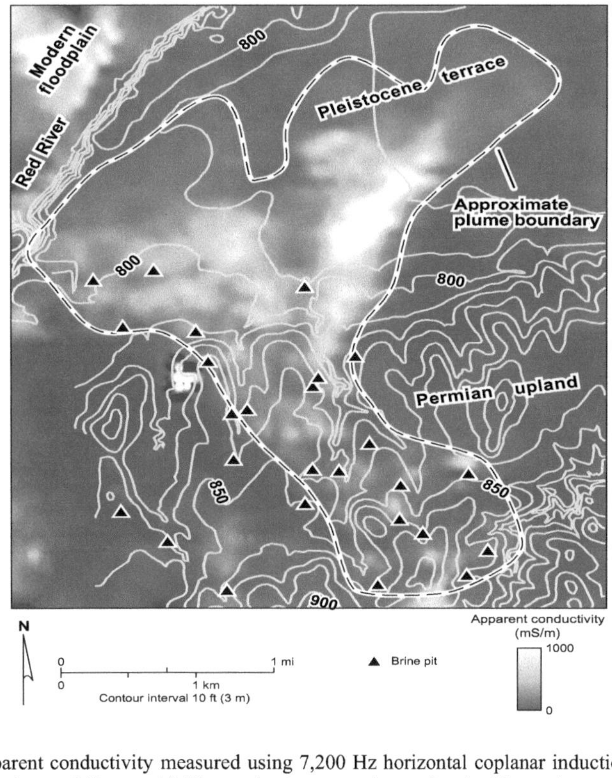

Figure 11.6. Apparent conductivity measured using 7,200 Hz horizontal coplanar induction coils towed by helicopter. Outlined area delineates highly conductive ground associated with a saltwater plume migrating from an oil field on the Permian upland. Brine pits are locations where produced saltwater was discharged during production. Pits where residual salinity remains are associated with elevated conductivities at this and higher frequencies. Modified from Paine (2003).

Apparent conductivity images at constant depths (or limited depth ranges) produced from airborne TDEM data can be useful in both salinization and groundwater resource investigations. Areas of low apparent conductivity identified on slices within the saturated zone of clastic aquifers can indicate favorable groundwater resources (Figure 11.7). In the search for groundwater, the goal is to find laterally and vertically extensive areas at aquifer depths having low conductivity relative to surrounding measurements. These areas delineate the most favorable groundwater resources where low conductivities indicate low salinity water. In clastic aquifers, low conductivity may also imply coarse-grained sediment. Well data can be used to establish an empirical relationship between host material, water saturation, and water quality. Airborne geophysical data can then be used to interpret resource quality between available wells, identify favorable groundwater resources, and reduce the cost of exploration by minimizing the number of wells necessary to assess a potential groundwater resource.

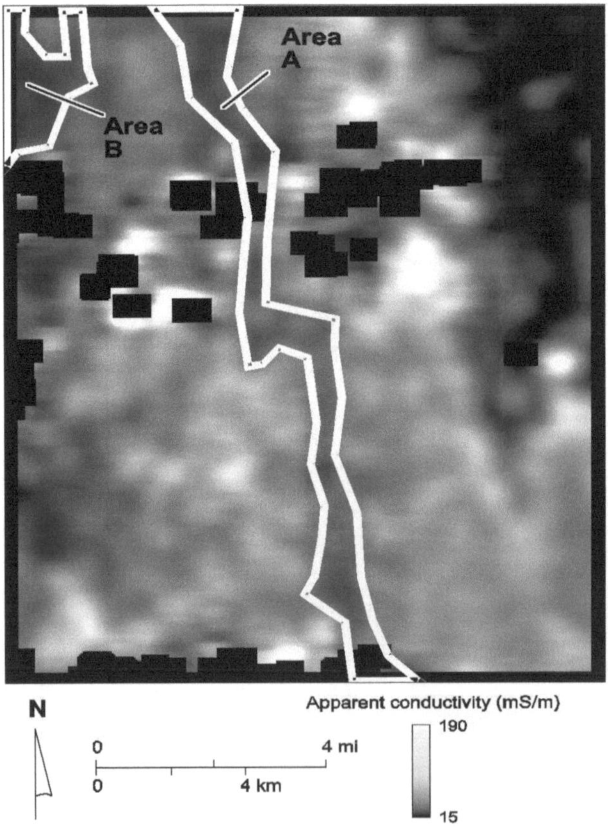

Figure 11.7. Depth slice at 120 m depicting apparent conductivity derived from airborne TDEM data and favorable groundwater exploration areas (A and B) within a West Texas basin. Modified from Paine and Collins (2003). Black boxes represent areas of no data.

The effectiveness of airborne TDEM and depth inversions in mapping water quality can be quantified by comparing water-well data with apparent conductivity derived from airborne measurements. We can compare TDS concentrations in samples from known depth ranges in wells near airborne measurement locations with apparent conductivity at that depth, calculated from the airborne TDEM data. Using data from a West Texas groundwater survey as an example (Paine and Collins, 2003), there is a reasonable relationship between TDS and apparent conductivity derived from airborne data (Figure 11.8). In this example, the relationship improves when fresh samples are excluded from the data set, exhibiting the increased dominance of water salinity over apparent conductivity as TDS concentrations rise. The change in slope between the TDS-water conductivity relationship (Figure 11.1) and the TDS-apparent conductivity relationship derived from airborne data can be used to estimate porosity in clastic aquifers, by recognizing that host materials with higher porosities will have higher slopes in the TDS-apparent conductivity relationship (Figure 11.9). For the example shown here, dividing the slope of the TDS to apparent ground conductivity line by the slope of the TDS to water conductivity line yields a reasonable porosity estimate of about 30% (Figure 11.9).

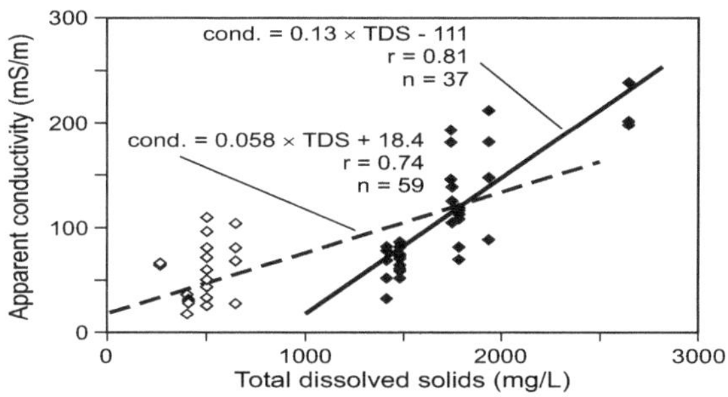

Figure 11.8. Relationship between TDS concentration in groundwater samples and apparent conductivities calculated from airborne TDEM data for depths below the water level and within the screened intervals for each available well within the survey area. Dashed line represents the least-squares fit to all data; solid line represents the fit for data points with TDS concentrations above 1,000 mg/L.

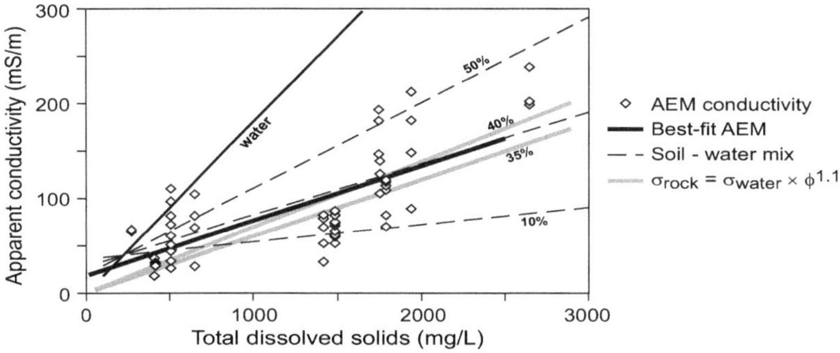

Figure 11.9. Relationship between the electrical conductivity of water and TDS concentration (thin solid line) and the calculated conductivity of poorly conductive host material, with porosities ranging from 10 to 50% (dashed lines), saturated with water at various TDS concentrations, as well as conductivity calculated using the Archie (1942) relationship (gray lines) at a porosity exponent of 1.1. Also shown are data (open diamonds) and the calculated best-fit (thick solid line) between groundwater samples and airborne survey-derived apparent conductivities from West Texas (Figure 11.8).

11.3 Gamma-Ray Spectrometry

All rocks and soil contain radioactive elements, and the decay of these elements gives rise to a natural gamma-ray flux. Almost all gamma radiation detected near or at the earth's surface derives from the natural radioactive decay of just three elements—potassium (K), uranium (U), and thorium (Th). Gamma-ray surveys map the distribution of these elements at the earth's surface; this distribution is a function of source-rock mineralogy and active surface processes such as erosion, pedogenesis, and sediment deposition. Source rock types and the processes that modify them affect hydrogeologically important issues such as runoff, moisture retention, recharge potential, and contaminant migration.

Gamma rays are the most penetrating form of radiation from the natural decay of K, U, and Th. They can penetrate some 35 cm of rock and several hundred meters of air. Gamma rays can thus be used to remotely sense terrestrial radioelement concentrations that vary with rock, sediment, and soil type. Airborne gamma-ray spectrometry, originally developed as a uranium exploration tool, is now widely used in geological and environmental mapping efforts having direct hydrogeological applicability.

Gamma-ray photons have an associated energy that is diagnostic of the emitting isotope. Some of the gamma rays from radioisotope disintegrations near the earth's surface penetrate through the earth and lower atmosphere and can be recorded with an airborne instrument. Most airborne gamma-ray detectors used today (Figure 11.10) consist of crystals of thallium-activated sodium iodide (NaI). Gamma rays absorbed in the crystals result in a scintillation of light being emitted—the intensity of which is proportional to the energy of the absorbed photon. Airborne detectors measure a gamma-ray spectrum—i.e., both the number of gamma rays recorded during a specific sample period, and the energy of each photon sorted into predefined energy windows, or bins. The number of gamma rays recorded is proportional to the concentration of the radioelements in the source, and the energies of the gamma rays can be used to diagnose the composition of the source isotopes.

Figure 11.10. Gamma-ray spectrometer with a thallium-activated sodium iodide detector (photograph courtesy of Exploranium)

In airborne surveys, a gamma-ray spectrum is typically recorded every second. Short counting periods are offset by large detector volumes. Ancillary data include positional information (Global Positioning System [GPS] navigation), temperature, pressure, and the height of the aircraft above the ground. The recorded data require substantial processing before accurate estimates of the ground concentrations of K, U, and Th can be made (Minty, 1997; Minty et al., 1998). The main corrections are for background radiation, the height of the aircraft above the ground, and the response and sensitivity of the detector.

11.3.1 METHODS, INSTRUMENTS, AND PLATFORMS

Survey design for airborne gamma-ray spectrometry (Pitkin and Duval, 1980; Duval, 1989; IAEA, 1991) is governed mainly by the precision required of the estimates of K, U, and Th concentrations, and by the spatial resolution required of these estimates. Spatial resolution is governed mainly by the flight line spacing, but also by the sample interval and speed of the aircraft. The line spacing is a trade-off between spatial resolution and survey cost. For geological mapping applications, line spacings between 100 and 400 m are typical. Sample spacing is governed by the speed of the aircraft—typically 50 to 60 m for a fixed-wing aircraft with a 1-second sample interval.

The greater the number of gamma rays recorded during each sampling interval, the better the precision of the data. Count rates can be increased by increasing the size of the detector, by flying closer to the ground, or by increasing the sample time. However, because the sample time also affects the spatial resolution of the survey, this is never increased beyond 1 second. The detector size is limited by the weight penalty any extra volume of detector imposes. Detectors of 33 to 50 liters are commonly used. For greater detector volumes, more expensive aircraft types would need to be considered to carry the extra payload.

Gamma-ray surveys are typically flown at less than 100 m above ground level on a regular grid of parallel flight lines. Above 500 m height, most of the gamma rays emitted from the ground would be absorbed in the intervening air. Because of their penetrating nature, gamma rays recorded at survey height originate from the top 30 to 35 cm of the earth's surface and from an area below the aircraft several hundred meters in diameter. The size of this "circle of investigation" depends on the survey height (Duval et al., 1971; Grasty et al., 1979; Pitkin and Duval, 1980). At 100 m height, about 80 percent of recorded photons originate from a circle below the aircraft with diameter of about 600 m. A single airborne estimate of radioelement concentrations represents an average over a fairly large area.

11.3.2 APPLICATIONS

Estimates of potassium (K), uranium (U), and thorium (Th) concentrations are commonly displayed as ternary images of K (red), U (blue), and Th (green) (Duval, 1983) that can be draped over a digital elevation model (Figure 11.11). Three-dimensional viewing of the data greatly assists interpretation because (a) geomorphic processes affect the surface concentrations of radioelements, and (b) topography strongly affects geomorphic processes. Because we only measure the radioelement concentrations in a thin layer at the earth's surface, topography has a significant effect on the distribution of the radioelements.

Figure 11.11. Perspective image of K concentrations in the Cootamundra area, New South Wales, Australia, draped over a digital elevation model. The width of the foreground is about 25 km with a 10 times vertical exaggeration. The image shows high K over thin soils in actively eroding areas, and low K associated with alluvial and colluvial sediments on the depositional plains. Image courtesy Geoscience Australia.

Gamma-ray spectrometric surveys have many geoscience applications. Those relevant to hydrogeology include:

Geologic mapping. Radioelement data are commonly used to map lithologic units that may control recharge, influence groundwater flow patterns, or help identify basin structure (Graham and Bonham-Carter, 1993; Jaques et al., 1997; Doll et al., 2000). There is significant overlap between the range of radioelement concentrations found in common rock types, but some general associations can be made. Igneous rocks are usually more radioactive than sedimentary rocks, although some slates and shales can have high concentrations of U and Th. Radioelement concentrations in igneous rocks tend to increase with increasing silica content, with the highest concentrations found in pegmatites. Felsic intrusions are the most radioactive. Mafic intrusions and volcanics tend to have lower concentrations of radioactive minerals. Mappable contrasts in the gamma spectrum caused by differing primary or secondary mineralogy exist among most rock types.

Regolith and soil mapping. Weathering modifies the concentration and distribution of the radioelements, and the degree of weathering depends on the climate to which the rocks and soils have been exposed. Differences in radioelement concentrations between fresh bedrock, weathered regolith, and transported material can be expected. U can be leached from rocks and soils and deposited in sediments some distance away. Th is less soluble but can be differentiated from the parent material through transport in solid mineral grains. Even where soils are *in situ*, radioelement concentrations can vary through the profile. Because weathering modifies the concentration and distribution of the radioelements, radioelement data can provide information on geomorphic processes and soil/regolith properties. Soil, geologic, and geomorphic information gleaned from gamma-ray data are used to map soils and surface geologic units (Cook et al., 1996; Wilford et al., 1997), identify potential recharge areas (Bierwirth and Welsh, 2000), estimate landscape stability, erosion and deposition rates, and soil development (Pickup and Marks, 2000; Pickup et al., 2002), delineate geologic features that might enhance or restrict groundwater flow, and identify and manage soil and water salinization (Coppa et al., 1998; Wilford et al., 2001).

Land-use and surface contaminant applications. Radioelement data are increasingly being used in decision-making processes related to land use and radioactive contaminant characterization. These include areas such as trace-element pollution (Schwarz et al., 1997; Aage et al., 1999; Doll et al., 2000), water quality, metal contents in foods, and natural levels of radiation—including the health risks associated with high levels of atmospheric radon.

11.4 Magnetics

Moisture content and water-quality variations in the shallow subsurface have no effect on the earth's magnetic field, yet magnetic field data are routinely acquired during many airborne EM and radiometric surveys. These data are commonly not the primary geophysical data acquired, but can be useful in hydrogeological studies through their ability to determine basin geometry, to delineate igneous intrusions that may affect groundwater flow, and to identify shallow, local magnetic anomalies caused by faults, paleochannels, eolian deposits, or man-made features such as wells and pipelines.

Earth's magnetic field is largely generated by electrical currents flowing in the liquid outer core (Telford and others, 1990). This field can be approximated by a magnetic dipole source (Figure 11.12) with its axis tilted a few degrees from the earth's rotational axis (Milsom, 1989). The strength of the main field increases from about 30,000 nanoteslas (nT; a unit equivalent to an older unit, the gamma) at the magnetic equator to about 60,000 nT at the magnetic north and south poles. Lines of force defining this field have shallow dips at low magnetic latitudes that increase to near 90 degrees approaching the magnetic poles. Significant deviations from ideal dipole behavior are attributed to core-fluid motion, electrical current flowing at the core-mantle boundary, and the presence of magnetically susceptible rock within the earth's crust and upper mantle. Short-term (and relatively large magnitude) variations in magnetic field strength are caused by episodic solar activity such as flares and sunspots, and by electrical current flowing in the ionosphere. Magnetic field data should be acquired during periods of low solar activity.

Magnetic susceptibility, the ability of a material to become magnetized in the presence of an external field, varies over several orders of magnitude. Very few minerals found in the earth's crust have magnetic susceptibilities high enough to measurably perturb the local magnetic field strength. These few minerals include magnetite, pyrrhotite, and maghemite, of which only magnetite is relatively common. Systematic changes in the abundance of magnetite in common rock and sediment types allow magnetic methods to be useful in hydrogeologic studies in which lithology and geologic structure are important, and in which a contrast exists in magnetic susceptibility or permanent (remanent) magnetization among the key lithologic or structural units.

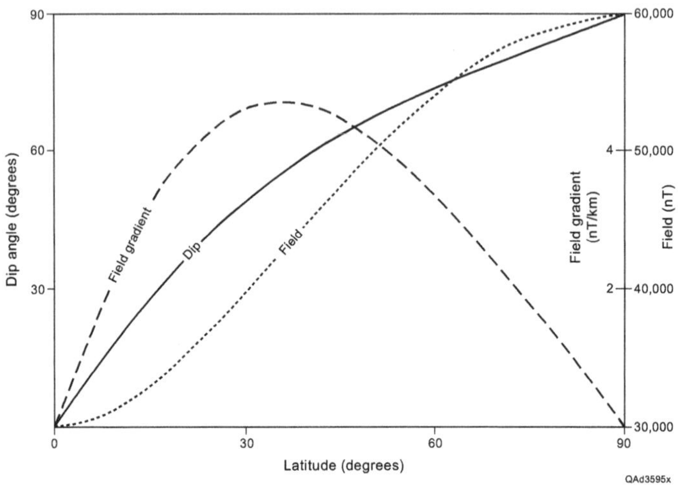

Figure 11.12. Magnetic field intensity, dip, and horizontal gradient for an ideal earth-dipole aligned with the earth's spin axis. Adapted from Milsom (1989).

11.4.1 METHODS, INSTRUMENTS, AND PLATFORMS

Magnetometers measure the strength and sometimes direction of the magnetic field at a given point. These measurements can be used in the context of adjacent measurements to identify local magnetic anomalies that may be hydrogeologically relevant. In airborne surveying, one or more magnetometers are towed by or affixed to a helicopter or fixed-wing aircraft (Figure 11.13). Cesium vapor magnetometers measure the total magnetic field strength by quantifying the influence that the magnetic field has on cesium-133 atoms contained within the sensor. These instruments operate over a field-strength range of about 20,000 to 100,000 nT and are sensitive to magnetic-field changes as small as 0.01 to 0.001 nT, or about 20 to 200 parts per billion. In addition to a single measurement of field strength made using one sensor, multiple sensors can be used at known lateral and vertical separations to measure magnetic-field intensity gradients.

Flight height and line spacing directly influence measured anomaly magnitude and detectability, because field strength drops off steeply with source distance (proportional to the inverse of the distance cubed) and because most materials have small bulk magnetic susceptibilities. Because magnetic field measurements may not be the primary geophysical data used in hydrogeologic surveys, survey flight height and line spacing may be dictated by the primary method and platform. Helicopter-based systems (Figure 11.2) are typically flown at nominal instrument heights of 30 m, but can be flown at terrain clearances of less than 10 m. At a 10 Hz sample rate, along-line sample spacing is governed by flight speed, which results in spacings of 3 m or less. Fixed-wing systems are flown at higher ground clearances of 50 to 150 m and at higher speeds, reducing the measured anomaly magnitude and decreasing the achievable spatial resolution.

Figure 11.13. Survey aircraft carrying a stinger-mounted cesium magnetometer. Photograph by David M. Stephens.

Total magnetic field strength as measured by the sensor is processed to produce maps and flight-line profiles for analysis and interpretation. The daily or diurnal magnetic-field change, established by monitoring field strength at a nearby ground base station, is removed from the airborne measurements acquired while the magnetic field strength varied with time. The diurnally corrected data are subtracted from the International Geomagnetic Reference Field (IGRF), a periodically updated numeric approximation of the earth's magnetic field strength, to emphasize local magnetic field anomalies caused by crustal features. Magnetic data can be spatially analyzed to enhance anomalies caused by sources at various depths, within the spatial frequency limits imposed by flight height and line spacing (Reid, 1980). Anomalies evident at high spatial frequencies (short wavelengths) are attributed to shallow sources, whereas anomalies evident at low spatial frequencies (long wavelengths) are attributed to deeper sources. Maps constructed from these data allow anomalies of a few to a few tens of nanoteslas, potentially caused by faults, dikes, intrusions, paleochannels, and man-made features, to be identified and enhanced.

11.4.2 APPLICATIONS AND EXAMPLES

Magnetic field data, acquired alone or combined with other types of geophysical data, have proven useful in establishing basin boundaries; mapping bedrock depth or basement topography (Babu et al., 1991; Maus et al., 1999), geologic structure, and lithologic boundaries (Plume, 1988); identifying important geologic features such as faults, shear zones, intrusions, and paleochannels that might influence groundwater flow (Mackey et al., 2000; Grauch, 2001; Rama Rao et al., 2002); and delineating local anomalies related to potentially significant infrastructure or contaminant sources such as wells, pipelines, and buried drums (Frischknecht et al., 1985; Wilson et al., 1997; Gamey et al., 1998; Paine et al., 1999; Doll et al., 2000).

For example, studies of groundwater salinization in oil fields (Paine et al., 1997; Smith et al., 1997) rely on EM methods to delineate salinized ground and on magnetic field data to identify small local magnetic field anomalies caused by oil or gas well casings (Figure 11.14). Relationships between the two anomaly types can help to discriminate among natural, agricultural, and oil-field salinization sources. Closely spaced flight lines and residual or derivative magnetic anomaly maps constructed from aeromagnetic data enhance the identification of these anomalies, which may be only tens of meters in width and tens of nanoteslas in magnitude.

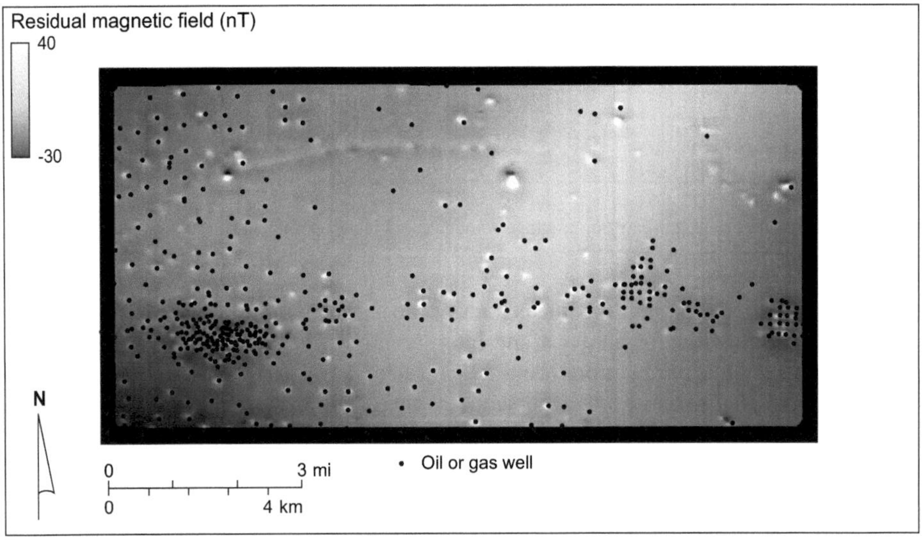

Figure 11.14. Residual magnetic field intensity map of an oil field produced by subtracting the IGRF from total magnetic intensity and minimizing long wavelengths. Local features are enhanced, including small anomalies associated with oil and gas wells and linear anomalies along pipelines. Well locations from the Railroad Commission of Texas.

Applying high-pass spatial filters to magnetic field data removes long-wavelength, large-amplitude magnetic field effects resulting from deep sources or major tectonic features, which are of little use in most hydrogeologic studies beyond establishing basin geometry. Maps depicting higher spatial frequencies enhance local anomalies arising from minor and relatively shallow sources that influence subsurface hydrogeological properties more directly. These images cover a more restricted range of field strength values that reveal near-surface faults (Figure 11.15), paleochannels associated with increased concentrations of magnetically susceptible minerals, dikes and minor intrusions, and infrastructure such as wells or pipelines.

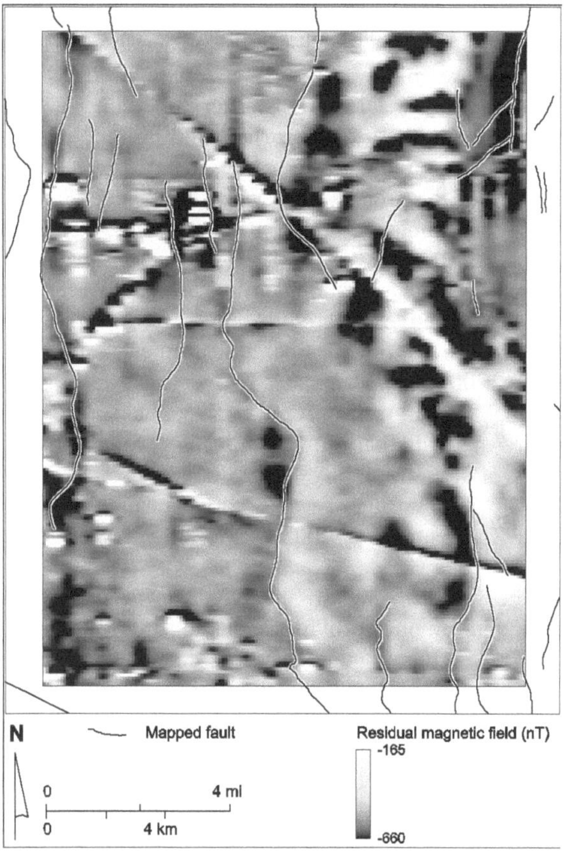

Figure 11.15. Shallow-depth transform of magnetic field data of intrabasinal alluvial deposits in a fault-bounded basin near El Paso, Texas, U.S.A. Prominent features include linear anomalies associated with pipelines and curvilinear anomalies that coincide with mapped intrabasinal faults.

11.5 Other Relevant Airborne Systems

Numerous other instruments have been mounted in aircraft (or spacecraft) and acquired data that have been widely used in hydrogeological investigations (Meijerink, 2000; Jackson, 2002). But generally, these instruments only provide information about the land surface or upper few centimeters of soil and are largely beyond the scope of this chapter. These include such systems as visible and infrared cameras, multispectral and thermal infrared scanners, active and passive microwave sensors (including synthetic-aperture radar), and lidar. Aerial photographs have long been used in soils and geologic mapping and land-use studies, and still provide critical ground-truth information for airborne geophysical surveys. Multispectral imaging systems passively record signal intensity within multiple channels of the ultraviolet to thermal infrared electromagnetic spectrum, providing images of land or water surfaces that can be analyzed to reveal soil and geologic information, elemental and mineral distribution, vegetation characteristics, land-use practices, soil-moisture variations, groundwater discharge sites, and water-quality parameters (Huntley, 1978; Taylor and Dehaan, 2000; Hakvoort et al., 2002). Airborne synthetic-aperture microwave systems, such as the NASA Jet Propulsion Laboratory's AIRSAR, employ active transmitters operating at high EM frequencies (C, L, and P band) to image the ground surface and provide digital elevation models using interferometry. Because radar scattering is governed by surface roughness and by the electrical properties of the ground, it is useful for detecting surface features and is being used to quantify soil moisture (Giacomelli et al., 1995; Bolognani et al., 1996). Lower radar frequencies (100 MHz or less) have been employed to image the shallow subsurface, but the achievable exploration depth is limited to a few meters or less where ground is electrically conductive (soils with high clay, moisture, or salinity content). ESTAR, a passive airborne radiometer operating in the L microwave frequency band (Le Vine and others, 2001), has been used to detect soil-moisture changes at the watershed and regional scales (Jackson et al., 1995; Jackson et al., 1999). In a novel application of gravity methods that are typically used for basin delineation, satellite-based measurements arising from NASA's GRACE (gravity recovery and climate experiment) might allow monitoring of water-storage changes in major aquifers (Rodell and Famiglietti, 2002; Swenson et al., 2003).

All major airborne geophysical methods benefit from accurate topography. Airborne lidar (light detection and ranging) provides elevation information of higher resolution and greater accuracy than topographic maps or derivative digital elevation models. Airborne lidar (Wehr and Lohr, 1999; Gutiérrez et al., 2001) combines three technologies: a laser, an inertial measurement unit (IMU), and geodetic-quality GPS receivers. Range measurements are produced by multiplying the speed of light by the time required for a laser pulse to travel from an aircraft to the earth's surface and a reflection to be detected. The IMU measures the rotations and accelerations of the aircraft, and GPS provides its position. The ranges, platform motion, and positions are processed into a series of x, y, and z ground points having vertical accuracies of about 10 to 15 cm over smooth, open surfaces. Lidar systems can detect ground reflections from laser energy that penetrates small gaps in foliage, giving lidar the capability to map terrain beneath trees.

11.6 Summary

Many hydrogeologically important physical properties can be measured from an airborne platform. Advantages common to all airborne methods include rapid data acquisition over large areas and dense spatial coverage. In general, airborne instruments cannot match the lateral and vertical resolution obtained from surface and borehole instruments, but rather complement them by providing data that allow typically sparse ground or subsurface data to be extended laterally. In many cases, airborne instruments are the only practical means of acquiring quantitative information over significant areas.

Successful airborne surveys depend upon the presence of a measurable contrast in a physical property that is a reasonable proxy for the hydrogeological parameter of interest. Because airborne geophysical surveys can be costly, proper survey design is especially important. Most successful airborne surveys are preceded by modeling efforts and ground-based surveys that serve to demonstrate the likely success of the method and help determine optimal airborne survey parameters.

Hydrogeologic applications are a developing frontier for airborne geophysical methods, many of which have been borrowed from other original purposes and are being modified to suit new applications. With this in mind, we can expect continued improvements in signal strength, detector sensitivity, system bandwidth, and processing capability, which will in turn improve the resolution, lower the detection limits, and increase the accuracy of remotely acquired information—to apply to an ever-growing list of significant hydrogeological issues.

References

Aage, H.K., U. Korsbech, K. Bargholz, and J. Hovgaard, A new technique for processing airborne gamma ray spectrometry data for mapping low level contaminations, *Applied Radiation and Isotopes, 51*(6), 651–662, 1999.

Archie, G.E., The electrical resistivity log as an aid in determining some reservoir characteristics, *SPE-AIME Transactions, 146*, 54–62, 1942.

Avdeev, D.B., A.V. Kuvshinov, O.V. Pankratov, and G.A. Newman, Three-dimensional frequency-domain modeling of airborne electromagnetic responses, *Exploration Geophysics, 29*, 111–119, 1998.

Babu, H.V.R., N.K. Rao, and V.V. Kumar, Bedrock topography from magnetic anomalies: An aid for groundwater exploration in hard-rock terrains, *Geophysics, 56*, 1051–1054, 1991.

Barongo, J.O., Selection of an appropriate model for the interpretation of time-domain airborne electromagnetic data for geological mapping, *Exploration Geophysics, 29*, 107–110, 1998.

Beamish, D., An assessment of inversion methods for AEM data applied to environmental studies, *Journal of Applied Geophysics, 51*(2–4), 75–96, 2002.

Beard, L.P., J.E. Nyquist, and P.J. Carpenter, Detection of karst structures using airborne EM and VLF, *Proc. of the 64th Ann. Internat. Mtg: Soc. of Expl. Geophys.*, 555–558, 1994.

Becker, Alex, Airborne resistivity mapping, *IEEE Transactions on Antennas and Propagation, 36*(4), 557–562, 1988.

Bierwirth, P.N., and W.D. Welsh, *Delineation of Recharge Beds in the Great Artesian Basin Using Airborne Gamma Radiometrics and Satellite Remote Sensing*, Bureau of Rural Sciences, Kingston, Australia, 2000.

Bishop, J., D. Sattel, J. Macnae, and T. Munday, Electrical structure of the regolith in the Lawlers District, Western Australia, *Exploration Geophysics, 32*, 20–28, 2001.

Bolognani, O., M. Mancini, and R. Rosso, Soil moisture profiles from multifrequency radar data at basin scale, *Meccanica 31*(1), 59–72, 1996.

Brodie, R.C., Investigating salinity using airborne geophysics, *Preview, 82*, 13–16, 1999.

Bromley, J., B. Mannstrom, D. Nisca, and A. Jamtlid, Airborne geophysics; Application to a ground-water study in Botswana, *Ground Water, 32*(1), 79–90, 1994.

Christensen, N.B., K.I. Sorensen, A.V. Christiansen, T.M. Rasmussen, and L.H. Poulsen, The use of airborne electromagnetic systems for hydrogeological investigations, *Proceedings, Symposium on the Application of Geophysics to Engineering and Environmental Problems*, Arlington, VA, U.S.A., 73–82, 2000.

Christiansen, A.V., and N.B. Christensen, A quantitative appraisal of airborne and ground-based transient electromagnetic (TEM) measurements in Denmark, *Geophysics 68*(2), 523–534, 2003.

Cook, P.G., and S. Kilty, A helicopter-borne electromagnetic survey to delineate groundwater recharge rates, *Water Resour. Res., 28*(11), 2953–2961, 1992.

Cook, S.E., R.J. Corner, P.R. Groves, and G.J. Grealish, Use of airborne gamma radiometric data for soil mapping, *Australian Journal of Soil Research, 34*(1), 183–194, 1996.

Coppa, I., P. Woodgate, and A. Webb, Improving the management of dryland salinity in Australia through the national airborne geophysics project, *Exploration Geophysics, 29*, 230–233, 1998.

Doll, W.E., J.E. Nyquist, L. P. Beard, and T. J. Gamey, Airborne geophysical surveying for hazardous waste site characterization on the Oak Ridge Reservation, Tennessee, *Geophysics, 65*, 1372–1387, 2000.

Duval, J.S., B. Cook, and J.A.S. Adams, Circle of investigation of an airborne gamma-ray spectrometer, *Journal of Geophysical Research, 76*, 8466–8470, 1971.

Duval, J.S., Composite color images of aerial gamma-ray spectrometric data, *Geophysics, 48*(6), 722–735, 1983.

Duval, J.S., Radioactivity and some of its applications in geology, *Proceedings, Symposium on the Application of Geophysics to Engineering and Environmental Problems*, Golden, Colorado, U.S.A., pp. 1–61, 1989.

Ellis, R.G., Inversion of airborne electromagnetic data, *Exploration Geophysics, 29*, 121–127, 1998.

Fitterman, D.V.,ed., Developments and applications of modern airborne electromagnetic surveys, *U.S. Geological Survey Bulletin* 1925, 1990.

Fitterman, D.V., and M. Deszcz-Pan, Helicopter EM mapping of saltwater intrusion in Everglades National Park, Florida, *Exploration Geophysics, 29*, 240–243, 1998.

Fountain, D., Airborne electromagnetic systems—50 years of development, *Exploration Geophysics, 29*, 1–11, 1998.

Fraser, D.C., Resistivity mapping with an airborne Multicoil electromagnetic system, *Geophysics, 43*(1), 144–172, 1978.

Frischknecht, F.C., R. Gette, P.V. Raab, and J. Meredith, Location of abandoned wells by magnetic surveys-acquisition and interpretation of aeromagnetic data for five test areas, *U.S. Geological Survey Open-File Report 85-614-A*, 1985.

Frischknecht, F.C., V.F. Labson, B.R. Spies, and W.L. Anderson, Profiling using small sources, In: *Electromagnetic Methods in Applied Geophysics—Applications, Part A and Part B*, M.N. Nabighian, ed., Society of Exploration Geophysicists, pp. 105–270, 1991.

Gamey, T.J., J.S. Holladay, J. Nyquist, and W. Doll, An airborne multisensor characterization of an active nuclear waste site, In: *Geologic Applications of Gravity and Magnetics: Case Histories*, R.I. Gibson and P.S. Millegan, eds., American Association of Petroleum Geologists, *Studies in Geology 43*, pp. 120–122, 1998.

Giacomelli, A., U. Bacchiega, P.A. Troch, and M. Mancini, Evaluation of surface soil moisture distribution by means of SAR remote sensing techniques and conceptual hydrological modeling, *Journal of Hydrology, 166*(3–4), 445–459, 1995.

Graham, D.F., and G.F. Bonham-Carter, Airborne radiometric data: A tool for reconnaissance geological mapping using a GIS, *Photogrammetric Engineering and Remote Sensing 59*, 1243–1249, 1993.

Grasty, R.L., Direct snow-water equivalent measurement by airborne gamma-ray spectrometry, *Journal of Hydrology, 55*(1–4), 213–235, 1982.

Grasty, R.L., K.L. Kosanke, and R.S. Foote, Fields of view of airborne gamma-ray detectors, *Geophysics, 47*, 1737–1738, 1979.

Grauch, V.J.S., High-resolution aeromagnetic data, a new tool for mapping intrabasinal faults; Example from the Albuquerque Basin, New Mexico, *Geology, 29*(4), 367–370, 2001.

Grauch, V.J.S., Aeromagnetic mapping of hydrologically important faults, Albuquerque Basin, New Mexico, *Proceedings, Symposium on the Application of Geophysics to Engineering and Environmental Problems*, Denver, Colorado, U.S.A., 2001.

Gutiérrez, R., J.C. Gibeaut, R.C. Smyth, T. Hepner, J.R. Andrews, C. Weed, W. Gutelius, and M. Mastin, Precise airborne LIDAR surveying for coastal research and geohazards applications, *International Archives of Photogrammetry and Remote Sensing, 34*(3/W4), 185–192, 2001.

Hakvoort, H., J. de Haan, R. Jordans, R. Vos, S. Peters, and M. Rijkeboer, Towards airborne remote sensing of water quality in The Netherlands; Validation and error analysis, *Journal of Photogrammetry and Remote Sensing, 57*(3), 171–183, 2002.

Hill, R.A., G.M. Smith, R.M. Fuller, and N. Veitch, Landscape modelling using integrated airborne multi-spectral and laser scanning data, *International Journal of Remote Sensing, 23* (11), 2327–2334, 2002.

Hunter, D., and J. Macnae, Subsurface conductivity structure as approximated by conductivity-depth transforms, *Australian Society of Exploration Geophysics Extended Abstracts*, Brisbane, 2001.

Huntley, D., On the detection of shallow aquifers using thermal infrared imagery, *Water Resour. Res., 14*(6), 1075–1083, 1978.

International Atomic Energy Agency, Airborne gamma ray spectrometer surveying, International Atomic Energy Agency, Vienna, *Technical Report Series, No. 323*, 1991.

Jackson, T.J., Remote sensing of soil moisture; implications for groundwater recharge: *Hydrogeology Journal, 10*(1), 40–51, 2002.

Jackson, T.J., D.M. Le Vine, A.Y. Hsu, A. Oldak, P.J. Starks, C.T. Swift, J.D. Isham, and M. Haken, Soil moisture mapping at regional scales using microwave radiometry: The Southern Great Plains Hydrology Experiment, *IEEE Transactions on Geoscience and Remote Sensing, 37*(5), 2136–2151, 1999.

Jackson, T.J., D.M Le Vine, C.T. Swift, T.J. Schmugge, and F.R. Schiebe, Large area mapping of soil moisture using the ESTAR passive microwave radiometer in Washita '92, *Remote Sensing of Environment, 54*(1), 27–37, 1995.

Jaques, A. L., P. Wellman, A. Whitaker, and D. Wyborn, High-resolution geophysics in modern geological mapping, *AGSO Journal of Australian Geology and Geophysics, 17(2)* , 159–173, 1997.

Kaufman, A.A., and G.V. Keller, Frequency and transient soundings, *Methods in Geochemistry and Geophysics, No. 16*, Elsevier, Amsterdam, The Netherlands, 1983.

Le Vine, D.M., C.T. Swift, and M. Haken, Development of the synthetic aperture microwave radiometer, ESTAR, *IEEE Transactions on Geoscience and Remote Sensing 39*(1), 199–202, 2001.

Mackey, T., K. Lawrie, P. Wilkes, T. Munday, N. de Souza Kovacs, R. Chan, D. Gibson, C. Chartres, and R. Evans, Palaeochannels near West Wyalong, New South Wales; A case study in delineation and modelling using aeromagnetics, *Exploration Geophysics, 31*(1–2), 1–7, 2000.

Macnae, J.C., R. Smith, B. D. Polzer, Y. Lamontagne, and P. S. Klinkert, Conductivity-depth imaging of airborne electromagnetic step-response data, *Geophysics, 56*(1), 102–114, 1991.

Macnae, J., A. King, N. Stolz, and P. Klinkert,, 3-D EM inversion to the limit, In: *Three-Dimensional Electromagnetics*, M. Oristaglio, B. Spies, and M.R. Cooper, eds., Tulsa, Oklahoma, U.S.A., Society of Exploration Geophysicists, Geophysical Development Series, *7*, 489–501, 1999.

Maus, S., K.P. Sengpiel, B. Röttger, B. Siemon, and E.A.W. Tordiffe, Variogram analysis of helicopter magnetic data to identify paleochannels of the Omaruru River, Namibia, *Geophysics, 64*(3), 785–794, 1999.

McNeill, J.D., Electrical conductivity of soils and rocks, Geonics Ltd., Mississauga, Ont., *Technical Note TN-5*, 1980a.

McNeill, J.D., Electromagnetic terrain conductivity measurement at low induction numbers, Geonics Ltd., Mississauga, Ont., *Technical Note TN-6*, 1980b.

Meijerink, A.M.J., Groundwater, In *Remote Sensing in Hydrology and Water Management*, G.A. Schultz and E.T. Engman, eds., Springer, Berlin, Heidelberg, New York, pp. 305–325, 2000.

Milsom, J., *Field Geophysics*, Geological Society of London Handbook, Open University Press, Milton Keynes, 1989.

Minty, B.R.S., Fundamentals of airborne gamma-ray spectrometry, In: *Airborne Magnetic and Radiometric Surveys*, P.J. Gunn, ed., Australian Geological Survey Organisation, *Journal of Australian Geology and Geophysics, 17*(2), 39–50, 1997.

Minty, B.R.S., P. McFadden, and B.L.N. Kennett, Multichannel processing for airborne gamma-ray spectrometry, *Geophysics, 63*(6), 1971–1985, 1998.

Morrison, H.F., A. Becker, and G.M. Hoversten, The physics of airborne EM systems, *Exploration Geophysics, 29*, 97–102, 1998.

Newman, G.A. and D.L. Alumbaugh, Frequency-domain modeling of airborne electromagnetic responses using staggered finite differences, *Geophys. Prosp., 43*, 1021–1042, 1995.

Paine, J.G., Determining salinization extent, identifying salinity sources, and estimating chloride mass using surface, borehole, and airborne electromagnetic induction methods, *Water Resour. Res., 39*(3), 3-1–3-10, 2003.

Paine, J.G., and E.W. Collins, Applying airborne electromagnetic induction in groundwater salinization and resource studies, West Texas, *Proceedings, Symposium on the Application of Geophysics to Engineering and Environmental Problems*, San Antonio, Texas, U.S.A., 2003.

Paine, J.G., A.R. Dutton, and M.U. Blüm, Using airborne geophysics to identify salinization in West Texas, *Bureau of Economic Geology Report of Investigations No. 257*, The University of Texas at Austin, 1999.

Paine, J.G., A.R. Dutton, J.S. Mayorga, and G.P. Saunders, Identifying oil-field salinity sources with airborne and ground-based geophysics: A West Texas example, *The Leading Edge, 16*(11), 1603–1607, 1997.

Palacky, G.J., The airborne electromagnetic method as a tool of geological mapping, *Geophysical Prospecting, 29*(1), 60–88, 1981.

Palacky, G.J., Airborne resistivity mapping, Geological Survey of Canada, *Paper 86-22*, 1986.

Palacky, G.J., and G.F. West, Airborne electromagnetic methods, In: *Electromagnetic Methods in Applied Geophysics—Applications, Part A and Part B*, M.N. Nabighian, ed., Tulsa, Oklahoma, Society of Exploration Geophysicists, pp. 811–879, 1991.

Parasnis, D.S., *Principles of Applied Geophysics*, Chapman and Hall, 1986.

Peltoniemi, M., Depth of penetration of frequency-domain airborne electromagnetics in resistive terrains, *Exploration Geophysics, 29*, 12–15, 1998.

Pickup, G., and A. Marks, Identifying large-scale erosion and deposition progresses from airborne gamma radiometrics and digital elevation models in a weathered landscape, *Earth Surface Processes and Landforms, 25*(5), 535–557, 2000.

Pickup, G., A. Marks, and M. Bourke, Paleoflood reconstruction on floodplains using geophysical survey data and hydraulic modeling, In: *Ancient Floods, Modern Hazards: Principles and Applications of Paleoflood Hydrology*, P. K. House, R. H. Webb, V. R. Baker, and D. R. Levish, eds., *Water Science and Application, 5*, 47–60, 2002.

Pitkin, J.A., and J.S. Duval, Design parameters for aerial gamma-ray surveys, *Geophysics, 45*(9), 1427–1439, 1980.

Plume, R.W., Use of aeromagnetic data to define boundaries of a carbonate-rock aquifer in East-Central Nevada, *U.S. Geological Survey Water Supply Paper 2330*, 1988.

Puranen, R., H. Säävuori, L. Sahala, I. Suppala, M. Mäkilä, and J. Lerssi, Airborne electromagnetic mapping of surficial deposits in Finland, *First Break, 17*(5), 145–154, 1999.

Rama Rao, C., M.P. Lakshmi, and H.V. Ram Babu, Delineation of new structural controls using aeromagnetics and their relation with occurrence of ground water in an exposed basement complex, in a part of South Indian Shield, In: *Proceedings of the International Groundwater Conference on Sustainable Development and Management of Groundwater Resources in Semi-Arid Regions, with Special Reference to Hard Rocks,* M.S. Thangarajan, N. Rai, and V.S. Singh, eds., Lisse, The Netherlands, A.A. Balkema, 2002.

Reid, A.B., Aeromagnetic survey design, *Geophysics, 45*(5), 973–976, 1980.

Robinove, C.J., R.H. Langford, and J.W. Brookhart, Saline-water resources of North Dakota, *U.S. Geological Survey Water-Supply Paper 1428*, 1958.

Rodell, M., and J.S. Famiglietti, The potential for satellite-based monitoring of groundwater storage changes using GRACE: The High Plains Aquifer, Central U.S., *Journal of Hydrology, 263*(1–4), 245–256, 2002.

Sattel, D., and L. Kgotlhang, Modeling aquifer structures with airborne EM in the Boteti area, Botswana, *Proceedings, Symposium on the Application of Geophysics to Engineering and Environmental Problems*, San Antonio, Texas, U.S.A., 36–44, 2003.

Schwarz, G.F., L. Rybach, and E. E. Klingele, Design, calibration, and application of an airborne gamma spectrometer system in Switzerland, *Geophysics, 62*(5), 1369–1378.

Sengpiel, K.P., Resistivity depth mapping with airborne electromagnetic survey data, *Geophysics, 48*(2), 181–196, 1983.

Sengpiel, K.P., Approximate inversion of airborne electromagnetic data from a multilayered ground, *Geophysical Prospecting, 36*, 446–459, 1988.

Sengpiel, K.P. and B. Siemon, Advanced inversion methods for airborne electromagnetic exploration, *Geophysics, 65*, 1983–1992, 2000.

Siemon, B., Improved and new resistivity-depth profiles for helicopter electromagnetic data, *Journal of Applied Geophysics, 46*(1), 65–76, 2001.

Smith, B.D., K.P. Sengpiel, J.L. Plesha, and R.J. Horton, Airborne electromagnetic mapping of subsurface brine, Brookhaven Oil Field, Mississippi, Tulsa, Oklahoma, Society of Exploration Geophysicists, *Annual Meeting Expanded Technical Program Abstracts with Biographies, 62*, 340–343, 1992.

Smith, B.D., R. Bisdorf, L.J. Slack, and A.T. Mazzella, Evaluation of electromagnetic mapping methods to delineate subsurface saline waters in the Brookhaven Oil Field, Mississippi, *Proceedings of the Symposium on the Application of Geophysics to Engineering and Environmental Problems*, Reno, Nevada, U.S.A., pp. 685–693, 1997.

Smith, B.D., A.E. McCafferty, and R.R. McDougal, Utilization of airborne magnetic, electromagnetic, and radiometric data in abandoned mine land investigations: Reston, VA, U. S. A., *U.S. Geological Survey Open-File Report OFR-00-0034*, pp. 86–91, 2000.

Smith, B.D., D.V. Smith, P.L. Hill, and V.F. Labson, Helicopter electromagnetic and magnetic survey data and maps, Seco Creek area, Medina and Uvalde Counties, Texas: Reston, VA, U.S.A., *U. S. Geological Survey Open-File Report OFR-03-226*, 2003.

Smith, C.L., and A.P. Tribble, The utilization of side-looking airborne radar (SLAR) in the analysis of karst topography, *Remote Sensing Quarterly, 1*(1), 49–56, 1979.

Smith, R., and P. Annan, The use of B-field measurements in an airborne time-domain system: Part 1: Benefits of B field versus dB/dt data, *Exploration Geophysics, 29*, 24–29, 1998.

Smith, R.S., and P.B. Keating, The usefulness of multicomponent, time-domain airborne electromagnetic measurements, *Geophysics, 61*, 74–81, 1996.

Sorensen, K.I., M. Halkjaer, and E. Auken, SkyTEM–A new high resolution helicopter TEM system, *Proceedings, Symposium on the Application of Geophysics to Engineering and Environmental Problems*, Colorado Springs, Colorado, U.S.A., pp. 668–672, 2004.

Spies, B.R., and F.C. Frischknecht, Electromagnetic sounding, In: *Electromagnetic Methods in Applied Geophysics—Applications, Part A and Part B*, M.N. Nabighian, ed., Society of Exploration Geophysicists, Tulsa, Oklahoma, pp. 285–386, 1991.

Street, G.J. and A. Anderson, Airborne electromagnetic surveys of the regolith, *Exploration Geophysics, 24*(3–4), 795–800. 1993.

Street, G.J., and G. P. Roberts, SALTMAP—High resolution airborne EM for electrical conductivity profiling, *Proceedings of the Symposium on the Application of Geophysics to Engineering and Environmental Problems*, Orlando, Florida, U.S.A., pp. 273–278, 1995.

Swenson, S., J. Wahr, and P.C.D. Milly, Estimated accuracies of regional water storage variations inferred from the Gravity Recovery and Climate Experiment (GRACE), *Water Resour. Res., 39*(8), 1223, 2003.

Taylor, G.R., and R.L. Dehaan, Mapping soil salinity with hyperspectral imagery, *Proceedings of the Fourteenth International Conference on Applied Geologic Remote Sensing*, pp. 512–519, 2000.

Telford, W.M., L.P. Geldart, and R.E. Sheriff, *Applied Geophysics*, 2nd edition, New York, Cambridge University Press, 1990.

Toyra, J., Assessment of airborne scanning laser altimetry (lidar) in a deltaic wetland environment, *Canadian Journal of Remote Sensing, 29(6)*, 718–728, 2003.

Vrbancich, J., M. Hallett, and G. Hodges, Airborne electromagnetic bathymetry of Sydney Harbour, *Exploration Geophysics, 31*(1-2), 179–186, 2001.

Wehr, A., and U. Lohr, Airborne laser scanning—An introduction and overview, *ISPRS Journal of Photogrammetry and Remote Sensing, 54*(2–3), 68–82, 1999.

West, G.F., and J.C. Macnae, Physics of the electromagnetic induction exploration method, In: Electromagnetic Methods in Applied Geophysics—Applications, Part A and Part B, M.N. Nabighian, ed., Tulsa, Oklahoma, Society of Exploration Geophysicists, pp. 5–45, 1991.

Wilford, J.R., P.N. Bierwirth, and M.A. Craig, Application of airborne gamma-ray spectrometry in soil/regolith mapping and applied geomorphology, *Journal of Australian Geology and Geophysics, 17*(2), 201–216, 1997.

Wilford, J.R., D.L. Dent, T. Dowling, and R. Braaten, Rapid mapping of soils and salt stores using airborne radiometrics and digital elevation models, *Australian Geological Survey Organisation Research Newsletter, 34*, 33–40, 2001.

Wilson, C.R., G. Tsoflias, M. Bartelmann, and J. Phillips, A high precision aeromagnetic survey near the Glen Hummel Field in Texas: Identification of cultural and sedimentary anomaly sources: *Leading Edge, 16*(1), 37–42, 1997.

Witherly, K.E., H. Golden, and S. Medved, Safety in airborne geophysical surveying, *The Leading Edge, 18*(5), 635–637, 1999.

Wolfgram, P., and G. Karlik, Conductivity-depth transform of GEOTEM data, *Exploration Geophysics*, *26*, 179–185, 1995.

Wood, E.F., D.-S. Lin, M. Mancini, D. Thongs, P.A. Troch, T.J. Jackson, J.S. Famiglietti, and E.T. Engman, Intercomparisons between passive and active microwave remote sensing, and hydrological modeling for soil moisture, *Advances in Space Research, 13*(5), 167–176, 1993.

Wynn, J., Evaluating groundwater in arid lands using airborne magnetic/EM methods: an example in the Southwestern U.S. and northern Mexico, *The Leading Edge, 21*, 62–64, 2002.

Wynn, J., and M. Gettings, An airborne electromagnetic (EM) survey used to map the upper San Pedro River Aquifer, near Fort Huachuca in southeastern Arizona, *Proceedings, Symposium on the Application of Geophysics to Engineering and Environmental Problems*, Chicago, Illinois, U.S.A., pp. 993–998, 1998.

Hydrogeophysical Case Studies

12 HYDROGEOPHYSICAL CASE STUDIES AT THE REGIONAL SCALE

MARK GOLDMAN[1], HAIM GVIRTZMAN[2], MAX MEJU[3], and VLADIMIR SHTIVELMAN[1]

[1]*Geophysical Institute of Israel, Israel*
[2]*Hebrew University of Jerusalem, Israel*
[3]*Lancaster University, UK*

12.1. Introduction

12.1.1 GENERAL BACKGROUND

Geophysical methods are widely applied in both local and regional hydrogeological investigations. Examples of local hydrogeophysical studies will be given in the next two chapters of this book. In this chapter, several original, mostly unpublished case histories showing regional hydrogeophysical applications will be presented.

Typical regional-scale hydrogeophysical investigations are characterized by the specific dimensions of the study area (x and y), varying from several kilometers to several tens of kilometers; and by the depth to the target (z), varying from several tens of meters to several hundred meters. In some special cases, x and y can be on the order of hundreds of kilometers, and the depth to the target may exceed a few kilometers.

All major exploration geophysical methods are employed in these regional hydrogeophysical studies, although some of them have a limited application because of insufficient penetration depth (such as ground-penetrating radar) or a low sensitivity to major hydrogeophysical parameters of the target (such as magnetic methods). The main methods applied in the overwhelming majority of regional hydrogeophysical investigations are electrical/electromagnetic (EM) and seismic methods. Electrical and seismic regional surveys are carried out using ground-based techniques, whereas electromagnetic (as well as gravitational, magnetic, and radiometric) methods are applied in both surface and airborne modifications.

Based on the sensitivity of the methods to measured (or more precisely, interpreted) physical parameters, each geophysical technique has specific applications for which it is best suited. For example, seawater intrusions are most accurately delineated by electrical methods, particularly by EM methods, owing to the very close relationship that exists between the interpreted electrical/EM resistivity and groundwater salinities. Because of the temporal variability of the target, electrical/EM methods are also frequently applied to monitoring seawater intrusion and similar hydrogeological targets. In addition, some geometrical parameters of the aquifer—such as overall thickness, intercalation of aquiferous portions and aquicludes/aquitards—are successfully detected by seismic methods, as well by electrical/EM and gravitational methods.

Y. Rubin and S.S. Hubbard (eds.), Hydrogeophysics, 361–389.

In some cases, the combined application of relevant geophysical techniques may significantly increase the reliability and accuracy of regional hydrogeophysical surveys. This usually happens when the individual geophysical methods supplement each other with respect to different hydrogeological properties of the target. For example, the combined application of nuclear magnetic resonance (NMR) and transient electromagnetic (TEM) methods in the Mediterranean and Dead Sea coastal aquifers of Israel enabled us to quantitatively estimate the amount of fresh groundwater within the aquifers (Legchenko et al., 1998). This estimation became possible because NMR and TEM are very efficient in detecting two supplementary properties of the target, namely the depth to the water table and the water content within the aquifer (with NMR) and the depth to freshwater/seawater interface (with TEM).

Although the primary objective of regional hydrogeophysical surveys is mapping the largely geometrical features of hydrogeological targets, several attempts (admittedly less successful) have been made to investigate hydraulic/hydrological properties of aquifers, such as porosity, transmissivity, hydraulic conductivity, and clay content. We believe that local hydrogeophysical investigations (see Chapters 13 and 14 of this volume) are more suitable for solving this problem. Regional hydrogeophysical studies, such as those discussed in this chapter, can provide important information regarding the geometry of a specific hydrogeological target, as well as the necessary boundary conditions for more detailed local hydrogeophysical investigations.

12.1.2 SELECTED GEOPHYSICAL TECHNIQUES

Electrical (particularly DC resistivity) methods have been widely used to image the resistivity structure of near-surface targets (i.e., within a few tens of meters of the ground surface), owing to their ease of operation and low equipment cost (see Chapter 5 of this volume). In some environments, however, these methods are difficult or impossible to use for delineating deep targets, since they require large array dimensions in relation to the maximum depth at which useful information can be obtained (about 5–6 times the target depth of interest). The problem is particularly severe in the case of multidimensional resistivity imaging techniques, which require a large number of electrodes to be involved in the measurements. On the other hand, 1-D resistivity soundings that sample great depths usually deal with significant lateral resistivity variations, thus leading to essential problems in data interpretation. Controlled-source electromagnetic (EM) techniques, which play an important role in shallow-depth conductivity mapping (see Chapter 6), have limited frequency bandwidth and hence relatively poor depth sounding capability (e.g., Meju et al., 2001).

The methods most suitable for deep soundings are the transient (TEM) or time-domain electromagnetic (TDEM) and various magnetotelluric (MT) techniques, including audio-magnetotelluric (AMT) and controlled-source audio-magnetotelluric (CSAMT). These EM methods are particularly appropriate for hydrogeophysical studies in arid and semi-arid regions (see Bazinet and Legault, 1986; Meju, 2002). Note that EM methods can be combined with electrical methods for improved subsurface mapping (see Chapter 6 of this volume). The TEM method is particularly suitable for the characterization of groundwater salinity, because of its superior sensitivity to electrically conductive targets (Fitterman and Stewart, 1986).

For high-resolution stratigraphic mapping of deep aquifers and their confining formations, the seismic reflection method is emerging as the tool with the best potential in some sedimentary environments. This method can be combined with electromagnetic or electrical methods for improved target identification in large-scale hydrogeophysical studies.

12.1.3 TEM METHOD

In TEM surveys, a primary field is generated by a rectangular or square transmitter (Tx) loop, with dimensions of ten to a few hundred meters. Depending on ground conditions, a 50–200 m sided Tx loop could suffice for imaging depths of about 20–1,000 m (see, for example, Meju 1998; Meju et al., 2000). The primary field is not continuous, but rather consists of a series of pulses between which the primary field is switched off and measurements made. The rapid termination of the primary field in the Tx causes eddy currents to be produced in the subsurface, and the associated secondary field is measured on the surface using a receiver (Rx) coil. The receiver may be a small multi-turn coil (about 1 m diameter), or the Tx loop can serve as the Rx during transmitter-off times (an arrangement called the single- or coincident-loop method). The decaying transient signal is sampled at various time gates to yield a sounding curve. This transient signal can last from a few microseconds to several hundreds of milliseconds (i.e., broadband), and an observational bandwidth of about 0.005 to 30 ms is desirable for subsurface investigations covering 20 to 500 m depth. The character of the transient signal is a function of the subsurface resistivity structure, with the shallow structure influencing the transient at early times (soon after Tx turn-off) and measurement at later times, allowing the deeper electrical structure to be interpreted.

12.1.4 MAGNETOTELLURIC METHOD

12.1.4.1 *Basic Principles and Field Practice*

The magnetotelluric (MT) method is an EM technique for determining the resistivity distribution of the subsurface from measurements of natural time-varying magnetic and electric fields on the surface of the earth. Naturally occurring interminable variations in the earth's magnetic field induce eddy currents in the earth that are detectable as electric (or telluric) field variations on the surface. The ratio of the horizontal electric field to the orthogonal horizontal magnetic field (termed the EM impedance), measured at a number of frequencies, gives earth resistivity as a function of frequency or period, resulting in a form of depth sounding. This is a wide-band depth sounding technique, with frequencies ranging from about 10^4 to 10^{-6} Hz, and it is suited for shallow as well as deep geological investigations. The depth of investigation in MT is a function of subsurface resistivity and frequency (or the inverse period) of the EM signals. The depth of sounding can be roughly related to frequency by the skin depth, defined as

$$\delta \approx 0.503\ (\text{resistivity/frequency})^{1/2}\ \text{km} \tag{12.1}$$

where resistivity is in ohms per meter (ohm-m) and frequency in hertz (Hz). MT can probe very great depths (up to 600 km) by utilizing sufficiently low frequencies. The MT signals of frequencies greater than 1 Hz are of atmospheric origin and are caused by global thunderstorm activities. Some thunderstorm energy is converted to EM fields, which are propagated with slight attenuation in the earth-ionosphere interspace as a guided wave; at a large distance from the source, this is a plane wave of variable frequency. The MT signals of frequencies lower than 1 Hz are of magnetospheric origin (i.e., resulting from the interaction of solar wind and the earth's magnetic field).

In the scalar MT method, the electric field in one horizontal direction and the magnetic field in an orthogonal horizontal direction are measured simultaneously at a surface location. The more widely used tensor MT method requires simultaneous observation of two orthogonal horizontal (usually north-south and east-west) components of the magnetic field (Hx, Hy) and the electric field (Ex, Ey), and the vertical magnetic field component (Hz) at a surface location (see Figure 12.1). Magnetic sensors are squid magnetometers or induction coils (consisting of several hundred turns of fine copper wire wound over a ferrite of high magnetic permeability). Electric sensors are nonpolarizing electrodes, commonly copper rods inserted in supersaturated copper sulphate solution or lead rods in supersaturated lead chloride solution in porous pots. Traditional MT surveying involves tensor depth soundings at widely spaced stations

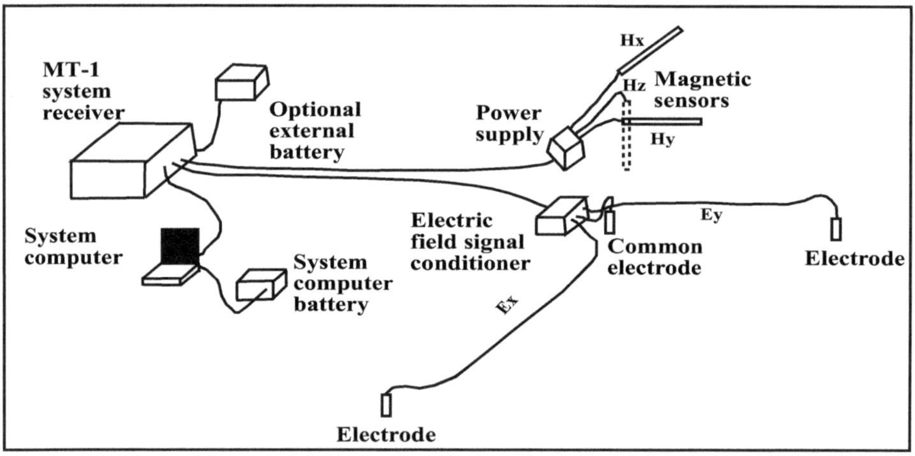

Figure 12.1. Typical field setup for a single station or discrete MT depth sounding

and herein dubbed discrete surveys. The electromagnetic array profiling (EMAP) method (Torres-Verdin and Bostick, 1992), a continuously dense sample variant of the traditional MT technique, provides good lateral resolution of short-wavelength features in the subsurface, but the survey cost approaches that of seismics.

12.1.4.2 *Data Processing and Characterization of Anisotropy*

Data processing involves converting the field records into EM impedance (Z) data and the determination of apparent resistivity and phase information from Z for the various observational frequencies or periods. The EM wave impedance is obtained as Z = E/H. This is a complex quantity (it has amplitude and phase). Since two mutually perpendicular horizontal components of E and H are measured in the field, Z can be defined for any given direction in tensor MT. For example, the electric field in the north-south direction and the magnetic field in the east-west direction may be used to define the impedance Zxy = Ex/Hy, while the electric field in the east-west direction and the orthogonal magnetic field component may be used to define the impedance Zyx = Ey/Hx. Zxy and Zyx are equal if the subsurface is homogeneous and isotropic. If the ground is heterogeneous, both impedances differ in magnitude and phase, as in the examples presented in Figure 6.15 (Chapter 6 of this volume) for the controlled-source variant (CSMT). In general, the magnetotelluric impedance is a tensor whose values depend on the measurement direction. The components of the impedance tensor can be rotated mathematically to attain maximum values in the direction of linear conductive subsurface structures (and minimum values in the perpendicular direction). The direction in which the impedance tensor elements are maximized during rotation is termed the electrical strike or azimuth of the impedance tensor in MT parlance, and usually coincides with the structural trend of a major geological feature. The magnitude of Z is used to define an interpretative quantity called the apparent resistivity, ρa, which for a given EM wave frequency or period (T) is given as:

$$\rho a = 0.2T\,|Z|^2 \qquad (12.2)$$

Apparent resistivity is in ohm-m when the wave period T is in seconds (the reciprocal of frequency in Hz), E is in mV/km, and H is in nanoTesla or gamma. The phase of Z (symbolically denoted by θ) is the ratio of the imaginary to real parts of the complex impedance, i.e., θ = imag Z / real Z, and is in degrees.

The relationship between the vertical magnetic field Hz and the horizontal magnetic components can also be used to estimate the electrical strike (and define directions of current concentrations or induction arrows in MT parlance). The differences in strike directions determined by impedance tensor rotation and vertical-horizontal magnetic-field relationships serve to indicate the three-dimensionality of subsurface conductivity structures and provide a tool for characterizing anisotropy in advanced data processing (see Bahr, 1988, 1991; Groom and Bailey, 1989; Groom and Bahr, 1992).

Mathematical modeling techniques exist for interpreting MT data, commonly apparent resistivities, and phases (e.g., deGroot-Hedlin and Constable, 1990; Madden and Mackie, 1993). When the subsurface is comprised of horizontally stratified rock sequences, the resistivity varies only with depth, and the resistivity distribution is said to have a one-dimensional (1-D) form. Parameters sought in 1-D interpretation are layer resistivities and thicknesses or (alternatively) conductivity-thickness products. When the geological sequences are not horizontally stratified, the resistivity structure is described as being three-dimensional (3-D) and may vary in all directions. However, there are many situations in which tectonic processes have imparted strong structural lineations in the subsurface rocks, such that there is a well-defined strike, and the gross electrical structure is approximately taken to be 2-D (i.e., the structure changes only across strike and at depth).

12.1.5 SEISMIC REFLECTION METHOD

The high-resolution seismic reflection method is by far the most effective geophysical technique for detailed and reliable imaging of the subsurface structure over a wide range of depths and for various applications. Because of the high vertical and lateral resolution of this method, one can perform very detailed mapping of various subsurface features, including layer geometry and fracture and fault zones. The theoretical bases of the method are well known; its basic principles are described briefly in Chapter 8 of this volume. The method has been widely used in various geological environments for groundwater-related studies (Birkelo et al., 1987; Geissler, 1989; Miller et al., 1989; Miller and Steeples, 1990; Bruno and Godio, 1997; Shtivelman and Goldman, 2000).

12.1.6 CASE STUDIES

Four case histories of regional investigations are presented in the following sections to demonstrate the usefulness of ground and marine TDEM (Sections 12.2 and 12.3, respectively), MT (Section 12.4) and seismic reflection (Section 12.5) methods in deep groundwater resource investigations in arid and semi-arid regions.

12.2 Characterization of Groundwater Salinity Using a Time-Domain Electromagnetic (TDEM) Method

12.2.1 INTRODUCTION

A schematic map of Israel's Mediterranean Coastal Aquifer is shown in Figure 12.2. The coastal aquifer is of paramount importance to Israel's water system, due to its close proximity to the most densely populated region of the country. As a result, the aquifer is heavily overexploited, despite the existence of a sophisticated water management system based on both a large number of the observation wells and extensive monitoring using TDEM. The overpumping, in turn, leads to the formation of hydrological depressions, in which the water table sinks below sea level, and this results in further encroachment of seawater into the aquifer (up to approximately 2 km inland).

Electrical and electromagnetic methods are particularly suitable for studying the salinity of groundwater, given the close relation that exists between the salinity and electrical resistivity (or conductivity, which is the reciprocal of resistivity) measured by these methods. It is well known that the more saline the groundwater to be detected, the more accurate and unique are the electrical/electromagnetic signatures. From a hydrological viewpoint, this results from the fact that, starting from certain salinities (e.g., normal sea water), the lithology of the aquifer plays a much less significant role, so that the resistivity of the aquifer is almost solely controlled by the water salinity (Goldman et al., 1988). It is important to emphasize, however, that the latter observation is mostly applicable to granular clastic aquifers rather than to fractured carbonate aquifers, because of the much greater variability of the porosity in the carbonates. As a result, the resistivity of seawater-saturated carbonate aquifers may vary over a much wider range than granular aquifers. Moreover, the resistivity of seawater-saturated carbonate aquifers can be similar to or greater than the resistivity of some dry or freshwater-saturated clastic formations (e.g., clays, marls). Thus, in the latter (freshwater) case, non-uniqueness may exist in hydrogeological interpretations of the geophysical results (see Chapter 17 of this volume for discussions on handling such non-uniqueness).

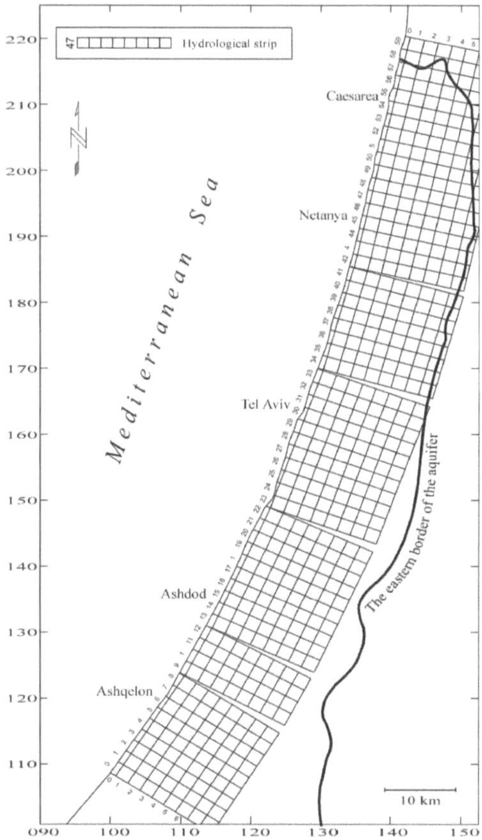

Figure 12.2. Schematic map of the Mediterranean Coastal Aquifer of Israel

From a geophysical viewpoint, seawater-saturated clastic aquifers, which are characterized by extremely low resistivities, can be more accurately determined than other electrical targets, using a properly selected electromagnetic technique. Both theoretical investigations and extensive practical studies (Kaufman and Keller, 1983; Fitterman and Stewart, 1986; Nabighian and Macnae, 1991; Goldman et al., 1991) show that the superior method in detecting salt-water-saturated layers in the subsurface is the TDEM method. This method possesses the highest sensitivity to electrically conductive targets, provides excellent vertical and lateral resolution, and can be easily employed in terrains with electrically difficult surface conditions, such as dry sands and hard rocks. The method suffers, however, from an extremely low sensitivity to thin resistive horizons and possesses relatively low noise-protection capabilities.

The TDEM method has been extensively applied for detecting saline groundwater in practically all clastic aquifers of Israel. A total of approximately 800 soundings have been carried out here during the last 15 years. Most of the soundings were conducted in the Mediterranean coastal plain (roughly 400) and in the Sea of Galilee and its close vicinity (roughly 300). The remaining approximately 100 TDEM measurements were performed in the Dead Sea coastal area, the Arava desert, and the Gulf of Eilat coastal aquifer. A number of deep TDEM soundings have been recently carried out to detect saline groundwater within the regional carbonate aquifer. The most recent results of the TDEM surveys in the Mediterranean Coastal Aquifer and those of the marine TDEM survey in the Sea of Galilee are presented in the first two sections of this chapter.

12.2.2 THE HYDROGEOLOGICAL SETTING

The Mediterranean Coastal Aquifer is approximately 110 km long and extends from the Gaza Strip in the south to Mount Carmel in the north. As a convenience to hydrogeological studies, the aquifer is divided into strips perpendicular to the coast, each of these being 2 km wide (Figure 12.2). The lateral extension of the aquifer varies from approximately 5 km in the north to about 30 km in the south. Figure 12.3 shows a typical hydrogeological cross section of the aquifer, including the geometry of lithological structures, the water table, and seawater intrusions into different subaquifers. The average thickness of the aquifer is approximately 150 m, decreasing in both eastern and northern directions. The aquifer consists of a Quaternary sequence of marine and continental deposits composed predominantly of calcareous sandstone resting on impermeable black shales and clays of the Saqiye group formed in the Pliocene-Miocene age. The western part of the aquifer, extending from the shoreline to about 4 km inland, consists of impermeable (mostly clay) horizons, which subdivide the aquifer into generally four subaquifers (Figure 12.3) herein denoted A (phreatic), B (phreatic or confined), and C and D (both primarily confined).

To date, the exploitation of the aquifer is being carried out solely from the upper sub-aquifers (A and B), which are also exposed to heavy agricultural and urban/industrial pollution. The existence of fresh or brackish water in lower subaquifers (C and D), often below seawater intrusion (this phenomenon is referred to in hydrogeological literature as hydrological reversal or inversion), was generally known from a few deep observation wells, in which the salinity measurements were conducted at appropriate depths. However, this water has never been exploited, despite the permanently growing deficit of fresh water within the upper subaquifers, and despite the fact that the water in C and D is almost free of any agricultural or industrial pollution. There are two main reasons for such discrimination, both of which result from the extremely sparse data regarding water salinity within the lower subaquifers:

1. The true dimensions of fresh/brackish water bodies within C and D are practically unknown.
2. The hydrological connection between the sea and the lower subaquifers is unclear.

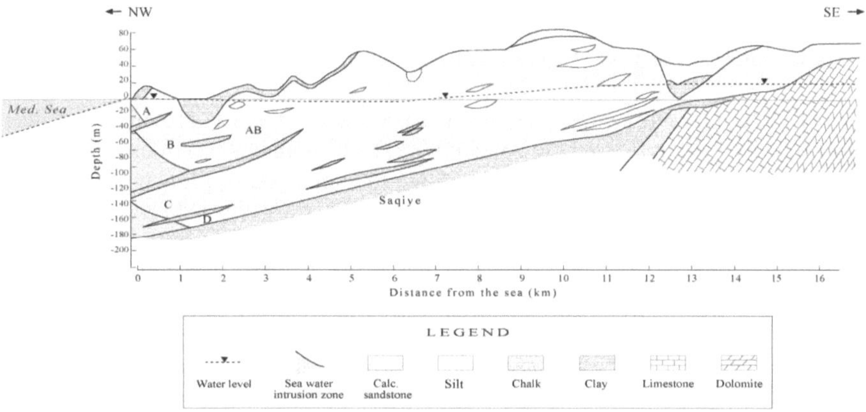

Figure 12.3. Typical hydrogeological cross section of the Mediterranean Coastal Aquifer.

Both problems are crucial for rational management of the lower subaquifers, the natural recharge of which is far slower than that in the upper ones. It is clear that the exploitation of deep subaquifers becomes efficient only where the amount of fresh/brackish water is significant and where, in addition, the subaquifers are hydrologically disconnected from the sea (e.g., as a result of the known thickening of clay layers towards the sea—Kolton, 1988).

To resolve both of the above-mentioned problems, an extensive TDEM survey, including almost 100 central loop soundings, was recently carried out along the Mediterranean coastal plain. In contrast to all previous surveys, most of the measurements were conducted as close as possible to the shoreline, using 100 m by 100 m transmitter (Tx) loops. In the previous surveys, the Tx size was dictated mainly by the estimated depth to fresh/sea water interface in the upper subaquifers. Therefore, for the most part, those few measurements, carried out very close to the seashore in the previous surveys, did not provide reliable information about the deep subaquifers because of their limited penetration depth. Although the required information has been obtained in numerous measurements remote from the seashore, it was insufficient to prove whether C and D are connected to the sea or not. Indeed, if the water in lower subaquifers is fresh far away from the sea, it does not necessarily mean that there is no seawater intrusion closer to the seashore. However, if this is the case near the sea, it is reasonable to assume that C and D are blocked towards the sea in the vicinity of the measurement point. In any case, the geophysical data must be supported by relevant hydrogeological/geochemical information (Merkado, 1989, Yechieli et al., 1997).

To evaluate the amount of fresh/brackish water in lower subaquifers, we carried out a number of TDEM soundings along selected lines perpendicular to the seashore. The results of these measurements are presented below, in the form of quasi-2-D resistivity cross sections.

12.2.3 COMPARISON OF TDEM RESISTIVITY AND BOREHOLE CONDUCTIVITY/ SALINITY MEASUREMENTS

Fourteen TDEM measurements have been carried out in close proximity to deep observation wells. Salinity (borehole conductivity) measurements were performed in all four subaquifers. In all cases, the interface within the upper subaquifers was accurately detected by TDEM. Figure 12.4 is a typical example, showing a drop in resistivity below the threshold value of 2 ohm-m at a depth of 50 m. At exactly the same depth, the borehole electrical conductivity shows the top of the seawater intrusion. It should be noted, however, that because of limitations in the layered TDEM interpretation, the drop in TDEM resistivity sometimes occurs within the diffusion zone and not exactly at the top of the seawater intrusion, as in the case under consideration. Note that the compared data are shown in reciprocal units, namely borehole conductivities (σ) in mS/cm and TDEM resistivities (ρ) in ohm-m. Such a presentation is standard for both techniques, and the transition from one unit to the other is very simple: $\rho = 10/\sigma$.

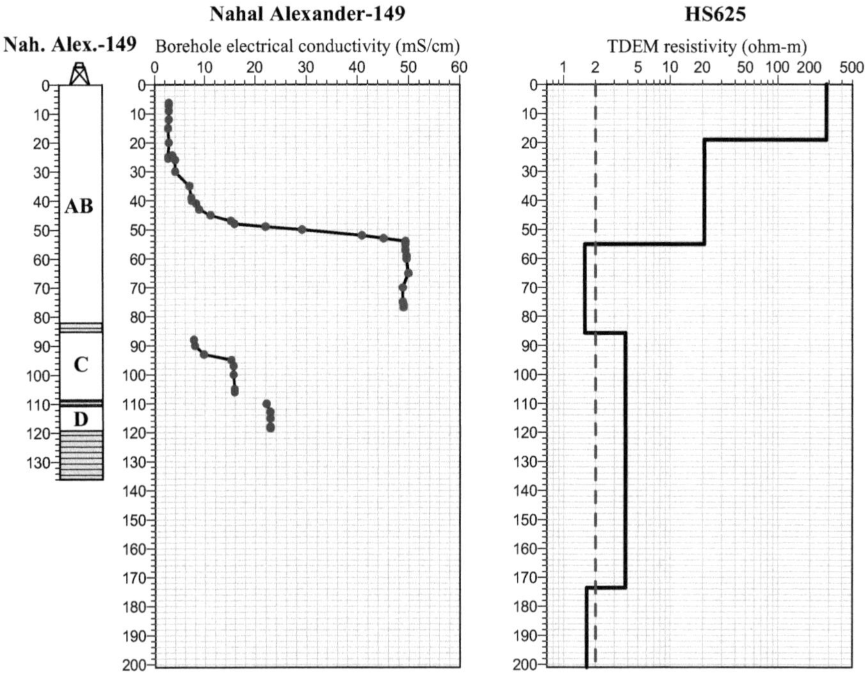

Figure. 12.4. Comparison of TDEM resistivities and borehole conductivity measurements. Dashed line roughly represents the threshold resistivity value for seawater. The TDEM resistivities perfectly fit borehole conductivities within the entire aquifer.

In all comparisons, the resistivity of seawater-bearing parts of the upper subaquifers (formation resistivity) varied within a fairly narrow range, between 1.5 to 1.9 ohm-m. The average resistivity value of 1.7 ohm-m is thus identical to that obtained in the first feasibility study survey, in which the TDEM results where correlated with 39 mostly shallow (within subaquifers A and B only) observation wells (Goldman et al., 1991).

The appropriate borehole conductivity (fluid conductivity) is approximately 50 S/cm, which corresponds to a resistivity of 0.2 ohm-m. Thus the formation resistivity is roughly an order of magnitude greater than the fluid resistivity. This difference is mainly dictated by the porosity of the aquifer and is quantitatively described by Archie's equation (see Chapter 4 of this volume for more details).

Figure 12.4 also shows an increase in the resistivity below seawater intrusion, an observation that completely fits the distribution of salinities measured in the observation well. Indeed, the salinities within C and D decrease roughly 2–3 times, and the TDEM resistivities increase approximately 2 times at appropriate depths (below 86 m). Unfortunately, unlike the upper subaquifers, the TDEM resistivities do not always fit the borehole conductivities measured within the lower subaquifers. Figure 12.5 shows an example in which the borehole salinities/conductivities within the lower subaquifers contradict the appropriate TDEM resistivities (note that the coincidence in the upper subaquifers is still good). According to the borehole conductivity values, the water in the upper subaquifers is fresh up to a depth of about 70 m. Then the salinity increases somewhat at the bottom of B, and afterwards decreases again within deep subaquifers C and D. Thus, according to the borehole conductivity measurements, no interface is detected within the entire aquifer. The TDEM resistivity accurately characterizes the salinity distribution within A and B, but then dramatically drops somewhere in the middle of CD to the values typical of seawater-saturated lithologies. A similar picture was observed in approximately 20% of the comparisons made. It is important to emphasize that in all these cases, the TDEM resistivities within C and D consistently indicated much higher salinities than those actually measured in the observation wells.

The most reasonable explanation of this phenomenon may be related to the fact that the total thickness of clays within the lower subaquifers is usually much greater than that within the upper ones. Because of the very low hydraulic conductivity of clays, they can be saturated with fossil saline water, even if the water within the aquiferous parts of the subaquifer is fresh. Experience shows that resistivities of aquicludes and aquitards saturated with seawater are similar, if not identical, to those of the seawater-saturated aquifers, and both are completely different from any freshwater-saturated or dry geological units (see Chapter 4 of this volume). Also, taking into account the extremely low sensitivity of the TDEM method to the existence of thin resistive layers, it is reasonable to assume that, in the case under consideration, the TDEM resistivities in fact characterize clay layers within deep subaquifers rather than freshwater-saturated aquiferous horizons.

The ability of TDEM to accurately characterize the water salinity within C and D thus strongly depends on the aquifer/aquiclude thickness ratio, as well as on the depth to the lower subaquifers and on the salinity distribution within the upper subaquifers. All these parameters are rarely known *a priori* and can only be partially determined by TDEM. Therefore, the application of TDEM for detecting groundwater salinity within the lower subaquifers at separate locations seems questionable in this environment. However, since TDEM consistently overestimates salinities in C and D, which typically occur only relatively rarely (at approximately 20% of the comparison measurements), the method is expected to be feasible for regional salinity mapping, where a few outliers can be easily recognized.

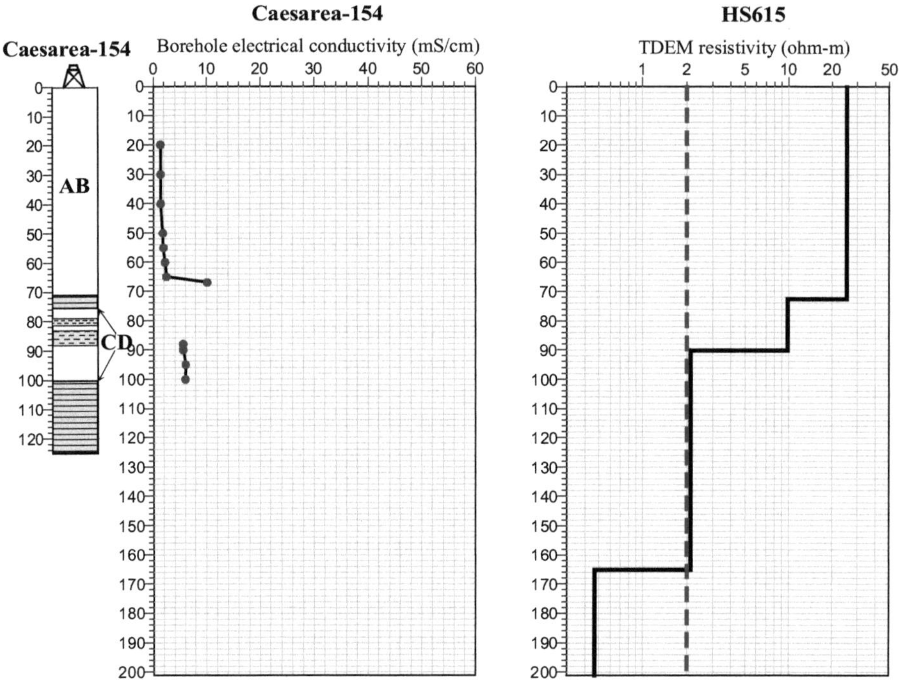

Figure 12.5. Comparison of TDEM resistivities and borehole conductivity measurements. Dashed line roughly represents the threshold resistivity value for seawater. In this example, the TDEM resistivities fit borehole conductivities in the upper subaquifers (AB) only.

12.2.4 MAPPING GROUNDWATER SALINITIES IN LOWER SUBAQUIFERS

An attempt to characterize groundwater salinity within the lower subaquifers at the regional scale has recently been made, along a line approximately 70 km long, between the city of Ashdod in the south and the city of Caesarea in the north (Figure 12.2). The survey included 60 TDEM measurements, most of which were carried out fairly close to the shoreline. The measurements clearly distinguished between two large areas: the southern area extending approximately between coordinates 140 and 187 and the northern area between coordinates 190 and 206 (Figure 12.2). The overwhelming majority of the TDEM measurements in the southern area clearly delineated fresh-to-brackish groundwater within the lower subaquifers, whereas all the measurements in the northern area showed significantly greater salinity (lower resistivity) within the lower subaquifers. Despite the presence of a few outliers in the southern area, which can be easily explained by the above-mentioned limitation of the TDEM method, the area is recommended for exploitation of the lower subaquifers, both for a direct use (provided the water is sufficiently fresh) or for a further desalination (if the water is brackish) (Kafri et al., 2003). It is important to emphasize that, because of the slow recharge rate into the lower subaquifers, particular caution would need to be exercised both before and during any extensive exploitation of the subaquifers.

To evaluate the expected amount of fresh/brackish groundwater within the lower subaquifers in the area, a number of TDEM traverses were collected perpendicular to the shoreline. Three such traverses, which have been measured to date, indicate a significant amount of relatively fresh water, comparable to that in the overexploited upper subaquifers A and B (see Figure 12.6 as an example). Unfortunately, no validation of this observation is presently available, due to the very limited number of deep observation wells in which salinity measurements have been carried out within C and D.

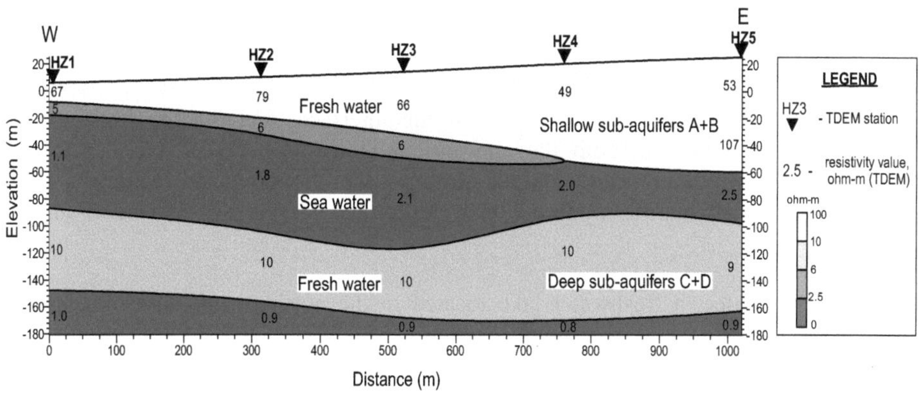

Figure 12.6. Quasi-2-D resistivity cross section along the TDEM traverse at the Ziqim site.

12.2.5 SUMMARY AND CONCLUSIONS

TDEM resistivities below 2 ohm-m uniquely characterize salinity associated with seawater intrusion into upper subaquifers A and B. The depth to the interface within the upper subaquifers is detected by TDEM measurements alone, with accuracy similar to that of direct borehole measurements. In most cases, TDEM is also able to characterize groundwater salinity within the lower subaquifers. However, if the subaquifers are fully or partially saturated with fresh or brackish water, the thickness of which is relatively small compared to the thickness of clays within C and D, the resolution of TDEM may be insufficient to delineate the fresh/brackish parts of the subaquifers. Since this phenomenon is relatively rare (found in approximately 20% of the calibration measurements), and since the TDEM measurements consistently overestimate the actual salinities within the lower subaquifers, TDEM can still be used to characterize groundwater salinity in C and D at the regional scale. The appropriate regional mapping of the Mediterranean coastal plain of Israel indeed allowed the delineation of large areas with potentially fresh/brackish groundwater within the lower subaquifers, despite a few outliers showing apparently saline water in zones C and D in these areas.

12.3 Delineating Saline Groundwater within Sub-bottom Sediments in the Sea of Galilee, Israel, Using a Marine TDEM System

12.3.1 INTRODUCTION

In contrast to seawater intrusion into coastal aquifers, where the hydrodynamics are well understood and thus the geometry of freshwater/seawater interface is predictable, the interface existing beneath the Sea of Galilee (a freshwater lake) and its vicinity has a complicated geometry, one that cannot be easily explained by groundwater hydrodynamics. In this area, ancient brine has entrapped in the impermeable sediments beneath the lake, but has washed out from adjacent permeable sediments. Groundwater hydrodynamics at the lake vicinity, while depending mainly on a topography-driven flow system, are also influenced by compaction-driven and density-driven flow systems.

To delineate the complicated geometry of the interface, and thereby to study the groundwater flow system, both the conventional terrestrial TDEM (similar to that applied in the Mediterranean Coastal Aquifer—see the previous section) and its specially developed marine modification were applied for the Sea of Galilee and its vicinity.

The Sea of Galilee, a freshwater lake located in the northern part of Israel, has an average salinity, expressed in chloride concentration, of 230 mgCl/L. This is an order of magnitude higher than the concentration in the Jordan River and other streams entering the lake. This order-of-magnitude difference is a result of saline water sources entering the lake. The saline sources supply less than 10% of the lake water, but almost 90% of its salts. The saline sources are the onshore and offshore saline springs, located mostly at the northwestern coast, and seepage of saline water from the lake's bottom sediments. The objective of this study was to delineate the saline groundwater beneath the lake. The surface-deployed marine modification of the TDEM method was used to map the spatial distribution of brines in the sediments below the lake. As will be shown below, saline groundwater was detected within low-permeability sediments at a very shallow depth (about 10 m) under 85% of the lake area. It was hypothesized that the saline water is a relic of a former saline lake that existed in this area about 20,000 years ago (during the last glacial period).

12.3.2 SITE DESCRIPTION

The Dead Sea rift is a left-lateral, strike-slip transform, separating the Sinai-Levant sub-plate from the Arabian plate (Garfunkel, 1981). The rift includes several rhomb-shaped grabens, one of which contains the Sea of Galilee, which is the world's lowest freshwater lake (210 m below MSL). This lake supplies 500 million cubic meters of water per year, which is about 25% of Israel's annual consumption. The lake quality is degraded by saline water sources that enter the lake. These include onshore and offshore springs with chloride concentrations in the range of 300–18,000 mg/L (Gvirtzman et al., 1997a; Rimmer et al., 1999), as well as areal seepage from sediments beneath the lake (Stiller, 1994).

A gradual increase in chloride concentration with depth was observed in three 5.0 m cores drilled to sediments within the lake basin. At the water-sediment interface, chloride concentrations are 230 mg/L, increasing to 2,000-3,500 mg/L at 5 m depth (Stiller, 1994). No deeper sediment cores are available. This study's purpose was to use the time-domain electromagnetic (TDEM) geophysical method as a "remote sensing" technique to detect the spatial distribution of brines in the sediments below the lake.

12.3.3 MARINE TDEM SYSTEM

To the best of our knowledge, all previous marine TDEM surveys have been carried out using a transmitter and/or receiver antenna located on the sea floor. Taking advantage of the fact that the water in the Sea of Galilee has low salinity (and therefore low conductivity), Goldman et al. (1996) developed a new effective marine TDEM system (Figure 12.7).

This system was operated for the first time in the Sea of Galilee. The array consists of a 25 m diameter circular transmitter loop with the receiver antenna located in the center. The system floats on the water surface and is towed by a motorboat. Navigation was carried out with a highly accurate differential global positioning system (GPS). About 300 onshore and offshore soundings were performed within the lake and its surroundings (Figure 12.8). The density of the soundings was determined by the observed areal variations in the apparent resistivity during measurement.

12.3.4 GEOELECTRIC RESULTS

Results are presented as 2-D resistivity cross sections at selected profiles (Figure 12.9) and as resistivity maps at six different depths below the lake's floor (Figure 12.10). Resistivity values may be divided into two major units, distinguished by high and low resistivity. The high resistivity region is plotted in white and light gray colors and has values greater than 2.5 ohm-m; the low resistivity is plotted in black and dark gray colors and has resistivities lower than 2.5 ohm-m. High resistivity was found at the upper few meters of the sediments in the internal part of the lake, at a major part of the explored section along the margin of the lake, and obviously the water of the lake itself (Figures 12.9 and 12.10). In fact, the lake water has typical resistivity values of 7 to 12 ohm-m, usually closer to 10 ohm-m. Along the northeastern and northwestern margins of the lake, high resistivity was detected down to the entire explored depth (approximately 100 m), whereas along the northern and eastern margins of the lake, high resistivity is found only in the upper 70 to 90 m of the profile. In the southeast part of the lake, high resistivity was not detected below the lake bottom.

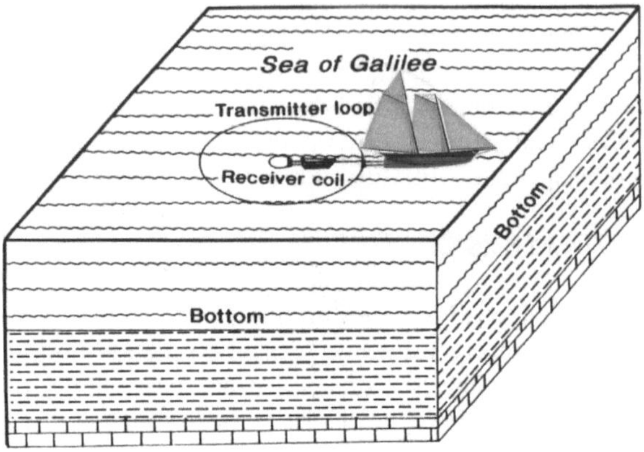

Figure 12.7. Schematic diagram showing the marine TDEM system

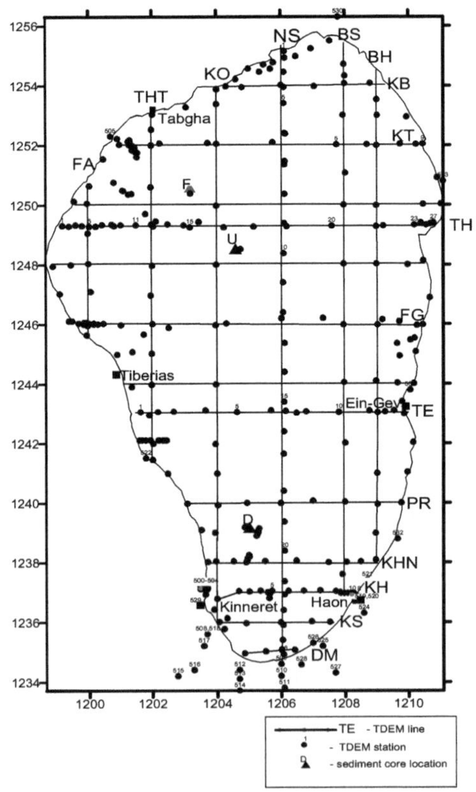

Figure 12.8. Location map of the off- shore and on-shore TDEM measurements

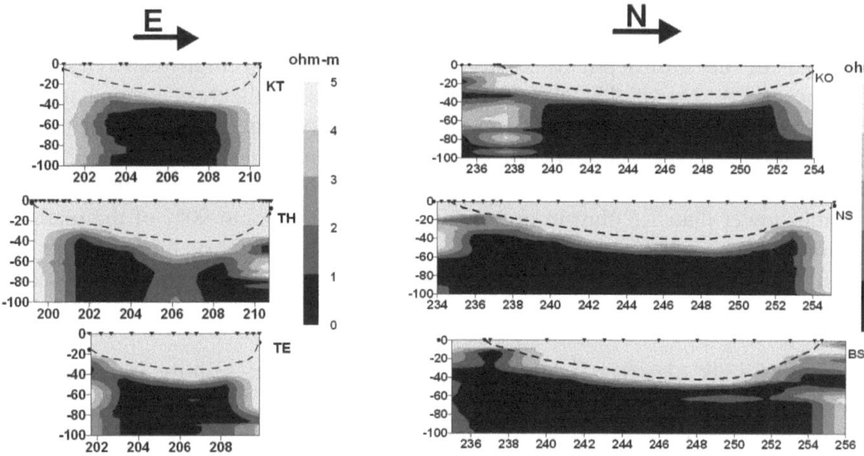

Figure 12.9. Pseudo-2-D resistivity cross sections in the W-E direction (left picture) and in the S-N direction (right picture). Dashed line in each cross section represents the lake bottom.

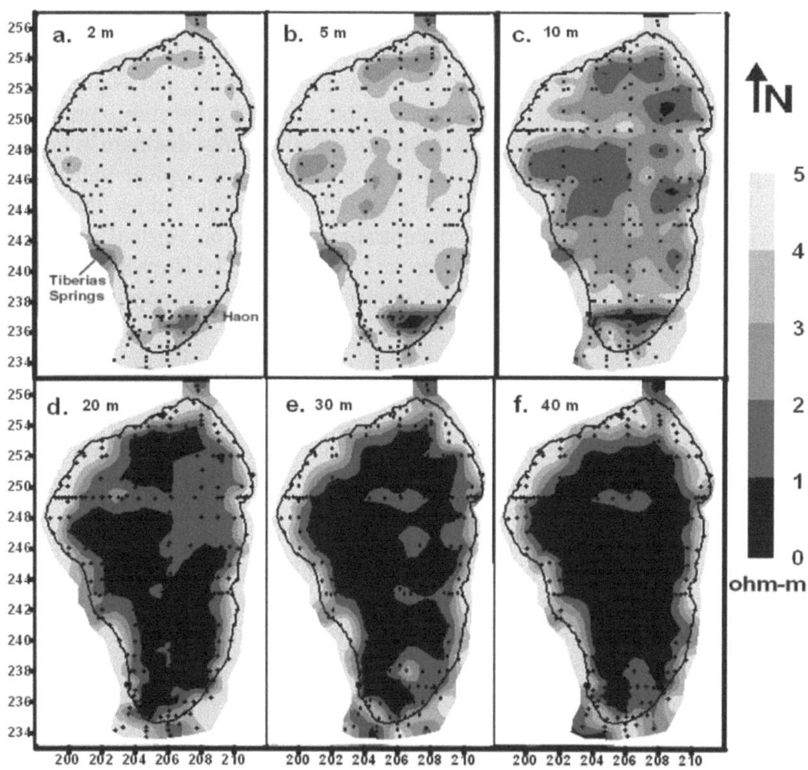

Figure 12.10. True resistivity maps at different depths beneath the lake bottom

A prominent feature of most 2-D cross sections is the transition between high and low resistivity zones, which are relatively parallel to the lake bottom within the central part of the lake. In the 2 m depth resistivity map (Figure 12.10), high resistivity covers almost the entire area of the lake. The exceptions are small areas located near the Tiberias Springs and Haon. In the 10 m depth resistivity map, nearly half of the area under the lake is black and dark grey, indicating resistivities less than 2.5 ohm-m (Figure 12.10). In the 20 to 40 m depth maps, areas on the center of the lake with resistivities lower than 1.5 ohm-m occupy approximately 70% to 90% of the total lake area. Note that, except for the southern and southwestern parts of the lake, the high resistivity is located along the entire perimeter of the lake to a distance of approximately 2 km from the shore.

Calibration of bulk electrical resistivity against groundwater salinity is required for analyzing the spatial distribution of saline groundwater. Previous studies have suggested that a resistivity value of 1 ohm-m corresponds to a TDS (total dissolved solid) concentration of approximately 20,000 mg/L (Goldman and Neubauer, 1994). Although resistivity values are related to the TDS content, only chloride concentration data were available for calibration in the Sea of Galilee. The proportion of chloride out of the TDS, as measured in onshore springs around the lake, is about one third.

Resistivity values in the sub-bottom sediments of the Sea of Galilee lower than 1.5 ohm-m may be uniquely attributed to saline water, because no lithology can have such a low resistivity unless saturated with saline water (Goldman and Neubauer, 1994). Resistivity values greater than 1.5 ohm-m can be caused by varying combinations of pore-water salinity and lithology. Thus, resistivity alone cannot distinguish between the brackish groundwater and some lithologies (e.g., clays with typical values of 2.5–5 ohm-m). Therefore, the calibration correlated only resistivity values lower than 1.5 ohm-m with the appropriate salinity concentrations. Based on vertical cores from the lake-floor (Stiller, 1994) samples taken at Ha'on-2 well, Hurwitz et al. (1999) concluded that values of 1.5, 1, and 0.5 ohm-m correspond to approximately 7,000, 11,000, and 22,000 mgCl/L, respectively.

12.3.5 DISCUSSION

Previous geological studies (Stein et al., 1997) have suggested that between 70,000 to 18,000 years ago, Lake Lisan covered a large area of the Dead Sea rift valley. This former lake extended from the Sea of Galilee in the north to the Dead Sea in the south. The salinity of the lake varied between 100,000 to 300,000 mg/L TDS throughout most of its history. After the retreat of the lake, two residual lakes remain: the flow-through, freshwater Sea of Galilee in the north and the terminal, hypersaline Dead Sea in the south. It has been proposed that the saline groundwater detected in the low-permeability sediments beneath the Sea of Galilee is a relic of this former lake (Stiller, 1994). Since the Sea of Galilee formed, approximately 18,000 years ago, freshwater has covered the sediment-hosted saline groundwater. Consequently, diffusion of salts from the sediment into the lake has been initiated. This study verifies that a saline lake, probably Lake Lisan, once covered the entire area covered by the Sea of Galilee.

The resistivity values measured within the present lake basin show the present horizontal and vertical distribution of the interstitial saline groundwater, which is probably different from its original spatial distribution. The most significant changes, both in brine concentration and in depth distribution, are concentrated at a distance of 1 to 2 km from the shore, where the low resistivity values are either absent or located at depths of a few tens to a few hundreds of meters. Investigators have proposed that the sharp lateral contrasts between high and low resistivities at some sites along the margins of the lake reflect faults separating the clastic and carbonate units. Indeed, these interfaces coincide with major faults (Hurwitz et al., 2002).

It is hypothesized that groundwater advection from the regional aquifers is the dominant transport mechanism along the lake's margins. High resistivity values found at the marginal part of the lake indicate that if highly concentrated brines were originally located there, they have been leached. The different transport mechanisms are also confirmed by flow measurements. In the central part of the lake, measured upward fluxes are two orders of magnitude lower than those measured off the western coast (Stiller et al., 1975; Nishri et al., 1997).

12.3.6 SUMMARY AND CONCLUSIONS

The subsurface electrical resistivity distribution beneath the Sea of Galilee suggests that at the margins of the lake, groundwater is much less saline than beneath the central part of the lake. We suggest that this is a result of different mass-transport mechanisms. In the margins, advection of diluted saline groundwater by regional flow discharging into the lake basin is the major transport mechanism; whereas in the central part of the lake, salt flux into the lake is mainly by diffusion. In the southern part of the lake, resistivity measurements suggest that the saline unit is located at shallower depths and is probably more concentrated. The subsurface vertical interfaces between fresh and saline groundwaters near the lake margin are probably fault controlled, placing high-resistivity, permeable carbonate units in contact with low-resistivity, low-permeability sediments. Results also indicate that a saline water body, Lake Lisan, has covered most if not all of the area underlying the present lake.

12.4 Magnetotelluric Imaging to Locate a Deep Aquiferous Graben, for a Large-Scale, Urban-Water-Supply Scheme in a Semi-arid Region of Northeastern Brazil

12.4.1 INTRODUCTION

The deep structural control on groundwater distribution in the semi-arid northeastern margin of Parnaiba Basin in Brazil (Figure 12.11) is poorly understood. Parnaiba Basin is mainly Palaeozoic in age and filled with clastic siliceous sediments of mostly continental origin, deposited in five major depositional cycles from the Upper Ordovician to the Cretaceous. Three of the megacycles of deposition are represented by the Serra Grande (Silurian), Caninde (Devonian) and Balsas (Permian-Carboniferous-Triassic) Groups. The total sedimentary thickness is about 3,400 m near the center of the basin, with the Serra Grande and Caninde Groups reaching up to 2,900 m in thickness and consisting mainly of sandstones with subordinate siltstones and shales (see Meju et al., 1999, Figure 2). The Serra Grande Group contains the main aquifers in the region and is exposed only in the southern and eastern parts of the oval-shaped basin. In the semi-arid marginal parts of the basin, the major problem is how to identify those areas with thickened aquiferous formations, which may be developed for groundwater supply to nearby urban populations.

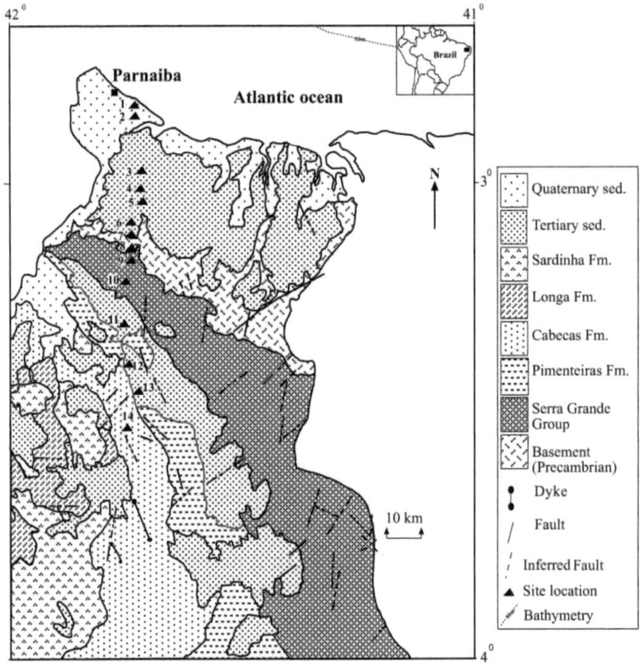

Figure 12.11. Geological map of the northern margin of Parnaiba Basin, Brazil. The distribution of TDEM-MT stations in the area of study is also shown (Mohamed et al., 2002).

The deep aquifers and their confining formations in this basin have contrasting electrical resistivities (see Meju et al., 1999), enabling them to be mapped successfully with a combined transient electromagnetic (TDEM) and magnetotelluric (MT) approach. Joint TDEM-MT investigation of the southeastern margin of the basin had previously identified linear trough-like zones of enhanced conductivity, interpreted as major grabens, close to areas with significant human populations (Fontes et al., 1997; Meju et al., 1999). Subsequent follow-up drilling by a Brazilian government agency confirmed this interpretation (see, for example, Fontes et al., 1997) and encouraged the regional (Piaui) state government to commission further regional investigations for any such occurrences of potentially aquiferous grabens that may serve for large-scale urban groundwater supply. Since the joint TDEM-MT profiling approach has been successfully used to guide deep drilling (ca. 1 km) for groundwater for urban supply schemes in the region (Fontes *et al.* 1997), the adopted exploration model is a conductive trough-like feature of regional extent that may coincide with known fault zones or hitherto undiscovered grabens and contain thick enough aquiferous materials (at least 1 km deep) to warrant development by deep drilling. The survey of the northern margin is described here, since it illustrates the cost-effective target identification capability of TDEM-MT necessary for deep groundwater resource development.

12.4.2 FIELD SURVEY AND DATA ANALYSIS TECHNIQUES

Joint TDEM and MT soundings were conducted at 14 locations along a N-S line extending from the Atlantic coast onto the hinterland for 95 km (Figure 12.11), across the northeastern margin of the Parnaiba basin (Mohamed et al. 2002). The average station spacing was 7 km. The TDEM and MT soundings were centered on the same positions. The MT data were measured in the magnetic east-west and north-south directions. The electric dipoles were 50 or 100 m long, depending on site constraints. Magnetic field components were recorded using induction coils. The recording bandwidth was 0.01 to 176 Hz. TDEM data were necessary to constrain the structure of the top 100 m and also served to correct MT data on static shift problems arising from near-surface heterogeneities (see Berdichevsky and Dmitriev 1976; Jones 1988), so as to obtain reliable resistivity-depth estimation during 2-D MT imaging.

The MT data were processed using standard tensorial techniques, and the relevant interpretative parameters were computed for the sounding frequency range 176 to 0.01 Hz. In terms of continuity of data points and size of data errors, the data are of good quality at all the stations. The main lithological structural trend in the field area is WNW-ESE. However, it is conventional practice to gauge the structural dimensionality or "electrical strike" directly from the data recorded at each station, using the Groom and Bailey (1989) tensor decomposition method. The strike directions determined at measurement frequencies higher than 0.3 Hz (using the Groom-Bailey method) showed a dominant electrical strike that is fairly consistent with the main WNW-ESE lithological trend. The MT data were therefore rotated to this direction, yielding the appropriate dual polarization data required for 2-D imaging (see Mohamed et al., 2002, Figure 4). As a final preparation for 2-D inversion, these assumed dual-polarization apparent resistivity data sets were first corrected for static shift, using the TDEM data (e.g., Meju et al., 1999; Meju, 2002; Mohamed et al., 2002), and then projected onto a profile perpendicular to the adopted strike (i.e., trending N37°E) so that the distances between the stations are different from those shown on the actual site location map (Figure 12.11).

12.4.3 TWO-DIMENSIONAL MT IMAGING AND INTERPRETATION

Simultaneous inversion of dual-polarization MT data (Mackie et al. 1997) was done using several initial models with the design of the 2-D calculation mesh optimized for the top 30 km of the subsurface. On the ocean-side, the 2-D grid made use of the available bathymetric data and incorporated a surface conductive slab of 2.5 ohm-m, representing the seawater layer (cf. Arora et al., 1999). Initial trial inversions were done to determine the best regularization parameters (in terms of model smoothness and data fit) for this particular data set, which helped to guide the subsequent inversion studies (see Mohamed et al., 2002). An example of an optimal least-squares 2-D inversion model is presented in Figure 12.12 (only the top 5 km is shown for convenience).

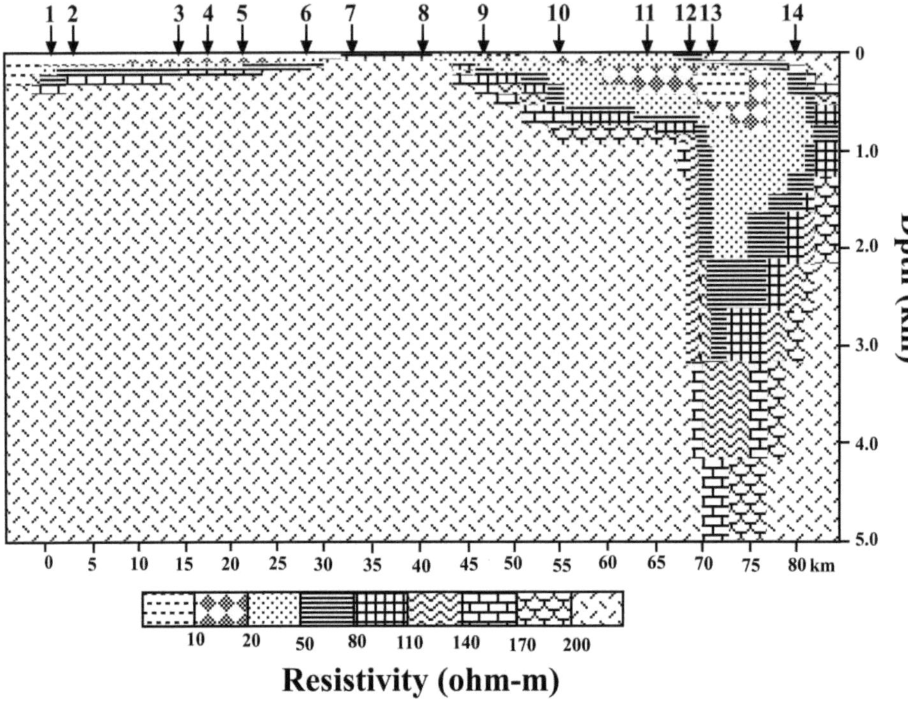

Figure 12.12. A resistivity model derived from a 2-D inversion for the northern margin of Parnaiba Basin (after Mohamed et al., 2002)

The match between the calculated model responses and the observed data is good, and has been shown elsewhere (see Mohamed et al., 2002). It was found that the unweathered crystalline basement in the area of study is highly resistive, with a minimum value of 200 ohm-m, and may be traced to the north and south from that part of our model coinciding with the zone of basement outcrop on the geological map (Stations 7 and 8). Relatively conductive sedimentary cover rocks appear to increase in thickness away from the zone of basement outcrop, as expected from the geology; note that stations 1 to 6 are believed to lie in what is conveniently dubbed by Goes et al. (1993) as the Barreirinhas basin. The cover units extending from stations 1 to 6 probably correspond to Quaternary and Tertiary sediments, with the uppermost section being very conductive ($\leq$ 10 ohm-m) underneath Stations 1 to 3 (a feature constrained by TDEM data). Southward from Station 8, the basement cover units show considerable thickening, starting from Station 12 and attaining a maximum thickness of about 2 km underneath Station 13. This is anomalous, since the projected sedimentary cover thickness (based on the studies conducted further south by Góes et al., 1993) would be about 1 km, just beyond our Station 14. These cover units exhibit some stratification, consisting of an uppermost resistive (>150 ohm-m) sliver starting near Station 12 (and correlating with the zone of outcropping Cabecas sandstone on the geology map); an underlying conductive (<20 ohm-m) unit (correlated with the Pimenteiras shale, which is draped over by Quaternary sediments at Station 11); and a basal unit of moderate resistivity (ca. 20–50 ohm-m). This basal unit may represent thickened segments of the Serra Grande formations, which outcrop at Stations 9 and 10, or a combination of the Serra Grande group and pre-Silurian sediments, as suggested for deep grabens found elsewhere in the basin (see De Sousa, 1996; Ulugergerli, 1998; Arora et al., 1999; Meju et al., 1999).

12.4.4 SUMMARY AND CONCLUSIONS

The main features of this model are the relatively conductive trough-like feature in the top 2 km of the subsurface at Stations 12 to 14 and the highly resistive arch-like feature underneath Stations 6 to 8. The highly resistive body beneath Stations 6 to 8 coincides geographically with the zone of basement outcrop on the geological map (cf. Figure 12.11). The trough-like feature may possibly be related to a basin-bounding fault or graben (Mohamed et al., 2002). Based on the studies conducted further south by Góes et al. (1993)—using data gathered from surface geology, geochemical surveys, exploratory wells, and seismic sections—the projected sedimentary cover thickness would be about 1 km just beyond our Station 14. The presence of thickened (ca. 2 km) conductive electrical units near Stations 12 and 13 in our MT models (inferred to be sedimentary cover rocks) is therefore anomalous; it is best explained by the existence of a hitherto undiscovered graben-like structure. The Serra Grande group contains the most important aquifer in Parnaiba basin, and this anomalously thickened zone may thus have hydrogeological implications. It was recommended that this graben-like feature should be investigated by drilling, since it may contain aquiferous materials as found elsewhere in the basin (Mohamed et al., 2002).

12.5 Using High-Resolution Seismic Reflection Data for Studying the Aquifer Structure in the Western Mesaoria Area of Cyprus

12.5.1 INTRODUCTION

The investigated site is located in the vicinity of the village of Meniko, southeast of Nicosia in Cyprus (Figure 12.13). The upper part of the geological section in the area is composed mainly of Pliocene to Middle Miocene marls of the Nicosia Formation. The formation is subdivided into two units. The upper unit is composed of marls with inclusions of thin beds of gravels and fine sandstones. The lower unit consists of gray clays and marls, and includes a thin chalk layer. Between the two units, a phreatic aquifer is developed in a clastic horizon consisting of gravels and sands. The depth to the aquifer layer is about 200 m, and its thickness varies from 0 to 80 m. The Nicosia Formation lies unconformably on the Pillow Lavas of the Troodos Ophiolite Complex. The area suffers from an insufficient supply of fresh water. A number of wells drilled in the area produced reasonable quality groundwater used for agriculture. At the same time, other wells drilled in the area proved to be dry or produced water of poor quality.

This case study reports the use of seismic reflection data for mapping the aquifer structure to aid in the water resource assessment.

12.5.2 SEISMIC SURVEY

To understand the hydrological situation in the area, a seismic reflection survey was carried out at the Meniko site (Figure 12.13). The survey was performed within the framework of the INCO-DC program (Grant # 951076) with the objective of studying the general structure of the aquifer in the area and, in particular, detecting and mapping the water-bearing layers and delineating the contact between the sediments of the Nicosia formation and the igneous rocks.

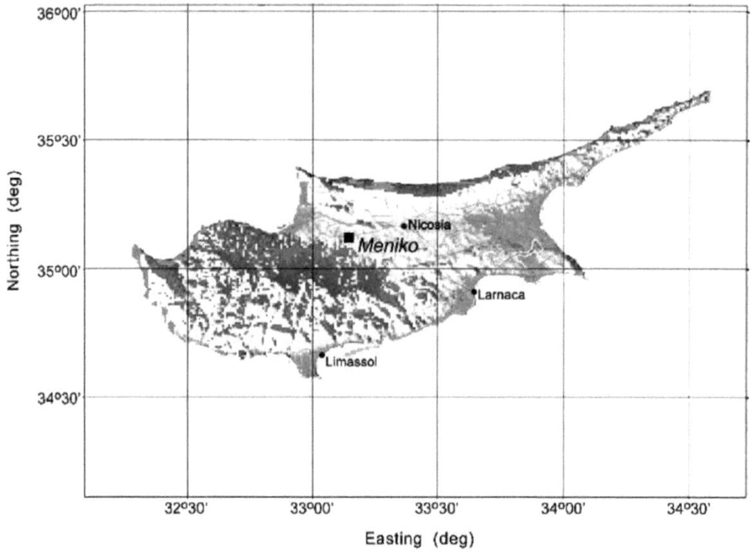

Figure 12.13. Schematic location of seismic reflection survey at the Meniko site in the Western Mesaoria area of Cyprus

12.5.3 DATA ACQUISITION AND PROCESSING

The survey included acquisition of two seismic reflection lines, which were shot with a 48-channel recorder using explosives as signal sources. The receivers were single 10 Hz geophones; the source and receiver spacing was 10.0 m. Figure 12.14 shows a seismic time section along one of the lines. The section presents a rather complicated structural picture of the subsurface. The continuity of the reflection sequence appearing on the section is clearly interrupted at many locations, apparently by a system of faults, as marked on the section. Owing to the fragmented character of the reflections, it is very difficult to trace them along the line. Since no velocity information from boreholes was available in the investigated area, no attempt at depth conversion of the section was made; however, rough estimates of the depth of the reflected events were made on the basis of stacking velocities obtained from the seismic data.

12.5.4 HYDROSTRATIGRAPHIC INTERPRETATION

The seismic line passes in close proximity to three water wells (EB18A, 34/99 and 40/99—Figure 12.14). In borehole EB18A, drilled to the depth of 460 m, a sand and gravel aquifer layer with good water quality was penetrated within the depth range of 225–300 m. Borehole 34/99 was drilled down to the depth of 393 m; brackish water was found in a thin sand layer penetrated at the depth of 236 m. Borehole 40/99 reached the depth of 386 m. The clastic horizon of the Nicosia formation, consisting of fine sandstone, was encountered within a depth range of 190–220 m; the water in this aquifer was of poor quality. Locations of the three wells are marked above the section in Figure 12.14, and the borehole information was used for correlation of reflections appearing on the section.

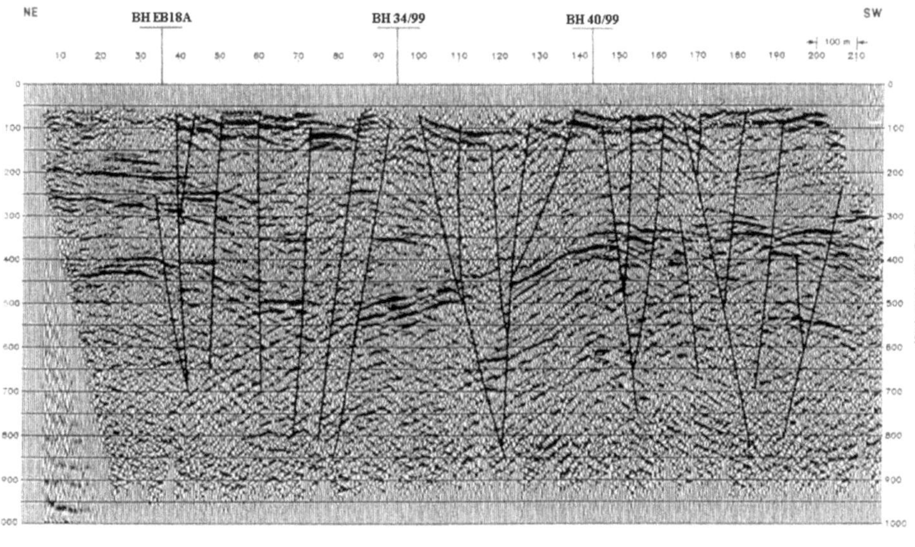

Figure 12.14. Seismic time section along one of the reflection lines at Meniko site. The lines on the section designate the interpreted faults.

In the vicinity of Station 35 the reflections can be correlated to the geological layers penetrated in borehole EB18A (Figure 12.15). The two shallow reflectors, appearing at times of about 180 ms and 210 ms, are interpreted to be two thin gravel layers that were encountered in the boreholes at depths of 134 and 165 m, respectively.

The reflector appearing at about 280 ms may be related to the top of the main aquifer layer encountered at a depth of 225 m. To the SW of the borehole, continuity of the aquifer layer is interrupted, apparently by a system of faults. The two deeper reflectors, appearing at times of 400 ms and 440 ms, can possibly be identified with the chalk and clay layers encountered at depths of 363 m and 372 m, correspondingly. The reflector appearing in the vicinity of Station 143 at about 350 ms may be related to the contact of the sediments with the igneous rocks encountered in borehole 40/99 at a depth of 317 m. This reflector has a clear NE inclination. In the vicinity of borehole 34/99 (Station

95), this reflector appears at about 500 ms (about 600 m depth) and therefore was not reached by the borehole.

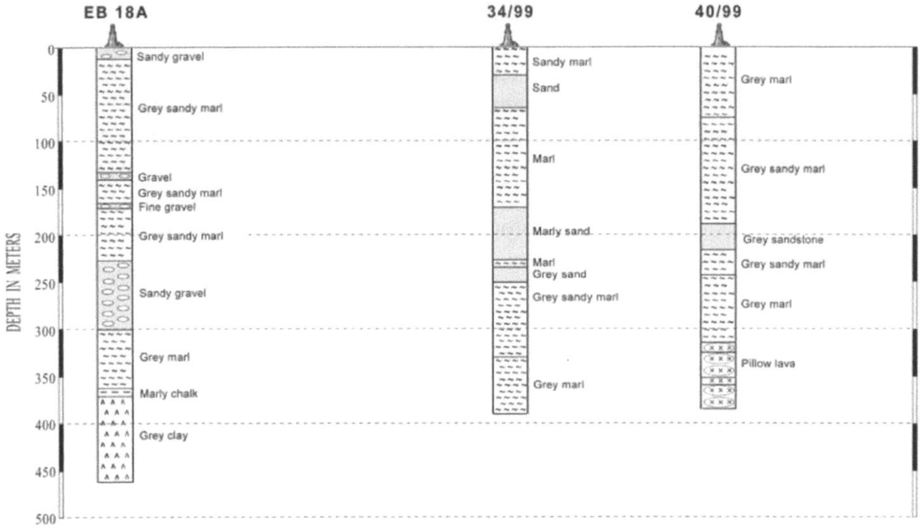

Figure 12.15. Lithological logs in boreholes drilled at Meniko site

The water-bearing sandstone layer encountered in borehole 40/99 at the depth of 190 m, and the thin sand layer encountered in borehole 34/99 at the depth of 236 m, cannot be detected on the seismic section. A possible reason for this may be that the acoustic properties (seismic velocities) of these layers are similar to those of the surrounding marls and, therefore, their contact does not produce any significant reflection.

12.5.5 SUMMARY

The results of the high-resolution seismic reflection survey carried out in the Western Mesaoria area of Cyprus demonstrate how the reflection technique can be used for studying the aquifer structure in the area. The seismic section obtained by the survey represents a detailed two-dimensional structural image of the subsurface, thus providing a good basis for the hydrogeological interpretation. Correlation of the reflections appearing on the section with the borehole data enabled us to relate them to various geological layers penetrated by the wells. In particular, the section clearly shows the lateral extent of the gravel layer related to the main fresh water aquifer. The section also reveals a complicated fault pattern interrupting the continuity of the entire layer system.

Acknowledgments

The geophysical surveys in Israel were supported by the Israeli Water Commission and by the Earth Science Administration of the Ministry of National Infrastructures. The borehole salinity/conductivity data were kindly provided by the Hydrological Service of Israel. The hydrogeological interpretation of the TDEM results in the Mediterranean Coastal Aquifer of Israel was carried out by Uri Kafri. M. Goldman wishes to thank Michael Gendler, Vladimir Fridman, and Israel Gev for invaluable assistance and useful discussions. The MT survey in northeast Brazil was conducted jointly with the National Observatory in Rio de Janeiro; special thanks to Sergio Fontes and Adel Mohamed for their contributions to the field study. We thank P. Styles and A. Binley for their critical reviews, which helped improve the clarity of this chapter.

References

Arora, B.R., A.L. Padilha I. Vitorello, N.B. Trivedi, S.L. Fontes, A. Rigoti, and F.H. Chamalaun, 2-D geoelectrical model for the Parnaiba Basin conductivity anomaly of northeast Brazil and its tectonic implications, *Tectonophysics, 302,* 57–69, 1999.

Bahr, K., Geological noise in magnetotelluric data: A classification of distortion types, *Phys. Earth Planet. Int., 60,* 24–38, 1991.

Bahr, K., Interpretation of the magnetotelluric impedance tensor: Regional induction and local telluric distortion, *J. Geophys., 62,* 119–127, 1988.

Berdichevsky, M.N., and V.I. Dmitriev, Distortion of magnetic and electrical fields by near-surface inhomogeneities, *Acta Geodaet. Geophys. et Montanist. Hung., 11,* 447–483, 1976.

Birkelo, B.A., D.W. Steeples, R.D. Miller, and M. Sophocleous, Seismic reflection study of a shallow aquifer during a pumping test, *Ground Water, 25,* 703–709, 1987.

Bruno, P.P.G., and A. Godio, Environmental risk assessment of a shallow aquifer in Piana Campana (Italy): A field comparison between seismic refraction and reflection methods, *European Journal of Environmental and Engineering Geophysics, 2,* 61–76, 1997.

DeGroot-Hedlin, C., and S. Constable, Occam's inversion to generate smooth, two-dimensional models from magnetotelluric data, *Geophysics, 55,* 1613–1624, 1990.

DeSousa, M.A., Regional gravity modeling and geohistory of the Parnaiba Basin (NE Brazil), Ph.D Thesis, University of Newcastle, 1996.

Fitterman, D.V., and M.T. Stewart, Transient electromagnetic soundings for groundwater, *Geophysics, 53,* 118–128, 1986.

Fontes, S.L., M.A. Meju, J.P.R. Lima, R.M. Carvalho, E.F. La-Terra, C.R. Germano, and M. Metelo, Geophysical investigation of major structural controls on groundwater distribution, north of Sao Raimundo Nonato, Piaui, *5th Internat. Congr. Brazilian Geophys. Soc., Expanded Abstract, 2,* 766–769, Sao Paolo, Brazil, 1997.

Garfunkel, Z., Internal structure of the Dead Sea leaky transform (Rift) in relation to plate kinematics, *Tectonophysics, 80,* 81–108, 1981.

Geissler, P.E., Seismic profiling for groundwater studies in Victoria, Australia, *Geophysics, 54,* 31–37, 1989.

Góes, A.M.O., W.A.S. Travassos, and K.C. Nunes, Projeto Parnaiba: Reavaliacão da bacia e perspectivas exploratórias, *Internal Report,* Petrobras/ DEXNOR-DINTER, 1993.

Goldman, M., A. Arad, U. Kafri, D. Gilad, and A. Melloul, Detection of freshwater/seawater interface by the time-domain electromagnetic (TDEM) method in Israel, *Naturwet. Tijdsehr., 70,* 339–344, W. De Breuck, ed. 1988.

Goldman, M., D. Gilad, A. Ronen, and A. Melloul, Mapping of seawater intrusion into the coastal aquifer of Israel by the time-domain electromagnetic method, *Geoexploration, 28,* 153–174, 1991.

Goldman, M., S. Hurwitz, H. Gvirtzman, B. Rabinowich, and Y. Rotshtein, Application of the marine time domain electromagnetic method in lakes: the Sea of Galilee, Israel, *European J. of Environmental and Engineering Geophysics, 1,* 125–138, 1996.

Goldman, M., and F.M. Neubauer, Groundwater exploration using integrated geophysical techniques, *Surv. Geophys., 15,* 331–361, 1994.

Groom, R.W., and K. Bahr, Corrections for near-surface effects: Decomposition of the magnetotelluric impedance tensor and scaling corrections for regional resistivities: A tutorial, *Surv. Geophys., 13,* 341–379, 1992.

Groom, R.W., and R.C. Bailey, Decomposition of magnetotelluric impedance tensors in the presence of local three-dimensional galvanic distortion, *Journal of Geophysical Research, 94,* 1913–1925, 1989.

Gvirtzman, H., G. Garven, and G. Gvirtzman, Hydrogeological modeling of the saline hot springs at the Sea of Galilee, Israel, *Water Resour. Res., 33,* 913–926, 1997a.
Gvirtzman, H., G. Garven, and G. Gvirtzman, Thermal anomalies associated with forced and free convective groundwater systems at the Dead-Sea Rift Valley, *Geological Society of America Bulletin, 109,* 1167–1176, 1997b.
Hurwitz, S., M. Goldman, M. Ezersky, and H. Gvirtzman, Geophysical (TDEM) delineation of a shallow brine beneath a fresh-water lake, the Sea of Galilee, Israel, *Water Resour. Res., 35,* 3631–3638, 1999.
Hurwitz, S., V. Lyakhovsky, and H. Gvirtzman, Transient salt transport modeling of shallow brine beneath a fresh water lake, the Sea of Galilee, *Water Resour. Res., 36,* 101–107, 2000a.
Hurwitz, S., E. Stanislavsky, V. Lyakhovsky, and H. Gvirtzman, Groundwater interactions with an alternating fresh/saline-water lake in a continental rift: Sea of Galilee, Israel, *Geological Society of America Bulletin, 112,* 1694–1702, 2000b.
Hurwitz, S, Z. Garfunkel, Y. Ben-Gai, M. Reznikov, Y. Rotstein, and H. Gvirtzman, The tectonic framework of a complex pull-apart basin: Seismic reflection observations in the northern Kinarot-Beit-Shean Basin, Dead Sea Transform, *Tectonophysics, 359,* 289–306, 2002.
Jones, A.G., Static shift of magnetotelluric data and its removal in a sedimentary basin environment, *Geophysics, 53,* 967–978, 1988.
Kafri, U., M. Goldman, M. Gendler, and H. Ben-Arie, Detection of freshwater and brackish water bodies in the deep sub-aquifers of the coastal aquifer using geophysical methods (TDEM), *Joint GSI-GII Report* GSI/27/03-914/210/02 (in Hebrew), 2002.
Kolton, Y., Analysis of the connection between groundwater and sea water in the Pleistocene coastal aquifer in the continental shelf of the Mediterranean coastal plain in central Israel, *Tahal Report* 01/88/31 (in Hebrew), 1988.
Legchenko, A., M. Goldman, and A. Beauce, A combined use of the NMR and TDEM methods for evaluating the amount of fresh groundwater in coastal aquifers of Israel, Presented at the SEG Summer Workshop on NMR Imaging of Reservoir Attributes, Park City, Utah, U.S.A., 1998
Mackie, R.L. and T.R. Madden, Three-dimensional magnetotelluric inversion using conjugate gradients, *Geophys. J. Int., 115,* 215–229, 1993.
Mackie, R.L, S. Rieven, and W. Rodi, Users manual and software documentation for two-dimensional inversion of magnetotelluric data, *Earth Resources Lab. Report,* Massachusetts Institute of Technology, 1997.
Meju, M.A., A simple method of transient electromagnetic data analysis, *Geophysics, 63,* 405–410, 1998.
Meju, M.A., S.L. Fontes, M.F.B. Oliveira, J.P.R. Lima, E.U. Ulugergerli, and A.A. Carrasquilla, Regional aquifer mapping using combined VES-TEM-AMT/EMAP methods in the semi-arid eastern margin of Parnaiba Basin, Brazil, *Geophysics, 64,* 337–356, 1999.
Meju, M.A., P.J. Fenning, and T.R.W. Hawkins, Evaluation of small-loop transient electromagnetic soundings to locate the Sherwood Sandstone aquifer and confining formations at well sites in the Vale of York, England, *J. of Applied Geophysics, 44,* 217–236, 2000.
Meju, M.A., S.L. Fontes, E.U. Ulugergerli, E.F. La Terra, C.R. Germano, and R.M. Carvalho, A joint TEM-HLEM geophysical approach to borehole siting in deeply-weathered granitic terrains, *Ground Water, 39,* 554–567, 2001.
Meju, M.A., Geoelectromagnetic exploration for natural resources: Models, case studies and challenges, *Surveys in Geophysics, 23,* 133–205, 2002.
Merkado, A., Geochemical examination of the hydraulic connection between the coastal aquifer and the Mediterranean Sea, *Tahal Report 01/89/18* (in Hebrew), 1989.
Miller, R.D., D.W. Steeples, and M. Brannan, Mapping a bedrock surface under dry alluvium with shallow seismic reflections, *Geophysics, 27,* 1528–1534, 1989.
Miller, R.D., and D.W. Steeples, A shallow seismic reflection survey in basalts of the Snake River Plain, Idaho, *Geophysics, 55,* 761–768, 1990.
Mohamed, A.K., M.A. Meju, and S.L. Fontes, Deep structure of the north-eastern margin of the Parnaiba Basin, Brazil, from magnetotelluric imaging, *Geophysical Prospecting, 50,* 589–602, 2002.
Nabighian, M., and J. Macnae, Time domain electromagnetic prospecting methods, In: *Electromagnetic Methods in Applied Geophysics,* M. Nabighian, ed., SEG, 427–478, 1991.
Raiche, A. P., The effect of ramp function turnoff on the TEM response of a layered earth, *Bulletin of the Australian Society of Exploration Geophysicists, 15,* 37–42, 1984.
Rimmer, A., S. Hurwitz, and H. Gvirtzman, Spatial and temporal characteristics of saline springs, Sea of Galilee, Israel, *Ground Water, 37,* 663–673, 1999.

Shtivelman, V., and M. Goldman, Integration of shallow reflection seismics and time domain electromagnetics for detailed study of the coastal aquifer in the Nitzanim area of Israel, *Journal of Applied Geophysics, 44,* 197–215, 2000.
Stein, M., A. Starinsky, A. Katz, S.L. Goldstein, M. Machlus, and A. Schram, Strontium isotopic, chemical, and sedimentological evidence for the evolution of Lake Lisan and the Dead Sea, *Geochim. Cosmochim. Acta, 61,* 3975–3992, 1997.
Sternberg, B.K., J.C. Washburne, and L. Pellerin, Correction for the static shift in magnetotellurics using transient electromagnetic soundings, *Geophysics, 53,* 1459–1468, 1988.
Stiller, M., The chloride content in pore water of Lake Kinneret sediments, *Israel J. Earth Sci., 43,* 179–185, 1994.
Torres-Verdin, C., and F.X. Bostick, Principles of spatial surface electric field filtering in magnetotellurics: Electromagnetic array profiling (EMAP), *Geophysics, 57,* 603–622, 1992.
Ulugergerli, E.U., Development and application of 2-D magnetotelluric inversion in complex domain, PhD Thesis, University of Leicester, U.K., 1998.
Yechieli, Y., D. Ronen, and A. Vengosh, Isotopic measurements and groundwater dating at the fresh-saline water interface region of the Mediterranean Coastal Plain aquifer of Israel, *Geol. Surv. Israel, Report GSI/28/96*, 1997
Yilmaz, O., Seismic data analysis, In: *Investigations in Geophysics, 1,* S. M. Doherty, ed., M. R. Cooper, Series ed., Society of Expl. Geophys., Tulsa, Oklahoma, 2001

13 HYDROGEOPHYSICAL CASE STUDIES AT THE LOCAL SCALE: THE SATURATED ZONE

DAVID HYNDMAN[1] AND JENS TRONICKE[2]

[1]*Dept. of Geological Sciences, Michigan State University, East Lansing, Michigan 48824, U.S.A.*

[2]*Institute of Geophysics, Swiss Federal Institute of Technology, ETH-Hoenggerberg, CH-8093 Zurich, Switzerland*

13.1 Introduction

Modern geophysical methods provide significant promise for estimating subsurface aquifer properties of the saturated zone in a minimally invasive manner. Mapping aquifer boundaries and internal stratification, estimating spatial distribution of hydrogeologic parameters, or monitoring tracer and contaminant plumes are examples of geophysical tools successfully applied. A general benefit of geophysical methods is the ability to collect high-resolution data in the horizontal dimension, where core/borehole data is nearly always limited. The complementary nature of core/borehole data and 2-D or 3-D geophysical data promises to help improve the accuracy and resolution of aquifer characterization at a variety of scales. However, one significant remaining difficulty is transforming geophysical parameters into flow and transport properties. A series of approaches and petrophysical models have been developed to help in this transformation (e.g., see Chapter 4 and Chapter 9 of this volume), but often complex and non-unique parameter relationships complicate data analysis and interpretation.

The uncertain and potentially non-unique relation between geophysical and hydrogeologic properties is one of the main reasons that geophysical data are not regularly used to estimate hydrogeologic properties (such as hydraulic conductivity). There is little reason to expect a fundamental relation between these properties (e.g., seismic velocities do not directly depend on the ability of sediment to transmit fluids). For particular frequencies of seismic energy, however, the ability of fluids to flow is important to the stiffness of the media, which is related to seismic velocity and attenuation (Bourbie et al., 1987). The effect of fluid moving through pores to accommodate the stress imposed by a sound wave was also explored by Dvorkin et al. (1995) as an alternative to Biot theory. Estimating the relation between geophysical and hydrogeologic parameters is a site-specific endeavor, since no general relation is expected.

This chapter provides an overview of several recent case studies that use geophysical information to characterize aquifers at the scale of typical contaminant site investigations, or the "local" scale. The chapter focuses on seismic and ground-penetrating radar (GPR) methods for different saturated zone hydrogeological objectives, although a few electrical-method examples are mentioned. Several of the

Y. Rubin and S.S. Hubbard (eds.), Hydrogeophysics, 391–412.

presented examples demonstrate the application of individual geophysical techniques for specific hydrogeologic problems, while others integrate multiple geophysical datasets as well as geophysical and hydrogeologic information.

As was described in Chapter 1 of this volume, geophysical methods have been applied to solve several classes of subsurface problems. One of the most common classes involves inferring stratification or general structural properties. This can be achieved via surface reflection seismic and GPR techniques as well as electrical methods, depending on the scale and depth of interest. The goal of such methods is commonly to map the geologic architecture of the aquifer, which can be used to help generate a hydrostratigraphic model that better explains hydrologic observations or processes. A second class of methods focuses on estimating hydrogeologic aquifer parameters, such as hydraulic conductivity and porosity, using geophysical measurements constrained by hydrologic data. Many of these approaches rely on petrophysical relationships to transform geophysical properties into hydrogeologic properties. This class also includes the estimation of spatial correlation structures from geophysical data. A third class of methods attempts to map temporal variations in subsurface geochemistry. This can be especially enlightening at sites where point-source contaminants have a geophysical signature that differs from the surrounding fluids. This class of methods also involves dynamic imaging of tracer or contaminant movement to estimate aquifer properties. Such methods are especially well suited to estimate hydrogeologic properties in the subsurface, because time-lapse estimates of solute movement are more easily related to aquifer hydraulic properties than static measurements. A final class deals with studying appropriate aquifer analog sites (e.g., in gravel pits or quarries). Such studies aim to explore the potential of geophysical tools in different environments, e.g., by comparing the geophysical results to closely located outcrops. Below, we review these classes of geophysical investigation, using specific illustrative case studies.

13.2 Structural Characterization of Aquifers

A variety of geophysical techniques can provide important information about subsurface architecture. For example, electrical and electromagnetic methods have successfully distinguished between different subsurface lithologies (e.g., Pellerin, 2002), and detailed gravity surveys have been used to estimate depth to bedrock (e.g., Kick, 1985). Although these techniques can also be applied at the local scale, their main field and strength of application is found at larger scales (cf. Chapter 12 of this volume and Ward, 1990). Thus, these techniques are not discussed in detail here.

An important step toward defining aquifer hydraulic parameters is identifying a site's stratigraphic zonation, as well as important interfaces such as the water table. For such purposes, seismic and GPR techniques are often used, although other methods such as DC resistivity (e.g., Baines et al., 2002; Yaramanci et al., 2002) can also be used for this purpose. Within the last decade, a number of successful studies have been published that provide detailed two- or three-dimensional insights into the subsurface zonation and stratigraphy of different depositional environments using seismic (e.g., Miller et al., 1990; Büker et al. 2000; Baker et al., 2000; Jarvis and Knight, 2002) or

GPR reflection methods (e.g., Beres and Haeni, 1991; Smith and Jol, 1992; Beres et al., 1995; Young and Sun, 1996; McMechan et al., 1997; Tronicke et al., 1999; Lesmes et al., 2002; Lunt et al., 2003). Applying these techniques to characterize a specific site clearly requires consideration of both technical and physical limitations of such approaches in the environment of interest (cf. Chapters 7 and 8 of this volume). As an example, Figure 13.1 illustrates the potential of high-resolution 3-D GPR surveying for delineating subsurface stratigraphic features of saturated Quaternary deposits in the Reuss delta, Switzerland (Nitsche et al., 2002). In this example, "time slices" (which can be easily converted to "depth slices" for a homogenous or layered subsurface velocity distribution) from a 3-D data cube show prominent channel structures. At this site, total penetration depths of ~10 m were achieved using GPR antennas with a nominal center frequency of 100 MHz, in an environment where the water table was located ~1 m below ground surface. Such a depth of investigation is common for 100 MHz GPR surveys in gravel- and sand-dominated sites.

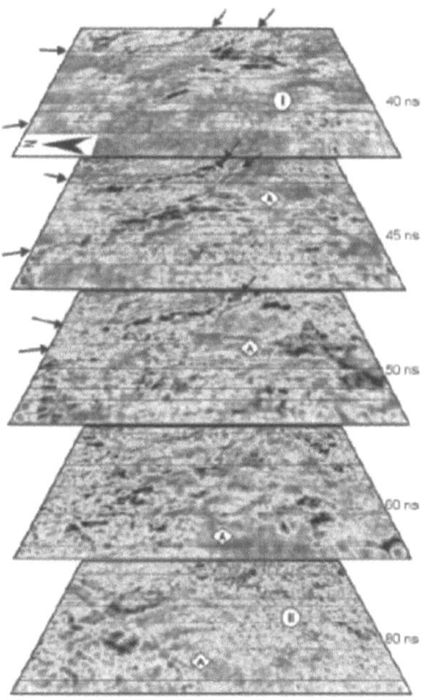

Figure 13.1. Example of detailed stratigraphic features obtained using 3-D GPR surveying (adapted from Nitsche et al., 2002). Time slices were extracted from a 3-D data cube covering a total area of 80 × 40 m, located in saturated Quaternary deposits of the Reuss delta, Switzerland. Arrows identify boundaries of prominent channel structures and labels identify characteristic radar units as interpreted by Nitsche et al. (2002).

For detailed stratigraphic analysis and interpretation, concepts of seismic and radar facies analysis have been developed (e.g., Sangree and Widmier, 1979; Beres and Haeni, 1991; Huggenberger, 1993; Beres et al., 1999; Regli et al., 2002; Lunt et al., 2003). As an example, Figure 13.2 shows a radar facies chart compiled by van Overmeeren (1998) for various sedimentary depositional environments in The Netherlands. This figure shows characteristic patterns obtained for several common sedimentological settings. Such a collection of representative reflection patterns allow researchers and practitioners to identify sedimentary sequence types and help distinguish, for example, between glacial and aeolian settings.

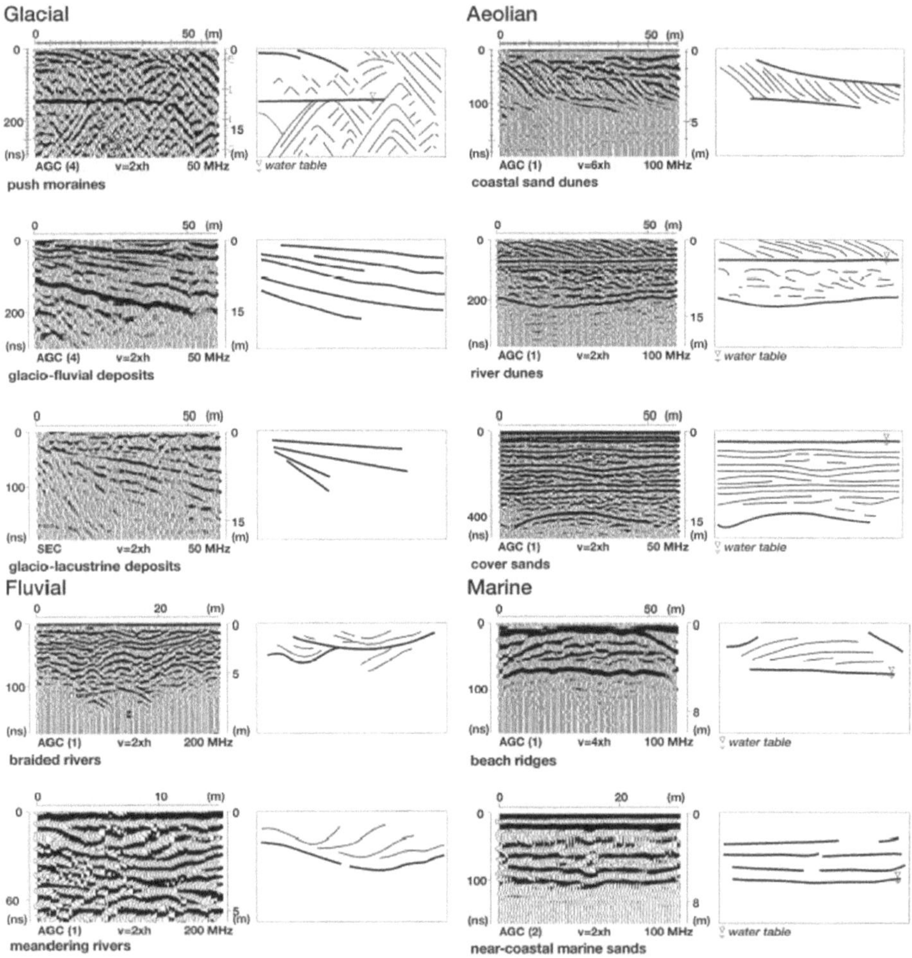

Figure 13.2. Radar facies chart for various characteristic sedimentary environments in The Netherlands (modified after van Overmeeren, 1998)

In many cases, the geometry of lithologic zones (e.g., sand, gravel, or clay) is the most important information that can be provided from a geophysical survey. Hyndman and Harris (1996) parameterized a crosswell seismic inversion, using a limited number of velocity values, to estimate the geometry of the primary lithologies for the shallow Kesterson aquifer in California's San Joaquin Valley (Figure 13.3). The approach used for this site simultaneously inverts travel times between all available well pairs for the spatial distribution of a limited number of seismic slowness (the reciprocal of seismic velocity) populations. This method can provide images that capture the dominant scale of subsurface heterogeneity, which can be an advantage over common tomographic parameterizations that provide smooth slowness fields, but can be difficult to translate into hydrogeologic parameters. The hydrogeologic properties for the estimated zones can then be inferred from core data, hydraulic testing, or some type of pump test or tracer test inversion.

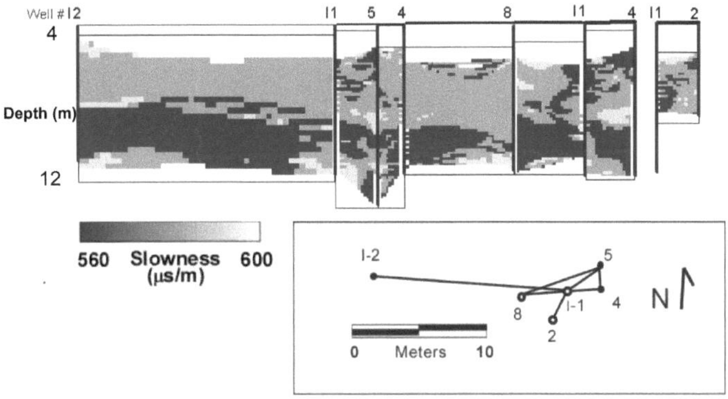

Figure 13.3. Zonal estimates of seismic slowness for the Kesterson site in California's San Joaquin Valley, developed using the multiple-population-inversion approach of Hyndman and Harris (1996)

Zonal geophysical estimates have also been developed by integrating multiple geophysical methods. For example, Fechner and Dietrich (1997) and Dietrich et al. (1998) used a combination of seismic and electric tomography to delineate major lithological units in a sedimentary aquifer. Tronicke et al. (2004) systematically investigated the potential of integrating crosshole GPR velocity and attenuation tomography to characterize highly heterogeneous sand and gravel aquifers. They found that this combination is a promising tool for delineating the major subsurface zonation as well as estimating hydrogeologic parameters within each zone. Figure 13.4 gives an example of such a zonal data integration (adapted from Tronicke et al., 2004). The crosshole tomographic GPR data set was gathered at the Boise Hydrogeophysical Research Site, Idaho, and first-arrival travel times and amplitudes were used to tomographically reconstruct the velocity and attenuation structure between two boreholes (Figures 13.4a and 13.4b). These two parameter fields were then integrated using *k*-means cluster analysis, a well-known unsupervised classification technique. Tronicke et al. (2004) conclude that the resulting zonal image of the interborehole plane (Figure 13.4c) outlines the main structural features visible in the two tomographic images and is consistent with well logs from the site. Well logs (such as neutron porosity and flowmeter logs) or other available information on the hydraulic properties (e.g., from pump or tracer tests) can then be used to develop a zonal hydrogeological model of the subsurface.

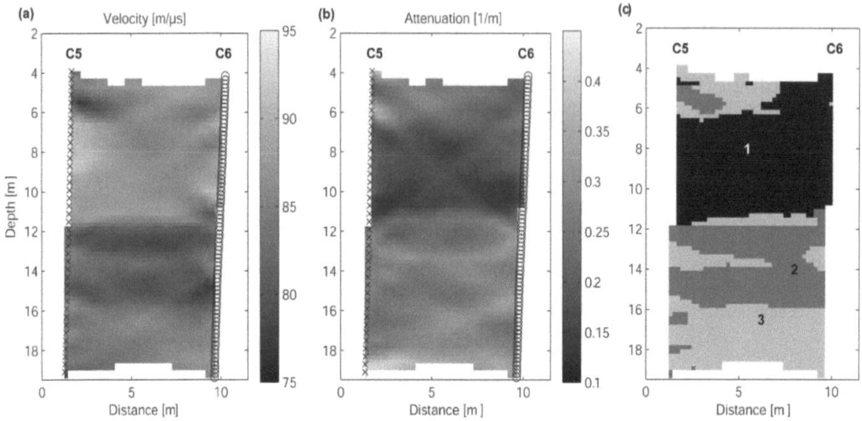

Figure 13.4. Example of integrating different crosshole tomographic parameter distributions into one zonal subsurface model (modified after Tronicke et al., 2004): (a) GPR velocity tomogram, (b) GPR attenuation tomogram, and (c) zonal model generated using cluster analysis. In (a) and (b), stars and circles denote transmitter and receiver positions in boreholes C5 and C6, respectively. In (c), numbers and gray scales represent three characteristic zones that can be identified using cluster analysis.

Estimating Aquifer Flow and Transport Parameters

There are a number of studies in the recent literature that focus on estimating hydrogeological parameters and their spatial distributions directly from geophysical measurements. Often, petrophysical relationships and models are used (as presented for example in Chapters 4 and 9 of this volume) to transform a resulting geophysical parameter into the hydrogeological parameter of interest. Hydrogeologic properties are most commonly estimated using crosshole geophysical methods, because it is difficult to resolve such properties with sufficient accuracy using current surface-based geophysical methods.

13.3.1 ESTIMATING HYDROGEOLOGICAL PROPERTIES

A significant effort over the last two decades has focused on estimating aquifers from the complementary nature of different types of data collected at multiple overlapping scales. Sediment cores provide information at the point scale, whereas pump- and slug-test data provide information in the vicinity of a well bore. Crosswell geophysical tomography provides dense information for a cross section between wells, whereas some surface geophysical data can provide three-dimensional information about the geometry of shallow lithologies. Measured tracer concentrations and changes in hydraulic heads during pumping events provide information about the average hydraulic properties between wells. Since the scale measured by each type of data is different but overlapping, combinations of these data types provide information that could not be obtained using any single data type alone. Several approaches have been used to combine such disparate datasets. Below, we review several approaches and give examples of how they have been used to estimate hydrogeological parameters in the saturated zone at the local scale.

13.3.1.1 *Geostatistical Methods*

Geostatistical methods can be used to combine diverse data types such as those obtained from geophysical and hydrogeologic data. These methods interpolate between available data points using weights that represent modeled spatial correlation structures and the uncertainty in the different measurement methods. Weights are generally assigned through a covariance matrix that incorporates the inferred spatial correlation of each variable and the interrelationships between variables, as described in Chapter 3 of this volume.

Several geostatistical methods that use surface seismic and GPR data sets to provide information about the heterogeneous structure of aquifers and reservoirs have been developed. The simplest approach is to *cokrige* the different data sets based on the inferred correlation structures of each variable and cross-correlation between datasets. The benefit of adding high-resolution soft seismic data to hard well-log data has been demonstrated in a number of studies. For example, Araktingi and Bashore (1992) cokriged three-dimensional surface-seismic-velocity estimates with porosity measurements, and found that even poor-quality seismic estimates provided information about reservoir properties. However, traditional cokriging cannot incorporate the likely nonlinear and non-unique relations between a hydrogeologic parameter of interest and the accompanying soft geophysical estimates.

Sequential Gaussian co-simulation (Deutsch and Journel, 1998) provides an additional geostatistical method for simulating hydraulic conductivity realizations based on hydraulic conductivity measurements (hard, or direct data) and geophysical attributes (soft, or indirect data). Sequential Gaussian co-simulation honors all estimated geophysical attributes, as well as the sample probability distribution and variograms of these attributes. One limitation of this approach is the reliance on a linear relation between soft and hard data (such as seismic slowness and hydraulic conductivity), which limits its application for many sites. For the Kesterson aquifer in California's San Joaquin Valley, Hyndman et al. (2000) found significant linear correlation between seismic slowness estimates obtained from crosshole seismic tomography and natural log hydraulic measurements from permeameter analysis of core samples and pump tests in the wells. Figure 13.5 is a log conductivity realization generated using sequential Gaussian co-simulation of *hard* hydraulic conductivity data, and *soft* seismic slowness estimates obtained from seven crosshole tomography planes for the site (modified from Hyndman et al., 2000). This study also demonstrated that the addition of crosswell seismic tomography data improved the accuracy of simulated tracer transport through the Kesterson aquifer. This example illustrates a simple approach for incorporating dense geophysical data and sparse measurement of the hydrogeologic properties of interest, in cases where reasonable correlation exists between these properties.

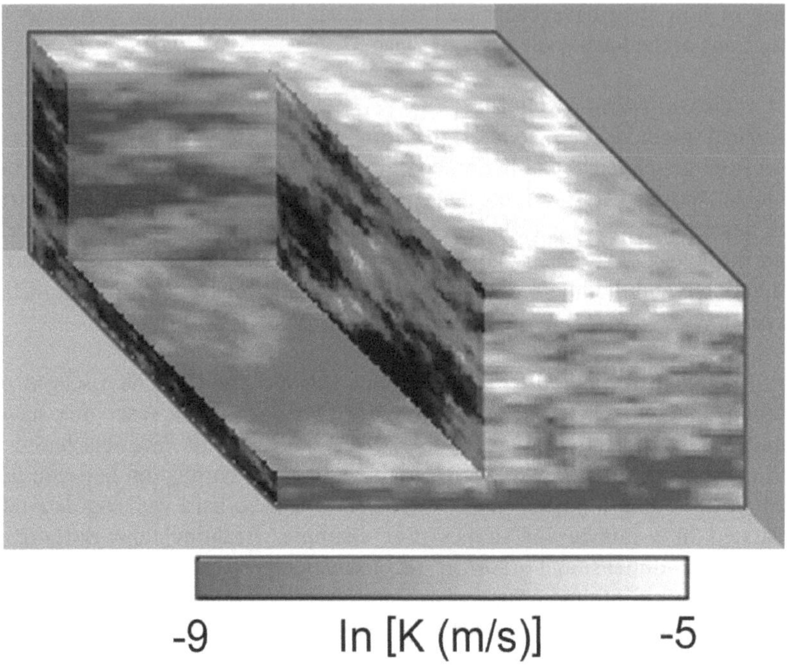

Figure 13.5. Realization of ln(K) obtained using sequential Gaussian co-simulation of crosswell seismic tomograms and local values of hydraulic conductivity at boreholes for the Kesterson site in California (modified after Hyndman et al., 2000)

13.3.1.2 *Bayesian Approaches*

Another suite of approaches for integrating geophysical and hydrogeological data involves Bayesian updating principles. These approaches develop prior estimates of hydraulic properties as described in Chapter 17 of this volume, and then update these prior estimates using more densely sampled geophysical data. The first step in the Bayesian estimation method is the development of a prior probability distribution function (pdf). This can be accomplished by (for example) interpolating the direct estimates of hydraulic conductivity from well bores using kriging (e.g., Deutsch and Journel, 1998). The second step is to estimate the joint distribution between the log of hydraulic conductivity and a geophysical parameter (such as seismic or GPR velocity) based on collocated data. This distribution is then used within Bayes' formula to update the prior pdf based on the estimated geophysical parameters, providing a posterior distribution. This inversion process is schematically shown in Figure 13.6 (modified from Hubbard et al., 2001) and is discussed in Chapter 17 of this volume.

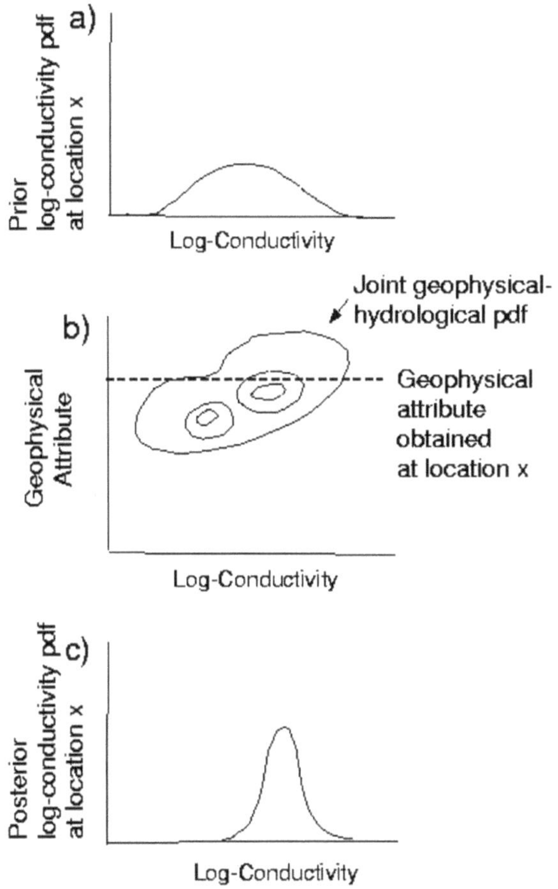

Figure 13.6. Schematic diagram of the Bayesian estimation approach: (a) example of a prior log K pdf for a point; (b) joint distribution of log hydraulic conductivity and a particular geophysical attribute such as GPR velocity; and (c) example of a posterior log (K) pdf obtained using the Bayes theorem to update a using b (modified after Hubbard et al., 2001)

This general approach was applied to estimate the log-hydraulic conductivity values at the US Department of Energy bacterial transport site near Oyster Point, Virginia (Hubbard et al., 2001). Chen et al. (2001) demonstrated that the addition of crosshole GPR tomograms reduced the estimation errors relative to the prior estimates, based on hydraulic conductivity data alone. Figure 13.7 (modified from Hubbard et al., 2001) illustrates the hydraulic conductivity estimates obtained using the Bayesian approach, along with an overlay of the hydraulic conductivity values estimated at a borehole using a flowmeter. As expected, geophysical information was most beneficial in the parameter estimation process far from locations where direct estimates of hydraulic conductivity existed. Hubbard et al. (2001) also examined tracer concentrations relative to the tomographic estimates and found that both spatial moments of the plume and the visual aspects of the project were reasonably consistent with the geophysical data.

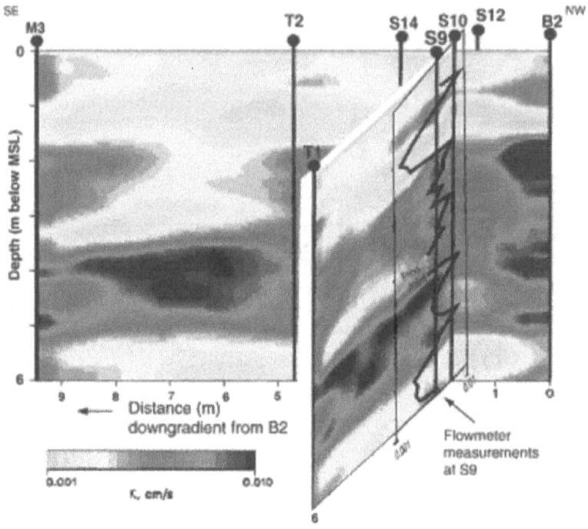

Figure 13.7. Estimated mean log (K) values estimated for the Oyster Point, VA, site, based on crosshole GPR tomograms and conductivity data obtained from borehole flowmeter tests, one of which is shown for well S9 (modified after Hubbard et al., 2001)

13.3.1.3 *Zonal Inversion Methods*

A different philosophy for combining geophysical and hydrogeologic data involves estimating the geometry of lithologic zones and the effective hydrogeologic properties for each zone that are consistent with geophysical, tracer, and hydraulic data sets. This philosophy was implemented in the split inversion method (SIM) (Hyndman et al., 1994), which co-inverts independently collected datasets (e.g., crosswell seismic travel times and tracer concentration histories) for the zonation of lithologies, effective hydraulic conductivity values for each zone, and an effective dispersivity for the region. The motivation for developing such an approach was that the combined analysis of crosswell geophysical data and tracer concentrations, which have independently sampled portions of the same environment, should provide better estimates of the geometry of subsurface lithologies and effective zonal properties than obtained using either data set alone. A particular advantage of this approach is that it does not rely on knowledge of the relationship between the geophysical attribute and the hydraulic parameter, although it assumes some relationship exists for large-scale lithologic zones. The relationship can be nonlinear and non-unique. For example, Hyndman et al. (1994) found that this approach could accurately identify the properties of two synthetic aquifers that had the same seismic response but different distributions of hydraulic conductivity.

Hyndman and Gorelick (1996) also applied the SIM to the Kesterson aquifer discussed in Sections 13.2 and 13.3.1.1 above. The seismic slowness was first estimated along the vertical planes between wells using seismic travel-time tomography (updated from zonal tomograms shown in Figure 13.3; the tomogram planes were below the white lines in Figure 13.8a). Sequential Gaussian simulation (Deutsch and Journal, 1998) was then used to develop three-dimensional conditional seismic slowness realizations (e.g., Figure 13.8a) for a region surrounding the tomograms, using the modeled correlation structure from the tomograms. The SIM approach was then used to split each slowness realization into lithologic classes (e.g., Figure 13.8b) and to estimate the effective hydraulic conductivity for each lithology as well as a regional dispersivity value.

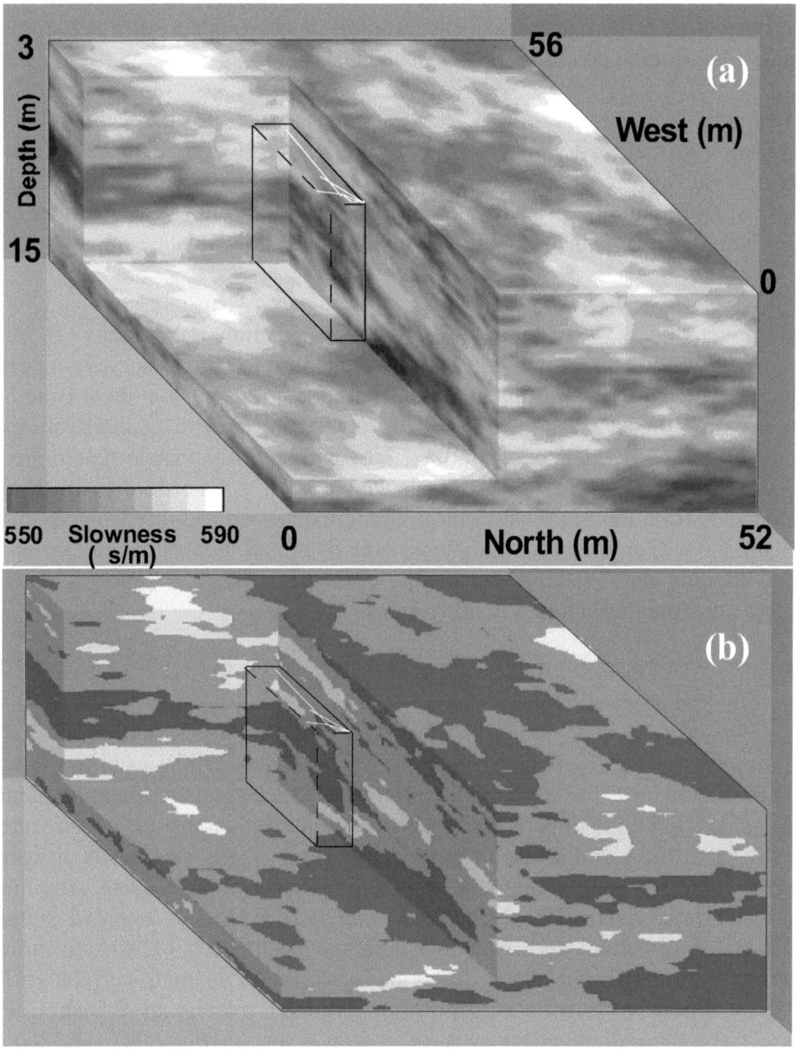

Figure 13.8. (a) Three-dimensional seismic slowness realization generated using sequential Gaussian simulation, which preserves slowness estimates and correlation structure from tomograms along planes below white lines. (b) Three-dimensional estimate of log hydraulic conductivity developed using the SIM. Hydraulic conductivity estimates are 1.4×10^{-4}, 3.6×10^{-4}, and 5×10^{-4} (m/s) for the black, dark gray and light gray zones respectively. Modified from Hyndman and Gorelick (1996).

The objective function of this inversion was to minimize the squared residual between observed and simulated tracer arrival-time quantiles, and drawdown during the forced gradient tracer test. This study again demonstrated the value of seismic tomography for estimating the hydraulic conductivity structure of a heterogeneous aquifer. The tomography estimates provided information about the geometry of lithologic zones and the correlation structure of the aquifer lithologies. Drawdown values measured during the tracer test provided critical information about the average regional hydraulic conductivity at the site, and tracer concentration data provided information about the continuity of hydraulic conductivity across the test region.

The parameter estimation approaches described above illustrate the potential of geophysical information to increase the resolution of hydraulic property estimates. Changes in subsurface lithologies at some scales should commonly change both geophysical and hydraulic properties—consequently, combining these diverse datasets provides better estimates than can be obtained with any single dataset.

13.3.1.4 *Theoretical Methods*

Yamamoto et al. (1995) employed crosshole seismic tomography to estimate permeability within a limestone aquifer. They used tomographic surveys repeated for different signal frequencies to obtain velocity-frequency dispersion data. Based on the Biot-squirt flow theory, resulting dispersion data were used to qualitatively estimate permeability values, which compared well to permeability values determined from pumping test data. Yamamoto (2001) also developed a procedure to quantitatively estimate the distribution of porosity, shear strength, and permeability from seismic data. This technique was applied to crosshole seismic data, and the results were compared to available logging data. It gave reasonable agreement in some cases, whereas only qualitative agreement was found in other cases.

13.3.2 ESTIMATING SPATIAL CORRELATION PARAMETERS USING GEOPHYSICAL DATA

Several studies have recognized that geophysical methods can be used to estimate the spatial correlation structure of hydraulic properties. Such spatial correlation parameters are used to develop three-dimensional estimates of aquifer parameters and are needed to create stochastic flow and transport models. However, these parameters are generally difficult to estimate in the horizontal direction because of inadequate sampling with typical well and core data. Geophysical data can provide densely sampled estimates in the horizontal direction and can thus improve estimates of spatial correlation parameters. Rea and Knight (1998) used surface GPR to estimate correlation lengths and the maximum correlation direction for a gravel pit site in British Columbia, Canada. They then digitized images of a gravel pit vertical face and compared correlation lengths from this image to the geophysical estimates, and found that the results were similar. Tercier et al. (2000) found significant differences in the correlation structures obtained from GPR reflection images of deltaic and barrier-spit depositional environments. They concluded that the correlation structure of GPR images can be closely related to the processes that formed the imaged geological section, and thus this information can be useful for characterizing subsurface spatial heterogeneity. More recently, Oldenborger et al. (2003) used a similar approach in a gravel pit with comparisons to hydraulic conductivity measured on cores and found that the correlation structure of stacked velocities from a common midpoint (CMP) survey compared reasonably well with those from the measured hydraulic conductivity field. In contrast, they found that geostatistical analyses of the more common reflection images provided quite different estimates of correlation parameters. Using crosshole tomographic data, Hubbard et al. (1999) investigated the importance of considering the scale of the measurement relative to the scale of heterogeneity when using geophysical data (in this case, seismic and radar tomographic data) to obtain structural correlation parameters.

13.4 Mapping Temporal Changes

Perhaps the most exciting advance in hydrogeophysics is the ability of geophysical methods to provide time-lapse images of tracers moving through the subsurface. Such approaches provide the potential to overcome the unknown and non-unique relations between geophysical and hydrologic properties discussed in the introduction to this chapter. Methods that are able to geophysically image the movement of fluids or solutes can be more directly related to the hydrogeologic properties of interest than passive imaging methods.

13.4.1 IMAGING CHANGES IN WATER TABLE ELEVATION

Conventionally, the hydraulic response of an aquifer is measured in a series of monitoring wells during a pump test. Often, the number of observation wells is limited, and their spatial distribution is not sufficient to study the response of the experiment in detail. In this section, we explore the potential of seismic and GPR methods to provide high-resolution estimates of water table fluctuations. Electrical Resistance Tomography (ERT) and other electrical methods can also be used for this type of analysis, as discussed in Chapter 14 of this volume.

Birkelo et al. (1987) reported the use of time-lapse seismic reflection data, gathered during a pumping test, for investigating an unconfined alluvial aquifer. They demonstrated that seismic reflection data can image the top of shallow (<2.5 m) saturated zones, but that there was little change in the seismic response during eight days of pumping, even though drawdowns of several meters were measured in nearby wells. They speculated that a near-surface clay layer inhibited drainage in the upper part of the studied aquifer. Baker et al. (2000) used surface seismic reflection data to monitor seasonal water table fluctuations in an alluvial aquifer, with a reported accuracy of ± 12 cm. They concluded that detailed seismic velocity information, which can be obtained from common midpoint surveys, was critical in providing accurate estimates of water table depths from seismic reflection data.

Endres et al. (2000) investigated the potential of time-lapse surface GPR reflection data to map the drawdown associated with a pumping test carried out at the Canadian Forces Base (CFB) test site in Borden, Ontario. The aquifer at this site is composed of clean, well-sorted, medium- to fine-grained Pleistocene sand. They observed a reflection in the transition zone between the overlying residually saturated material and the capillary fringe below, rather than at the water table depth. They used the arrival-time delay of this reflection to infer the drawdown during the pumping test. Figure 13.9 shows distance- and time-drawdown curves based on GPR and piezometer data collected during this experiment. The comparison shows that the drawdown estimated from GPR is both smaller and delayed compared to the drawdown directly measured from piezometers. Endres et al. (2000) interpreted the difference to result from an increase of the combined thickness of the transition zone and capillary fringe during the pump test. They concluded that GPR-derived information can be useful in imaging a pumping test, but that there is a need for future research. Bentley and Trenholm (2002) also investigated the accuracy of GPR for estimating water table elevations in unconsolidated sediments in Canada. They used theoretical models and field data to quantify sources of uncertainty for estimating water table elevations (uncertainty in the velocity model, uncertainty in surface elevation, and uncertainty in the capillary fringe height) using GPR frequencies of 100 and 200 MHz. They concluded that the shallow water table elevation can be estimated with an accuracy of roughly 0.2 m in their particular test environment. Extensive discussions of water table mapping using GPR data are given in Chapter 14 of this volume.

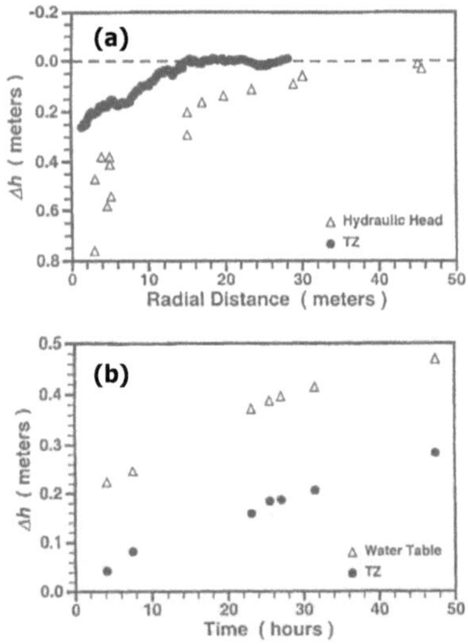

Figure 13.9. Comparison of drawdown during a pumping experiment measured with piezometers and GPR (modified after Endres et al., 2000): (a) Distance-drawdown after 23 hours determined from hydraulic head measurements and from transition zone reflection times (TZ); (b) Time-drawdown curves from hydraulic head measurements and from transition zone reflection times (TZ) for a radial distance of 5 m.

In another interesting experiment, Tsoflias et al. (2001) studied the GPR response during a pumping test in a fractured aquifer. First, they employed forward electromagnetic modeling of thin layers (simulating the fractures) to investigate the GPR waveform changes associated with changes in fracture saturation. GPR surveys carried out before and during a pump test provided data that were consistent with piezometric head measurements. Using these data, Tsoflias et al. (2001) interpreted a drainage pattern for a prominent fracture. They concluded that GPR surveying combined with hydraulic data can provide improved understanding of fluid flow within fractured formations at the local scale.

The studies discussed above indicate that geophysical data can provide insight into variations in water table and capillary fringe elevations during pumping tests or natural seasonal changes. However, these studies also indicate that additional research is needed to improve both the resolution of the estimates and the interpretation of measured changes. Additional examples of the use of geophysical methods for mapping the water table are given in Chapter 14 of this volume.

13.4.2 PLUME MAPPING AND DYNAMIC IMAGING

There is a significant need for new methods that can either infer contaminant concentrations across plumes or remotely image the size and shape of contaminant plumes. Even limited direct characterization of plumes is extremely expensive. Without remote technology, contaminant plumes are always undersampled. New remote plume imaging methods could reduce the cost of contaminant remediation systems, because they could be placed in locations that would maximize the contaminant treatment with the minimum installation and operation costs.

A number of controlled experiments at the laboratory or field scale indicate the potential of geophysical techniques, such as electric and electromagnetic methods, to detect and map nonaqueous phase liquids (NAPLs) and other contaminants within the subsurface (e.g., Olhoeft, 1992; Brewster and Annan, 1994; Endres and Redman, 1996; Grumman and Daniels, 1996; Sandberg et al., 2002). Because of the complexity and heterogeneity of near-surface environments, it is especially difficult to relate a single geophysical parameter (e.g., electrical resistivity or GPR reflectivity) to a specific contaminant. Additional information can be provided through the use of multiple geophysical techniques constrained by various borehole-based data, as for example presented by Atekwana et al. (2000). Deeds and Bradford (2002) report the direct detection of light non-aqueous phase liquid (LNAPL) using advanced processing techniques for multi-offset GPR data. They used detailed velocity information and AVO (amplitude versus offset) analysis assisted by systematic coring.

Another approach for detecting and mapping contamination plumes is to observe subsurface changes with time. In the Borden experiment (Greenhouse et al., 1993), a controlled DNAPL (dense NAPL) release was monitored with various geophysical techniques. In this experiment, 770 L of perchloroethylene (PCE) were released over a 70-hour period into a $9 \times 9 \times 3.3$ m cell containing medium- to fine-grained sand. Brewster and Annan (1994) present the results of detecting and monitoring this release with 200 MHz GPR. Under these ideal conditions, it was possible to relate certain reflectors to plume movement. Brewster and Annan (1994) also found that lateral and/or temporal variations in GPR velocity from CMP gathers can be helpful in reliably interpreting the data. The referenced publications and other recent studies indicate that in uncontrolled field settings, assisting information (e.g., direct sampling by systematic coring) is indispensable for obtaining a reliable link between geophysical data and contaminant plume detection. Identifying DNAPL pools at contaminated sites is a daunting task, because it can be difficult to separate variations in fluid properties from changes in subsurface geology, unless the site is already well characterized.

Dynamic imaging of changes in fluid properties is another exciting area of hydrogeophysics. Many of the traditional problems of interpreting geophysical images in terms of hydraulic properties can be reduced or eliminated in this class of problems. The potential of electrical methods was recognized decades ago, especially to monitor the movement of saline tracers and to derive hydrological parameters such as groundwater velocity (Fried, 1975; White, 1988; Bevc and Morrison, 1991; White 1994). Ramirez et al. (1993) discussed an experiment in which electrical resistance tomography (ERT) was used to image a plume of injected steam designed to remediate volatile organic contaminants at the Lawrence Livermore National Laboratory site. They found that ERT did a reasonable job of imaging the measured changes in sediment resistivity, and they also found that the steam flood was likely constrained by a gravel layer across the site. More recently, developments in inversion techniques allow tomographic imaging of the tracer plume and its development with time. This has increased the popularity of dynamic electrical imaging. Good examples of this technique for various applications include Slater et al. (1997), Daily and Ramirez (2000), and Kemna et al. (2002).

In a carefully designed field experiment, the background conditions can be measured before a tracer is injected, and then a series of differenced geophysical images can be used to image the moving tracer plume. One of the keys to this type of experiment is choosing tracer concentrations that provide enough geophysical contrast while still allowing some energy to propagate through the tracer plume. Details of the design of this type of test and some preliminary results for the dynamic imaging of a bromide salt tracer plume injected into the Boise Hydrogeophysical Research Site, Idaho, are given by Barrash et al. (2003) and Goldstein et al. (2003).

Figure 13.10 illustrates the results of a 3-D saline tracer experiment monitored using ERT (adapted from Mohrlock and Dietrich, 2001). In a gravel- and sand-dominated aquifer located in the Rhine Valley in southwest Germany, these authors used arrays of borehole electrodes installed in six boreholes. In a long-term experiment (i.e., over a period of several weeks), the use of borehole electrodes improved data quality, because these measurements were less affected by the significant near-ground-surface resistivity variations, e.g., caused by changes in water content. Using a 3-D inversion technique and relating certain iso-surfaces of electrical resistivity to the tracer plume, Mohrlock and Dietrich (2001) were able to generate 3-D images of the tracer plume and its movement with time, as illustrated in Figure 13.10.

Day-Lewis et al. (2002) illustrated the utility of dynamic tracer imaging using time-lapse crosshole GPR tomography. Using a synthetic GPR data set, the authors developed a sequential inversion approach to image multiple time slices of an electrically conductive tracer moving through a fracture zone. Day-Lewis et al. (2003) then applied this approach to evaluate a saline tracer test at the U.S. Geological Survey Fractured Rock Hydrology Research Site in Grafton County, New Hampshire. They found that their sequential time-lapse inversion allowed them to identify a preferential flowpath through an identified fracture and estimate the time of peak tracer arrival at the geophysically imaged planes. Such field tests need to be carefully designed to avoid temporal aliasing, which occurs when the time step of the GPR acquisition is greater than the time step of the tracer plume movement.

Time-lapse imaging methods, such as presented in the cases discussed above, provide the potential for significant insight into flow and transport dynamics in heterogeneous aquifers. Water levels and solute concentrations can only be measured at a limited number of well points across an aquifer, whereas geophysical methods have the potential to provide 2- to 3-dimensional estimates of solute concentrations through time between such wells.

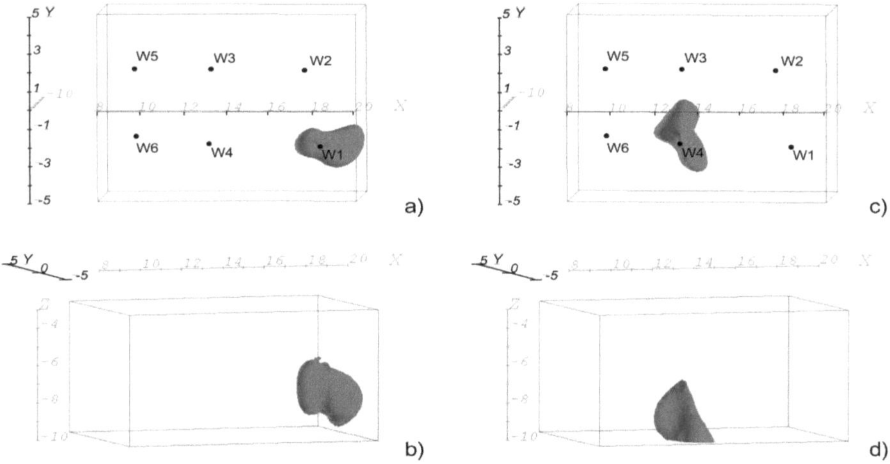

Figure 13.10. Three-dimensional electrical imaging of a salt tracer plume and its propagation in a sand- and gravel-dominated aquifer (adapted from Mohrloch and Dietrich, 2001). The images show selected iso-surfaces of changes in electrical resistivity for different times and perspectives: (a) Top view after twelve days of tracer propagation; (b) Side view after twelve days; (c) Top view after 24 days; and (d) Side view after 24 days. All axes dimensions are in meters.

13.5 Aquifer Analog Studies

Outcrop analog approaches have been applied mainly in the field of petroleum exploration to characterize the properties of petroleum reservoirs (e.g., Miall, 1988). The concept has also been used in aquifer studies to investigate geometries and hydraulic properties of various sedimentological units (e.g., Macfarlane et al., 1994; Heinz et al., 2003). Such analog approaches involve the detailed investigation of accessible outcrops of a representative sedimentary formation. It is assumed that the geometric and physical parameter distributions are comparable to those of the inaccessible aquifer. Such models can then be used to simulate geophysical measurements (Dietrich et al., 1998; Kowalsky et all, 2001) as well as groundwater flow and transport (Whittaker and Teutsch, 1999) on an analog model of an aquifer that is as close as possible to reality.

Figure 13.11 illustrates an analog study presented by Tronicke et al. (2002). The authors carried out geophysical measurements (GPR reflection surveying, crosshole GPR tomography, and natural gamma activity logging) in braided stream deposits. Their test site was located in a gravel pit in the upper Rhine Valley of southwest Germany. After geophysical surveying, they excavated their test site and photographed the outcrop plane corresponding to their geophysical measurement plane, using a special wide-angle lens to ensure that the photographs were free of distortion. The digitized photographs were then interpreted in terms of different lithologies, resulting in a map of the 2-D lithofacies distribution in the surveyed plane. They found reasonable agreement between their structural interpretation of the combined geophysical results and the main sedimentary units observed in the outcrop. In addition, they transformed the obtained lithofacies distribution into hydraulic conductivity and porosity values using the procedure of Klingbeil et al. (1999). This enabled an impact analysis of different structures detected by geophysical tools using a series of groundwater flow and transport models. Tronicke et al. (2002) concluded that the incorporation of integrated high-resolution geophysical information improved the hydraulic characterization of the heterogeneous deposit.

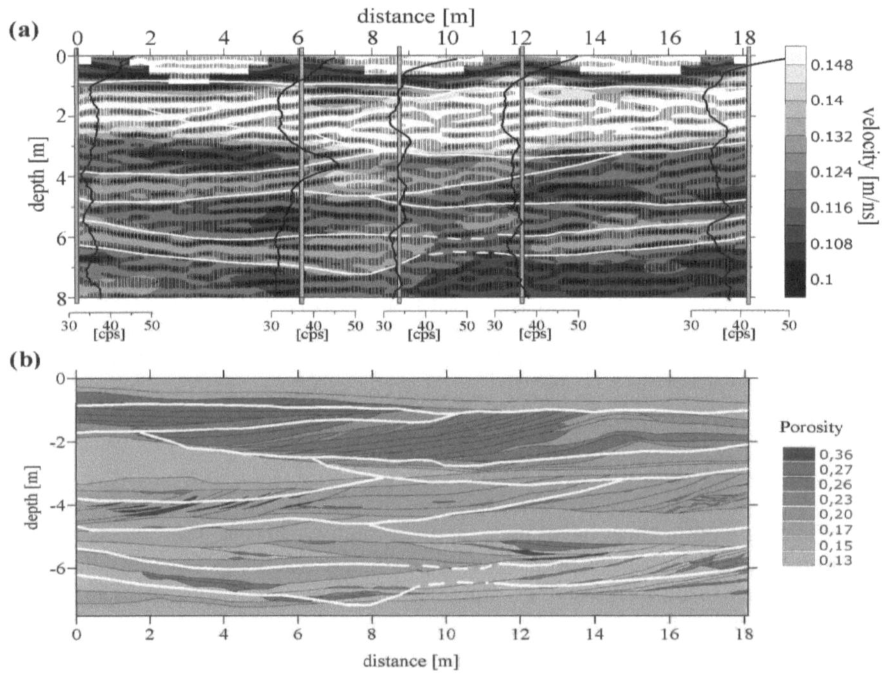

Figure 13.11. Example of a detailed aquifer analog study for evaluating geophysical techniques and their potential in braided stream deposits (modified after Tronicke et al., 2002): (a) Combined visualization of GPR reflection image, GPR velocity distribution tomographically reconstructed from three crosshole surveys, and natural gamma activity logging (borehole locations are denoted by grey rectangles); (b) Lithofacies distribution converted into porosities in the geophysical measuring plane interpreted from a photograph taken after excavation. The white lines represent a structural interpretation based on all shown geophysical results. Note the reasonable agreement between the structures detected by geophysical techniques and the main sedimentary units in (b).

13.6 Concluding Remarks

The case studies presented in this chapter illustrate the benefit of using geophysical techniques for a variety of subsurface characterization and monitoring objectives within saturated aquifers at the local scale. These objectives include delineation of structural aquifer features, direct estimation of hydraulic relevant parameters, integrated hydrogeophysical approaches to characterize flow and transport parameters, and dynamic imaging of variations in tracer or contaminant concentrations. The ability of geophysical methods to provide densely sampled information, relative to traditional hydrogeological methods, makes them attractive for subsurface characterization at the local scale, such as for investigating a contaminated site. While some tasks involve standard applications of geophysical methods (e.g., mapping depth to bedrock), other problems require novel techniques and analysis methods that are still in the development stages (e.g., relating and integrating geophysical and hydrological data). Integration of various soft geophysical data and various hard data, such as provided by borehole-based hydrogeological methods, may help to reduce uncertainties and non-uniqueness during processing and data analysis, and may help to generate more reliable flow and transport models at specific sites. The presented studies show the benefit of integrated hydrologic and geophysical approaches. One of the most exciting areas of research is the emerging field of dynamic imaging, which allows for passive contaminant site monitoring, active plume migration imaging, and more accurate hydraulic property estimation than with static measurements.

References

Araktingi, U.G., and W.M. Bashore, Effects of properties in seismic data on reservoir characterization and consequent fluid flow predictions when integrated with well logs, Paper presented at 67th Annual SPE Conference, Soc. Pet. Eng., Washington, D.C., Oct. 4–7, 1992.

Atekwana, E.A., W.A. Sauck, and D.D. Werkema, Investigations of geoelectrical signatures at a hydrocarbon contaminated site, *Journal of Applied Geophysics*, *44*, 167–180, 2000.

Baker, G.S., D.W. Steeples, C. Schmeissner, and K.T. Spikes, Ultrashallow seismic reflection monitoring of seasonal fluctuations in the water table, *Environmental and Engineering Geoscience*, *6*(3), 271–277, 2000.

Baines, D., D.G. Smith, D. Froese, P. Bauman, and G. Nimeck, Electrical resistivity ground imaging (ERGI): a new tool for mapping the lithology and geometry of channe-belts and valley-fills, *Sedimentology, 49*, 441–449, 2002.

Barrash, W., M.D. Knoll, D.W. Hyndman, T. Clemo, E.C. Reboulet, and E.M. Hausrath, Tracer/Time-Lapse Radar Imaging Test at the Boise Hydrogeophysical Research Site, In: *Proceedings of SAGEEP03, The Symposium on the Application of Geophysics to Engineering and Environmental Problems*, pp. 163–174, San Antonio, TX, 2003.

Bentley, L.R., and N.M. Trenholm, The accuracy of water table elevation estimates determined from ground penetrating radar data, *Journal of Environmental and Engineering Geophysics*, *7*(1), 37–53, 2002.

Beres, M., and F.P. Haeni, Application of ground-penetrating-radar methods in hydrogeologic studies, *Ground Water*, *29*(3), 375–387, 1991.

Beres, M., A.G. Green, P. Huggenberger, and H. Horstmeyer, Mapping the architecture of glaciofluvial sediments with three-dimensional georadar, *Geology*, *23*(12), 1087–1090, 1995.

Bevc, D., and H.F. Morrison, Borehole-to-surface electrical monitoring of a salt water injection experiment, *Geophysics*, *56*, 769–777, 1991.

Birkelo, B.A., D.W. Steeples, R.D. Miller, and M. Sophocleous, Seismic reflection study of a shallow aquifer during a pumping test, *Ground Water*, *25*(6), 703–709, 1987.

Bourbie, T., O. Coussy, and B. Zinszner, *Acoustics of Porous Media*, Gulf Publishing Company, Houston, 1987.

Brewster, M.L., and A.P. Annan, Ground-penetrating radar monitoring of a controlled DNAPL release: 200 MHz radar, *Geophysics*, *59*(8), 1211–1221, 1994.

Büker, F., A.G. Green, and H. Horstmeyer, 3-D high-resolution reflection seismic imaging of unconsolidated glacial and glaciolacustrine sediments: Processing and interpretation, *Geophysics*, *65*(1), 18–34, 2000.

Chen, J., S. Hubbard, and Y. Rubin, Estimating hydraulic conductivity at the South Oyster Site from geophysical tomographic data using Bayesian techniques based on the normal regression model, *Water Resour. Res.*, *37*(6), 1603–1613, 2001.

Daily, W.D., and A.L. Ramirez, Electrical imaging of engineered hydraulic barriers, *Geophysics*, *65*(1), 83–94, 2000.

Day-Lewis, F.D., J.M. Harris, and S.M. Gorelick, Time-lapse inversion of crosswell radar data, *Geophysics*, *67*(6), 1740–1752, 2002.

Day-Lewis, F.D., J.W. Lane, Jr., J.M. Harris, and S.M. Gorelick, Time-lapse imaging of saline-tracer transport in fractured rock using difference-attenuation radar tomography, *Water Resour. Res.*, *39*(10), *1290,* 10.1029/ 2002WR001722, 2003

Deeds, J., and J. Bradford, Characterization of an aquitard and direct detection of LNAPL at Hill Air Force Base using GPR AVO and migration velocity analyses, In: *Ninth International Conference on Ground Penetrating Radar (GPR 2002)*, edited by S. Koppenjan, and H. Lee, SPIE, 4758, pp. 323–329, Santa Barbara, California, 2002.

Deutsch, C.V., and A.G. Journel, *GSLIB: Geostatistical Software Library and User's Guide*, Oxford University Press, 1998.

Dietrich, P., T. Fechner, J. Whittaker, and G. Teutsch, An integrated hydrogeophysical approach to subsurface characterization, In: *Groundwater Quality: Remediation and Protection*, edited by M. Herbert, and K. Kovar, pp. 513–519, IAHS Publ., 250, Tübingen, Germany, 1998.

Dvorkin, J., G. Mavko, and A. Nur, Squirt flow in fully saturated rocks, *Geophysics*, *60*(1), 97–107, 1995.

Endres, A.L., W.P. Clement, and D.L. Rudolph, Ground penetrating radar imaging of an aquifer during a pumping test, *Ground Water*, *38*(4), 566–576, 2000.

Fechner, T., and P. Dietrich, Lithological inversion of tomographic data, In: *3rd Ann. Mtg., Environ. and Eng. Geophys. Soc., Euro. Section*, pp. 355–358, Aarhus, Denmark, 1997.

Fried, J.J., *Groundwater Pollution: Developments in Water Science 4*, Elsevier, 1975.

Goldstein, S.E., T.C. Johnson, M.D. Knoll, W. Barrash, and W.P. Clement, Borehole radar attenuation-difference tomography during the tracer/time-lapse test at the Boise Hydrogeophysical Research Site, In: *Proceedings of SAGEEP03, The Symposium on the Application of Geophysics to Engineering and Environmental Problems*, pp. 147–162, San Antonio, TX, 2003.

Greenhouse, J., M. Brewster, G. Schneider, J.D. Redman, A.P. Anna, G.R. Olhoeft, J. Lucius, K. Sander, and A. Mazzella, Geophysics and solvents: The Borden Experiment, *The Leading Edge of Exploration*, *12*, 261–267, 1993.

Grumman, D.L., and J.J. Daniels, Experiments on the detection of organic contaminants in the vadose zone, *Journal of Environmental and Engineering Geophysics*, 0(1), 31–38, 1995.

Heinz, J., S. Kleineidam, G. Teutsch, and T. Aigner, Heterogeneity patterns of Quaternary glaciofluvial gravel bodies (SW-Germany): application to hydrogeology, *Sedimentary Geology*, *158*, 1–23, 2003.

Hubbard, S.S., J. Chen, J.E. Peterson, E.L. Majer, K.H. Williams, D.J. Swift, B. Mailloux, and Y. Rubin, Hydrogeological characterization of the South Oyster Bacterial Transport Site using geophysical data, *Water Resour. Res.*, *37*(10), 2431–2456, 2001.

Hubbard, S.S., Y. Rubin, and E. Majer, Spatial correlation structure estimation using geophysical and hydrogeological data, *Water Resour. Res.*, *35*, 1809–1825, 1999.

Huggenberger, P., Radar facies: Recognition of facies patterns and heterogeneities within Pleistocene Rhine gravels, NE Switzerland, In: *Geological Society Special Publications No 75*, pp. 163–176, 1993.

Hyndman, D.W., and S.M. Gorelick, Estimating lithologic and transport properties in three dimensions using seismic and tracer data, *Water Resour. Res.*, *32*(9), 2659–2670, 1996.

Hyndman, D.W., J.M. Harris, and S.M. Gorelick, Coupled seismic and tracer test inversion for aquifer property characterization, *Water Resour. Res.*, *30*(7), 1965–1977, 1994.

Hyndman, D.W., and J.M. Harris, Traveltime inversion for the geometry of aquifer lithologies, *Geophysics*, *61*(6), 1996.

Hyndman, D.W., S.M. Gorelick, and J.M. Harris, Inferring the relationship between seismic slowness and hydraulic conductivity in heterogeneous aquifers, *Water Resour. Res.*, *36*(8), 2121–2132, 2000.

Jarvis K.D. and R.J. Knight, Aquifer heterogeneity from SH-wave seismic impedance inversion, *Geophysics, 67*(5), 1548–1557, 2002.

Kemna, A., J. Vanderborght, B. Kulessa, and H. Vereecken, Imaging and characterization of subsurface solute transport using electrical resistivity tomography (ERT) and equivalent transport models, *Journal of Hydrology, 267*(3–4), 125–146, 2002.

Kick, J.F., Depth to bedrock using gravimetry, *The Leading Edge of Exploration, 4*(4), 38–42, 1985.

Klingbeil, R., S. Kleineidam, U. Asprion, T. Aigner, and G. Teutsch, Relating lithofacies to hydrofacies: Outcrop-based hydrogeological characterization of Quaternary gravel deposits, *Sedimentary Geology, 129*(3–4), 299–310, 1999.

Kowalsky, M.B., Dietrich, P., Teutsch, G., and Y. Rubin, Forward modeling of GPR data usingdigitized outcrop images and multiple scenarios of water saturation, Water Resour. Res., 37(6), 1615-1625, 2001.

Lesmes, D.P., S.M. Decker, and D.C. Roy, A multiscale radar-stratigraphic analysis of fluvial aquifer heterogeneity, *Geophysics, 67*(5), 1452–1464, 2002.

Lunt, I.A., J.S. Bridge, and R.S. Tye, Development of a 3-D depositional model of braided river gravels and sands to improve aquifer characterization, In: *Aquifer Characterization*, SEPM Concepts in Paleontology and Sedimentology Series, J.S. Bridge and D.W Hyndman, eds., 2003 (in press).

Macfarlane, P.A., J.H. Doveton, H.R. Feldmann, J.J. Butler, J.M.J. Combes, and D.R. Collins, Aquifer/aquitard units of the Dakota aquifer system in Kansas: Methods of delineation and sedimentary architecture effects on groundwater flow and flow properties, *Journal of Sedimentary Research, B 64/4*, 464–480, 1994.

McMechan, G.A., G.C. Gaynor, and R.B. Szerbiak, Use of ground-penetrating radar for 3-D sedimentological characterization of clastic reservoir analogs, *Geophysics, 62*(3), 786–796, 1997.

Miller, R.D., D.W. Steeples, R.W. Hill, and B.L. Gaddis, Identifying intra-alluvial and bedrock structures shallower than 30 meters using seismic reflection techniques, In: *Geotechnical and Environmental Geophysics, Volume 2: Environmental and Groundwater*, edited by S. Ward, pp. 75–88, Society of Exploration Geophysics, 1990.

Miall, A.D., Reservoir heterogeneities in fluvial sandstones: Lessons from outcrop studies, *AAPG Bulletin, 72*(6), 682–697, 1988.

Mohrlok, U. and Dietrich, P., Exploration of preferential transport paths using geoelectrical salt tracer tests. In: Field Screening Europe 2001—Proceedings of the Second International Conference on Strategies and Techniques for the Investigation and Monitoring of Contaminated Sites, W. Breh, J. Gottlieb, H. Hötzl, F. Kern, T. Liesch, and R. Niessner, eds., pp. 327–330, 2001.

Nitsche, F.O., A.G. Green, H. Horstmeyer, and F. Büker, Late Quaternary depositional history of the Reuss delta, Switzerland: Constraints from high-resolution seismic reflection and georadar surveys, *Journal of Quaternary Science, 17*(2), 131–143, 2002.

Oldenborger, G.A., R.A. Schincariol, and L. Mansinha, Radar determination of the spatial structure of hydraulic conductivity, *Ground Water, 41*(1), 24–32, 2003.

Olhoeft, G.R., Geophysical detection of hydrocarbon and organic chemical contaminants, In: *Proceedings of SAGEEP92, The Symposium on the Application of Geophysics to Engineering and Environmental Problems*, pp. 587–595, Oakbrook, IL, 1992.

Pellerin, L., Application of electrical and electromagnetic methods for environmental and geotechnical investigations, *Surveys in Geophysics, 23*, 101–132, 2002.

Ramirez, A.L., W.D. Daily, D. Labrecque, E. Owen, and D. Chesnut, Monitoring an underground steam injection process using electrical-resistance tomography, *Water Resour. Res., 29*(1), 73–87, 1993.

Rea, J., and R. Knight, Geostatistical analysis of ground-penetrating radar data: A means of describing spatial variation in the subsurface, *Water Resour. Res., 34*(3), 329–339, 1998.

Regli C., P. Huggenberger, and M. Rauber, Interpretation of drill core and georadar data of coarse gravel deposits, *Journal of Hydrology, 255*(1–4), 234–252, 2002.

Sangree, J.M., and J.M. Widmier, Interpretation of depositional facies from seismic data, *Geophysics, 44*, 131–160, 1979.

Sandberg, S.K., L.D. Slater, and R. Versteeg, An integrated geophysical investigation of the hydrogeology of an anisotropic unconfined aquifer, *Journal of Hydrology, 267*(3–4), 227–243, 2002.

Slater, L., A. Binley, and D. Brown, Electrical imaging of fractures using ground-water salinity change, *Ground Water, 35*, 436–442, 1997.

Smith, D.G., and H.M. Jol, Ground-penetrating radar investigation of a Lake Bonneville Delta, Provo level, Brigham City, Utah, *Geology, 20*(12), 1083–1086, 1992.

Tercier, P., R. Knight, and H. Jol, A comparison of the correlation structure in GPR images of deltaic and barrier-spit depositional environments, *Geophysics, 65*(4), 1142–1153, 2000.

Tronicke, J., N. Blindow, R. Groß, and M.A. Lange, Joint application of surface electrical resistivity- and GPR-measurements for groundwater exploration on the island of Spiekeroog, northern Germany, *Journal of Hydrology*, *223*(1–2), 44–53, 1999.

Tronicke, J., P. Dietrich, U. Wahlig, and E. Appel, Integrating surface georadar and crosshole radar tomography: A validation experiment in braided stream deposits, *Geophysics*, *67*(5), 1495–1504, 2002.

Tronicke, J., K. Holliger, W. Barrash, M.D. Knoll, 2004, Multivariate analysis of crosshole georadar velocity and attenuation tomograms for aquifer zonation, *Water Resour. Res., 40(1), W01519,* 10.1029/2003WR002031.

Tsoflias, G.P., T. Halihan, and J.M. Sharp, Monitoring pumping test response in a fractured aquifer using ground-penetrating radar, *Water Resour. Res.*, *37*(5), 1221–1229, 2001.

van Overmeeren, R.A., Radar facies of unconsolidated sediments in The Netherlands: A radar stratigraphy interpretation method for hydrogeology, *Journal of Applied Geophysics*, *40*, 1–18, 1998.

Ward, S. H., editor, Geotechnical and Environmental Geophysics, Volumes I-III, Investigations in Geophysics #5, Society of Exploration Geophysicists, 1990.

White, P.A., Measurement of groundwater parameters using salt-water injection and surface resistivity, *Ground Water*, *26*, 179–186, 1988.

White, P.A., Electrode arrays for measuring groundwater flow direction and velocity, *Geophysics*, *59*(2), 192–201, 1994.

Whittaker, J., and G. Teutsch, Numerical simulation of subsurface characterization methods: application to a natural aquifer analogue, *Advances in Water Resources*, *22*(8), 819–829, 1999.

Yamamoto, T., T. Nye, and M. Kuru, Imaging the permeability structure of a limestone aquifer by crosswell acoustic tomography, *Geophysics*, *60*, 1634–1645, 1995.

Yamamoto, T., Imaging the permeability structure within the near-surface sediments by acoustic crosswell tomography, *Journal of Applied Geophysics*, *47*, 1–11, 2001.

Yaramanci, U., G. Lange, and M. Hertrich, Aquifer characterisation using surface NMR jointly with other geophysical techniques at the Nauen/Berlin test site, *Journal of Applied Geophysics*, *50*, 47–65, 2002.

Young, R.A., and Y. Sun, 3-D ground-penetrating radar imaging of a shallow aquifer at Hill Air Force Base, Utah, *Journal of Environmental and Engineering Geophysics*, *1*(2), 97–108, 1996.

14 HYDROGEOPHYSICAL CASE STUDIES IN THE VADOSE ZONE

JEFFREY J. DANIELS[1], BARRY ALLRED[2], ANDREW BINLEY[3], DOUGLAS LABRECQUE[4], and DAVID ALUMBAUGH[5]

[1]*Dept. Geol. Sciences, The Ohio State University, Columbus, OH, 43210*
[2]*USDA/ARS-SDRU , The Ohio State University, Columbus, OH, 43210*
[3]*Lancaster University, Department of Environmental Science, Lancaster, LA1 4YQ, UK*
[4]*Multi-Phase Technologies, LLC, Sparks Nevada*
[5]*University of Wisconsin, Madison, Wisconsin*

14.1 Overview: Detection and Monitoring in the Vadose Zone

The focus of this chapter is the characterization of the vadose zone, or the unsaturated section of the subsurface, using hydrogeophysical techniques. The regions of water saturation as they relate to the physical properties are shown for reference in Figure 14.1. Characterization below the water table (in the saturated section) is described in Chapter 13 of this volume and will not be discussed in detail here. From a physical properties perspective, the zones of variable saturation above the water table are transitional, and depend upon the soil or rock type and the lateral heterogeneity of the materials. In the vertical direction, the boundaries between all of these zones are dependent upon the types of soil, regolith, or rock that are present; the current and historical climatic conditions; and the regional and local geomorphology of the site. These same factors affect the heterogeneity of the vadose zone in the horizontal (lateral) direction and generally compound the problems of defining the different regions of moisture in the vadose zone.

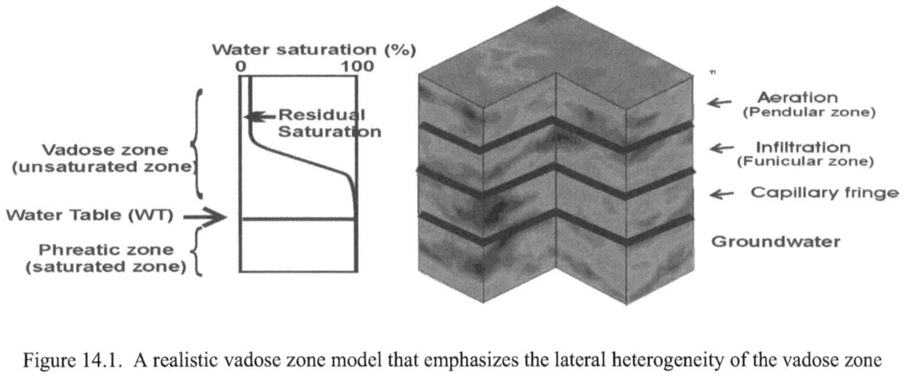

Figure 14.1. A realistic vadose zone model that emphasizes the lateral heterogeneity of the vadose zone

Y. Rubin and S.S. Hubbard (eds.), Hydrogeophysics, 413–440.

There is very little information in the literature regarding direct measurements of variations in geophysical properties (electrical conductivity, permittivity, seismic velocity, or density) in the unsaturated zone. Clearly, this is a region of fluctuations and seasonal variations in moisture content, and a region of changing geochemical regimes and bulk physical properties. Estimates of moisture within the vadose zone have primarily been studied through the use of electrical and nuclear measurements. The Topp equation (Topp et al., 1980) has been applied to measurements of the dielectric constant to obtain estimates of the moisture content, as was discussed in Chapter 4 of this volume. Estimates of dielectric constant from ground penetrating radar (GPR) have been used by several investigators to estimate water content, including Lesmes et al. (1999), Huisman et al. (2001), Huisman et al. (2003), and Grote et al. (2003). In clay-free environments, the electrical conductivity can be related to the amount of pore space occupied by fluids through Archie's law , as was also described in Chapter 4 of this volume.

Nuclear methods for estimating moisture in the near surface include the nuclear magnetic resonance (NMR) method and neutron-neutron techniques. Neutron-neutron methods are used routinely to determine the relative amount of moisture variation in the zone of aeration, but they are rarely used in a quantitative manner because of the difficulty in calibrating the instrument. NMR methods hold some promise of obtaining improved direct measurements of moisture and porosity, but field implementation of NMR is difficult, as will be discussed in Chapter 16 of this volume. Other, more common geophysical methods (magnetics, gravity, seismic, etc.) are sometimes used to determine general bulk properties of the subsurface, and to locate objects possibly associated with contaminants and the engineering aspects of the subsurface. However, these methods are not generally applicable to directly determining the hydrogeological parameters in the vadose zone.

In the past, the characterization of the vadose zone has been achieved using point measurement techniques, such as neutron probes, TDR, and tensiometry, but these methods have restricted measurement scales (typically on the order of centimeters). Since heterogeneity of the subsurface occurs at much larger scales, more appropriate measurement techniques are required. This has led to a growing interest in the use of geophysical methods for vadose zone characterization. Vadose zone hydrogeological objectives that can be facilitated using geophysical methods include: (1) defining the general hydrogeologic setting (vertical boundaries within the vadose zone), (2) determining the lateral heterogeneity of moisture content within the vadose zone, (3) monitoring and tracking infiltration through the vadose zone and into the saturated zone below the water table, and (4) locating sources of potential contamination and objects within the subsurface. In the following sections, field and experimental examples address the use of hydrogeophysical approaches for investigating these vadose zone objectives. In particular, the following case studies address some of the most fundamental problems encountered in the vadose zone: (1) estimation of hydrological boundaries, (2) soil property estimation, and (3) moisture monitoring.

14.2 Estimation of Hydrogeological Boundaries in the Vadose Zone

Determining the depth of the water table has been one of the fundamental objectives of hydrogeophysical investigations. In theory, this objective should be a simple thing to achieve with geophysical techniques, since the water table is a physical property boundary for seismic shear and compression waves, density, electrical permittivity (or the dielectric constant), and electrical conductivity (reciprocal of resistivity). The following discussion of a recent experiment that investigated the effect of moisture on the response of GPR will serve to illustrate some of the problems pertaining to interpreting geophysical measurements in terms of near-surface hydrologic properties.

A tank experiment was run at Ohio State University to determine the effect of changes in fluid levels (water and gasoline) on the GPR response. The details of this experiment were presented by Kim (2001). The tank model and experimental arrangement are shown in Figure 14.2. Fluid intake and outtake were located at the bottom of the tank to facilitate raising and lowering it without percolation of fluid from the surface. The tank was covered throughout the experiment to prevent the influx of rainwater. The overall project consisted of filling the bottom of the tank with fluid and making GPR measurements at closely spaced time intervals (approximately 30 minutes) after the water level was changed. GPR measurements were made using a GSSI SIR-10 system and a 500 MHz antenna, which was a time-domain system with co-pole antennas (parallel transmit and receive transducers) oriented perpendicular to the direction of the line traverse. Seventeen lines were run over the grid, as shown in Figure 14.2b. Consistent survey and processing parameters were maintained throughout the experiment.

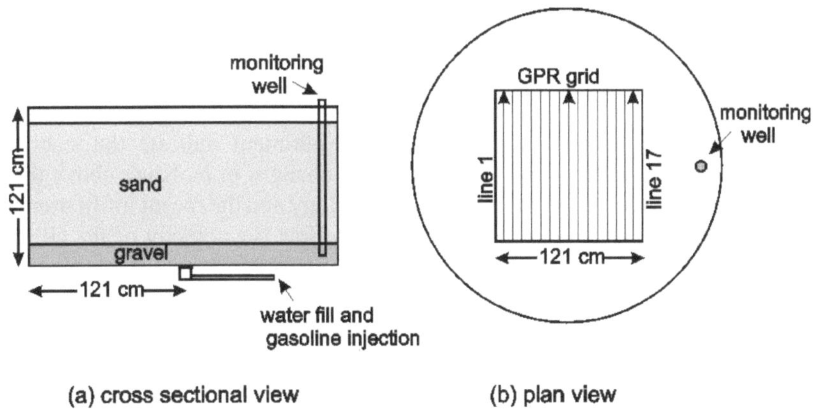

Figure 14.2. Configuration of tank and GPR survey grid in vadose zone experiment (modified from Kim, 2001): (a) cross sectional view, and (b) plan view. The porosity of the sand, gravel, and sand-gravel mix was 31%, 36%, and 11%, respectively. The capillary rise measured in the laboratory for the sand was 11.4 cm. The relative permeability for the sand, gravel, and sand-gravel interface was 7.3, 1460, and 1, respectively.

GPR 2-D profiles and 1-D traces extracted from the 2-D profiles for each monitoring stage are shown in Figure 14.3, along with the major reflection boundaries. A monitoring well inside of the tank was used to establish the water level. The water level was raised to 66 cm above the bottom of the tank in three stages (25.4 cm, 35.6 cm, and 66 cm, respectively), and subsequently was lowered in three stages (30.5 cm, 13.8 cm, and completely drained). A water level of 7.6 cm was observed in the bottom of the tank 16 hours after completely draining the tank. The residual water apparently came from residual moisture that slowly percolated down to the bottom of the tank. GPR measurements over the grid were repeated every 30 minutes to monitor changes in the GPR response. These measurements were repeated at each stage until no further changes occurred in the response on the GPR records. The stabilized results are shown for each stage in Figure 14.3. The water saturation, dielectric constant, and GPR velocity profiles determined from one-dimensional models are shown in Figure 14.4 for the 25.4 cm water level and after water drained from the tank. These models were generated using a one-dimensional modeling-program mixing model (or the complex refractive index method—CRIM; refer to Chapter 4 of this volume), which was introduced by Birchak et al. (1974) and further discussed by Wharton (1980) and Sihvola (1999).

The presence of a boundary between the zone of aeration and infiltration (refer to Figure 14.1) is not clear on the GPR records, and the GPR records (particularly the 1-D trace records) and 1-D numerical models (Figure 14.4) of the GPR data indicate that the GPR reflection interpreted as the water table actually came from the top of the capillary fringe. The capillary fringe is in tension saturation (close to 100% saturation), and it is effectively the GPR water table. The two-way travel time is nearly the same when the water was raised to the 25.4 cm level as it is when the water was lowered from 66 cm to 30.5 cm, even though the vertical distance was different. This shows that there was a significant amount of residual water in the pore space, which caused a decrease in velocity of the GPR wave after the water was drained from the tank.

Overall, the GPR data and models from this experiment indicate the stability and sensitivity of GPR measurements for determining changes in moisture above the water table. The residual water is a function of the mineralogy and the recent infiltration history of a site, and the residual water can significantly affect the velocity of the GPR wave. These measurements also show the complexity of interpreting the water table boundary from GPR data. The water table determined from GPR measurements is usually not the true hydrologic water table, but it is likely to be the upper boundary of the capillary fringe, which is close to 100% saturation. The ability to detect any hydrologic boundary on GPR records is highly dependent on the complexity of the local geology.

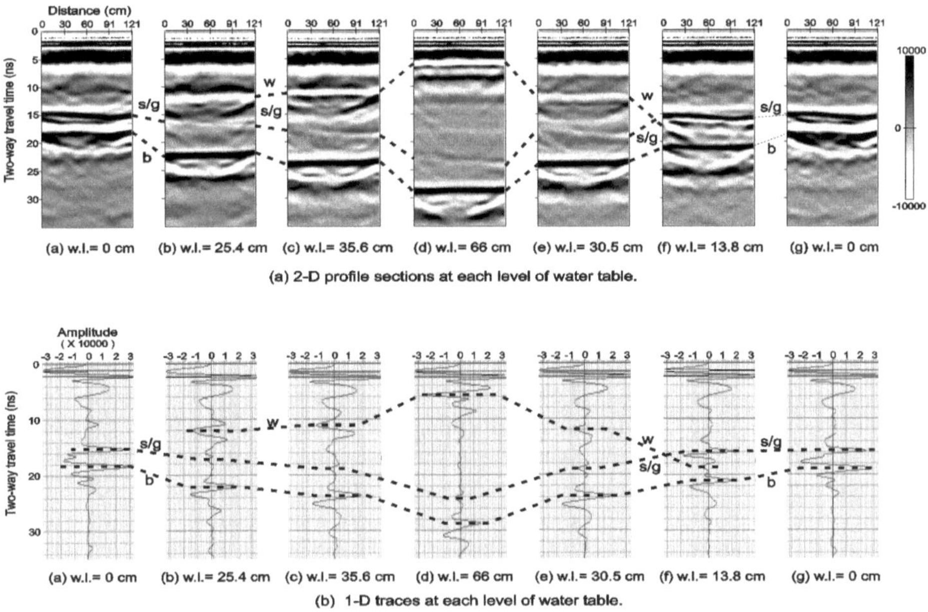

Figure 14.3. 2-D profile lines and 1-D traces after each change in the water level (modified from Kim, 2001) Water levels (wl) are given as elevation above the bottom of the tank. The water table reflection, the reflection from the sand/gravel boundary, and the reflection from the bottom of the tank are indicated by the symbols "w," "s/g," and "b," respectively.

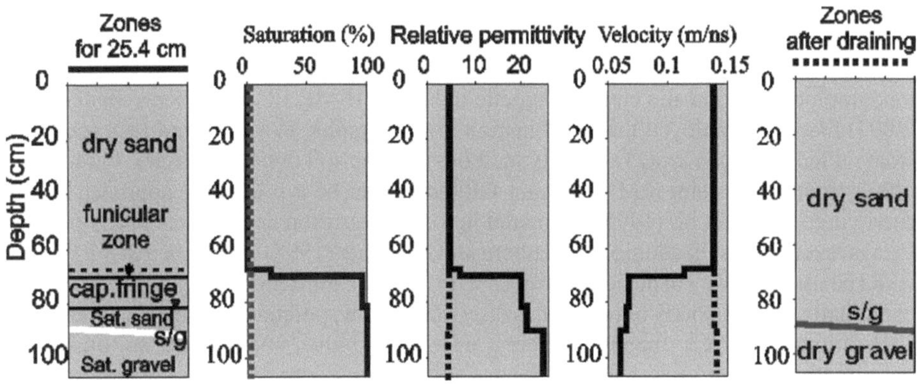

Figure 14.4. Hydrogeologic zones, water saturation, dielectric permittivity, and GPR velocity with depth at water level of 25.4 cm and after draining. The boundaries were obtained from 1-D modeling using the complex refractive index method (CRIM), as described by Wharton et al. (1980), to calculate the relative dielectric permittivity and the interval velocity of each hydrogeologic layer for a three-phase mixture of sand (quartz at a permittivity of 7 and feldspar assigned a relative permittivity of 7.5), water (relative permittivity of 80), and air (relative permittivity of 1). The sand-gravel interface is indicated by s/g. (modified from Kim, 2001).

14.3 Soil Property Estimation

The use of geophysical methods for providing hydrogeological property estimates in the vadose zone have received increased attention over the past few years for various applications—as ecological, environmental, engineering, and agricultural studies require denser and more precise estimates of soil and pore fluid properties. Near-surface geophysical methods, particularly those capable of mapping soil electrical conductivity and dielectric constant, have been used to estimate soil moisture and salinity. We review some of these studies in this section.

A substantial amount of study to date has focused on demonstrating that electromagnetic apparent conductivity (ECa) mapping is an effective tool to gauge the magnitude and spatial variability of soil salinity (Lesch et al., 1992; Hendrickx et al., 1992; Doolittle et al., 2001). Research results are mixed concerning the value of using ECa geophysical measurement techniques to monitor soil moisture. Scanlon et al. (1999) evaluated ECa measured with electromagnetic induction methods (EMI) as a reconnaissance technique to characterize unsaturated flow in an arid setting, and determined that the magnitude of the impact of moisture content on ECa was dependent on the geomorphic setting. An investigation conducted by Sheets and Hendrickx (1995) in an arid region of southern New Mexico discovered that a linear relationship existed between ECa and moisture content in the top 1.5 m of the soil profile. However, in a field study near Quebec City, Canada, carried out with traditional resistivity methods, Banton et al. (1997) found that the ECa mean and spatial patterns did not change significantly between wet and dry soil conditions. The study by Banton et al. (1997) also determined that ECa was moderately correlated with soil texture and organic matter, but not with porosity, bulk density, or hydraulic conductivity. Doolittle et al. (1994) determined a way to estimate clay pan depths in a Missouri soil, based on ECa values obtained with EMI methods. Furthermore, Fraisse et al. (2001) were able to define claypan soil management zones with a combination of topographic elevation and electromagnetic induction (EMI) ECa data. Kravchenko et al. (2002) likewise employed this combination of topographic elevation and ECa (obtained from pulled electrode array resistivity methods) to map soil drainage classes. Inman et al. (2002) found that using EMI ECa and GPR data can be a promising approach to soil surveying. Jaynes et al. (1995) estimated herbicide partition coefficients based on EMI ECa measurements. In addition, Eigenberg and Nienaber (1998) established that EMI ECa could be used as a way to detect field areas with high soil nutrient buildup. Consequently, a continually growing body of research is discovering new, potentially valuable agricultural applications for ECa mapping. Many of these studies may apply to other soil investigations.

As is apparent from this discussion of prior research, soil electrical conductivity can be affected by a number of different factors, some of which are more dominant than others depending on location, climate, etc. These factors include the soil type as determined by the mineralogy, chemistry, and grain distribution. The live vegetation can also impact the electrical property values. Electrical methods alone will not provide all of the answers for soil property analysis, but in many cases they can provide a relative indication of variations in the basic properties of salinity and moisture as a function of time and space. The following examples illustrate some of these applications.

14.3.1 EFFECT OF VARYING THE WATER TABLE ON CONDUCTIVITY

A recent investigation focused on determining the relative impact on apparent soil electrical conductivity (ECa) resulting from soil profile properties versus the factors associated with agricultural field operations or rainfall. EMI surveys were conducted under various field conditions, including different controlled shallow water table depths, changes in surface moisture content caused by rainfall or sprinkler irrigation, before and after fertilizer application and before and after tillage operations. This portion of the project took place at a test plot located behind the Ohio State University (OSU) ElectroScience Laboratory (ESL) in Columbus, Ohio (Figure 14.5). A subsurface drainage pipe system with two riser pipes connected up to the surface was installed at the site, allowing a shallow water table to be maintained at any desired level. The EMI ECa measurements for this part of the study were collected along lines oriented north-south, separated from one another by 1.5 m.

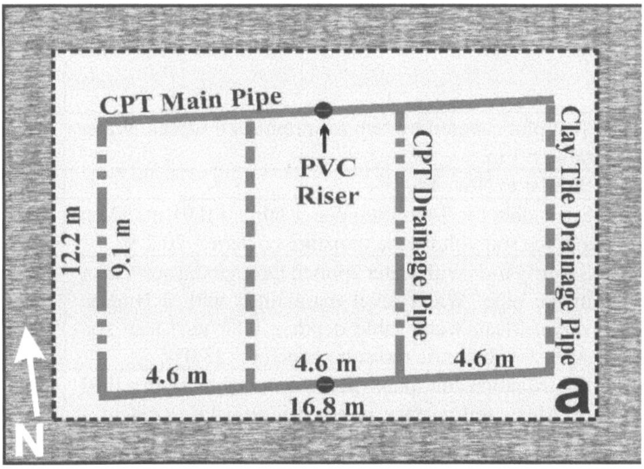

Figure 14.5. ESL #1 test plot schematic

Field conditions at the test plot utilized in this research project were monitored with water table observation wells, a time-domain reflectometry device for measuring soil surface moisture content, soil thermometers, an infrared thermometer for aboveground readings, and rain gauges. The type of soil present was a silty clay. A summary of test plot field conditions and corresponding geophysical survey results for the project are provided in Allred et al. (2003). The summary results presented in Table 14.1 for the ESL site are briefly described herein.

EMI apparent soil electrical conductivity ECa results were quite similar regardless of instrument frequency; therefore, our discussion will concentrate on data obtained at 14610 Hz using a GEM II electromagnetic induction system. Contour plots of electrical conductivity during the raising of the water table are shown in Figure 14.6. The shallow hydrologic-condition impacts were assessed through a linear regression analysis between average ECa versus average water table depth, and average ECa versus average soil surface volumetric moisture content. The ESL #1 test area EMI survey data incorporated into this statistical analysis included data obtained on days when there was sufficient water table or soil surface moisture information available. The coeffiecient of determination from linear regression analysis (R^2) for ECa versus water table depth was 0.00, and for ECa versus soil surface volumetric moisture content, it was 0.67. Although the correlation between ECa and soil surface moisture is definitely significant, it is probably not strong enough to warrant using ECa as a direct predictor of volumetric moisture content at the ground surface. This same ECa data exhibited only minor correlation to either soil temperature ($R^2 = 0.15$) or air temperature ($R^2 = 0.09$).

Table 14.1. ESL #1 field conditions and EMI results for the part of the project focused on determining the relative impacts on ECa due to soil profile properties versus factors associated with raising the water table. ECa values are in mS/m.

DATE	CONDITIONS	RESULTS (Volts × 10^{-3})	
		Mean ECa	St Dev ECa
11/14/01	Test plot covered by tarp for previous 6 weeks. Water table > 1 m.	14.16	3.09
11/16/01	Similar to Nov. 14	13.77	2.98
07/08/02	Site stable for 7 months. Water table > 0.91 m. Mean surface soil volumetric moisture content = 20.2 %	10.4	1.65
07/ 12/02	Subirrigation with water applied through the north riser intake pipe. Water level maintained with a Hudson valve. Mean water table depth = 0.84 m, Mean soil surface volumetric moisture content = 23.0%.	7.37	1.75
07/15/02	Subirrigation continued. Mean water table depth = 0.62 m, Mean soil surface volumetric moisture content = 22.7%.	8.98	2.16
07/18/02	Subirrigation continued. Mean water table depth = 0.35 m, Mean soil surface volumetric moisture content = 41.8%.	10.36	2.86
07/19/02	Subirrigation continued. Mean water table depth = 0.27 m, Mean soil surface volumetric moisture content = 52.1%.	13.33	2.87
07/26/02	Subirrigation had been discontinued seven days prior to this date. Mean water table depth = 0.78 m	13.8	2.6
08/7/02	Subirrigation had been discontinued 19 days prior to this time	13.49	2.51
09/8/02	As a supplement to the 45.5 kg of fertilizer already applied, 22.7 kg of 21-28-7 starter fertilizer were added to the test plot eight days before this date and test plot evenly watered. Mean water table depth =0.87 m. Mean soil volumetric moisture = 29.2%.	13.15	2.38

This ECa spatial-pattern consistency, evident regardless of the field conditions, is a strong indication that soil profile properties tend to dominate the ECa response measured by near-surface geophysical methods. Higher ECa numbers are found within a tongue-shaped area that extends westward from the east boundary for almost three quarters of the test plot length. On July 19, 2002, the water table was mounded to the surface over the central portion of the test plot directly above the subsurface drainage pipe system (compare Figure 14.5 and Figure 14.6), while along the periphery of the ESL #1 test area, with the exception of the southeast corner, the soil surface was dry and the water table much lower.

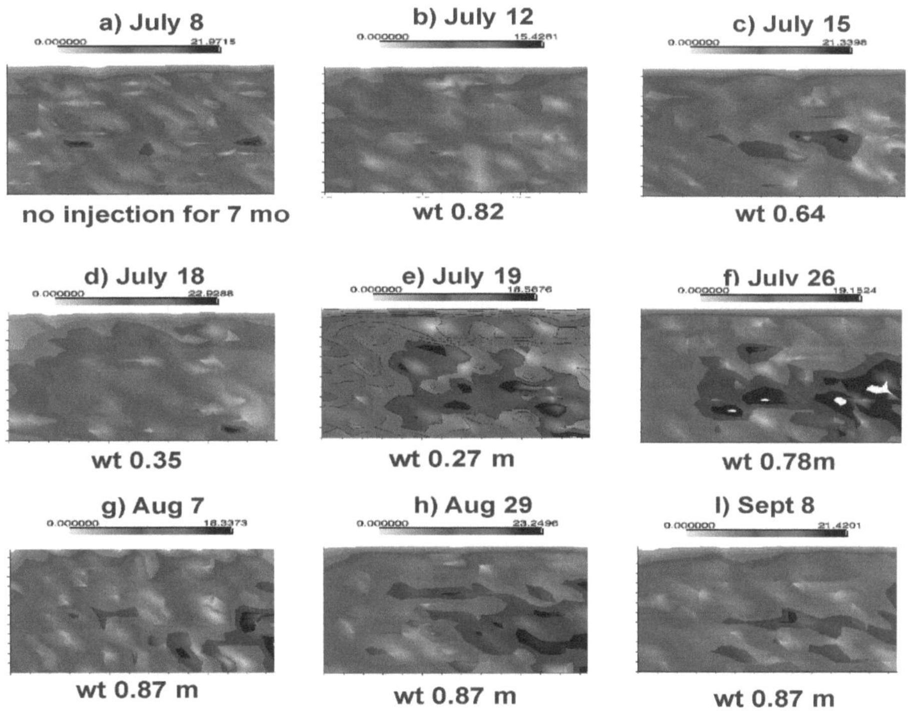

Figure 14.6. ECa contour maps for (a) July 8, 2002, (b) July 12, 2002, (c) July 15, 2002, (d) July 18, 2002, (e) July 19, 2002, (f) July 26, 2002, (g) August 7, 2002, (h) August 29, 2002, (I) September 8, 2002. Water table measurements were not taken on August 7 and August 29. Lighter colors are lower conductivity, and darker colors represent higher values of electrical conductivity.

14.3.2 ESTIMATION OF VOLUMETRIC WATER CONTENT USING SURFACE GPR GROUNDWAVE TECHNIQUES

Several recent studies have investigated the accuracy and resolution of using GPR groundwave travel-time measurements to estimate near-surface soil water content; the use of groundwave, reflected, and transmitted GPR arrivals for estimating soil water content was recently reviewed by Huisman et al. (2003b).

The groundwave is the part of the radiated energy that travels between the transmitter and receiver through the top of the soil. Advantages of the groundwave approach over conventional approaches (GPR reflection approaches such as those discussed in Section 14.1) is that the groundwave does not require the presence of reflecting horizons or calibration information, such as the depth to a reflector, to interpret the groundwave travel time in terms of water content. As such, it has the potential to provide very-high-resolution estimates of shallow soil water content in a relatively straightforward way and in a completely noninvasive manner. Although limited to the very-near-surface soil layer, the depth sampled by the GPR groundwaves is often the critical zone for many engineering, ecological, agricultural, and environmental studies.

Using the known distance between the transmitting and receiving antenna and the measured signal travel time, the GPR velocity can be calculated and converted to dielectric constant, as was discussed in Chapter 7 of this volume. Using a petrophysical relationship between dielectric constant and volumetric water content, either developed for the particular study site or borrowed from literature (such as Topp's Equation [Topp et al., 1980]), the dielectric constant values obtained from GPR groundwave data can be used to estimate near-surface water content. Two different acquisition modes can be used to obtain dielectric constant from surface GPR data: common midpoint (CMP) or common (or single) offset, both of which are described in Chapter 7. Both Huisman et al. (2001) and Grote et al. (2003) reported that water-content estimates, from GPR groundwave CMP data, agreed well with colocated TDR estimates of volumetric water content. Huisman et al. (2001) reported a volumetric water-content accuracy of 0.024 m^3m^{-3} using 225 Mhz antennas, and Grote et al. (2003) reported root mean squared errors of the GPR-obtained volumetric water-content estimates of 0.022 and 0.015 m^3m^{-3} using 450 and 900 MHz antennas, respectively.

However, multi-offset GPR acquisition is cumbersome and time consuming, and thus not well suited for rapid estimation of water content at the field scale. As mentioned above, dielectric constant can also be obtained using common-offset GPR acquisition approaches, provided that the approximate arrival time of the groundwave is known from a multi-offset GPR measurement. The idea of using GPR groundwave data collected using the quick common-offset acquisition geometry was first suggested by Du (1996). Recent experiments by Lesmes et al. (1999), Hubbard et al. (2002), Huisman et al. (2002), Huisman et al. (2003a), and Grote et al. (2003) have confirmed that soil-water-content estimation using common-offset groundwave travel-time data is a viable approach for quickly providing soil-water-content estimates at the field scale. In an irrigation experiment, Huisman et al. (2002) used 225 MHz common-offset GPR groundwave data to estimate soil water content over a 60 × 60 m area and found that the GPR estimates agreed very well with those measurements obtained using conventional TDR methods, but that the GPR approach provided much-higher-resolution information in a noninvasive manner.

Using 450 and 900 MHz common offset GPR groundwave data, Grote et al. (2003) estimated the spatial and temporal variations in near-surface water content over a large-scale agricultural site during the course of a year. They found that soil texture exerted an influence on the spatial distribution of water content over the field site. In addition, although the mean values of water content varied as a function of season and precipitation, the spatial pattern of moisture content within the natural field was temporally persistent. Compared with soil sample gravimetric measurements of water content, they reported a root mean squared error of 0.011 for the 900 MHz data and 0.017 for the 450 MHz data using the common-offset GPR groundwave approach.

Figure 14.7 illustrates a time series of volumetric water-content estimates obtained at a California winery study site using 900 MHz surface common offset GPR data (modified from Hubbard et al., 2002). Each image in this figure includes over 20,000 estimates of volumetric water content obtained using GPR data, and although the absolute estimates of water content change with time, the spatial pattern clearly remains the same.

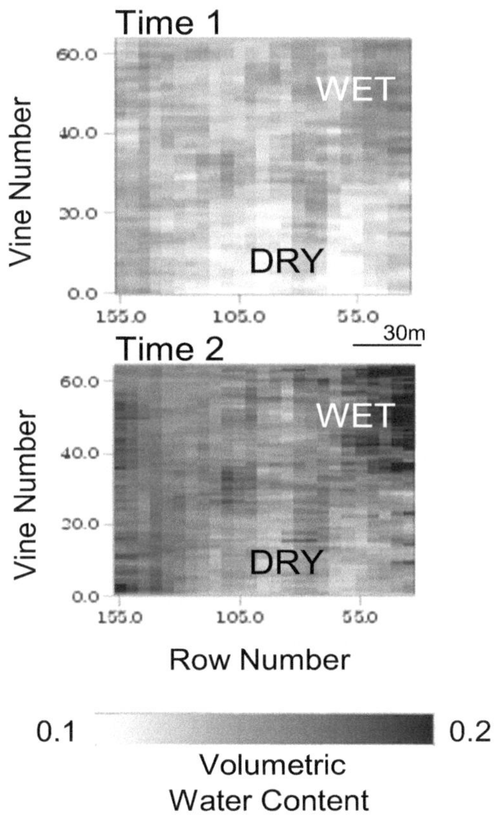

Figure 14.7. Soil moisture estimated over a winery study site using surface 900 MHz GPR groundwave data at various times during the year. Each image = ~20,000 data points. Modified from Hubbard et al. (2002).

The effective measurement volume of the GPR groundwave is still a topic of research. Huisman et al. (2001) concluded that soil-water-content measurements based on groundwave data are similar to measurements with 0.10 m long TDR probes for both the 225 MHz and 450 MHz GPR antennas. Grote et al. (2003) compared estimates of water content obtained using 450 MHz and 900 MHz common-offset GPR groundwave data with measurements collected using gravimetric techniques, in soils having various textures and at depths between 0 and 0.10 m, 0.10 and 0.20 m, and 0 and 0.20 m below ground surface. They found that the estimates obtained using these frequencies showed the highest correlation with the soil-water-content values averaged over the 0–0.20 m range and the least correlation with the water-content measurements taken from the 0.10–0.20 m interval.

As described by Huisman et al. (2003b), there are some drawbacks to using the GPR groundwave to estimate soil water content. These drawbacks include the potential difficulties associated with: (1) clearly distinguishing the groundwave from other arrivals, (2) choosing an antenna separation for which the arrival times of the air and groundwave can consistently be picked, despite moving the antennas across a field with varying soil water content, and (3) collecting groundwave data at far antenna separations, (since the groundwave is attenuated more quickly than other waves.) In spite of these potential limitations, GPR groundwave travel-time data have provided accurate, very-high-resolution estimates of shallow soil water content.

14.4 Moisture Monitoring in the Vadose Zone Using Resistivity

Because electrical resistivity is a volumetric measurement of volumetric moisture content (see Chapter 4 of this volume), resistivity surveys have proven to be of great value in aiding hydrological studies of the vadose zone. Most previous applications have concentrated on monitoring forced loading of the system via tracer tests (see, for example, Section 14.5 below). Few studies, however, have used resistivity methods to examine changes in moisture content of soils caused by natural loading. The focus here is on examples of how resistivity has been used to study such conditions.

Several investigations of vadose zone processes have used surface-deployed resistivity surveys for monitoring moisture. Kean et al. (1987) described the use of vertical electrical soundings (VES; see Chapter 5 of this volume) for studying changes in the vadose zone under natural loading. At one of their survey sites, Kean et al. (1987) observed changes over a 10-week period, albeit in a shallow water table environment. Frohlich and Parke (1989) also used VES to infer changes in moisture content in the unsaturated zone. In their case, measurements were made over a period of 9 weeks; changes in resistivity to the water table depth (approximately 3 m) were observed and compared with the conventional neutron probe measurements.

Surveys based on VES are somewhat limited, given the non-uniqueness of inversion methods for soundings (as demonstrated, for example, by Simms and Morgan, 1992). Perhaps more restrictive is that the measurement volume (lateral and vertical variations) for soundings increases with depth. This will create significant errors if any lateral variability in resistivity exists (caused by lithology and/or moisture content). Benderitter and Schott (1999) have shown how surface resistivity surveys can be used to reveal such lateral variability, by analyzing changes in moisture content during rainfall events over a 100-day period. Benderitter and Schott used a short (7.5 m) traverse with 15 electrodes and measured apparent resistivity using a dipole-dipole configuration (see Chapter 5 of this volume). Consequently, the depth of sensitivity for their surveys was limited to the top meter. In a more recent study, Zhou et al. (2001) used a combined surface-borehole electrode array to map natural changes in resistivity caused by rainfall inputs. In their investigation, only short-duration events (several days) and near-surface (top 1.5 m) changes were studied. Zhou et al. (2001) claim that their approach, at their site, is capable of measuring moisture contents with 0.1 m^3 m^{-3} errors, although changes in moisture content should be more accurately resolved.

Two other examples are presented here, in which the resistivity method has been used to study changes in moisture content within the vadose zone under natural loading. In the first case study, surface resistivity arrays are used to monitor the spatial variability of infiltration resulting from snowmelt in a small field plot in Norway. In the second case study, borehole electrode arrays are utilized to monitor small moisture-content changes to a depth of 10 m at a sandstone aquifer in the U.K.

14.4.1 MONITORING SNOWMELT USING SURFACE ELECTRICAL METHODS

Experiments have been carried out in Norway to study flow and transport processes during snowmelt. The experiments were driven by the need to determine rate estimates for transport and retardation of pollutants from diffuse source pollutants, such as de-icing chemicals. At the Morreppen field site, adjacent to Oslo Airport, an extensive series of hydrological and biogeochemical investigations were completed, and a series of geophysical surveys were taken during 2001 to supplement earlier studies. French et al. (2002) report results from cross-borehole electrical resistivity surveys, designed to monitor the movement of salt tracers applied beneath the snow cover. The results from French et al. (2002) support earlier observations of preferential transport through the sand and gravel sediments in the vadose zone. The following case study summarizes results from surface resistivity surveys carried out during the snowmelt period of 2001. These surveys were performed to assess the spatial variability of infiltration caused by snowmelt, and thus determine the validity of assuming diffuse sources within any predictive hydrological model applied at the site.

Figure 14.8 illustrates the change in groundwater level during the 2001 snowmelt period, when the response of the water table to onset of snowmelt during April 2001 was rapid.

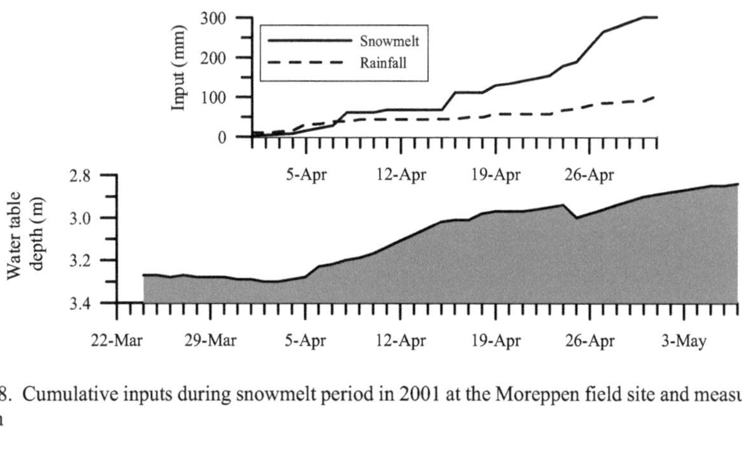

Figure 14.8. Cumulative inputs during snowmelt period in 2001 at the Moreppen field site and measured water table depth

During 2000, to determine the spatial variability of infiltration, investigators installed 80 electrodes at 0.2 m depth (to avoid near-surface ground-frost effects) in a small field plot 1 m wide by 3.75 m long. The electrodes were arranged in a grid at 0.25 m spacing. Using a Wenner configuration (see Chapter 5 of this volume), investigators carried out electrical resistivity surveys prior to snowmelt and during the snowmelt period. The data were inverted using a 3-D inverse model to produce images of resistivity (to a depth of 0.6 m) during the snowmelt event.

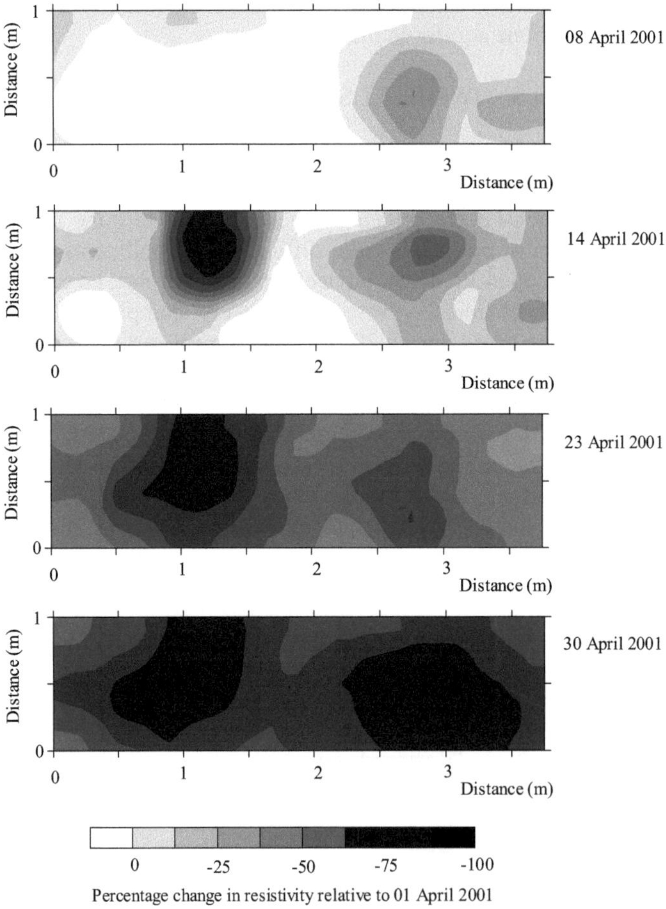

Figure 14.9. Changes in resistivity at the ground surface due to snowmelt at the Moreppen field site

Figure 14.9 shows a selection of results from the resistivity survey. Here, changes in resistivity relative to April 1, 2001, are shown for the horizontal plane corresponding to the top 0.1 m of the soil profile. The images reveal significant localized changes in resistivity during the initial phase of the snowmelt event, suggesting preferential infiltration. Towards the end of the event, more widespread infiltration appears to exist. The observed response is likely to be a result of localized thawing of the near-surface soil horizon. Prior to any such thaw, the soil surface impedes any infiltration. Note that significant changes in resistivity are observed, as a result of intensive loading on an initially dry soil. The results from the geophysical survey support a hypothesis that preferential infiltration occurs at the site. These findings must be considered alongside evidence of preferential transport in deeper soil sections and will hopefully lead to the development of more appropriate regulatory models of flow and transport at the site.

14.4.2 MONITORING SEASONAL CHANGES IN SOIL MOISTURE USING BOREHOLE ELECTRICAL METHODS

Given the limited depth sensitivity of surface resistivity surveys, Binley et al. (2002) demonstrated how borehole-based electrode arrays may be used to characterize seasonal variation in moisture content to depths of 10 m at a sandstone site in the UK. This study supplemented earlier work using borehole radar tomographic methods to determine changes in moisture content caused by forced (tracer) and natural hydraulic loading at the site (Binley et al., 2001). The study of seasonal variation by Binley et al. (2002) was carried out over two years, allowing comparison of changes with estimated net rainfall inputs.

For the study by Binley et al. (2002), electrical resistivity tomography (ERT) surveys (see Chapter 5 in this volume) from a 32-electrode array in a single borehole were converted to vertical resistivity profiles using data inversion techniques, by discretizing the vertical profile into 0.82 m thick units. Using site-specific petrophysical relationships between moisture content and resistivity, these electrical profiles were converted to moisture content profiles. Figure 14.10 shows example results from the work of Binley et al. (2002). In Figure 14.10, the monthly net rainfall at the site between July 1998 and November 2000 is shown. Changes in the bulk electrical conductivity for the near-surface unit (i.e., representing the top 0.82 m soil zone) are shown in Figure 14.10b. The correlation between electrical conductivity and net rainfall amounts is clearly seen. Figure 14.10c reveals changes in moisture content estimated from electrical conductivity throughout the unsaturated sandstone profile (2 to 10 m depth). The delay in response owing to the winter wetting periods, caused by the hydraulic impedance of the near-surface sediments, is seen to be several months. Once the sandstone responds to such inputs, migration through the profile to a depth of 6 to 7 m is rapid. Hydraulic resistance at this depth is then apparent from the step change in moisture content with depth. (Notice how small the increase is in moisture content at depths beyond 7 m during March and April 1999, despite continual increases at shallower depths.) The inferred hydraulic resistance is consistent with the observed location of fine-grained sandstone recorded from cores taken from the site. Further note, in Figure 14.10c, that the drying fronts are also visible to significant depths in the profile. The small moisture-content changes at depth inferred from these surveys could not be inferred from surface-deployed resistivity arrays.

Binley et al. (2002) also compared moisture-content changes inferred from electrical resistivity to those derived from borehole radar profiles—and demonstrated consistency in their responses. Binley and Beven (2003) used these data to examine the potential to calibrate hydraulic flow models for the site, with the aim of using geophysical data to constrain predictions of pollutant transport through the vadose zone.

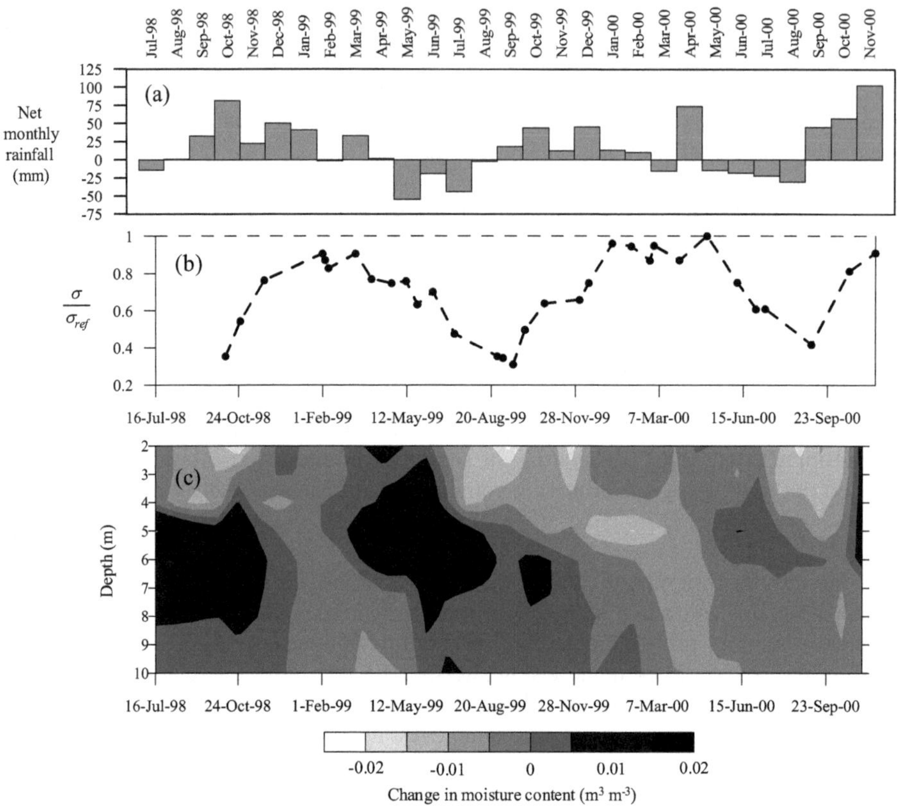

Figure 14.10. (a) Net monthly rainfall estimates at the Hatfield site (supplied courtesy of the UK Environment Agency); (b) fractional change in bulk electrical conductivity in 0–0.82 m layer relative to May 3, 2000; (c) change in moisture content throughout unsaturated zone (values computed relative to February 1, 1999)

14.5 Monitoring Forced Infiltration at a Single Site Using Multiple Methods

The Sandia/Tech vadose zone (STVZ) site in New Mexico was constructed to study the combined use of geophysical methods and hydrological modeling to monitor unsaturated flow processes within a heterogeneous sedimentary site. The instrumented portion of the STVZ site was approximately 10 m × 10 m (Figure 14.11) × 13 m deep. A 3 m × 3 m infiltrometer was installed at the center of the site to infiltrate water at a known, constant rate. The infiltrometer was divided into nine 1 m^2 arrays, each array containing 100 equally spaced medical needles mounted on 50 mm PVC pipes. The remainder of the test area was covered by a polyethylene tarp to create a no-flow boundary along the earth's surface. The top part of the infiltrometer was covered with insulating foam and a second polyethylene tarp to minimize evaporation or additional infiltration through the top (Paprocki, 2000).

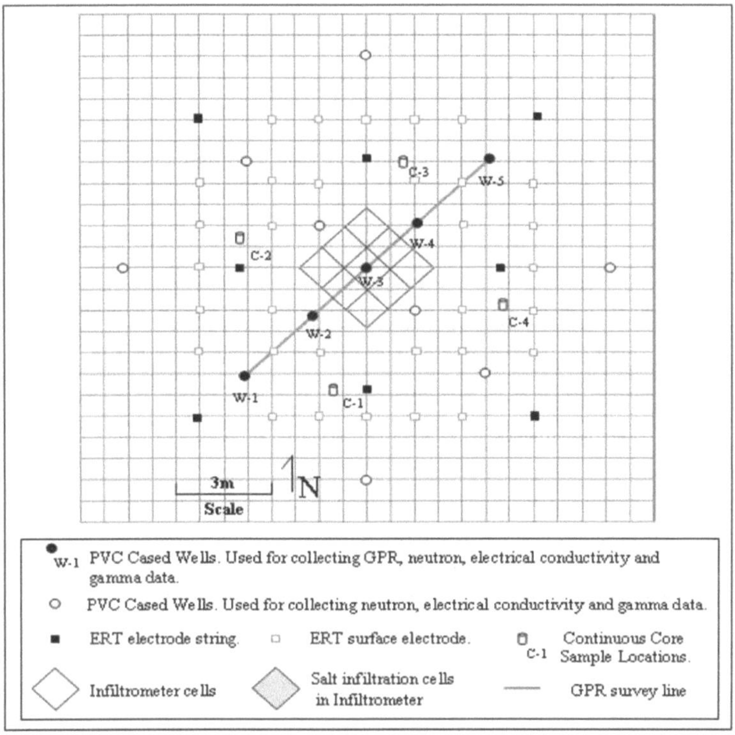

Figure 14.11. Arrangement of boreholes at the STVZ site

Site stratigraphy (Figure 14.12) was determined from four continuous core samples (see Figure 14.11). Note that the clay content of the poorly sorted gravel layer, from 4 to 7 m in depth, increased towards the northwest corner of the site. This layer had a lower hydraulic conductivity than the sand layers and served as a barrier to the downward migration of water.

Thirteen PVC-cased wells were installed at the site, with a maximum depth of 13 m. Neutron measurements and induction logs were collected along these wells. Cross-borehole ground penetrating radar (GPR) surveys were also conducted in five PVC-cased wells along an 11 m diagonal line from the southwest to the northeast end of the site (Figure 14.16). Nested time-domain reflectometry probes (TDR), tensiometers, and suction lysimeters (not shown) were placed along the outer perimeter of the infiltrometer and between the PVC-cased wells.

In addition to neutron measurements and GPR, a three-dimensional (3-D) ERT system was installed at the STVZ site to provide 3-D images of electric conductivity. The ERT system used a combination of surface and borehole electrodes. Surface electrodes consisted of 30 cm lengths of copper-plated steel rod driven into the surface of the site (Figure 14.10). Electrodes were placed in eight boreholes. Each borehole contained 17 stainless-steel electrodes placed at 0.76 m intervals, with the uppermost electrode approximately 0.5 m below the surface.

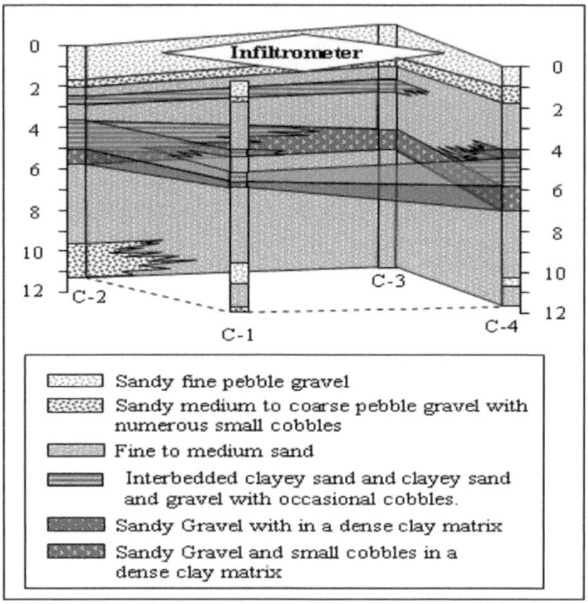

Figure 14.12. Stratigraphy of the STVZ site

14.5.1 GPR DATA COLLECTION AND PROCESSING AT THE STVZ SITE

A Sensors and Software PulseEKKO 100 system, with an antenna frequency of 100 MHz, was employed to obtain cross-borehole GPR measurements. Data were acquired by positioning the transmitting antenna within a borehole, and then collecting data at several receiver depths in another borehole 2 to 3 m away. The transmitting antenna was then moved to a new position, and data were collected in the same way. Repeating this process creates a dense array of intersecting ray paths, where each ray path represents the shortest path perpendicular to the EM wave front that passes through the region between the two boreholes. During data collection, the angle of the transmitting antenna and the receiving antenna to the horizontal was limited to 45° to avoid the influences of wave reflections from high angles of antenna offset (Peterson, 2001), low signal-to-noise ratios at high angles, and problems with the presence of wires on the ground (Paprocki, 2000; Alumbaugh, 2002). To assure that the output radar signal of the GPR system did not fluctuate during the data collection, a time-zero calibration was conducted every 10 transmitter depths. This involved placing the transmitting and receiving antennae in the air at a known distance. The correct time-zero can be determined from the travel time calculated by the speed of light in air and the known distance.

The inversion code GeotomCGTM (GeoTom, LLC, 1998) was used to produce images of velocity and attenuation from travel time and normalized amplitude measurements. GeotomCGTM is based on the simultaneous iterative reconstruction technique (Jackson and Tweeton, 1996) and uses ray tracing methods to generate tomographic images. Straight ray inversion, which assumes the ray path between the transmitter and receiver is a straight line, was chosen for both velocity and attenuation inversion. Theoretically, EM waves tend to travel through high-velocity regions rather than low-velocity regions. This makes the ray paths in an inhomogeneous media tend to curve around the low-velocity regions, and thus they are no longer straight lines. Although in theory it is more accurate to use curved-ray inversion (Alumbaugh et al., 2002), there were difficulties with the curved-ray inversion providing stable solutions for the attenuation data. Investigators chose to use the straight ray inversion because it seemed to be more robust for the attenuation tomograms in this study (see Chang et al., 2003).

The inversion code provides images of GPR velocity and attenuation on a discrete grid of cells, with an interval of 0.25 m in both the horizontal and vertical direction. The inverted images were then converted into estimated volumetric water content and electrical conductivity. Given volumetric water content, θ_v, the value of dielectric constant, k, can be estimated from the empirical relation developed for this site by Alumbaugh et al. (2003):

$$\theta_v = 0.0136\kappa - 0.033 \tag{14.1}$$

For GPR data where the frequency ω is high enough that $\omega\varepsilon >> \sigma$, the conductivity can be determined from

$$\sigma \cong 2\sqrt{\frac{\varepsilon_0}{\mu}}\sqrt{\kappa}\alpha \cong \frac{\sqrt{\kappa}}{188.5}\alpha \tag{14.2}$$

where α is attenuation in Nepers/m, σ is conductivity in S/m, ε_0 is the free-space permittivity, κ is dielectic constant, and μ is magnetic permeability.

14.5.2 ERT DATA COLLECTION AND PROCESSING STVZ SITE

The ERT data were collected using the dipole-dipole array with combinations of the vertical electrode arrays and surface electrodes. The data were collected using a system that utilized 3 receiver channels and a multiplexer capable of connecting up to 120 electrodes at a time. Reciprocal measurements were used to remove noisy data. Reciprocal pairs that differed by more than 10% of the data value were removed from the data. The remaining pairs were averaged. Typically, about 30% of the data were removed. The final averaged data sets usually contained about 20,000 points. The results presented here used the 3-D anisotropic inverse algorithm described by LaBrecque and Casale (2002). The code also implements a differencing inversion scheme similar to that described by LaBrecque and Yang (2001) to allow effective imaging of small changes in background resistivity. The region to be imaged was discretized into a mesh of 44 × 44 × 40, for a total of 77,440 elements. Element size ranged from 0.38 m in the center of the mesh, to 6 m along the boundaries. The inversion of a complete background data set took about 9 hours on a Pentium 4, 1.8 GHz PC.

14.5.3 BOREHOLE LOGGING DATA AT THE STVZ SITE

Neutron measurements were collected using a Campbell Pacific Nuclear International Inc. Model 503-DR Hydroprobe. Measurements were taken at 0.25-meter intervals. The probe was calibrated by collecting TDR and neutron data simultaneously at three depths in a small pit adjacent to the test area. Prior to emplacement, the TDR probes in the pit were calibrated in the laboratory using sediments from the test area (Paprocki, 2000). The calibrated TDR data were used to create a linear calibration curve for the neutron data. The same boreholes used for neutron measurements were also logged using a Geonics EM-39 borehole induction tool at 0.25-meter intervals.

14.5.4 RESULTS OF GEOPHYSICAL SURVEYS AT THE STVZ SITE

Prior to the start of water infiltration, multiple sets of background data were completed for each of the four techniques. Starting in March 1999, fresh water with a conductivity of about 70 mS/m was infiltrated at a rate equivalent to 2.7 cm/day over the area of the 3 m × 3 m infiltrometer. It should be noted that conductivity of the background water was around 200 to 400 mS/m, somewhat higher than the infiltrating water. Although the sediments contained some buffering capability such that the salinity (and thus the conductivity) of the infiltrating water increased, this buffering capability was depleted during the coarse of the experiment. Measurements of pore-water samples taken after two years of infiltration showed that the water in the principal flow paths was close to the conductivity of the pore water. There were two competing effects during the experiment: (1) increased pore-water content, tending to increase the bulk conductivity of the soil, and (2) decreasing salinity of the pore water, tending to decrease the bulk resistivity.

14.5.4.1 *3-D ERT Results at the STVZ site*

Figure 14.13 shows 3-D views of the background conductivity distribution for ERT and the changes in conductivity at three times after the start of infiltration. In the background images, the strongly resistive zones (Figure 14.13a) have been made transparent. The images show a relatively conductive zone corresponding to the clay bearing zones between 2 and 6 m in depth.

During the early phase of infiltration (Figure 14.13b), there is a small elliptical zone of increased conductivity in the fine sands below the infiltrometer. Very little water is retained within coarse sands and gravels immediately below the infiltrometer; thus, the water moves quickly through this layer and the bulk conductivity of this layer increases only slightly from the pre-infiltration value.

After just over one month of infiltration (Figure 14.13c), the water moved downward and westward, and some of the fluid was located outside the image zone. The conductive zone was large within the finer-grained sediment layers and smaller in the intervening beds.

After a year of infiltration (Figure 14.13d), the size of the conductive zone decreased from that shown in Figure 14.13c. This apparent decrease is at least in part caused by the removal of salts from the subsurface, such that the pore water within the zone has a lower conductivity. In addition, the water was significantly colder in March 2000. Because of the rapid infiltration, thermocouple data (not shown) from the site show annual changes in temperature as large as 18°C at a depth of 2 m and 11°C at 4 m. The decrease in temperature can significantly decrease the pore water electrical conductivity.

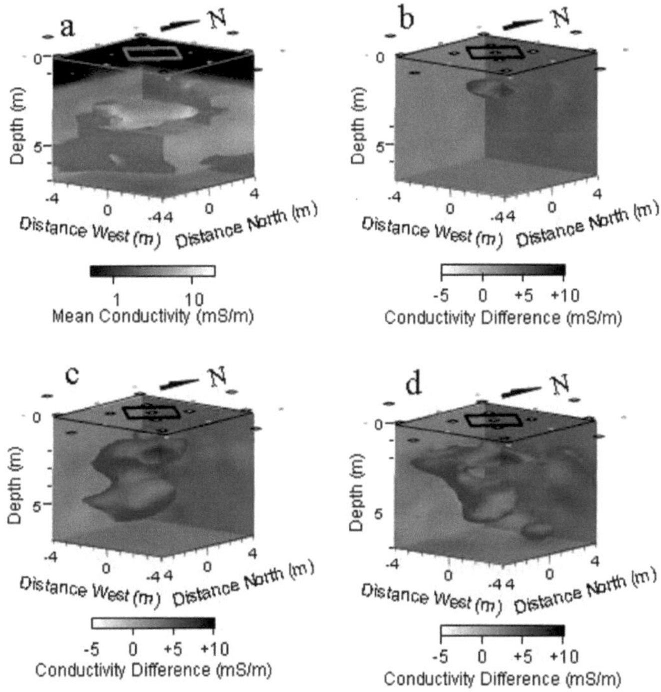

Figure 14.13. Composite showing images of 3-D ERT results of conductivity or change in conductivity for (a) pre-infiltration background data collected February 16, 1999; (b) data collected April 20, 1999, approximately 10 days after the start of infiltration, (c) data collected, March 20, 1999, approximately 40 days after the start of infiltration; and (d) data collected on March 28, 2000, approximately 1 year after the start of infiltration.

14.5.4.2 *Comparison of ERT, GPR, Neutron, and EM-39 Results*

Figure 14.14a shows an image of soil moisture content along the SW-NE profile as derived from GPR velocities. The values of moisture content estimated from calibrated borehole neutron measurements are superimposed on the GPR images. Both GPR and neutron results show water content as the percent of the total soil volume and are plotted on a scale of 0 to 10%.

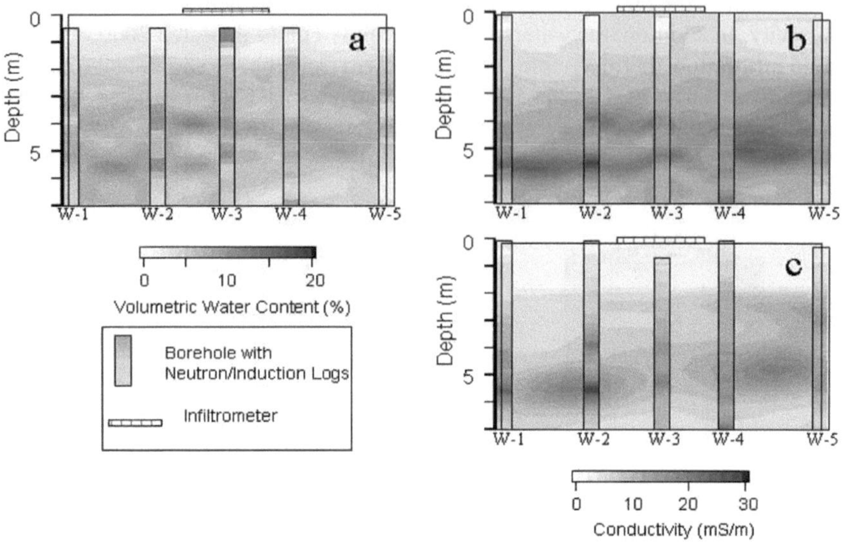

Figure 14.14. Pre-infiltration images of (a) volumetric water content from GPR and borehole neutron data; (b) conductivity data from crosshole GPR attenuation and borehole induction logs; and (c) ERT and borehole induction logs. Data were collected in February 1999.

The region of high moisture content seen at the top of the neutron log in borehole W-3 is caused by bentonite placed in the borehole annulus just below the infiltrometer. The bentonite was placed at the top of this borehole to prevent water from moving down the borehole annulus below the infiltrometer. The remaining portions of this borehole and all of the other boreholes were backfilled with native sediments. There are also some discrepancies between the radar and neutron results in the outer boreholes, particularly borehole W-1. Note that both methods have limitations, and the true moisture content probably lies somewhere between the two results. GPR images will tend to have lower resolution than neutron logs, particularly when the distance between the boreholes increase. On the other hand, the neutron data are more susceptible to near-borehole effects.

In Figure 14.14b, GPR attenuation data, obtained from GPR amplitudes, were used with the GPR dielectric information (obtained from GPR travel time data) to estimate electrical conductivity in mS/m, following Equation 14.2. Induction logging data collected using the EM-39 are superimposed on the GPR images. For most of the wells, there is very good correlation between the EM-39 results and the conductivity derived from GPR data. Although the electrical conductivity of sediments is a function of a number of factors—including clay content, pore-water salinity, and porosity—the similarities between the moisture content (Figure 14.14a) and conductivity (Figure 14.14b) indicate that conductivity and moisture content may be strongly correlated at this site.

The final panel, Figure 14.14c, shows conductivities extracted from the 3-D ERT images. Although the ERT images agree with the overall pattern seen in the GPR images, a highly resistive layer at the top and a conductive zone from 3.5 m to 6 m in depth, the ERT images have lower resolution than those of the GPR and EM-39.

After the onset of infiltration, neutron, GPR, and ERT data were collected periodically. Figures 14.15 and 14.16 compare the use of the three geophysical methods for following the infiltration front after approximately 10 days and 40 days of infiltration respectively. All of the methods show changes in volumetric moisture content or electrical conductivity from the background images shown in Figure 14.14.

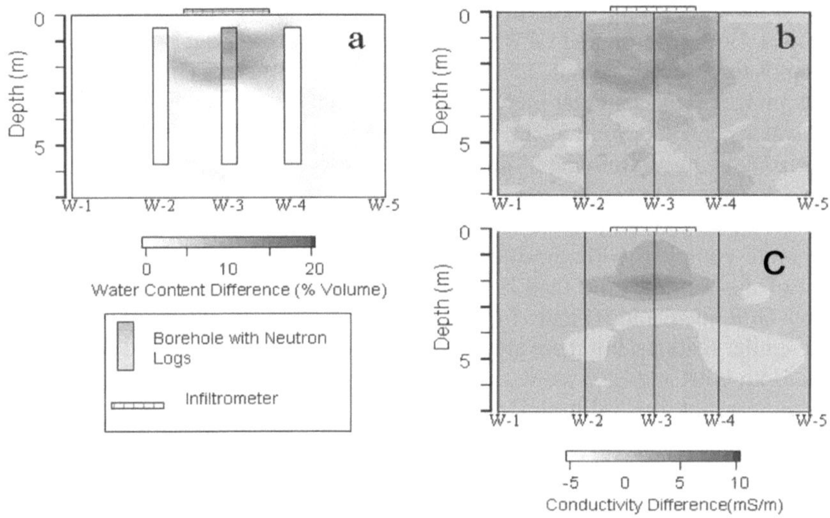

Figure 14.15. Images of (a) the change in volumetric water content from GPR and borehole neutron data, (b) conductivity data from crosshole GPR attenuation, and (c) ERT data. Data were collected between March 18

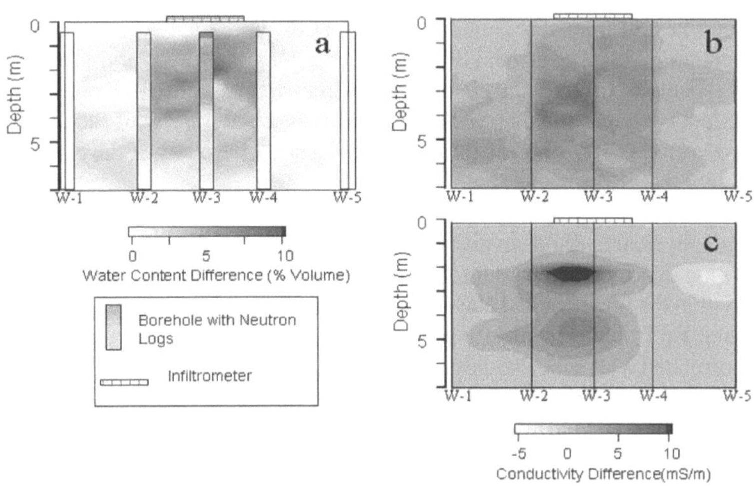

Figure 14.16. Images of (a) change in volumetric water content from GPR and borehole neutron data, (b) conductivity data from crosshole GPR attenuation, and (c) ERT data. The data were collected between April 20 and April 22, 1999, approximately 40 days after the start of infiltration.

After only 10 days of infiltration, the wetting front moved to a depth of about 2.5 m, as shown by the substantial increase in water content in the GPR images and neutron logs (Figure 14.15a). Unfortunately, because of the bentonite in the borehole annulus around W-3, the soil moisture content could not be accurately measured in the top meter of the borehole. The accuracy of the GPR was also limited in the upper meter, since there was a tendency for much of the signal to propagate up and along the air-earth interface rather than through the soil. Taking these limitations into account, the methods all show very similar results. Both neutron and GPR velocity images show a roughly tabular layer of increased moisture content about 2.5 m below the infiltrometer. The ERT and GPR attenuation images show a similar increase in conductivity corresponding to the zone of increased water content. None of the methods indicates that the infiltrating water has moved horizontally away from the infiltrometer.

After 40 days of infiltration (Figure 14.16), the images and the apparent movement of water are more complex. GPR moisture content images show more extensive lateral movement of the wetting front than is apparent in the neutron logs. The conductivity images from GPR attenuation (Figure 14.16b) data and the ERT data (Figure 14.16c) show somewhat similar features, but there are discrepancies in the magnitude and depth of the anomalies. Both GPR and ERT show strong increases in conductivity at shallow depths slightly southeast of the infiltrometer, and a deeper zone of modestly increased conductivity around 5 m extending almost to the boundary of the image region. GPR tends to show these two features as one continuous dipping layer; ERT shows them as a pair of distinct zones. In the GPR images, the anomalous conductive zones are slightly deeper than the zones of increased water content apparent in the neutron logs; in the ERT, they tend to be shallower. The ERT images agree fairly well with the site geology derived from soil cores that shows distinct layers at these two depths (Figure 14.16). The large differences in the GPR attenuation images always seem to fall between the large changes in moisture content. The GPR moisture content image and ERT image agree better than the ERT and GPR attenuation. It appears that GPR attenuation imaging is less robust than the velocity imaging, especially for tracking changes over time. Because the ERT represents an average that extends beyond the plane, it probably gives a better view of the overall large-scale processes on the site, whereas the GPR provides a better view of the complex local processes below the infiltrometer.

In summary, the geophysical methods in this study showed that the infiltration followed a complex path, moving rapidly downward through the coarse-grained layers, collecting and moving laterally within the finer-grained clay-bearing layers. All of the methods correlated well with each other. ERT data were able to provide fully 3-D images of the conductivity structure, which correlated well with changes in moisture content. This is probably because moisture is a volumetric parameter and ERT is a volumetric measurement. GPR data showed superior resolution to the ERT, but the results were limited to two-dimensional images along a single plane.

14.6 Conclusions

The studies presented in this chapter illustrated the use of many different geophysical methods for improving vadose zone characterization and monitoring. Resistivity (ERT), electromagnetic induction, and GPR methods are the primary tools for investigating the vadose zone, since they are the most sensitive of all of the conventional geophysical methods (excluding NMR and neutron-neutron scattering methods) to changes in water saturation. As was discussed in Chapter 1 of this volume, there is a natural ambiguity in using any single geophysical method to investigate the hydrogeologic properties in the vadose zone, and the use of multiple methods helps to confirm the location of major variations in the distribution of moisture.

The tank model study illustrated the sensitivity of GPR to subtle changes in the water saturation, and showed how 3-D mapping can provide valuable information concerning the distribution of regions of higher water saturation in pockets of higher water compaction. In addition, this study shows the complexities of using GPR to define the water table, since the geophysical water table determined by GPR is often different from the hydraulic water table defined in an observation well.

The potential of mapping soil water content in the presence of a varying water table using electrical measurements was discussed in Section 14.3.1, and the volumetric water content estimated from GPR groundwave studies in Section 14.3.2. These studies further illustrate the sensitivity of electrical conductivity and dielectric constant to soil water content.

Experimental results were shown that examine the changes in moisture content of soils due to natural loading as inferred from resistivity measurements. In one case study, surface resistivity arrays were used to monitor the spatial variability of infiltration resulting from snowmelt in a small field plot in Norway. In the second case study, borehole electrode arrays were utilized to monitor small changes in moisture content to a depth of 10 m at a site in a sandstone aquifers.

The studies at the STVZ site illustrate the use of crosshole GPR and ERT methods to infer or estimate the amount of water infiltration in the vadose zone. Although electrical conductivity is a function of a number of variables, its spatial pattern tended to be very strongly correlated to changes in moisture content within the vadose zone. Infiltrating water was observed to move rapidly through intervening sand and gravel layers, and tended to accumulate in clay-bearing layers. The changes as indicated in a qualitative sense by the geophysical images showed substantial lateral movement of the infiltrating water, illustrating that the infiltrating water moved beyond the image regions within a few months of the start of the experiment.

Each of the geophysical methods applied in this study, including the borehole methods, has strengths and weaknesses. The greatest strengths of ERT are its ability to monitor 3-D changes in moisture inferred from changes in electrical conductivity, and its ability to detect small changes in subsurface electrical properties. The method can be adapted to a range of scales; thus, the images in this study could be made from boreholes placed at convenient distances around the periphery of the site. As such, the method has less resolution than the techniques to which it was compared.

The cross-borehole GPR method provided very good comparisons with other techniques. Allowing for the differences in resolution of the methods, the GPR-velocity-derived images of moisture content showed similar values to results obtained with the calibrated neutron data. Electrical conductivity derived from GPR attenuation data also showed patterns similar to both the EM-39 logs and the ERT-derived conductivities.

Overall, these case studies illustrate a broad range of geophysical methods that can be applied to detecting and monitoring variations in vadose zone electrical properties. These studies also serve to illustrate that geophysical measurements provide an indirect means to monitor changes in hydraulic properties, and should always be interpreted in light of local geologic variations.

Acknowledgments

Jeff Daniels and Barry Allred would like to acknowledge funding support from the USDA, EPA, and technical Support Unit of Region V. They are grateful for the dissertation work of Changryol Kim (which was supported by the USEPA) and give special thanks to Mark Vendl, Jim Ursic, and Steve Ostrodka of the USEPA for their continued support of environmental testing. David Alumbaugh and Doug LaBrecque would like to acknowledge the Department of Energy's Environmental Science Program, and Lee Paprocki for collecting the field data and doing the lab analysis. Andrew Binley acknowledges financial support from the UK Natural Environment Research Council (Grant GR3/11500) and the UK Environment Agency for the sandstone vadose zone study. Field assistance to Binley was provided by Peter Winship at Lancaster University. The snowmelt study was carried out with assistance from Helen French and Leif Jakobsen (Agricultural University of Norway) and Peter Winship and Carol Hardbattle (Lancaster University).

References

Abdul, A.S., Migration of petroleum products through a sandy hydrogeologic system, *Ground Water Monitoring Review, 8*(4), 73–81, 1988.

Alumbaugh, D., L. Paprocki, J. Brainard, and C.A. Rautman, Monitoring infiltration within the vadose zone using cross borehole ground penetrating radar, *Proceedings of the Symposium on the Application of Geophysics to Engineering and Environmental Problems*, 273–281, Environmental and Engineering Geophysical Society, Wheat Ridge, Colorado, United States, 2000.

Alumbaugh, D., P.Y. Chang, L. Paprocki, J.R. Brainard, R.J. Glass, and C.A. Rautman, Estimating moisture contents using cross-borehole ground penetrating radar: A study of accuracy and repeatability in context of an infiltration experiment, Tentatively Accepted by *Water Resour. Res.*, 2002.

Banton, O., M. K. Seguin, and M. A. Cimon, Mapping field-scale physical properties of soil with electrical resistivity, *Soil Sci. Soc. Am. J., 61*, 1010–1017, 1997.

Bedient, P.B., H.S. Rifai, and C.J. Newell, *Ground Water Contamination*, Prentice Hall, Englewood Cliffs, New Jersey, 1994.

Benderitter, Y., and J.J. Schott, Short time variation of the resistivity in an unsaturated soil: The relationship with rainfall, *European J. Env. Eng. Geophys., 4*, 37–49, 1999.

Binley, A. and K. Beven, Vadose zone flow model uncertainty as conditioned on geophysical data, *Ground Water, 41*(2), 114–127, 2003.

Binley, A., P. Winship, R. Middleton, M. Pokar, and J. West, High resolution characterization of vadose zone dynamics using cross-borehole radar, *Water Resour. Res., 37*(11), 2639–2652, 2001.

Binley, A., P. Winship, L.J. West, M. Pokar, and R. Middleton, R., 2002a, Seasonal variation of moisture content in unsaturated sandstone inferred from borehole radar and resistivity profiles, *J. Hydrology, 267*, 160–172, 2002a.

Chang, P., D. Alumbaugh, J. Brainard, and L. Hall, The application of ground penetrating radar attenuation tomography in a vadose zone infiltration experiment, Submitted to *Journal of Contaminant Hydrology*, 2003.

Doolittle, J.A., K.A. Sudduth, N.R. Kitchen, and S.J. Indorante, Estimating depths to claypans using electromagnetic induction methods, *J. Soil and Water Cons., 49*(6), 572–575, 1994.

Doolittle, J., M. Petersen, and T. Wheeler, Comparison of two electromagnetic induction tools in salinity appraisals, *J. Soil and Water Cons., 56*(3), 257–262, 2001.

Du, S. Determination of water content in the subsurface with the groundwave of ground penetrating radar, PhD-Thesis for Ludwig-Maximilians-Universität, München, Germany, 1996.

Eigenberg, R.A., and J.A. Nienaber, Electromagnetic survey of cornfield with repeated manure applications, *J. Environ. Qual., 27*, 1511–1515, 1998.

Fraisse, C.W., K.A. Sudduth, and N.R. Kitchen, Delineation of site-specific management zones by unsupervised classification of topographic attributes and soil electrical conductivity, *Trans. ASAE, 44*(1), 155–166, 2001.

French, H.K., C. Hardbattle, A. Binley, P. Winship, and L. Jakobsen, Monitoring snowmelt induced unsaturated flow and transport using electrical resistivity tomography, *J. Hydrology, 267*, 273–284, 2002.

Frohlich, R.K. and C.D. Parke, The electrical resistivity of the vadose zone—Field survey, *Ground Water, 27*(4), 524–530, 1989.

GeoTom, LLC, *User Mannual for GeotomCG and GeoTom3D*, 1998.

Grote, K., S.S. Hubbard, and Y. Rubin, Field-scale estimation of volumetric water content using GPR groundwave techniques, *Wat. Resour. Res. 39*(11), 1321, 10.1029/2003WR002045, 2003

Kean, W.F., M.J. Waller, and H.R. Layson, Monitoring moisture migration in the vadose zone with resistivity, *Ground Water, 27*(5), 562–561, 1987.

Hendrickx, J.M.H., B. Baerends, Z.I. Rasa, M. Sadig, and M. Akram Chaudhry, Soil salinity assessment by electromagnetic induction of irrigated land, *Soil Sci. Soc. Am. J., 56*, 1933–1941, 1992.

Hubbard, S.S., K. Grote, and Y. Rubin. Mapping the volumetric soil water content of a California vineyard using high-frequency GPR groundwave data, *Leading Edge of Expl. 21*, 552–559, 2002.

Huisman, J.A., and W. Bouten, Accuracy and reproducibility of measuring soil water content with the groundwave of ground penetrating radar. *J. of Env. and Eng. Geophysics 8*(2), 65–73, 2003a.

Huisman, J.A., S.S. Hubbard, J.D. Redman, and A.P. Annan, Measuring soil water content with ground penetrating radar: A review, *Vadose Zone Journal, 4*(2), 476–491, 2003b.

Huisman, J.A., J.J.J.C. Snepvangers, W. Bouten, and G.B.M. Heuvelink, Mapping spatial variation in surface soil water content: Comparison of ground-penetrating radar and time domain reflectometry, *J. of Hydrol. 269*, 194–207, 2002.

Huisman, J.A., J.J.J.C. Snepvangers, W. Bouten, and G.B.M. Heuvelink, Space-time characterization of soil water content: Combining ground-penetrating radar and time domain reflectometry, Accepted by *Vadose Zone J.*, 2003.

Huisman, J.A., C. Sperl, W. Bouten, and J.M. Verstraten, Soil water content measurements at different scales: Accuracy of time domain reflectometry and ground-penetrating radar, *J. of Hydrol., 245*, 48–58, 2001.

Inman, D.J., R.S. Freeland, J. T. Ammons, and R. E. Yoder, Soil investigations using electromagnetic induction and ground penetrating radar in southwest Tennessee, *Soil Sci. Soc. Am. J., 66*, 206–211, 2002.

Kim, C.R., A physical model experiment on the hydrogeologic applications of GPR, Ph.D. Dissertation, The Ohio State University, 2001.

Kravchenko, A.N., G.A. Bollero, R.A. Omonode, and D.G. Bullock, Quantitative mapping of soil drainage classes using topographical data and soil electrical conductivity, *Soil Sci. Soc. Am. J., 66*, 235–243, 2002.

Jackson, M.J., and D.R. Tweeton, 3DTOM: Three-dimentional geophysical tomography, *Report of Investigations 9617*, Bureau of Mines, United Stated Department of the Interior, 1996.

Jaynes, D.B., J. M. Novak, T. B. Moorman, and C. A. Cambardella, Estimating herbicide partition coefficients from electromagnetic induction measurements, *J. Environ. Qual., 24*, 36–41, 1995.

LaBrecque, D.J., and X. Yang, Difference inversion of ERT data: A fast inversion method for 3D *in situ* monitoring, *Journal of Environmental and Engineering Geophysics, 5*, 83–90, 2001.

LaBrecque, D.J., and Casale, Experience with anisotropic inversion for electrical resistivity tomography, *Proceedings of the Symposium on the Application of Geophysics to Engineering and Environmental Problems* (SAGEEP) '02, 2002.

Lesch, S.M., J.D. Rhoades, L.J. Lund, and D.L. Corwin, Mapping soil salinity using calibrated electronic measurements, *Soil Sci. Soc. Am. J., 56*, 540–548, 1992.

Paprocki, L., Characterization of vadose zone in-situ moisture content and an advancing wetting front using cross-borehole bround penetrating radar, Masters Thesis, New Mexico Institute of Mining and Technology, 2000.

Peterson, J.E., Pre-inversion correction and analysis of radar tomographic data, *Journal of Environmental & Engineering Geophysics, 6*, 1–18, 2001.

Sasaki, Y., Resolution of resistivity tomography inferred from numerical simulation, *Geophysical Prospecting, 40*, 453–464, 1992.

Scanlon, B.R., J.G. Paine, and R.S. Goldsmith, Evaluation of electromagnetic induction as a reconnaissance technique to characterize unsaturated flow in an arid setting, *Ground Water, 37*(2), 296–304, 1999.

Simms, J.E., and F.D. Morgan, Comparison of four least-squares inversion schemes for studying equivalence in one-dimensional resistivity interpretation, *Geophysics*, *57*(10), 1282–1293, 1992.

Sheets, K.R., and J.M.H. Hendrickx, Noninvasive soil water content measurement using electromagnetic induction, *Water Resour. Res., 31*(10), 2401–2409, 1995.

Topp, G.C., J.L. Davis, and A.P. Annan, Electromagnetic determination of soil water content: Measurements in coaxial transmission lines, *Water Resour. Res., 16*(3), 574–582, 1980.

Wharton, R.P., G.A. Hazen, R.N. Rau, and D.L. Best, Advancements in electromagnetic propagation logging, *Paper SPE 9041*, Society of Petroleum Engineers of AIME, Rocky Mountain Regional Meeting, 1980.

Zhou, Q.Y., J. Shimada, and A. Sato, Three-dimensional spatial and temporal monitoring of soil water content using electrical resistivity tomography, *Water Resour. Res.*, 37(2), 273–285, 2001.

15 HYDROGEOPHYSICAL METHODS AT THE LABORATORY SCALE

TY P.A. FERRÉ[1], ANDREW BINLEY[2], JIL GELLER[3], ED HILL[4], and TISSA ILLANGASEKARE[5]

[1]*University of Arizona, Hydrology and Water Resources, 1133 E. North Campus Drive, Tucson, AZ, 85721-0011 USA*

[2]*Lancaster University, Department of Environmental Science, Lancaster, LA1 4YQ, UK.*

[3]*Lawrence Berkeley National Laboratory, Earth Sciences Division, 1 Cyclotron Road, Berkeley, CA, 94720 USA*

[4]*Massachusetts Institute of Technology, Dept. of Earth, Atmospheric, and Planetary Sciences, 77 Massachusetts Ave, Cambridge, MA 02139-4307 USA*

[5]*Colorado School of Mines, Environmental Science and Engineering, 1500 Illinois St., Golden, CO, 80401 USA*

15.1 Unique Contributions and Challenges of Laboratory-Scale Hydrogeophysics

Hydrogeophysics relies on the inference of hydrologically important properties based on the measurement of other properties that are more easily obtained. This inference requires, first, the definition of petrophysical relationships such as the dependence of the bulk dielectric permittivity of a medium on its volumetric water content (see Chapters 4 and 9 of this volume). Second, confounding effects must be defined or controlled to allow for appropriate corrections. For example, if electrical resistance tomography (ERT) is to be used to infer water-content changes, the change in measured electrical conductivity as a function of temperature must be accounted for, or measurements must be made under isothermal conditions. Third, if the measured property or the property of interest varies within the measurement sample volume, then the manner in which the measurement method averages these heterogeneous values must be considered.

Under controlled laboratory conditions, petrophysical parameters can often be defined more completely and more accurately throughout the measurement domain than is possible under field conditions. In addition, the confounding effects of multiple properties that may affect the geophysical response can be controlled more closely than under field conditions. As a result, when applied carefully and appropriately in the laboratory, geophysical methods can produce high-resolution (in space and time), nondestructive, and inexpensive measurements to monitor transient hydrologic processes. In addition, some of these methods can be applied in entirely noninvasive modes. This final quality is of particular interest for laboratory measurements made in relatively small domains, which may be more susceptible to impacts caused by measurement devices.

Y. Rubin and S.S. Hubbard (eds.), Hydrogeophysics, 441–463.

Despite the promise of hydrogeophysics for laboratory-based investigations, it is critical that geophysical measurements be interpreted with both hydrologic and geophysical understanding to avoid systematic errors. This chapter presents background material and case studies to demonstrate the contributions and limitations of laboratory-scale hydrogeophysics. Among the large number of geophysical methods that have been applied to laboratory studies, four are discussed here: time-domain reflectometry, x-ray tomography, electrical resistance tomography, and seismics. This is not intended to be a complete discussion of the application of geophysics at the laboratory scale, as this is too large a topic to cover in a single chapter. Rather, the purpose of this chapter is to demonstrate some of the strengths and weaknesses of geophysical methods for small-scale monitoring. Primary attention is focused on the achievable spatial and temporal resolutions of the method, the manner in which properties are averaged within the sample volume of each instrument, and the way in which hydrologic understanding should be used in designing and interpreting hydrogeophysical monitoring networks.

15.2 Underlying Principles of Selected Geophysical Methods

15.2.1 TIME DOMAIN REFLECTOMETRY (TDR)

Time domain reflectometry (TDR) is one of the most widely used geophysical methods in soil science and vadose zone hydrology. The method is used to determine the apparent dielectric permittivity of a medium based on the measured velocity of a guided electromagnetic (EM) wave. For a wide range of soils, the dielectric permittivity can be uniquely related to the volumetric water content with little or no medium-specific calibration (Topp et al., 1980).

A TDR probe (Figure 15.1) typically includes a waveguide comprised of two or three parallel metal rods. These rods range in length from 10 cm to several meters, are typically less than 1 cm in diameter, and are separated by less than 10 cm. The rods are inserted into the medium of interest using an attached nonmetallic probe handle. The handle is connected to a fast-rise-time pulse generator and an oscilloscope through a coaxial cable. An EM plane wave travels along the rods in transverse electromagnetic mode (the electric and magnetic fields are located in a plane that is perpendicular to the long axes of the rods). The wave traverses the length of the rods and is completely reflected from the ends of the rods. The two-way travel time of the wave along the rods is measured. A practical guide to the use of TDR for water-content measurement is presented by Topp and Ferré (2002), and a recent critical review of the theory and application of TDR and other related dielectric permittivity methods is presented by Robinson et al. (2003).

TDR measurements require dedicated high-frequency electronic instrumentation. However, the broad use of this method has led to the availability of commercial TDR systems. These systems are designed for laboratory and field use, and many are amenable to automated monitoring and multiplexing. With multiplexing, TDR is well suited to very rapid monitoring of the volumetric water content distribution. However, the method is inherently intrusive because the rods are inserted into the medium of interest.

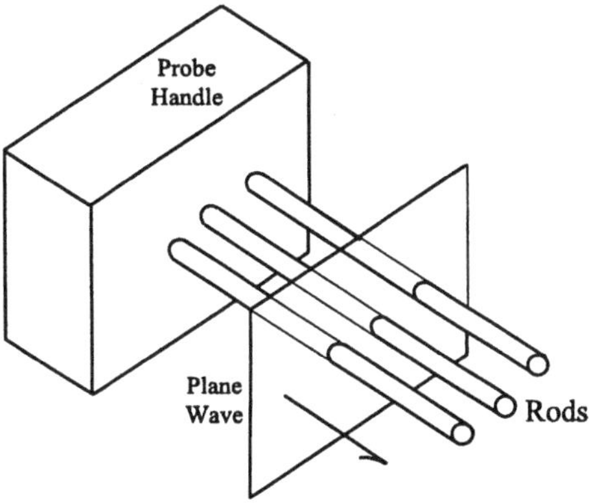

Figure 15.1. Schematic diagram of a TDR probe including the probe handle and three parallel metal rods. A guided EM plane waves travel along the rods.

15.2.2 ELECTRICAL RESISTANCE TOMOGRAPHY (ERT)

As described in detail in Chapter 4 of this volume, electrical resistivity varies spatially and temporally within a soil, owing to changes in porosity, saturation, fluid conductivity, temperature, and soil composition (e.g., clay content). Images of such variation within a volume of soil may thus provide useful hydrological information, particularly if these can be obtained in a noninvasive, or minimally invasive, manner. The general method of electrical resistivity imaging is now widely known as electrical resistance (or resistivity) tomography (ERT). The method has been used widely in the biomedical field (e.g., Webster, 1990; Brown, 2001) and chemical engineering (e.g., Wang et al., 2002), where it has also been referred to as electrical impedance tomography (EIT). ERT uses combinations of four-electrode measurements to derive the distribution of resistivity within a volume of interest. ERT may be applied at a wide range of spatial scales to determine hydrologic properties such as volumetric water content or solute concentration.

Adopting a four-electrode arrangement, electrical current is injected between two electrodes, and the potential difference is measured between another pair of electrodes. Figure 15.2 illustrates the approach when applied around the perimeter of a circular object, such as a soil core. Such measurements are repeated for different combinations of four electrodes, thus "scanning" different parts of the sample. It is then possible to determine the distribution of resistivity consistent with such a set of measurements. Because current flow lines are not linear, determination of the image of resistivity requires application of nonlinear inverse methods (see Chapter 5 of this volume). The distribution of resistivity is normally obtained using finite element or finite difference grid-based methods. Current flow lines are also not two-dimensional. In the schematic in Figure 15.2, measurements will be sensitive to variations in resistivity "off-plane," and thus true tomographic "slices" are not achievable with 2-D imaging. For a 64-

electrode arrangement shown in Figure 15.2, several hundred measurements are required for satisfactory image reconstruction. Using a typical low-cost multiplexed earth-resistance meter, data acquisition time may be of the order of several tens of minutes for one plane. This may limit the application of ERT for monitoring dynamic hydrological processes. Recent developments in multichannel earth-resistance meters are likely to prove valuable in studying time-varying processes in soils using ERT.

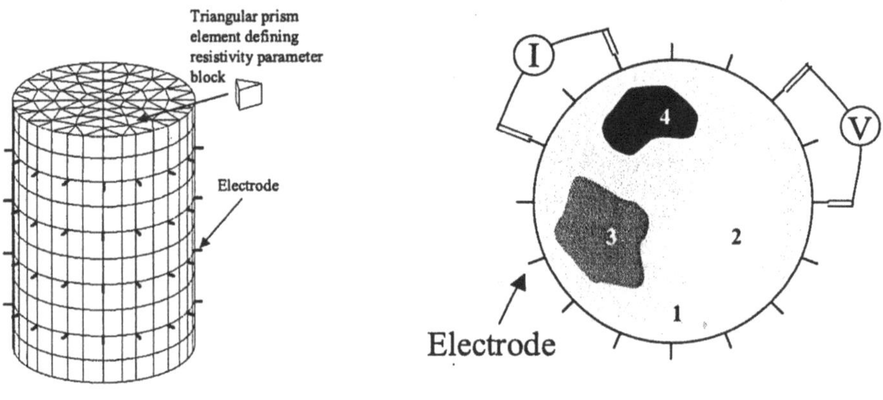

Figure 15.2. Electrical resistance tomography applied to a soil core (left). Within each horizontal slice, electrical current (*I*) is injected between two electrodes and the electrical potential (*V*) is measured between another pair of electrodes (right). The electrical conductivity distribution is then inferred within the slice.

15.2.3 X-RAY TOMOGRAPHY

Investigations of water content, porosity, nonaqueous phase liquid (NAPL) distributions, and NAPL mass in porous media can be accomplished through a range of different x-ray and gamma-ray attenuation techniques. For typical laboratory-scale natural-gamma and x-ray attenuation devices, the equations used for the measurement of porosities and fluid volume fractions within porous media are reviewed by Hill et al. (2002). These investigators have typically relied upon the Beer-Lambert relation,

$$\overline{IN}(\varepsilon) = \Delta t\ {}_0(\varepsilon)e^{-\rho u'(\varepsilon)l} = \Delta t\ {}_0(\varepsilon)e^{-u(\varepsilon)l} \qquad (15.1)$$

where $\overline{N(\varepsilon)}$ is the expected value of the number of observed photons at energy ε, Δt is the observation time, $I_o(\varepsilon)$ is the mean photon flux at energy ε, ρ is the material density, l is a differential path length, and $u'(\varepsilon)$ and $u(\varepsilon)$ are intrinsic and linear attenuation coefficients, respectively, with $\rho u'(\varepsilon) = u(\varepsilon)$. Applying this equation with observed values of photon counts and known attenuation coefficients, one can estimate apparent fluid path lengths, masses, and/or saturations. In all cases, differences in observed photon fluxes are related to changes in the hydrologic system from a background condition. In this chapter, densities are assumed constant, so that the linear attenuation coefficient may be used to infer the intrinsic attenuation coefficient.

Typical source/detector pairs are shown in Figure 15.3. The upper diagram shows a natural gamma source coupled with a sodium iodide detector. While this is a relatively inexpensive approach, it often suffers from poor spectral resolution and relatively long counting times. The x-ray and "high-performance germanium" configuration below is more expensive, but can provide higher overall photon observation rates, better spectral resolution, and, as a result, much lower measurement times.

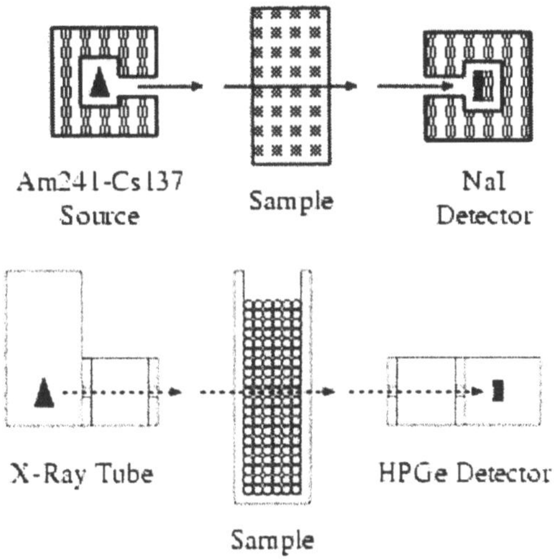

Figure 15.3. Typical source-detector configurations

In all systems, the number of photons observed tends to follow a Poisson distribution, meaning that the precision of any observation will increase with the square root of the observation time. Typical observation times (reported as detector "live" or counting times) for natural gamma systems are in the range of 100 to 500 seconds [Imhoff, 1992; Illangasekare et al., 1995], while x-ray and synchrotron sources can provide similar or better quality measurements in one to ten seconds [Imhoff et al., 1995; Okuda et al., 1996; Miller et al., 2000]. Practical issues such as motor positioning and equipment communication can increase the overall time required per location. Such time overheads can add anywhere from a few seconds to a few minutes to the total time. With a judicious choice of the number and scheduling of the sample locations, these measurement times have proven short enough to observe dynamic effects such as NAPL flow in sand-packed columns [Imhoff et al., 1995; Okuda et al., 1996; Miller et al., 2000; Hilpert et al., 2001] and dense NAPL (DNAPL) migration in large-scale tanks (Illangasekare et al., 1995). As improved electronics for photon observation and better analysis techniques (Hill et al., 2002) become available, measurement rates will increase, allowing more detailed views of DNAPL migration.

15.2.4 SEISMIC METHODS

The propagation characteristics of seismic waves depend upon the physical properties of the porous medium and its fluids. Chapter 8 of this volume describes the different types of seismic waves, their attributes (such as velocity, amplitude and frequency), and applications for characterization and imaging of the shallow subsurface. Most laboratory measurements of seismic properties are one-dimensional characterizations of core samples. In-depth discussions of the concepts and considerations for experimental design can be found in texts such as Bourbie et al. (1987) and Santamarina (2001). There are very few reports in the literature of two- and three-dimensional laboratory-scale seismic imaging (Chu et al., 2001; Li et al. 2001; McKenna et al., 2001; Geller et al., 2000), but these types of systems can be beneficial for process-monitoring, particularly for changes in fluid distribution. Air and common LNAPL (light NAPL) and DNAPL contaminants have acoustic velocities and bulk moduli less than those of water. As a result, their redistribution may be visible under certain conditions from the changes in seismic waveforms measured over time.

Laboratory-based seismic investigations must be designed to use the appropriate seismic source for the scale of the measurement domain and for the spatial scale of medium heterogeneities. The transmission distance must be at least several wavelengths long. The practical physical scale of laboratory experiments (centimeters to meters) therefore defines the useable frequency range for laboratory applications. Core-scale investigations use frequencies from 100–1000 kHz, which range from one to eight orders of magnitude greater than those used for seismic field investigations. Tank-scale measurements can be made at lower frequencies. Considering the range of seismic wave velocities for geologic media (refer to Table 8.2), wavelengths at laboratory frequencies can range from millimeters to centimeters, depending upon the media, the wave type (P or S), and the measurement technique. While the scale of the measurement domain defines the minimum frequencies that can be used in the laboratory, heterogeneities control the maximum frequencies. Scattering of seismic energy can be significant for objects that have length scales greater than one-tenth of the seismic wavelength. In samples having heterogeneities near the size of a wavelength or greater, energy may be lost to scattering and diffraction when the object has an acoustic impedance contrast with the surrounding media (e.g., Santamarina, 2001; also, Chapter 8 of this volume). This can apply to both solid-phase heterogeneities, such as gravel or clay inclusions, and fluid-phase heterogeneities, created by the presence of immiscible fluid phases.

A common way to generate and detect high-frequency seismic waves is to use piezoelectric elements that expand when subjected to a voltage. This transmits a wave that is sensed as a pressure pulse and converted to a voltage at the receiver. Waveforms are typically acquired digitally, and multiple waveforms are averaged (or stacked) to improve the signal-to-noise ratio. Typical measurement times are less than one minute and depend upon the length of the waveform and the number of waveforms stacked.

Wave transmitters and receivers (or transducers) can be positioned in a variety of locations to satisfy the coverage requirements for the desired type of imaging. Proper coupling of the transducer to the media ensures energy transmission through the sample without undesirable wave-conversion effects at the contact with the sample, or fast-path transmission around the sample. Because the piezoelectric transducers must be isolated from the sample, measurement is inherently noninvasive. For time-lapse imaging, transducer coupling must be reproducible, particularly if amplitude information is used. Transducers must also be located to avoid interference from reflections at sample boundaries.

Standard methods of seismic tomography, as described by Lo and Inderwiesen (1994), utilize the travel time of the wave and its amplitude, and invert for the two-dimensional distributions of velocity and attenuation. These methods are quite robust for low-resolution imaging, but some of the underlying assumptions do not necessarily hold in the presence of wavelength-scale heterogeneities. Development of tomographic methods to incorporate multidimensional wave propagation and full waveform inversion is an active area of research (Keers et al., 2001). Zero-offset monitoring is a straightforward method of imaging spatial and temporal changes in a sample. The zero-offset data are a subset of a tomographic dataset wherein the transmitter and receiver are directly opposite one another as they are moved vertically along the sample. The received waveforms plotted over the length of the sample provide an image of horizontally averaged properties, without waveform processing and inversion.

15.3 Spatial Averaging and Spatial Resolution

In forming petrophysical relationships to interpret hydrogeophysical measurements, steps are commonly taken to form homogeneous calibration samples. In contrast, most monitoring efforts are applied under conditions that give rise to spatially heterogeneous property distributions at critical times or locations, such as the time at which a wetting front reaches a measurement point or locations near the edges of NAPL pools. Correct interpretation of many hydrogeophysical measurements relies on an understanding of the concept of spatial averaging and the associated concepts of spatial sensitivity and sample volume. High-resolution laboratory measurements can be misinterpreted if the effects of spatial averaging are ignored.

15.3.1 SPATIAL SENSITIVITY AND SAMPLE VOLUME

The sensitivity of a measurement method can be defined as the change in the instrument response per unit change in the property of interest. The spatial sensitivity can then be considered to be a map of the change in instrument response per unit change in property of interest as a function of the location of that change. To better understand the effects of instrument design on spatial sensitivity, consider the two different configurations of electrodes shown in Figure 15.4: two circular electrodes or two parallel-plate electrodes. The electric flux density is uniform everywhere between the plate electrodes, as shown by the uniform spacing of the equipotentials. In contrast, for circular electrodes, the electric flux density is very high in the vicinity of the electrodes and lower farther from the electrodes, as shown by the nonuniform spacing of the equipotentials between the rods. As a result, a small change in the electrical conductivity at a point near one of the circular electrodes will cause a greater change in the measured electrical conductivity between the electrodes than the same conductivity change located farther from the electrodes. In contrast, a point change in conductivity at any location between the plate electrodes will have the same impact on the measured electrical conductivity. Note that Figure 15.4 was constructed assuming that the medium properties were uniform between the electrodes; the effects of heterogeneous medium properties are discussed with reference to TDR below.

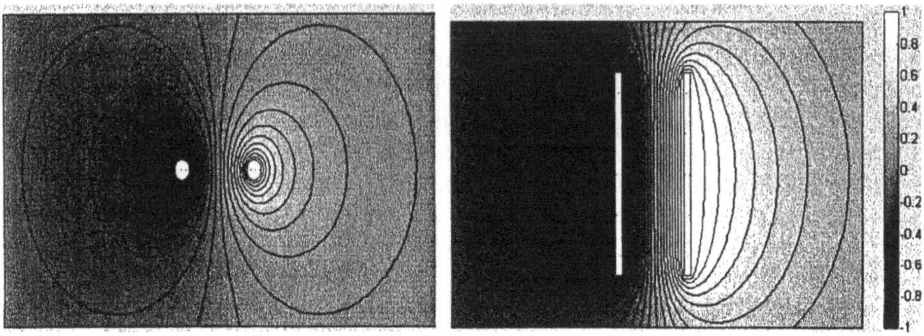

Figure 15.4. Equipotentials between two circular electrodes (left) and two plate electrodes (right) with constant potentials of 1 and –1 set on the left and right electrodes, respectively

15.3.2 TDR: EFFECTS OF HETEROGENEITY ON SPATIAL SENSITIVITY

Topp et al. (1982) showed experimentally that TDR measures the average water content along the rods even if the water content varies along the rods. It is more difficult to show directly how TDR responds to water-content variations in the plane perpendicular to the long axes of the rods. Knight (1992) presented an analytical expression describing the spatial sensitivity of TDR in the transverse plane if the water content is uniform. The analysis is based on the electrical potential distribution in the transverse plane. For a two-rod probe, this distribution is identical to the potential distribution around two circular electrodes, as shown in Figure 15.4. At any point in the plane, the sensitivity can be determined based on the square of the gradient of potential (Knight, 1992). As a result, the probe sensitivity is highest near the circular electrodes, especially in the region between the electrodes. In addition, if the water content is spatially uniform, the potential distribution, and therefore the sensitivity distribution, is independent of the water content.

Two approaches are commonly used to define the sample volume of an instrument. In the simplest and most common approach, measurements are made while the property of interest is varied at different distances from the instrument. The volume within which a given change in the property of interest gives rise to a measurable change in the instrument response is defined as the sample volume. While this approach gives a direct measure of the maximum volume in which a change can be sensed, it does not describe the relative contributions of the instrument response from different areas within this volume. A second approach to defining the sample volume relies on a description of the instrument spatial sensitivity.

For example, to define the sample area of TDR in the plane perpendicular to the long axis of the rods, Knight (1992) calculated the cumulative sensitivity within ellipses of varying sizes surrounding the rods. Then he identified the ellipse that contained some percentage of the total instrument sensitivity. For TDR, the sample volume can be defined by extending this ellipse over the length of the rods. This approach offers further insight into instrument sensitivity, but because it is based on an arbitrary choice of sample area shape (e.g., an ellipse), it does not define a unique sample area (or volume). Ferré et al. (1998) introduced a unique sample area definition that identifies the minimum area containing a given fraction of the total instrument sensitivity. This is achieved by summing areas within the domain, beginning with the areas of highest sensitivity and progressing to areas of continually decreasing sensitivity. Knight et al. (1997) presented a numerical approach to analyze TDR spatial sensitivity. The unique sample areas for four TDR probes based on this analysis are shown in Figure 15.5. The origin of the plot is located at the point equidistant from the centers of the outermost rods. Each quadrant represents a different probe design. The upper left quadrant shows the 90%, 70%, and 50% sample areas of a two-rod probe with relatively small diameter rods. Because of the symmetry of the probe design, all of the remaining quadrants surrounding this probe will have the identical fractional sample areas. The lower-left quadrant represents a two-rod probe with the same rod separation as that shown above, but with a larger rod diameter. The result demonstrates that the choice of rod size has little impact on the fractional sample areas of two-rod probes. The right quadrants represent three-rod probes. The outer rod separations and rod diameters are identical to those of the two-rod probes. These results show that the fractional sample areas of three-rod probes are much smaller than those of directly comparable two-rod probes. Furthermore, the 70% of the instrument response containing the areas of highest sensitivity is located in the region immediately adjacent to the probes, which is most susceptible to disturbance during rod insertion. Based on these results, three-rod probes are only recommended when measurements must be highly localized, and then only if medium disturbance during probe placement can be minimized.

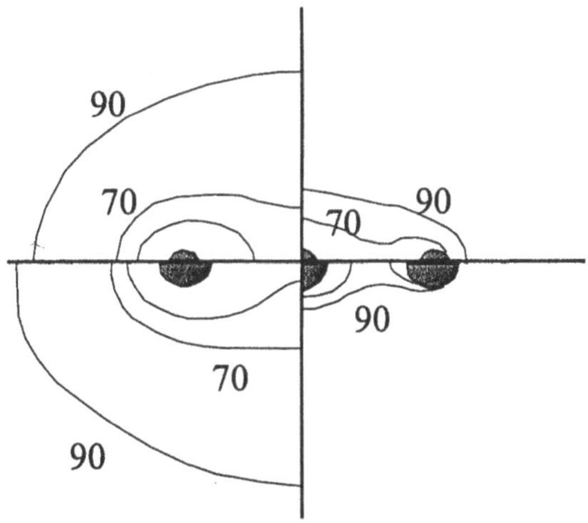

Figure 15.5. Unique fractional sample areas surrounding two-rod (left quadrants) and three-rod (right quadrants) TDR probes in a homogeneous medium. The upper and lower quadrants have the same rod separations, but smaller rods are considered in the upper quadrants. Labels show the percent of the total sensitivity contained within each contour.

Numerical analysis of TDR fractional sample areas can be applied to spatially heterogeneous media as well. Nissen et al. (2003) designed an experiment to test the effects of heterogeneous dielectric permittivity distributions on the spatial sensitivity of TDR in the transverse plane. Using multiple fluids, they achieved sharp boundaries between media of known dielectric permittivities. Two- and three-rod TDR probes were placed through the walls of a plastic box. Then, fluids (water, ethanol, oil) were added to the base of the box, causing the fluid-air interface to rise past the rods. Measurements were made continuously as the fluid level rose. At each measurement time, the height of the fluid interface was determined, allowing for the construction of a numerical model to represent the dielectric permittivity distribution. The predicted and measured dielectric permittivities showed very good agreement for horizontal two- and three-rod probes (oriented such that all probes were located in the same horizontal plane). The responses of vertical two-rod probes were also well predicted. However, for three-rod vertical probes, the results showed that the sharp dielectric permittivity contrast gave rise to two waves traveling simultaneously and at different speeds when the interface was between two of the rods. This behavior violates the assumption of plane wave propagation underlying the numerical analysis. As a result, the controlled laboratory experiments both validated the modeling approach for horizontal and vertical two-rod probes, and uncovered unexpected limitations to the use of three-rod TDR probes in heterogeneous media. The predicted dielectric permittivities were determined based on the calculated spatial sensitivities of the TDR probes for each air-fluid interface location. Therefore, this agreement provides validation of the spatial sensitivity determinations and allows for calculation of the fractional sample areas in heterogeneous media.

As the air-fluid interface rose past the TDR probe, the measured dielectric permittivity changed smoothly as a function of interface height from that of air to that of the fluid. This smoothing of the sharp interface is a function of the distributed spatial sensitivity of TDR probes. That is, when the interface is located at the probe midpoint, some of the probe sensitivity extends into each fluid. This is also true when the interface is located above or below the probe midpoint. The shape of the measured dielectric permittivity versus interface height can be explained based on the changing spatial sensitivities of the probes as a function of the dielectric permittivity distribution around the probe. The 70% and 90% sample areas for vertical two-rod probes are shown in Figure 15.6 for a series of air-water interface locations. There is significant distortion of the sample area, caused by medium heterogeneity. Specifically, most of the probe sensitivity is confined within the air until the interface reaches the upper rod. As a result, the measured dielectric permittivity is close to that of air until the interface reaches the upper rod. Once the interface passes the upper rod, the majority of the instrument sensitivity is located within the water. As a result of this changing spatial sensitivity, the most rapid change in measured dielectric permittivity occurs when the interface contacts the uppermost rod, not when the interface passes the probe midpoint. Ferré et al. (2002) show that this behavior can impact the accuracy of inversion of hydraulic properties from water-content measurements made during the advance of a wetting front.

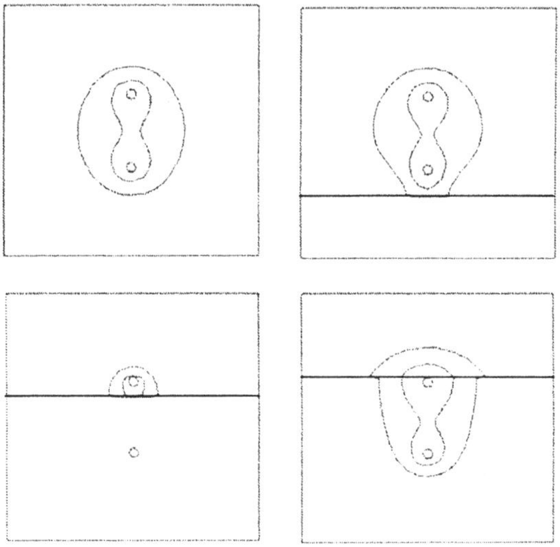

Figure 15.6. Unique fractional sample areas surrounding vertical two-rod TDR probes as water rises past the probe. The horizontal line shows the air-water interface: air overlies water in all cases. The outer contours contain 90% of the probe sensitivity and the inner contours contain 70%.

15.3.3 EFFECTS OF SAMPLE HOLDERS ON X-RAY METHOD AVERAGING

Spatial averaging of observations is largely determined by the beam geometry of the attenuation device employed. Typically, parameters such as the geometry of the x-ray focal spot (Knoll, 1979), the construction of the natural-gamma source, and the size and arrangement of collimators at the source and detector ends are used to control the beam geometry. In all instruments, the number of photons observed at a detector is a weighted average of the contributions of all the possible paths that the photons could travel from the source or sources to the detector.

To illustrate the physics of photon attenuation averaging, we will consider a simplified example involving a single source and detector combination as shown in Figure 15.3. Assuming that all photons travel in paths parallel to the z-axis, that the photon detector has perfect quantum efficiency, and that the contribution of Compton-scattered photons is negligible, the average rate of photon observation at the detector for any particular photon energy ε is the weighted average of photon flux over the projected beam area, according to:

$$\overline{N}(\varepsilon) = \int_{t=0}^{\Delta t} \int_{x=0}^{\Delta x} \int_{y=0}^{\Delta y} I_0(\varepsilon, x, y, t) e^{-\int_{z=z_d}^{z_s} \mu(\varepsilon, x, y, z, t) d} \, dydxdt \qquad (15.2)$$

where z_s and z_d are the source and detector locations (with the z axis parallel to the beam), Δx and Δy are the beam height and width, Δt is the observation time, $I_0(x, y, t)$ is the photon flux (production rate) per unit area at the source, $\upsilon(e, x, y, z, t)$ is the spatially and temporally distributed attenuation coefficient of the material(s) placed within the beam path, and $\overline{N}(\varepsilon)$ is the average number of photons observed at the detector. If photon production rate at the source is uniform over the area of the beam, then the following simplification may be applied as

$$I_0(\varepsilon, x, y, t) = \frac{1}{\Delta x \Delta y} I_0(\varepsilon, t). \tag{15.3}$$

The instrument response is an attenuation-weighted average of the photon observation rate. Since we are interested in employing this photon attenuation effect to measure densities or fluid saturations within porous media, it is useful to investigate how the attenuation average is related to a fluid mass average or a fluid volume average over the identical domain.

First, we consider a cuvette of constant and known thickness placed within the beam path, as shown in Figure 15.3. This technique is a commonly used method for the calibration of photon attenuation devices. Initially, the cuvette is scanned while it contains only air, which has a negligible attenuation coefficient. Then the cuvette is filled with water (or some other fluid with a known photon attenuation behavior) and scanned a second time. According to the Beer-Lambert equation (15.2), there is an exponential relationship between the thickness (or "path length") of the material and the number of observed photons. With this knowledge, one can perform calibration experiments to determine the extinction rate of the material (the "attenuation coefficient," μ) by simultaneously measuring path lengths and observed photon rates. Conversely, one can use previously calibrated attenuation coefficients and observed photon rates to measure material thicknesses.

In this case, the sample has a flat and uniform fluid thickness relative to the photon beam. It can be readily shown that the integrals along the temporal and all three spatial dimensions in Equation 15.2 can be reduced to a single instance of the Beer-Lambert equation. Thus, for a flat and constant absorber geometry, the attenuation weighted average is identical to a mass or a volume average of the fluid over the projected beam area (that is, the volume of the beam). However, for more complicated attenuator geometries, the equivalence between attenuation, mass, and volume averages does not, in general, hold. To prove there is a bias between attenuation and mass or volume averages, consider the hypothetical case of a flat plate, which has an "eggcrate"-like pattern of flat bumps on one side. The projected thickness, l, of this object at any x, y-coordinate can be specified as

$$l(x, y) = \bar{l} + \Delta l \operatorname{sgn}\left(\sin\left(\frac{2n\pi xy}{\Delta x \Delta y}\right)\right), \tag{15.4}$$

where $\bar{l}$ is the average plate thickness, $\Delta << \bar{l}$ is the protrusion length of the square "bumps," and the "sgn" function is defined as

$$\operatorname{sgn}(a) = \begin{matrix} -1, a < 0 \\ 0, a = 0 \\ 1, a > 0 \end{matrix} . \tag{15.5}$$

Given that the attenuation coefficient for the plate is constant over space and time, and that the plate geometry and the photon source strength are constant over time, Equations 15.2, 15.3, and 15.4 can be applied over the projected beam area to obtain:

$$\overline{N}(\varepsilon) = \frac{\Delta t I_0(\varepsilon)}{2}\left[e^{-(\bar{l}+\Delta l)\mu(\varepsilon)} + e^{-(\bar{l}-\Delta l)\mu(\varepsilon)}\right], \quad \ldots \tag{15.6}$$

which expresses the average number of observed photons at a particular energy ε as a function of both the average plate thickness $\bar{l}$ and the length Δ of the square "bumps" on its surface.

Applying either a mass or a volume average to the plate over the beam volume, one can readily show that the average thickness of the plate is $\bar{l}$,. since the high and low regions would exactly cancel each other out. For the attenuation-weighted case, the representative or "average" length ($l=\hat{l}$) as measured using observations of photon counts for a single energy would be from Equations 15.1 and 15.6:

$$\hat{l} = \bar{l} - \frac{1}{\mu(\varepsilon)}\ln\left(\frac{1}{2}e^{-\Delta l\mu(\varepsilon)} + \frac{1}{2}e^{\Delta l\mu(\varepsilon)}\right), \tag{15.7}$$

which gives a consistently negative bias. By examining the term within the logarithm, one can prove that

$$\begin{matrix} \hat{l} < \bar{l}, \Delta l \neq 0 \\ \hat{l} = \bar{l}, \Delta l = 0 \end{matrix} . \tag{15.8}$$

That is, the measured (attenuation-averaged) length will be consistently smaller than the mass-averaged or volume-averaged length for any non-zero "bump height" in this analysis. This result agrees with our previous description of a flat plate geometry and hints at a more general bias in measurements involving fluid saturations in porous media—since, as we have shown, an attenuation average over the beam area is not necessarily the same as a mass or volume average. A good example of such bias occurring within a laboratory experiment is the monitoring of oil saturations within a rock core sample or soil packed within a glass column (Imhoff, 1992, 1995). Both of these experimental domains have cylindrical geometries typically on the order of a few centimeters in diameter. If one uses a sufficiently small beam of radiation aligned to create a perpendicular intersection with the centerline of the cylindrical domain, then the assumption of a "flat" beam intersection area is probably justified. However, if the beam is not aligned with the center of the cylinder, or if the beam width approaches or exceeds the width of the cylindrical domain, then the bias can be large. In such cases, an attenuation measurement of the fluid mass at a point within the column may be dramatically less than the true mass within the beam intersection volume. In general, an experimentalist has no prior knowledge of the fluid distributions, so he or she cannot estimate the expected path length variations over the projected beam area as described by Equation 15.2. Thus it is, in general, impossible to derive *a priori* corrections. Furthermore, even if the fluid geometry were known, Equation 15.2 is itself only an approximation. This suggests that care should be taken to form experimental conditions that lead to near-uniform thicknesses for all measurements.

15.4 Sample Density

The spatial resolution of hydrologic properties and processes that can be achieved with geophysical methods depends on both the sample volume of each measurement and the spacing of measurements. As described in the preceding sections, the sample volume of each measurement depends on the instrument design and on the distribution of materials within the domain. Measurement spacing is defined by an investigator based on the expected relevant scale of the hydrologic process of interest, often in balance with a time constraint imposed by the time required for each measurement. The tradeoff between spatial and temporal resolution is most critical when geophysical methods are applied to monitor transient hydrologic processes. For example, this tradeoff between the measurement times, the number of locations, and the measurement accuracy/precision is often a critical practical concern for gamma- and x-ray systems, since observation times can be significant (1–500 s per location is typical) relative to the experimental time scales.

15.4.1 SAMPLE DENSITY EFFECTS ON X-RAY DNAPL MASS ESTIMATION

To demonstrate the impact of the choice of sample density on hydrologic interpretations, consider the estimation of DNAPL mass in a two-dimensional flow cell using an x-ray system (Moreno-Barbero, in submission). The container used had internal dimensions of approximately 25 cm long by 18 cm high by 2 cm thick (Figure 15.7). The cell was wet-packed with two uniform quartz sands: a coarse-grained sand to form the rectangular inclusion and a finer-grained sand surrounding it. A total of 27.0 ± 0.1 grams of tetrachloroethylene (PCE), dyed with a small mass of SudanIV to increase visibility, was slowly injected through a port in the back of the tank to create the high-saturation pool. Following injection, the tank was scanned with the x-ray system to measure the PCE mass at 256 locations (Figure 15.7). Because the box was wet-packed, the wet/dry differencing method for the x-ray measurement of the spatial distribution of porosities could not be applied. While having an unknown porosity distribution makes it difficult to determine DNAPL saturations precisely, it is not an impediment to the measurement of DNAPL masses. Assuming that the DNAPL has a constant density, each measurement is interpreted in terms of a representative DNAPL path length (or "effective thickness") at each measurement location. Statistical analyses of the error variance for this data set suggests that, at individual points, PCE path lengths were measured with a standard error of about 0.2 mm. Over the entire domain, the mass balance as calculated from all interpolated x-ray data from four complete scans is 27.7 ± 0.3 grams. This is in excellent agreement (less than 2.5% error) with the known total injected mass. Further examination of the results shows very high gradients in the PCE saturation between adjacent measurement locations at the lateral edges of the pool. In contrast, little variation in saturation exists within the pool. It is likely that a large fraction of the error in total mass estimation results from interpolation between the measurement points spanning the edge of the pool. A redesigned experiment could place measurement points closer together at the edges of the pool to capture the highly spatially variable saturations in this region; the lower limit on measurement spacing would be defined by the dimensions of the beam used.

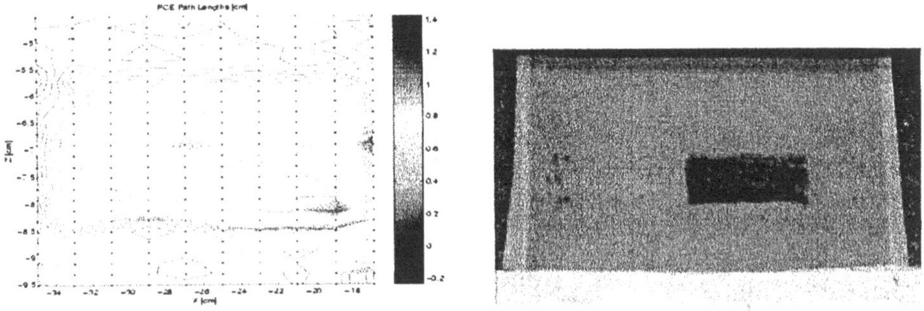

Figure 15.7. (left) A DNAPL pool (PCE dyed w/SudanIV) within a coarse sand inclusion; (right) DNAPL path lengths as measured by x-ray attenuation. The interior dimensions of the tank are approximately 25 cm long by 18 cm high by 2 cm thick.

15.5 Instrument Averaging

Every hydrogeophysical method has an associated sample volume (or sample volumes). This sample volume places a primary control on the achievable spatial resolution of the method. In addition, this sample volume describes the volume over which the medium properties are averaged by the instrument. That is, the measured property may depend on the distribution of the property of hydrologic interest within the sample volume. If this is the case, then there is a nonunique relationship between the measured property and the volume-average of the property of interest in the sample volume. This requires knowledgeable interpretation of hydrogeophysical results, as well as geophysical insight, to design effective hydrogeophysical measurement networks.

15.5.1 EFFECTS ON SEISMIC NAPL PROFILING

The distribution of immiscible fluid phases in porous media is extremely sensitive to how they are emplaced, and can be highly irregular, or patchy. Not only will values of fluid saturation vary with sample volume, but immiscible fluid distribution within a given sample volume can range from evenly dispersed to patchy. For measurement of NAPL saturation changes with *P*-waves at the laboratory scale, the millimeter- to centimeter-long seismic wavelengths are often of similar length scales as NAPL blobs or patches. Under these conditions, the seismic *P*-wave signature may respond to both the average medium properties, as well as to the geometry of NAPL distribution. The results from *P*-wave monitoring of a controlled NAPL spill in a laboratory-scale tank illustrate some of the issues that arise with measurement averaging.

An experiment was performed in a 60 cm diameter by 75 cm tall tank, with 3 cm diameter acrylic wells just inside the vessel wall for crosswell *P*-wave data acquisition. The *P*-wave source was placed in one well and the receiver in a second well on the opposite side of the tank. The bottom two-thirds of the tank was packed with a coarse, subrounded 0.85 mm to 1.7 mm grain diameter quartz sand, over which a 25 cm thick layer of 44 to 88-micron diameter glass beads was placed to form a capillary barrier. Then, n-Dodecane, dyed red, was injected from the bottom of the water-saturated tank, such that the lighter-than-water NAPL migrated upward, leaving residual NAPL segments along its flow paths, and accumulating as a lens in the coarse sand below the capillary barrier. The source and receiver were moved over the depth of the tank in increments of 1 cm, and the transmitted *P*-wave was recorded at each station, for the water-saturated tank. Over the course of three days, 6.7 L of n-dodecane was injected into the tank from the bottom. Displaced water was recovered from the top of the tank and the *P*-wave scan was repeated.

The record of the waveforms of each zero-offset scan collected before and after NAPL injection (Figure 15.8) shows a change in grayscale intensity (amplitude) associated with the first arrival of the *P*-wave. In zero-offset scans, the received waveforms represent horizontally averaged properties. The piezoelectric source produced waves with a central frequency of about 90 kHz, resulting in an approximately 2 cm wavelength. The reference scan (the water-saturated tank) in Figure 15.8a shows attenuation and diffraction caused by the glass bead/sand interface. This interface is not sharp, and includes a transitional region, indicated by the pair of dashed black lines in the figure, of about 2–3 cm where the smaller diameter beads fill the pore space of the coarse sand. In the postinjection scan (Figure 15.8b), negligible waveform changes occur in the glass bead layer, compared to the reference scan. In the 15 cm below the glass bead layer, delayed travel times and amplitude reductions of 60 to 95% result from the presence of the NAPL lens. The highly attenuated first-arrival times within the lens are indicated by a dashed white line. Below the lens, the residual NAPL causes amplitude reductions of 1 to 30% and changes in diffraction patterns. Amplitude decreases caused by the presence of NAPL are much greater than the variability of the water-saturated sand.

Following the post-injection scan, the three-dimensional distribution of the NAPL was determined by excavation. A thin-walled brass grid covering the horizontal cross-sectional area of the tank was pressed into the sand pack to guide the excavation of 5 cm × 5 cm × 5 cm samples. The NAPL volume in the samples taken from the scanning plane were measured by extracting the dyed NAPL into a known volume of clean n-dodecane and measuring the resulting dye concentration with a UV spectrophotometer. The measured porosity of each sample was used to calculate the NAPL saturation. No NAPL was detected within the glass bead layer (Figure 15.9a). Most of the NAPL accumulated in the top 15 cm of the coarse sand, where NAPL saturations ranged from 32% to 81% of the pore space. NAPL saturations are lower near the top of the lens, because of the smaller pore sizes where smaller glass beads partially fill the coarse sand pores. Part of the lens was excavated by freezing layers over the entire cross section of the tank with liquid nitrogen, without the brass grid, which facilitated the removal of large volumes of NAPL. This measurement produced horizontally averaged saturation values shown in Figure 15.9b. The other saturation values below the lens in Figure 15.9b were calculated from the 2-D distributions measured in Figure 15.9a. Below the lens, saturations range from 0 to 16%, with maximum values corresponding to the depths of maximum amplitude change seen in Figure 15.9c at 58 and 63 cm.

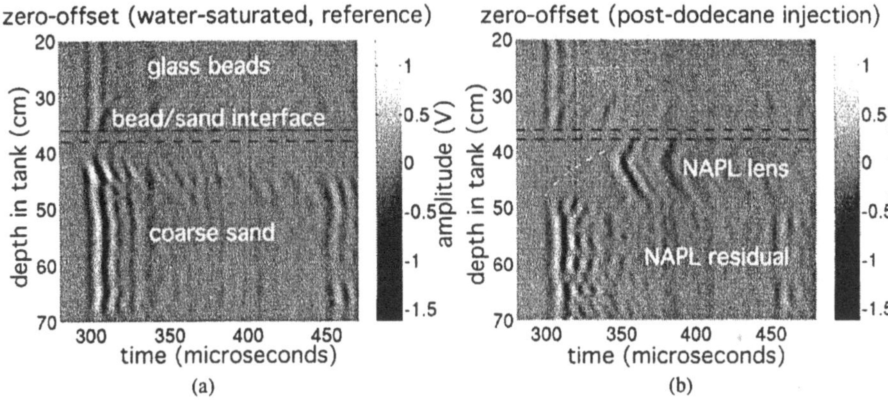

Figure 15.8. Comparison of zero-offset (source and receiver at same depths) *P*-wave travel time and amplitude data through tank for (a) water-saturated conditions (or reference scan), and (b) after injection of n-Dodecane from tank bottom. The arrival time of the *P*-wave is indicated by the first change in gray tone, moving along the time axis. The presence of the NAPL lens can be seen in the significant delay in arrival time (indicated by the dashed white line) and reduced amplitude over the 15 cm below the glass bead/sand interface. The NAPL residual below the lens causes velocity and amplitude changes relative to the reference scan. Data from Geller et al. (2000).

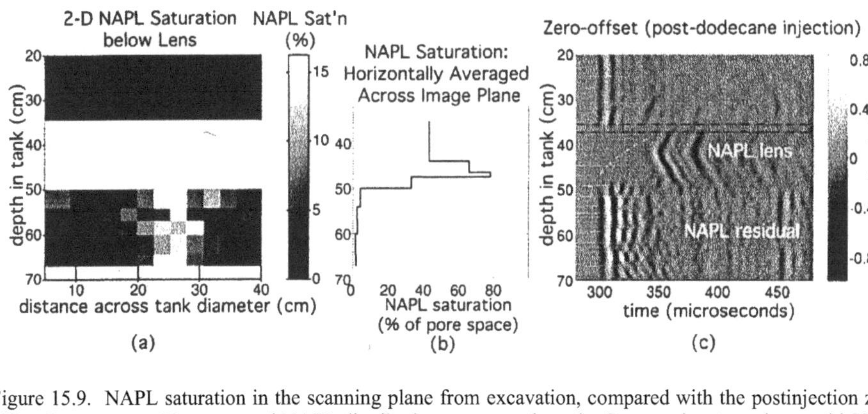

Figure 15.9. NAPL saturation in the scanning plane from excavation, compared with the postinjection zero-offset *P*-wave scan. The measured NAPL distribution corresponds to the *P*-wave signature change. (a) NAPL saturation below the NAPL lens, measured from samples excavated in 5 cm × 5 cm × 5 cm sections. (b) Horizontally averaged NAPL saturation across the image plane. The lens was excavated by freezing layers of sand with liquid nitrogen. Samples below the lens are calculated from the 2-D distributions in (a). (c) The post-injection scan. This amplitude scan has higher gain than in Figure 15.8b. Data from Geller et al. (2000).

Through the NAPL lens, the first-arrival time decreases with depth as NAPL saturation decreases. The ray path traveled by the *P*-wave may be affected by the complex liquid distribution within the lens and the glass-bead/sand interface. However, if the ray paths of the first arrivals through the lens are directly across the tank, then these data show an increase in *P*-wave velocity (the ratio of distance traveled to travel time) with decreasing NAPL saturation. This trend is consistent with column-scale measurements that showed that *P*-wave velocity decreased almost linearly with increasing NAPL saturation (Geller and Myer, 1995). The amplitudes within the region of the lens do not change significantly (seen by the constant gray-scale intensity in Figure 15.9c), while the average NAPL saturation changes from 81% to 32% (Figure 15.9b). This suggests a nonlinear relationship between saturation and attenuation. Towards the middle of the lens, the NAPL distribution is expected to be more homogeneous throughout the horizontal cross section than near the bottom of the lens. At the bottom of the lens, patches of water-saturated media occur between sections occupied by NAPL, which may be more attenuating compared to the same volume of NAPL evenly distributed throughout the cross section. This is consistent with other observations and analyses—Geller (1995); Seifert et al. (1998); Cadoret et al. (1998)—where increased energy attenuation occurs in the presence of irregular and patchy distributions of immiscible fluids compared to lower attenuation for more homogeneous fluid-phase distributions.

This experiment shows that attenuation is much more sensitive than velocity to the presence of NAPL. The effect of NAPL distribution on *P*-wave attenuation suggests that NAPL patches may scatter seismic energy at laboratory-scale frequencies. However, the dependence of *P*-wave attenuation on NAPL geometry indicates that while increased attenuation may be the more sensitive indicator of the presence of NAPL, it cannot be used to quantify the value of saturation. On the other hand, *P*-wave velocity may be used to estimate NAPL saturation because of the linearity of velocity response to NAPL saturation. The determination of valid relationships for *P*-wave velocity as a function of NAPL saturation in unconsolidated sands is an area of ongoing research. However, the available theories are consistent with a near-linear relationship under constant loads. Gassman's (1951) equations for pore-fluid substitution estimate the bulk modulus of a mixture as a function of its constituent solid and fluid properties, and have been applied to the problem of hydrocarbon detection in oil reservoirs (e.g., Mavko and Mukerji, 1998; Dominico, 1976). Geller and Myer (1995) predicted measured *P*-wave velocities through sands saturated with varying fractions of water and NAPLs using effective composition laws by Kuster and Toksöz (1974) and the average of the liquid properties. However, fitting the data to the prediction required lower porosities than were experimentally measured. It is possible that amplitude data may prove useful for screening for the presence of NAPL, velocity data may allow for quantitative analysis of NAPL saturations, and combined amplitude and velocity data could describe both the saturation and the distribution of NAPL.

15.6 Time-Lapse Monitoring

One of the most significant advantages of geophysical methods for hydrologic monitoring is their ability to measure with little or no disruption to the system under investigation. This allows for multidimensional monitoring of transient processes, which is highly limited, if not impossible, with conventional methods. The use of ERT for multidimensional, transient hydrologic monitoring is more advanced than other methods. An example of the use of ERT for monitoring solute transport through a structured column is presented.

15.6.1 BENEFITS OF ERT TIME-LAPSE MONITORING

ERT has limited value in characterizing soil or rock structure from static images and will certainly be unable to resolve millimeter scale macropore or fracture features. ERT may, however, be applied to study the spatial changes of pore fluid content over time and thus elucidate flow pathways within the system. By imaging changes in resistivity in an area or volume, ERT can be used to infer the proportion of pore volume contributing to flow, without necessarily imaging individual flow paths (as illustrated earlier for larger scale studies in Chapter 14 of this volume. Henry-Poulter et al. (1993), Binley et al. (1996a,b), Henry-Poulter (1996), and Olsen et al. (1999) have applied ERT to studies of solute transport in soil cores. This section presents some of the results and findings of these studies.

Binley et al. (1996a) used ERT to infer mobile and immobile pore fractions during the transport of a solute through a 32 cm diameter, 46 cm long undisturbed soil column. Four electrode planes were used, each with 16 electrodes, in a similar arrangement to that in Figure 15.2. A selection of ERT results at the four electrode planes monitored are shown in Figure 15.10, displayed as percentage changes in electrical conductivity relative to the pre-tracer conditions. Each horizontal plane is discretized into 104 triangular pixels, at each of which a resistivity value was computed for each frame.

During the early stages of the tracer test, changes in electrical conductivity are localized to a small volume of the soil core. After 19 hours, changes near the outflow are minor and yet breakthrough of the tracer was apparent. This is likely to be a result of transport through localized connected macropores within the core, which ERT is unable to resolve. Note that if one assumes purely advective transport at the pore-water velocity obtained from the model fit to the breakthrough curve, then the tracer "front" at 19 hours should be only approximately 6 cm from the base; and yet significant changes 28 cm from the base are clearly evident in the ERT images. After 34 hours, a marked zone of changes in soil conductivity is seen in the ERT image plane close to the outflow, at a time when the breakthrough curve showed a period of near-constant effluent concentration with time. This may result from early breakthrough and near-steady-state conditions within a small pore volume of highly transmissive macropores. After approximately 60 hours, the breakthrough curve showed a marked second rising limb of the curve, consistent with the increasing contributing volume observed in the ERT images near the outflow at 34 hours. After 94 hours, this volume has grown somewhat, but localized high conductivity changes are still significant. At 97 hours, a low-conductivity tracer was injected into the base of the column. Subsequent ERT images show that this low-conductivity tracer also influences the same zones that were influenced by the high-conductivity tracer. Other areas in which tracer has perhaps migrated laterally are slow to change.

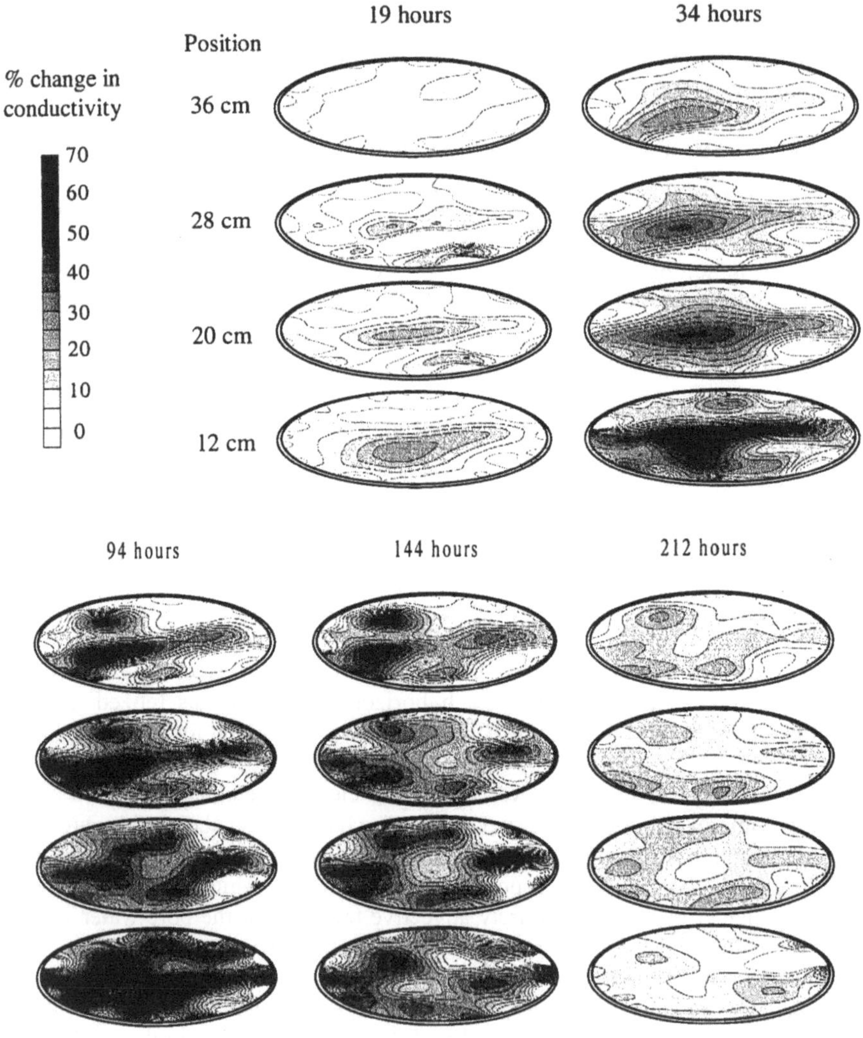

Figure 15.10. Sequence of ERT images of soil-core sections at four positions from the base of the core. Changes in electrical conductivity caused by saline tracer applied at base of core are shown. The high electrical conductivity tracer ends after 97 hours, followed by a low-electrical-conductivity tracer injection.

Binley et al. (1996a) introduced the simple concept of analyzing the ERT images as equivalent to results from pore water samplers located within the soil core (see also Slater et al., 2000; Slater et al., 2002). Figure 15.11 shows changes in conductivity, expressed as a relative concentration, for three pixels in the ERT plane 12 cm from the tracer injection (Figure 15.2). There is a marked difference between the three responses. Pixels 50 and 80 show symmetrical behavior, albeit with different magnitudes, similar to the observed effluent breakthrough curve (see Binley et al. [1996a] for details). Pixel 103, in contrast, shows a marked rise in conductivity until ~25 hours after tracer injection, after which time a comparably steady response is seen. Such behavior supports the notion of two distinct phases seen in the outflow breakthrough curve.

Dye staining of the core (see also Binley et al. [1996b] for application to different soil columns) revealed some consistency with the ERT results. However, most dye staining was confined to a very small fraction of the soil core volume. Such staining probably indicated the most transmissive fraction of the core and underestimates the true contributing pore fraction. Using 3-D ERT imaging of two soil columns, Olsen et al. (1999) extended the work of Binley et al. (1996a) by modeling the tracer breakthrough for two soil columns using transfer function techniques. With such models, Olsen et al. (1999) were able to quantify the significance of a preferential transport component in each system. Their ERT results clearly show supporting evidence of preferential flow in a soil column, with a significant bypass flow inferred from tracer breakthrough analysis.

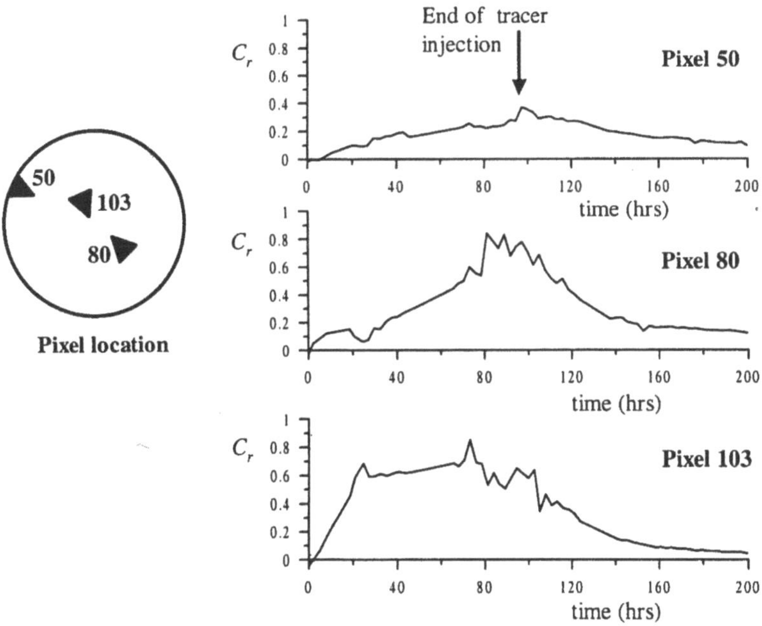

Figure 15.11. ERT pixel breakthrough curves for three selected locations in a plane 12 cm from tracer injection

15.7 Potential Contributions of Laboratory Hydrogeophysics When Properly Applied

Hydrogeophysical methods can make substantial contributions to laboratory studies of fluid flow and mass transport. These methods can offer minimally invasive, nondestructive, rapid, automated measurements that are relatively inexpensive compared with chemical analyses. These qualities are essential for conducting multidimensional monitoring of transient hydrologic processes. Case studies of the application of TDR, x-ray methods, seismic methods, and ERT are shown to demonstrate the unique abilities of hydrogeophysical methods for laboratory investigations. However, further discussion of case studies presents potential limitations to the use of hydrogeophysical methods. Specifically, TDR, x-ray, and seismics case studies demonstrate the need for the effects of spatial averaging within the measurement volume to be considered when making hydrologic interpretations. An x-ray case study is used to demonstrate the importance of the choice of sample locations when designing a hydrogeophysical monitoring network. Finally, a seismics case study demonstrates that many hydrogeophysical methods provide multiple measurements that can be used for hydrologic interpretation. Intelligent use of these measures, together with hydrologic understanding, can extend the use of hydrogeophysics for hydrologic investigations.

References

Binley, A., S. Henry-Poulter, and B. Shaw, Examination of solute transport in an undisturbed soil column using electrical resistance tomography, *Water Resour. Res., 32*(4), 763–769, 1996a.

Binley, A., B. Shaw, and S. Henry-Poulter, Flow pathways in porous media: Electrical resistance tomography and dye staining image verification, *Measurement Science and Technology, 7*(3), 384–390, 1996b.

Bourbie, T., O. Coussy, and B. Zinszner, *Acoustics of Porous Media,* Gulf Publishing Company, Houston, TX, 1987.

Brown, B.H., Medical impedance tomography and process impedance tomography: A brief review, *Measurement Science & Technology, 12*(8), 991–996, 2001.

Cadoret, T., G. Mavko, and B. Zinszner, Fluid distribution effect on sonic attenuation in partially saturated limestones, *Geophysics, 63*, 154–160, 1998.

Chu, D., D. Tang, T.C. Austin, A.A. Hinton, and R.I. Arthur, Jr., Fine-scale acoustic tomographic imaging of shallow water sediments, *IEEE Journal of Oceanic Engineering, 26*(1), 70–83, 2001.

Dominico, S.N., Effect of brine-gas mixture on velocity in an unconsolidated sand reservoir, *Geophysics, 41,* 882–894, 1976.

Ferré, T.P.A., H.H. Nissen, and J. Šimunek, The effect of the spatial sensitivity of TDR on inferring soil hydraulic properties from water content measurements made during the advance of a wetting front, *Vadose Zone Journal, 1*, 281–288, 2002.

Ferré, P.A., J.H. Knight, D.L. Rudolph, and R.G. Kachanoski, The sample area of conventional and alternative time domain reflectometry probes, *Water Resour. Res., 34*(11), 2971–2979, 1998.

Gassman, F., Uber die elasttizitat poroser medien, *Vier. der Natur Gesellschaft, 96,* 1–23, 1951.

Geller, J. T., M. B. Kowalsky, P. K. Seifert and K. T. Nihei, Acoustic detection of immiscible liquids in sand, *Geophysical Research Letters, 27*(3), 417–420, 2000.

Geller, J.T. and L.R. Myer, Ultrasonic imaging of organic liquid contaminants in unconsolidated porous media, *Journal of Contaminant Hydrology, 19*(3), 1995.

Henry-Poulter, S.A., An investigation of transport properties in natural soils using electrical resistance tomography, Unpublished PhD Thesis, Lancaster University, U.K., 1996.

Henry-Poulter, S.A., M.Z. Abdullah, A.M. Binley and F.J. Dickin, Electrical impedance tomography of tracer migration in soils, In: *Computational methods and Experimental Measurements IV, Vol 1: Heat and*

Fluid Flow, by Brebbia and Carlomagno, eds., Computational Mechanics Publications, pp. 101–115, 1993.
Hill III, E.H., L.L. Kupper, and C.T. Miller, Evaluation of path-length estimators for characterizing multiphase systems using polyenergetic x-ray absorption, *Soil Science*, *167*(11), 703–719, 2002.
Illangasekare, T.H., E.J. Armbruster III, and D.N. Yates, Non-aqueous-phase fluids in heterogeneous aquifers—Experimental study, *Journal of Environmental Engineering*, 121(8), 571- 579, 1995.
Imhoff, P.T., Dissolution of a nonaqueous phase liquid in saturated porous media, PhD Thesis, Princeton University, 1992.
Imhoff, P.T., S.N. Gleyzer, J.F. McBride, L.A. Vancho, I. Okuda, and C.T. Miller, Cosolvent-enhanced remediation of residual dense nonaqueous phase liquids: Experimental investigation, *Environmental Science & Technology*, *29*(8), 1966–1976, 1995.
Keers, H., D.W. Vasco and L.R. Johnson, Viscoacoustic crosswell imaging using asymptotic waveforms, *Geophysics, 66*(3), 861–870, 2001.
Knight, J.H. Sensitivity of time domain reflectometry measurements to lateral variations in soil water content, *Water Resour. Res.*, *28,* 2345–2352, 1992.
Knoll, G.F., Radiation Detection and Measurement, John Wiley and Sons, New York, NY, 1979.
Kuster, G.T., and M.N. Toksöz, Velocity and attenuation of seismic waves in two-phase media; Part I—Theoretical formulations, *Geophysics*, *39*(5), 587–606, 1974.
Li, X., L.R. Zhong, and L.J. Pyrak-Nolte, Physics of partially saturated porous media: Residual saturation and seismic-wave propagation, *Annual Review of Earth & Planetary Sciences*, *29*, 419 –460, 2001.
Lo, T-w., and P.L. Inderwiesen, *Fundamentals of Seismic Tomography,* Geophysical Monograph Series, #6, Society of Exploration Geophysicists, Tulsa, OK, 1994.
Mavko, G., and T. Mukerji, Bounds on low-frequency seismic velocities in partially saturated rocks, *Geophysics*, *63*(3), 918–924, 1998.
McKenna, J., D. Sherlock, and B. Evans, Time-lapse 3-D seismic imaging of shallow subsurface contaminant flow, *Journal of Contaminant Hydrology*, *53*, 133–150, 2001.
Miller, C.T., E.H. Hill, III, and M. Moutier, Remediation of DNAPL-contaminated subsurface systems using density-motivated mobilization. *Environmental Science & Technology*, *34*(4), 719–724, 2000.
Moreno-Barbero, E., Evaluation of partitioning inter-well tracer tests for character-ization of DNAPL pools, PhD Dissertation, Colorado School of Mines, Golden, Colorado, in submission.
Nissen, H.H., P.A. Ferré, and P. Moldrup, The transverse sample area of two- and three-rod time domain reflectometry probes: dielectric permittivity *Water Resour. Res.*, *38*(10), No. 1289, 2003.
Okuda, I., J.F. McBride, S.N. Gleyzer, and C.T. Miller, An investigation of physicochemical transport processes affecting the removal of residual DNAPL by nonionic surfactant solutions. *Environmental Science & Technology*, *30*(6), 1852–1860, 1996.
Robinson D.A., S.B. Jones, J.M. Wraith, D. Or, and S.P. Friedman, A review of advances in dielectric and electrical conductivity measurement in soils using time domain reflectometry, *Vadose Zone Journal*, *2*, 444–475, 2003.
Santamarina, J.C., *Soils and Waves*, in collaboration with K.A. Klein and M.A. Fam, John Wiley & Sons, Ltd., Chichester, West Sussex, England, 2001.
Seifert, P.K., J.T. Geller and L.R. Johnson, Effect of P-wave scattering on velocity and attenuation in unconsolidated sand saturated with immiscible liquids, *Geophysics*, *63*(1), 161–170, 1998.
Slater, L., A. Binley, W. Daily and R. Johnson, Cross-hole electrical imaging of a controlled saline tracer injection, *J. Appl. Geophysics*, *44,* 85–102, 2000.
Slater, L., A. Binley, R. Versteeg, R., G. Cassiani, R. Birken, and S. Sandberg, A 3D ERT study of solute transport in a large experimental tank, *J. Appl. Geophysics*, *49*(4), 211–229, 2002,.
Olsen, P.A., A. Binley, S. Henry-Poulter, and W. Tych, Characterising solute transport in undisturbed soil cores using electrical and x-ray tomographic methods, *Hydrological Processes*, *13,* 211–221, 1999.
Topp, G.C., J.L. Davis, and A.P. Annan, Electromagnetic determination of soil water content: measurement in coaxial transmission lines, *Water Resour. Res., 16*, 574–582, 1980.
Topp, G.C., J.L. Davis and A.P. Annan, Electromagnetic determination of soil water content using TDR: I. Applications to wetting fronts and steep gradients, *Soil Sci. Soc. Am. J.*, *46,* 672–678, 1982.
Topp, G.C., and P.A. Ferré, eds., Water content measurement methods, In: *Methods of Soil Analysis,* American Society of Agronomy, 2002.
Wang, M, W. Yin and N. Holliday, A highly adaptive electrical impedance sensing system for flow measurement, *Measurement Science & Technology, 13*(12), 1884–1889, 2002.
Webster, J.G., ed., *Electrical Impedance Tomography*, Adam Hilger, Bristol, England, 1990.

Hydrogeophysical Frontiers

16 EMERGING TECHNOLOGIES IN HYDROGEOPHYSICS

UGUR YARAMANCI[1], ANDREAS KEMNA[2], AND HARRY VEREECKEN[2]

[1] *Technische Universität Berlin, Germany*
[2] *Forschungszentrum Jülich, Germany*

Hydrogeological systems and hydrological processes are quite complex. In hydrogeological systems, as in any complex system, there is a great demand to improve investigative technologies. The motivation for such new technologies is mainly the need for improved assessment (i.e., understanding, access, and description) of properties controlling hydrogeological structure and dynamics. A better understanding of such properties will also improve exploration by enabling the complementary and joint use of hydrogeological methods.

Very often, advancements in hydrogeological technologies take the form of improvements in measurement and processing of currently used methods. These improvements include collection of more data in more accurate, faster, and cheaper ways, as well as faster processing. These kinds of improvements result from the continuous evolution of hardware and software technology. Besides these technical improvements, however, there is a need for better understanding of hydrogeological system properties.

Only very rarely in geophysics does a new technology or approach emerge that is completely different from those previously in existence. This chapter presents two such technologies, which have just passed the experimental stage to become promising and valuable tools for investigating hydrogeological structures and properties. The first technology, surface nuclear magnetic resonance (SNMR), enables, for the first time, the direct detection of water exclusively from surface measurements, with additional information about mobile water content and the pore-structure parameters controlling hydraulic conductivities. The second technology, the magneto-electrical resistivity imaging technique (MERIT)—based on a well-known electrical methodology but with new sensors and measurement parameters—enables better assessment of the subsurface resistivity structure than previously possible, and therefore a better assessment of hydrogeological characteristics. In this chapter, we review in detail SNMR and MERIT technologies and briefly discuss a few additional emerging methods that have great potential to improve hydrogeological investigations.

16.1 Surface Nuclear Magnetic Resonance

The SNMR method is a fairly new geophysical technique for directly investigating the existence, volume, and mobility of groundwater by surface geophysical measurements. The basic concept of SNMR is that hydrogen protons of water in the subsurface are

Y. Rubin and S.S. Hubbard (eds.), Hydrogeophysics, 467–486.

excited with an external electromagnetic wave at the resonance frequency of the protons. Subsequently, we can measure the electromagnetic field of relaxing protons, which depends on the spatial distribution, amount, and mobility of the water.

The first high-precision observations of nuclear magnetic resonance (NMR) signals from hydrogen nuclei were made in the 1940s. Since then, this technique has found wide application in chemistry, physics, tomographic imaging in medicine, and geophysics. It has become a standard investigative technology used on rock cores and in boreholes (Kenyon, 1992). The idea of making use of NMR in groundwater exploration from the ground surface was first developed as early as the 1960s, but only in the 1980s was effective equipment designed and put into operation for surface geophysical exploration (Semenov et al., 1988; Legchenko et al., 1990). Extensive surveys and testing have been conducted in different geological conditions, particularly in sandy aquifers but also in clayey formations and fractured limestone, as well as at special test sites (Schirov et al., 1991; Lieblich et al., 1994; Goldman et al., 1994; Legchenko et al., 1995; Beauce et al., 1996; Yaramanci et al., 1999; Meju et al., 2002; Supper et al., 2002; Vouillamoz et al., 2002; Yaramanci et al., 2002). These tests have revealed the power of this method, as well as its shortcomings.

16.1.1 BASICS OF SNMR

In SNMR measurement, an electromagnetic field is produced using a large loop on the earth's surface that works at the Larmor frequency (1–3 kHz, depending on the local earth magnetic field) with a defined intensity and duration. This field excites protons into a precession movement, which then emit an electromagnetic signal (Figure 16.1) that decays with time. The initial amplitude and decay envelope (Figure 16.2) are measured with the same loop.

Protons of hydrogen atoms in water molecules have a magnetic moment μ that is generally aligned with the local magnetic field B_0 of the earth. When another magnetic field B_1 (primary field or excitation field) is applied, the axes of the spinning protons are deflected, because of the applied torque (Figure 16.1). Thus, only the component of B_1 perpendicular to the static field B_0 acts as the torque force. When B_1 is removed, the protons generate a relaxation magnetic field as they become realigned along B_0 while precessing around B_0 at the Larmor frequency

$$f_L = \gamma B_0 / 2\pi, \tag{16.1}$$

where $\gamma = 0.2675$ Hz/nT is the gyromagnetic ratio of hydrogen protons. The expression $\omega_L = 2\pi f_L$ is called the angular Larmor frequency. Measurements are conducted using a loop with (typically) a circular or rectangular layout (Figure 16.1). An alternating current,

$$i(t) = i_0 \cos(\omega_L t), \tag{16.2}$$

with the angular Larmor frequency ω_L and strength i_0, is passed through the loop for a limited time τ, so that an excitation intensity (pulse moment) of $q = i_0 \tau$ is achieved. After the current in the loop is switched off, a voltage $e(t)$ with frequency ω_L and decaying amplitude is induced in the loop by the relaxation of the protons:

$$e(t) = \omega_L M_0 \int f(r)\, e^{-t/T(r)} \cos(\omega_L t + \varphi(r)) B_\perp(r) \sin(0.5\, \gamma B_\perp(r)\, q)\, dV, \quad (16.3)$$

where $M_0 = 3.29 \times 10^{-3} B_0$ J/(T m^3), is the nuclear magnetization for water at a temperature of 293 K. In a unit volume dV at location $r(x,y,z)$, the volume fraction of water is given by $f(r)$ and the decay time of protons by $T(r)$. $B_\perp(r)$ is the component of the exciting (primary) field (normalized to 1 A) perpendicular to the static magnetic field B_0 of the earth. In a conductive medium, $B_\perp(r)$ is composed of the primary field of the loop and the induced field. The induced field causes a phase shift $\varphi(r)$ with respect to the exciting field. The argument of the sine function, $\theta = 0.5\, \gamma B_\perp(r)\, q$, is the angle of deflection for the magnetic moment of the protons from the earth's magnetic field. The signal $e(t)$ is usually approximated by (Figure 16.2):

$$e(t) = E_0\, e^{-t/T} \cos(\omega_L t + \varphi). \quad (16.4)$$

The envelope of this voltage (i.e., $E_0\, e^{-t/T}$) is directly related to the water content $f(r)$ and to the decay time $T(r)$ of every volume element in the subsurface contributing to the signal. For nonconductive media (i.e., where no phase shift is produced), the initial amplitude E_0 at $t = 0$ is related only to the water content:

$$E_0 = \omega_0 M_0 \int f(r)\, B_\perp(r) \sin(0.5\, \gamma B_\perp(r)\, q)\, dV. \quad (16.5)$$

Using this equation, the initial amplitudes for various excitation intensities (Figure 16.2) and for water layers at different depths and thicknesses can be calculated. For deeper water layers, the maximum of E_0 occurs at higher q values, and the strength of E_0 is directly related to the thickness of the layer—i.e., the amount of water.

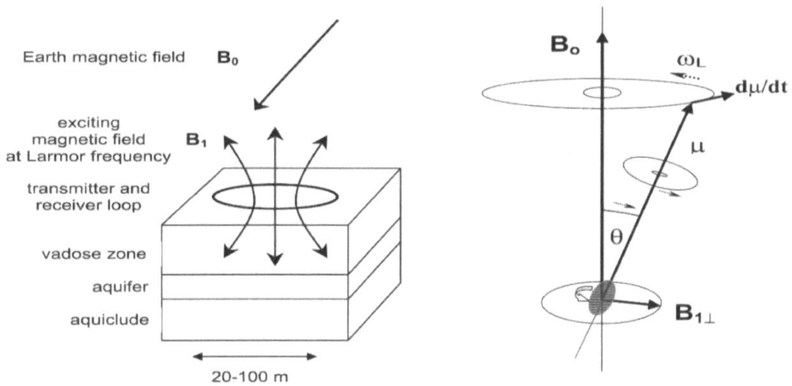

Figure 16.1. Layout of an SNMR measurement and dynamics of a hydrogen proton

The recorded decay time T (transversal relaxation denoted by T_2^* in NMR terminology) is related to the mean pore size and (therefore) grain size, as well as the hydraulic conductivity of the material. Decay times are shorter for materials with finer grains. Clay, including sandy clays, usually has a decay time of less than 30 ms, whereas sand has a decay time of 60–300 ms and gravel a decay time of 300–600 ms (Schirov et al., 1991; Yaramanci et al., 1999).

The resolution and accuracy of the method depend on $B_{\perp}(r)$ and decrease with depth. Higher i_0 and/or τ are needed to excite the protons at greater depth (as long as $\tau << T_2^*$). By increasing q, the depth of the investigation is increased. In fact, the choice of q focuses the excitation to a certain depth range. Currently, 100–150 m depth can be investigated, with a resolution of a few decimeters at shallow depths and a few meters at greater depths. The accuracy achievable for water content is roughly a few percent, depending on depth.

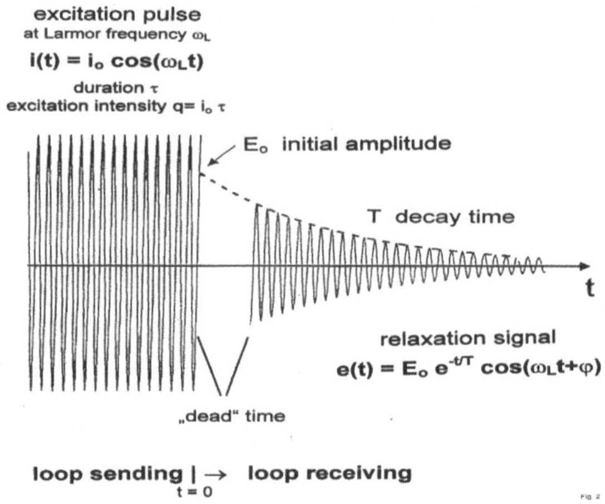

Figure 16.2. Input and output signals in SNMR measurements

The measurements are conducted for different excitation intensities (q), and the main parameters recorded for every q are the initial amplitude E_0 and the decay time T. This set of data, in the form of $E_0(q)$ and $T(q)$, is inverted to find the distribution of water content with depth $f(z)$ and of decay time with depth $T(z)$. Thus, a modified version of Equation (16.3) becomes the basis of the inversion for horizontally layered models. Two- or three-dimensional inversions and suitable measurement approaches for this are not available yet, but are currently being investigated. The usual inversion scheme employed in SNMR is based on a least-squares solution with regularization (Legchenko and Shushakov, 1998). New inversion schemes are being developed that use model optimization with simulated annealing; these schemes allow more flexibility in choosing layer thicknesses and also allow layer thicknesses to be optimized (Mohnke and Yaramanci, 1999, 2002).

16.1.2 INVESTIGATIONS AT THE HALDENSLEBEN TEST SITE

A demonstration of the use of SNMR in an integrated geophysical survey was carried out at a site near Haldensleben in Germany (Yaramanci et al., 1999). This site was quite suitable for testing SNMR, because many features were known from previous surveys. The geology of the area in which three boreholes (B2, B7, and B8) were drilled consisted of Quaternary deposits: well-sorted sands interbedded with cohesive glacial till and silt (Figure 16.3). The sands formed good aquifers with hydraulic conductivities of 10^{-4}–10^{-3} m/s. The glacial tills and silts, which act as aquicludes, were at points as thick as 20 m. Since the aquifers were discontinuous, local hydraulic links existed between them.

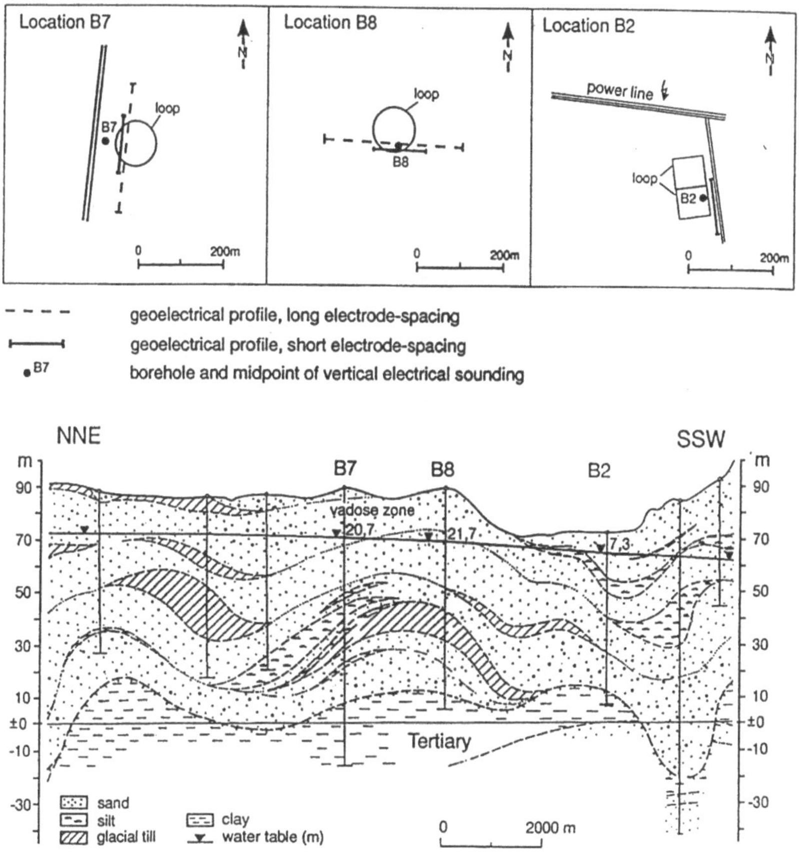

Figure 16.3. Geological section at the Haldensleben test site, according to borehole and geoelectrical measurements (Yaramanci et al. 1999)

The hydrogeological conditions at locations B7 and B8 were similar. The water table was at a depth of about 20 m and the depth to the base of the first aquifer varied between 40 to 50 m. At B7, there were interbedded impermeable tills and silts in the depth interval between 40 to 65 m. At B8, there was an impermeable till layer about 12 m thick at a depth of 47 m. At both locations there was a confined second aquifer below the till. The regional aquiclude, the Rupelian clay, occurred at a depth of 75 to 80 m, immediately below the Quaternary. The situation at B2 was somewhat different. The water table there was at a depth of 7.3 m, caused mainly by the lower elevation of the site. A 6 m thick silt layer was present at a depth of 16 m and acted as an aquiclude for a perched-water zone.

The NUMIS system of IRIS Instruments, which at the time this chapter was written was the only commercially available system for SNMR measurements, was employed for field surveys at these sites. Standard NUMIS software was also used for processing and inversion. The measurements at locations B7 and B8 were carried out using a circular loop 100 m in diameter. To reduce the influence of noise, a figure-eight loop shaped like two adjacent 37.5 m squares was used at location B2. The local magnetic field of the earth was $B_0 = 48{,}757$ nT, which corresponds to a Larmor frequency of $f_L = 2{,}076$ Hz. The duration of the excitation current was kept constant at 40 ms, corresponding to 80 cycles. Measurements were carried out at several current strengths between 5 A and 250 A, varying the excitation intensity from 200 A ms to approximately 8,000 A ms. The noise levels were about 500 to 700 nV at B7, 100 to 300 nV at B8, and 600 to 1,100 nV at B2.

The signal amplitudes measured at B7 and B8 show a typical shape for sounding curves obtained for an aquifer at moderate depth (Figure 16.4). Inversion shows that the water content gradually increases with depth from 5% in the unsaturated zone to 20–25% at 20 m, the depth of the water table. Below 30–35 m, the water content decreases. At location B2, the signal amplitudes and water content after inversion are almost constant and very low, with the maximum water content about 10%. The inversions have an rms-error less than 5%, which indicates good agreement between the observed amplitudes and the amplitudes reconstructed from the model.

Except for the two highest excitation intensities, the decay times observed at B7 were 140 to 220 ms and quite irregular. At B8, the observed decay times increased smoothly from 180 to 250 ms at 4,000 A ms excitation intensity, and then decreased slightly. The decay times at B2 changed from 150 to 100 ms, and then to 160 ms with increasing excitation intensity. The range of the inverted decay times was 200 to 250 ms for B7 and B8 at depths of higher water content, i.e., where the aquifer is composed of primarily a medium grained sand. The phases at B7 and B8 began with 0° and increased to 90° and 60°, respectively, indicating the existence of more conductive layers at depth.

In general, the SNMR results were in agreement with the borehole data and electrical data from Sites B7 and B8, at least down to a depth of 40 m, clearly confirming the presence of the aquifer. Resolution with depth became poorer because the thickness of the horizontal layers used in inversion increased. The inferred decrease in water content with depth was not necessarily reliable. At location B2, the distribution of the water content did not correspond to the known geology, but the noise level was remarkably high (despite use of the figure-eight loop).

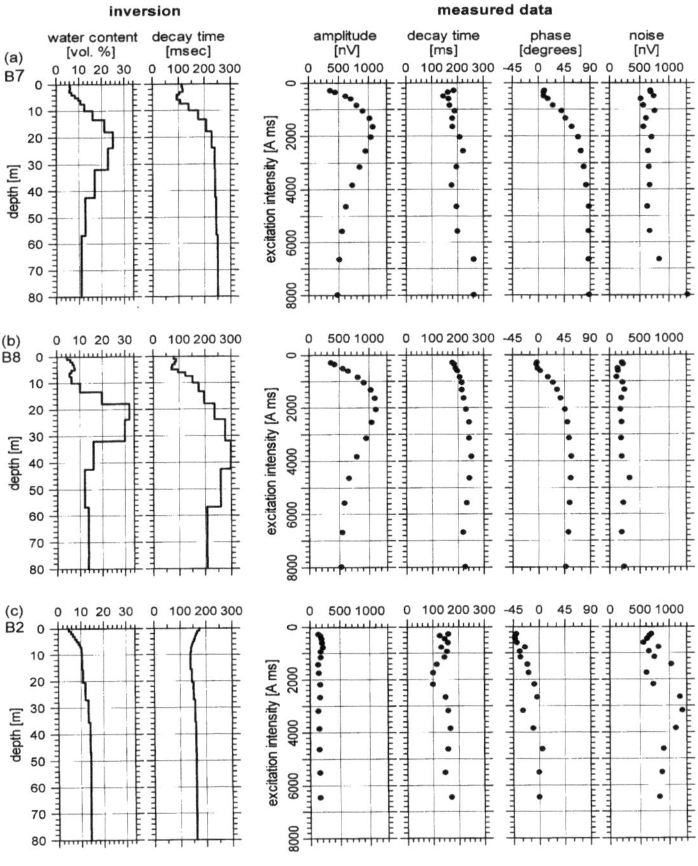

Figure 16.4. SNMR data and inversion at the test site at Haldensleben, Germany

At the test location, an extensive electrical survey was carried out using direct current measurements (see Chapter 5 of this volume). In general, the resistivities were high enough to produce negligible electromagnetic induction effects on the SNMR measurements. Multi-electrode 2-D resistivity measurements were made, using a basic electrode spacing of 10 m and a large array of 300 m or more. Resistivity was found to decrease distinctly and rapidly with depth, indicating the presence of groundwater and clayey layers. However, the inversion algorithms tended to "smear" the resistivity when large differences were present. Consequently, the depth to groundwater could not be precisely determined. Schlumberger soundings were then carried out at all locations, using 1-D inversion and very large electrode layouts (allowing sharp boundaries). At the three areas (B7, B8, and B2), a sharp decrease in resistivity very clearly indicated the presence of the water table at 23, 20, and 6 m, respectively, in good agreement with borehole data. The resistivity of the unsaturated zone was about 5×10^3 Ωm, and the aquifer had a resistivity of about 100 Ωm. The relatively high resistivities of the aquifers resulted from the low concentrations of total dissolved solids in the groundwater. Estimation of water content from these resistivities using standard relationships failed at Haldensleben. Note that resistivity not only depends on the water content, but also on the salinity of the water, the pore structure, and (to a large extent) the clay content, as was discussed in Chapter 4 of this volume. Therefore, uncertainty is typically associated with the relationship between water content and resistivity.

The SNMR measurements were centered on the boreholes at the test site (Figure 16.3) so that well logging gave independent data for verification. The SNMR and electrical results, as well as the parameters derived from grain-size analysis of cores, were checked against gamma-ray, induction, impulse neutron gamma (ING), and salinity logs. The lithology, conductivity, and water content of the various layers, as well as the conductivity of the groundwater, were derived from borehole data (Yaramanci et al., 1999). The logs indicated that the water content for boreholes B7 and B8 has approximately the same range and distribution. In the unsaturated zone down to a depth of 6 m, the total water content averaged about 15%, which increased slightly to 20% at about the depth of the water table. Experience indicates that the amount of adhesive (retained) water in the saturated zone should be less than this. The high values here were probably caused by the presence of percolation (i.e., seepage water). The water content of the capillary fringe just above the water table and the aquifer was about 35% on average. Slightly higher or lower values were caused by finer- or coarser-grained material. Clearly visible in the log was the effect of clay seals behind the casing, where the water content was 50–65%. In addition, sieve analyses carried out on aquifer material from boreholes allowed estimations of porosities and amounts of adhesive water using conventional methods. The reliability of these estimates is high, especially for well-sorted sands like those in the test area. The total porosities derived were in good agreement with the ING logs.

Comparison of the SNMR results with the ING log showed that the water content determined with the SNMR was at first sight lower by 5–10% in the unsaturated zone and by about 10–12% in the upper part of the aquifer (i.e., between the water table and a depth of 40 m). There was, however, a fundamental difference between the SNMR and ING: SNMR is sensitive only to the mobile water. Water bound to the pore surfaces by strong molecular attraction has very short relaxation times and cannot be detected by SNMR, because of the need for a delay (dead) time of 30 ms prior to the start of the measurements (Figure 16.2). The water content given by ING logs was obviously higher than that determined by SNMR, the difference being the bound water and some of the weakly bound seepage water.

The water content itself gave no clear indication whether the soil was saturated or unsaturated. Therefore, it was not a direct measure of well yield. In the unsaturated zone, many isolated pockets of capillary water may represent a considerable volume of free water, which, however, is not exploitable groundwater. A complete evaluation of water in a soil requires knowledge not only of the amount of water in the soil, but also its energy status. This is described by the soil-water retention curve. The SNMR results for the test area showed a relatively large amount of water in the unsaturated (vadose) zone. This was in good agreement with the ING log, which shows 12–15% total water on average.

To obtain hydraulic conductivities from SNMR measurements, the empirical relationship between decay time and average grain size observed in many SNMR surveys (Schirov et al., 1991) can be used. Combining this with the relationship between grain size and hydraulic conductivity often used in hydrogeology (e.g., Hölting, 1992) leads to a simple estimation of hydraulic conductivity, as proposed by Yaramanci et al. (1999):

$$k \approx T^4, \tag{16.6}$$

where k = hydraulic conductivity in meters per second and T = decay time in milliseconds. The decay times of about 100 to 200 ms from the Haldensleben survey suggest hydraulic conductivity values of about 0.8×10^{-4} to 1.4×10^{-3} m/s, which are in very good agreement with those conductivities derived from the grain-size analyses of the core material.

16.1.3 CONCLUSIONS AND DEVELOPMENTS

SNMR has passed the experimental stage to become a powerful tool for groundwater exploration and aquifer characterization. Some further improvements are still necessary and in progress. The most significant concerns currently are the induction effects, and their inclusion into the analysis and inversion as the amplitude and phase of the primary field is modified by the presence of conductive structures. Earlier considerations of this (Shushakov and Legchenko, 1992; Shushakov, 1996) have led to appropriate theoretical description and numerical handling of this problem (Valla and Legchenko, 2002; Weichmann et al., 2002). In fact, the incorporation of resistivities allows modeling of the phases in a reliable way (Braun et al., 2002), which is not only useful for understanding the phases measured, but also the basis for a successful inversion of phases to obtain resistivity information directly from SNMR measurements.

In an analysis of SNMR, relaxation is generally assumed to follow a single exponential, as in Equation (16.4). Even if decays of individual layers are single-exponential, the integration, as in Equation (16.3), results in a multi-exponential decay of the measured signal. The best way to thoroughly account for this behavior is to consider decay time spectra in the data as well as in the inversion (Mohnke et al., 2001). This leads to a pore-size distribution that is new information and allows improved estimation of hydraulic conductivities. The estimation of water content is also improved significantly, because the initial amplitudes are much more accurately determined using the decay-spectra approach.

Currently, SNMR is carried out with a 1-D working scheme. However, the errors in such a scheme might be very large, because of neglecting the 2-D or even 3-D geometry of the structures (Warsa et al., 2002), which have to be considered in any future analysis and inversion. Measurement layouts are to be modified to meet the multidimensional conditions, which is easier to accomplish for nonconductive structures. In multidimensional structures, the greatest difficulty involves numerical incorporation of the electromagnetic modeling for the exciting field.

With SNMR, as in any geophysical measurement, inversion plays a key role in interpreting the data. However, the limits of inversion and the imposed conditions (including geometrical boundary conditions and the differences in the physical model on which inversion is based) may lead to considerable differences in interpretation. The inversion of SNMR data may be problematic, since different regularizations in the inversion impose a certain degree of smoothness upon the distribution of water content (Legchenko et al., 1998; Yaramanci et al., 1998; Mohnke and Yaramanci, 1999), and the number and size of the layers forced in the inversion may considerably affect the results. The root-mean-squared error measure is not necessarily a sufficient one for assessing the quality of model fit to the observed data. Recent research suggests that a layered modeling with movable boundaries avoids the problems associated with regularization and accounts for the blocky character of the subsurface structure, where appropriate (Mohnke and Yaramanci, 2002).

Further improvement in inversion can be achieved when electrical measurements are available and can be incorporated into a joint inversion with SNMR. Examples of joint inversion of SNMR with vertical electrical sounding show considerable improvement in aquifer detectability and geometry, and allow use of appropriate petrophysical models to separate mobile and adhesive water (Hertrich and Yaramanci, 2002).

At sites where no information is available in advance, SNMR should always be carried out along with electrical methods (i.e., direct current electrical, electromagnetic, and even ground-penetrating radar). This will help to decrease the uncertainty in the results, and also allow hydrogeological parameters to be estimated (Yaramanci et al., 2002). In any case, the quality of geophysical exploration for groundwater and aquifer properties will have an increased degree of reliability wherever SNMR is used as a direct indicator of water and soil properties.

The importance of the SNMR method lies in its ability to detect water directly and to facilitate reliable estimation of mobile water content and hydraulic conductivity. In this respect it is unique, since all other geophysical methods at best yield indirect estimates, via resistivity, induced polarization, dielectric permittivity, or seismic velocity. Combining SNMR with other geophysical methods not only allows direct assessment, but also complements the information gathered by other geophysical methods.

16.2 Magneto-Electrical Resistivity Imaging

As highlighted in previous chapters of this volume, electrical and electromagnetic methods are among the most powerful tools for hydrogeophysical investigations at the field scale. Their potential arises from the fact that electrical properties measured with these methods are closely related to various parameters and state variables characterizing the structure and dynamics of hydrogeological systems (Chapter 4).

Whereas electromagnetic methods make use of induction phenomena to measure the full electromagnetic response of the subsurface to an electromagnetic excitation, standard electrical methods operate in a range where inductive effects are negligible and measure the electrical response to an impressed dc (or low-frequency) electric current only. However, any such current is also intimately associated with a (noninductive) magnetic field. Importantly, this magnetostatic response contains additional information on the subsurface electrical resistivity structure, reflecting (for instance) variations in lithology, water content, or water salinity. Moreover, magnetic measurements are possible in free space, which implies enormous practical advantages over standard electrical measurements that require galvanic contact with the ground.

Although the benefits of magnetic measurements have been recognized for a long time, they have been disregarded in the development of modern resistivity imaging techniques such as electrical resistance tomography (ERT; see Chapter 5 of this volume). Reasons for this include the increased instrumental requirements for magnetic-field sensing and the lack of appropriate, multidimensional modeling routines. Given the pertinent technological and computational advances in this area over the last few years, however, the idea of using magnetic-field measurements as a complement to standard electrical surveying has received new attention, and the magneto-electrical resistivity imaging technique (MERIT) has emerged. The emphasis of this present section is on the magnetic-field-related aspects of the new method. For more general aspects relevant to conventional electrical resistivity imaging, see Chapter 5 of this volume.

16.2.1 HISTORICAL DEVELOPMENT

The idea of measuring the noninductive magnetic field associated with a low-frequency (typically < 10 Hz) electric current, injected into the earth to determine the subsurface resistivity structure, is based on the well-known magnetometric resistivity (MMR) method. (For a comprehensive review, see Edwards and Nabighian, 1991.) Development of the method was motivated by the practical advantage that magnetic field sensing does not require galvanic contact with the ground (as do electric potential measurements). It thus enables very flexible data acquisition, making measurements in (for instance) dry (e.g., in the vadose zone) or plastic-cased boreholes possible. Another advantage of the method over conventional resistivity methods is that the surface magnetic response of a conductive target at depth (e.g., a clay layer) is significantly less reduced, due to the shielding effect of a present conductive overburden, than the surface electrical response (see Edwards, 1974). This phenomenon is based on the different sensitivity characteristics of electrical and magnetic measurements.

Interpretation of MMR data was originally based on surface anomaly maps of one or more measured magnetic-field components reduced by the theoretical response of a uniform earth. Unfortunately, such anomaly maps normally show quite complex patterns that are not easy to interpret. Despite the development of modeling tools for simple subsurface structures (e.g., Edwards et al., 1978; Gomez-Trevino and Edwards, 1979), interpretation generally remained difficult. This circumstance, together with the lack of low-cost magnetic sensors sufficiently sensitive to detect the relatively weak anomalous

MMR signals (typically < 1 nT), may be the main reason why the method has not yet become a routinely employed exploration technique. With the increased computational capabilities now available, MMR was recently revived as a modeling tool and, importantly, both 3-D modeling (Boggs et al., 1999; Haber, 2000) and inversion (Chen et al., 2002) routines were developed.

MMR applications for hydrological purposes have been rare, although Acosta and Worthington (1983) highlighted the potential of the method in this area. One possible application is mapping water flow along preferential pathways (e.g., in fractured aquifers). Here, electric current channeling is likely to occur, which in turn should yield a distinct MMR response. The idea of using the MMR method in an imaging framework to monitor subsurface fluid flow was pursued by Kulessa et al. (2002). Moreover, they proposed the magneto-electrical resistivity imaging technique (MERIT), i.e., the simultaneous acquisition and joint analysis of ERT and MMR data, to merge the benefits of both methods. Kulessa et al. (2002) state that combining both data types yields complementary information, and thus MERIT should have the capacity to resolve spatial structures better than ERT or MMR individually. This potential was also seen in synthetic experiments performed by LaBrecque et al. (2003). Independent from the above-mentioned developments in the geophysical field, magneto-electrical resistivity imaging has recently started to be developed for medical applications (Levy et al., 2002). This demonstrates the considerable interdisciplinary value of the technique.

16.2.2 PHYSICAL FUNDAMENTALS

Magneto-electrical resistivity imaging involves measurement of both the electric potential and the magnetic field as a response to a quasi-stationary electric current impressed into a volume conductor using a pair of electrodes. As in standard electrical methods (see Chapter 5 of this volume), the electric potential, $V(\mathbf{r})$, for a single current electrode idealized as a point source with strength I at the origin, is defined by the Poisson equation:

$$\nabla \cdot [\sigma(\mathbf{r}) \nabla V(\mathbf{r})] = -I\delta(\mathbf{r}), \tag{16.7}$$

where $\sigma(\mathbf{r})$ is the given electrical conductivity distribution and δ denotes the Dirac delta function. The current flow in the considered volume (Figure 16.5a), with current density

$$\mathbf{J}(\mathbf{r}) = -\sigma(\mathbf{r}) \nabla V(\mathbf{r}), \tag{16.8}$$

in turn creates a magnetic field (Figure 16.5b), $\mathbf{B}(\mathbf{r})$, according to the Biot-Savart law:

$$\mathbf{B}(\mathbf{r}) = \frac{\mu}{4\pi} \int \frac{\mathbf{J}(\mathbf{r}') \times (\mathbf{r} - \mathbf{r}')}{|\mathbf{r} - \mathbf{r}'|^3} d\mathbf{r}', \tag{16.9}$$

where μ denotes the magnetic permeability. For hydrogeophysical applications, magnetization effects can be generally neglected, and thus the magnetic permeability of free space can be assumed, i.e., $\mu \approx \mu_0 = 4\pi \times 10^{-7}$ Vs/(Am). Rewriting Equation (16.9), by inserting Equation (16.8) and applying Stokes' theorem (Edwards and Nabighian, 1991) as

$$\mathbf{B}(\mathbf{r}) = \frac{\mu}{4\pi} \int \frac{\nabla' V(\mathbf{r}') \times \nabla' \sigma(\mathbf{r}')}{|\mathbf{r} - \mathbf{r}'|} d\mathbf{r}' \tag{16.10}$$

reveals the important fact that the magnetic field requires information on gradients of the conductivity and not actual conductivity values.

In practice, the measured magnetic field also contains a contribution associated with the wires delivering the current to the electrodes. This overlapping magnetic field can be likewise calculated from the Biot-Savart law (16.9) by integration along the involved wires. Therefore, the wire layout in magneto-electrical resistivity imaging has to be carefully taken into account.

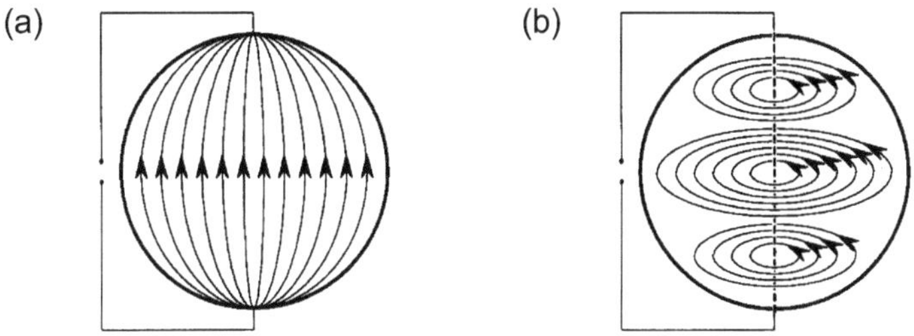

Figure 16.5. Schematic electric current flow (a) and associated magnetic field (b) in the cross section of a spherical or cylindrical volume conductor (e.g., soil column) for two diametrically opposed driving electrodes (adapted from Malmivuo and Pionsey, 1995)

16.2.3 SURVEY GEOMETRIES

For hydrological purposes, magneto-electrical resistivity imaging measurements may be collected in a variety of geometrical arrangements, depending on the scale of investigation. To achieve the highest possible spatial resolution, tomographic layouts involving large arrays of both electrodes and magnetic sensors can be deployed.

At the laboratory scale, cylindrical measurement geometries were suggested to monitor water flow and solute transport in soil columns and lysimeters (Kemna et al., 2003; Zimmermann et al., 2003). Here, electrodes and magnetic sensors are advantageously placed almost to circle the object (Figure 16.6a), similar to medical applications. At the field scale, typical cross-borehole ERT electrode arrangements may be complemented by a magnetic sensor array at the surface (e.g., LaBrecque et al., 2003) (Figure 16.6b). Such a setup combines the benefits of ERT and surface MMR measurements with respect to the delineation of (respectively) vertical and lateral resistivity variations. However, more complex layouts are also possible, (for instance) involving MMR measurements in boreholes (e.g., Nabighian et al., 1984) and current injection at the surface. The latter configuration is an attractive alternative for situations in which only dry or plastic-cased boreholes are available, and thus electrical borehole measurements are problematic. This flexibility in magnetic data acquisition makes magneto-electrical resistivity imaging a valuable tool for improved subsurface characterization and process monitoring.

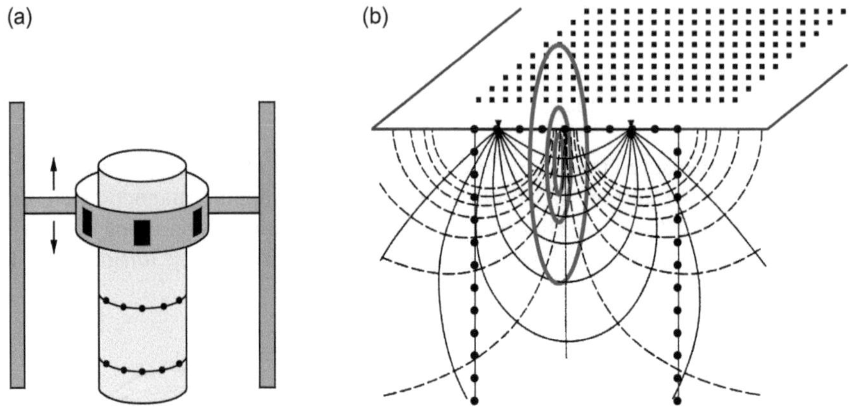

Figure 16.6. Proposed measurement layouts for data acquisition in magneto-electrical resistivity imaging: (a) Fixed electrode rings (circles) complemented by a movable scan ring of magnetic sensors (rectangles) for investigations of cylindrical objects; (b) Cross-borehole electrode setup (circles) complemented by a surface array of magnetic sensors (squares) for shallow subsurface investigations—current density (solid black line), electric potential (dashed black line), and associated magnetic field (thick grey line) distributions indicated for exemplary current injection at the surface (triangles).

16.2.4 IMAGING APPROACHES

Magneto-electrical resistivity imaging requires conversion of the measured electrical and, importantly, magnetic data into 2-D or 3-D distributions of electrical resistivity (or conductivity) by application of appropriate inversion algorithms. (For the general concept of inversion, see Chapter 5 of this volume.) Over the last decade, effective multidimensional inversion schemes have been developed for interpreting electrical data. Recently, however, imaging techniques have also been proposed to likewise interpret the magnetostatic response in terms of the underlying current density or conductivity distribution.

For 2-D current flow, $\mathbf{J}(\mathbf{r}) = [J_x(\mathbf{r})\,\delta(z), J_y(\mathbf{r})\,\delta(z), 0]^{\mathrm{T}}$, the current density may be deconvolved from a 2-D magnetic field map, $\mathbf{B}(x, y, z_0)$, measured at some level z_0, using inverse spatial filtering techniques (Roth et al., 1989) or singular value decomposition (Tan et al., 1995). For high data noise levels, so-called Richardson-Lucy type algorithms have also been successfully applied to similar deconvolution problems in statistical astronomy. These stable, iterative techniques are derived by constraining the Bayes' theorem (see Chapter 17 of this volume) on conditional probabilities using a Poisson statistic (e.g., Lucy, 1974). Adaptation for the present problem of 2-D current flow yields:

$$J_{x,y}^{(i+1)}(x, y, 0) = J_{x,y}^{(i)}(x, y, 0)\left[G(-x, -y, z_0) * \frac{B_{y,x}(x, y, z_0)}{G(x, y, z_0) * J_{x,y}^{(i)}(x, y, 0)}\right], \quad (16.11)$$

where i is the iteration index, $B_{y,x}(x, y, z_0)$ is the respective input horizontal component of the magnetic field, $G(x, y, z_0) = z_0/(x^2+y^2+z_0^2)^{3/2}$ is the kernel of the convolution integral resulting from Equation (16.9) for $B_{y,x}$, and $*$ denotes 2-D convolution with respect to x- and y-directions.

Kulessa et al. (2002) used Equation (16.11) with a method proposed by Sumi et al. (1996) to reconstruct the underlying 2-D conductivity distribution from two magnetic field maps corresponding to independent current density distributions. Note that, as demonstrated by synthetic and laboratory experiments (Kulessa et al., 2002), this relatively simple scheme may be effectively applied to hydrological problems at the field scale (for example, to map horizontal conductivity variations associated with subsurface solute flow by means of magnetic surface measurements—Figure 16.7). However, the lateral extension of the considered flow region should be large compared to its vertical extension, so that the assumed two-dimensionality represents a reasonable approximation.

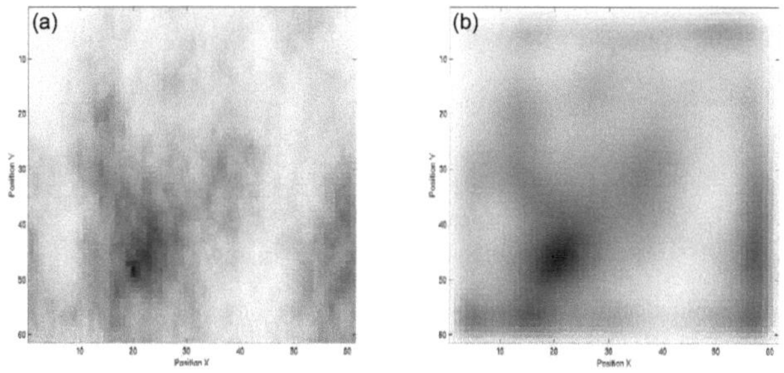

Figure 16.7. Synthetic input electrical conductivity distribution (a), and corresponding inverse solution (b) using the imaging scheme proposed by Kulessa et al. (2002). High and low conductivities are shaded in black and white, respectively; the maximum input conductivity contrast is 50 (after Kulessa et al., 2002).

In general, application of magneto-electrical resistivity imaging to hydrogeological problems will involve the reconstruction of a 3-D resistivity distribution from one- (e.g., borehole) and/or two-dimensionally (e.g., surface) distributed electrical and magnetic measurements (Figure 16.6). The associated inverse problem is fundamentally different from the one addressed above, in which a 2-D parameter distribution is deconvolved from a 2-D data map. It is generally characterized by strong inherent non-uniqueness, insufficient available information, and considerable data noise. Although relatively simple iterative reconstruction techniques were successfully applied to tomographic magnetic data for qualitative imaging purposes (Kemna et al., 2003), quantitative imaging requires adaptation of sophisticated inversion strategies such as those developed for conventional electrical resistivity imaging (see Chapter 5 of this volume). Here, the inverse problem is usually solved as a regularized optimization problem, which involves the minimization of an objective function comprising data misfit (measured versus predicted) as well as certain model characteristics (e.g., model roughness). This approach, however, can be extended to magneto-electrical resistivity imaging (LaBrecque et al., 2003; Levy et al., 2002), by merging both electrical and magnetic data vectors, $\mathbf{d}_e$ and $\mathbf{d}_m$, into a single vector $\mathbf{d} = (\mathbf{d}_e, \mathbf{d}_m)^T$, and applying the inversion scheme outlined in Chapter 5.

The feasibility of such a procedure was recently demonstrated by synthetic model studies representing typical lab-scale (Levy et al., 2002) and field-scale (LaBrecque et al., 2003) applications. Significantly, LaBrecque et al. (2003) compared the magneto-electrical imaging result with that obtained by conventional electrical imaging. Although only considering surface magnetic measurements, the comparison still revealed improvements in the recovery of a target at depth using combined electrical and magnetic data in the inversion (Figure 16.8). This finding is in accordance with actual resolution matrix analyses performed by Tillmann et al. (2003) for a cylindrical (lab-scale) measurement setup.

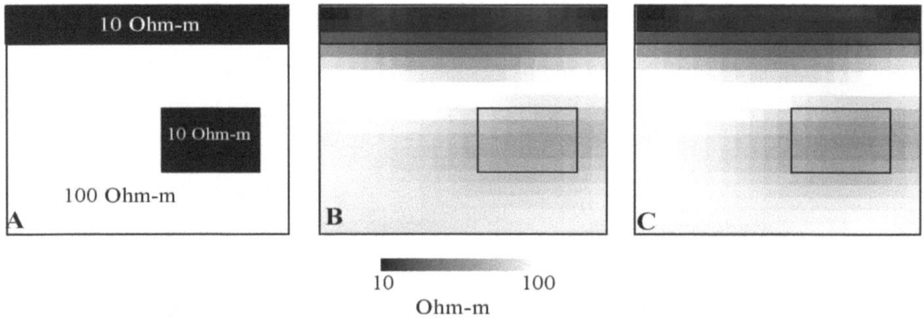

Figure 16.8. Synthetic input electrical resistivity distribution (A), and corresponding inverse solution using electrical resistivity imaging (B) and magneto-electrical resistivity imaging (C). Cross sections through a 3-D model are shown. Electrical and magnetic data were simulated for a field-type measurement layout, involving borehole and surface electrodes, plus, in (c), a surface array of magnetic sensors (after LaBrecque et al., 2003).

16.2.5 CURRENT DEVELOPMENTS AND RESEARCH NEEDS

While several recent studies clearly indicate the potential of magneto-electrical resistivity imaging for advanced structure characterization and process monitoring at different scales (e.g., Tillmann et al., 2004), certainly much more research is required to ultimately develop the method into an established investigation tool for hydrogeophysics. Methodologically, several issues need to be addressed in more detail. For instance, a deeper understanding of the interaction of electrical and magnetic data in the inversion process, and their respective impact on the convergence behavior, would provide the basis for the development of optimized inversion strategies. Moreover, future studies should focus on survey design aspects, such as optimum geometrical arrangements (layout of electrodes, magnetic sensors) and individual measurement configurations for the various potential applications. Most important, however, is the continued successful demonstration of the method in real laboratory and field experiments—an issue that intimately relates to the need for adequate tomographic measurement systems.

In developing large magnetic sensor arrays for tomographic data acquisition purposes, the demand for cost-effective yet sufficiently sensitive sensors arises. Fortunately, in the last few years, significant advances in magnetic-field sensing technology have taken place. These advances involve improved sensitivity, smaller size, and electronic system compatibility of magnetic sensors (Caruso et al., 1998). In particular, magnetoresistive sensors (Tumanski, 2001) are receiving increasing attention with regard to geophysical applications (McGlone, 1997; McGlone and Kutrubes, 2001). These sensors are relatively small (enabling their use in small-scale investigations), robust, and available at much lower cost than (for instance) the widely used fluxgate magnetometers. Zimmermann et al. (2001) investigated the performance of anisotropic magnetoresistive (AMR) sensors using different electronic implementations. Their key finding was that AMR sensors minimized the sensor noise to a level of 20–30 pT/$\sqrt{\text{Hz}}$ at frequencies above 30 Hz (Figure 16.9). This approaches the noise level of the highest quality fluxgate sensors (~5 pT/$\sqrt{\text{Hz}}$ at 1 Hz; e.g., Ripka, 2001) and is considered sufficient for typical magneto-electrical resistivity imaging applications (see, for example, Chen et al., 2002, and Tillmann et al., 2004).

Currently, tomographic measurement systems for laboratory applications of magneto-electrical resistivity imaging, involving AMR sensor arrays and modern lock-in technology for magnetic data acquisition, are under development (Zimmermann et al., 2003). These systems are specifically designed for monitoring flow and transport in soil columns and lysimeters. Svoboda et al. (2002) started the construction of a field-measurement system for hydrological applications. Here, established ERT instrumentation is complemented by a radio-linked, GPS-supported fluxgate magnetometer system, enabling flexible magnetic data collection at the earth's surface.

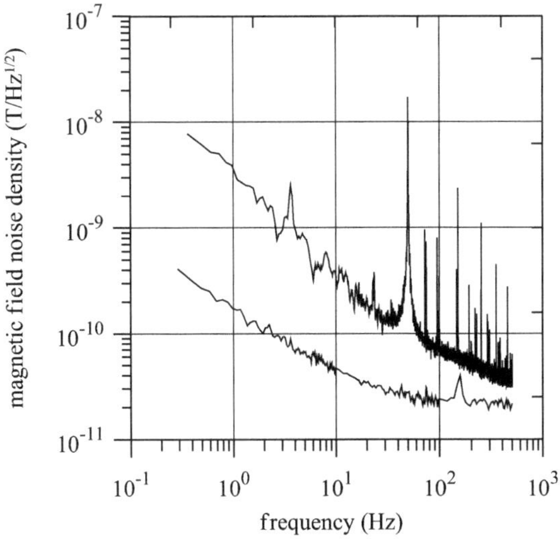

Figure 16.9. Noise spectrum of the anisotropic magnetoresistive sensor HMC1001 by Honeywell, operated with modulation technique (lower curve), compared to a typical external noise spectrum in a laboratory environment (upper curve)—in particular exhibiting characteristic power line and computer monitor frequencies and higher harmonics

16.3 Potential Emerging Methods

Besides the two methods reviewed in this chapter, there are two other methods still at an experimental stage that are likely to be useful for hydrogeological assessment. The first of these is the seismoelectric method (SEM), based on observing electromagnetic waves that both accompany and are generated by seismic waves passing through a geological boundary. The electromagnetic signal is influenced by porosity, hydraulic and electrical conductivity, dielectric constant, and zeta potential of the material, as well as by the contrasts in these parameters at boundaries (see Chapters 4 and 9 of this volume for basic rock physical aspects). Current research focuses on reliable measurements of signals and the extraction of information from these signals. The second possible new method for improved assessment of hydrogeological structures is the specific use and application of seismic surface waves (see Chapter 8 of this volume). The dispersive behavior of these waves is specifically influenced by the shear properties of the material that they pass through, which in turn depend on the presence, amount, and mobility of the water. Further research is needed to fully understand how dispersion of surface waves is related to rock physical parameters, and how this technology could be made useful for direct assessment of hydrological properties.

With the availability of adequate measurement systems, in conjunction with the ongoing optimization of multidimensional modeling and inversion algorithms, these emerging technologies are likely to play increasingly important roles as characterization and monitoring tools in the field of hydrogeophysics.

References

Acosta, J.E., and M.H. Worthington, A borehole magnetometric resistivity experiment, *Geophys. Prosp., 31,* 800–809, 1983.

Beauce, A., J. Bernard, A. Legchenko, and P. Valla, Une nouvelle méthode géophysique pour les études hydrogéologiques: L'application de la résonance magnétique nucléaire, *Hydrogéologie, 1,* 71–77, 1996.

Boggs, D.B., J.M. Stanley, and M.K. Cattach, Three-dimensional numerical modelling of sub-audio magnetic data, *Expl. Geophys., 30,* 147–156, 1999.

Braun, M., M. Hertrich, and U. Yaramanci, Modelling of phases in surface NMR, *Proceedings of 8th European Meeting on Environmental and Engineering Geophysics,* EEGS-ES, 2002.

Caruso, M.J., T. Bratland, C.H. Smith, and R. Schneider, A new perspective on magnetic field sensing, *Sensors, 15* (12), 34–46, 1998.

Chen, J., E. Haber, and D.W. Oldenburg, Three-dimensional numerical modelling and inversion of magnetometric resistivity data, *Geophys. J. Internat., 149,* 679–697, 2002.

Edwards, R.N., and M.N. Nabighian, The magnetometric resistivity method, In: *Electromagnetic Methods in Applied Geophysics, 2,* M.N. Nabighian, ed., *Application: Soc. Expl. Geophys.*, 47–104, 1991.

Edwards, R.N., The magnetometric resistivity method and its application to the mapping of a fault, *Can. J. Earth Sci., 11,* 1136–1156, 1974.

Edwards, R.N., H. Lee, and M.N. Nabighian, On the theory of magnetometric resistivity (MMR) methods, *Geophysics, 43,* 1176–1203, 1978.

Goldman, M., B. Rabinovich, M. Rabinovich, D.Gilad, I. Gev, and M. Schirov, Application of the integrated NMR-TDEM method in groundwater exploration in Israel, *Journal of Applied Geophysics, 31,* 27–52, 1994.

Gomez-Trevino, E., and R.N. Edwards, Magnetometric resistivity (MMR) anomalies of two-dimensional structures, *Geophysics, 44,* 947–958, 1979.

Haber, E., A mixed finite-element method for the solution of the magnetostatic problem with highly discontinuous coefficients in 3D, *Comput. Geosci., 4,* 323–336, 2000.
Hertrich, M., and U. Yaramanci, Joint inversion of Surface Nuclear Magnetic Resonance and Vertical Electrical Sounding, *Journal of Applied Geophysics, 50,* 179–191, 2002.
Hölting, B., *Hydrogeologie,* Enke Verlag, Stuttgart, 1992.
Kemna, A., A. Tillmann, A. Verweerd, E. Zimmermann, and H. Vereecken, MERIT—A new magneto-electrical resistivity imaging technique: (1) Modeling and tomographic reconstruction, *Proc. of the 3rd World Congress on Industrial Tomography,* 256–261, Univ. of Calgary/VCIPT/UMIST/Univ. of Leeds, 2003.
Kenyon, W.E., Nuclear magnetic resonance as a petrophysical measurement, *Int. Journal of Radiat. Appl. Instrum, Part E, Nuclear Geophysics, 6* (2), 153–171, 1992.
Kulessa, B., U. Jaekel, A. Kemna, and H. Vereecken, Magnetometric resistivity (MMR) imaging of subsurface solute flow: Inversion framework and laboratory tests, *J. Environ. Eng. Geophys., 7,* 111–118, 2002.
LaBrecque, D., R. Sharpe, D. Casale, G. Heath, and J. Svoboda, Combined electrical and magnetic resistivity tomography: Synthetic model study and inverse modeling, *J. Environ. Eng. Geophys., 8,* 251-262, 2003.
Legchenko, A.V., A.G. Semenov, and M.D. Schirov, A device for measurement of subsurface water saturated layers parameters (in Russian), *USSR Patent 1540515,* 1990.
Legchenko, A.V., and O.A. Shushakov, Inversion of surface NMR data, *Geophysics, 63,* 75–84, 1998.
Legchenko, A.V., O.A. Shushakov, J.A. Perrin, and A.A. Portselan, Noninvasive NMR study of subsurface aquifers in France, *Proceedings of 65th Annual Meeting of Society of Exploration Geophysicists,* 365–367, 1995.
Levy, S., D. Adam, and Y. Bresler, Electromagnetic impedance tomography (EMIT): A new method for impedance imaging, *IEEE Trans. Med. Imag., 21,* 676–687, 2002.
Lieblich, D.A., A. Legchenko, F.P. Haeni, and A.A. Portselan, Surface nuclear magnetic resonance experiments to detect subsurface water at Haddam Meadows, Connecticut, *Proceedings of the Symposium on the Application of Geophysics to Engineering and Environmental Problems, Boston, 2,* 717–736, 1994.
Lucy, L., An iterative technique for the rectification of observed distributions, *Astr. J., 79,* 745–754, 1974.
Malmivuo, J., and R. Plonsey, *Bioelectromagnetism: Principles and Applications of Bioelectric and Biomagnetic Fields,* Oxford Univ. Press, Inc., 1995.
McGlone, T.D., The use of giant magneto-resistance technology in electromagnetic geophysical exploration, *Proc. Symp. Application of Geophysics to Engineering and Environmental Problems,* Environ. Eng. Geophys. Soc., 705–713, 1997.
McGlone, T.D., and D.L. Kutrubes, The use of a giant magneto-resistance (GMR) based magnetometer for differentiation of subsurface electrical and non-electrical material, *Proc. Symp. Application of Geophysics to Engineering and Environmental Problems,* Environ. Eng. Geophys. Soc., 2001.
Meju, M.A., P. Denton, and P. Fenning, Surface NMR sounding and inversion to detect groundwater in key aquifers in England: Comparisons with VES–TEM methods, *Journal of Applied Geophysics, 50,* 95–111, 2002.
Mohnke, O., and U. Yaramanci, A new inversion scheme for surface NMR amplitudes using simulated annealing. *Proceedings of 60th Conference of European Association of Geoscientists & Engineers,* 2–27, 1999.
Mohnke, O., and U. Yaramanci, Smooth and block inversion of surface NMR amplitudes and decay times using simulated annealing, *Journal of Applied Geophysics, 50,* 163–177, 2002.
Mohnke, O., M. Braun, U. Yaramanci, Inversion of decay time spectra from Surface-NMR data, *Proceedings of 7th European Meeting on Environmental and Engineering Geophysics,* EEGS-ES, 2001.
Nabighian, M.N., G.L. Oppliger, R.N. Edwards, B.B.H. Lo, and S.J. Cheesman, Cross-hole magnetometric resistivity (MMR), *Geophysics, 49,* 1313–1326. 1984.
Ripka, P., *Magnetic Sensors and Magnetometers,* Artech House, Inc., 2001.
Roth, B.J., N.G. Sepulveda, and J.P. Wikswo, Jr., Using a magnetometer to image a two-dimensional current distribution, *J. Appl. Phys., 65,* 361–372, 1989.
Schirov, M., A. Legchenko, and G. Creer, A new direct non-invasive groundwater detection technology for Australia. *Exploration Geophysics 22,* 333-338, 1991.

Semenov, A.G., A.I. Burshtein, A.Y. Pusep, and M.D. Schirov, A device for measurement of underground mineral parameters (in Russian). USSR Patent 1079063. 1988.

Shushakov, O.A., Groundwater NMR in conductive water, *Geophysics, 61,* 998–1006, 1996.

Shushakov, O.A., and A.V. Legchenko, Calculation of the proton magnetic resonance signal from groundwater considering the electroconductivity of the medium (in Russian), *Russian Acad. of Sci., Inst. of Chem. Kinetics and Combustion, Novosibirsk, 36,* 1–26, 1992.

Sumi, C., A. Suzuki, and K. Nakayama, Determination of the spatial distribution of a physical parameter from the distribution of another physical variable—A differential inverse problem, *J. Appl. Phys., 80,* 1–7, 1996.

Supper, R., B. Jochum, G. Hübl, A. Römer, and R. Arndt, SNMR test measurements in Austria, *Journal of Applied Geophysics, 50,* 113–121, 2002.

Svoboda, J.M., B. Canan, J.L. Morrison, G.L. Heath, and D. LaBrecque, Advanced technology for mapping subsurface water conductivity, *Proc. Symp. Application of Geophysics to Engineering and Environmental Problems,* Environ. Eng. Geophys. Soc., 2002.

Tan, S., N.G. Sepulveda, and J.P. Wikswo, Jr., A new finite-element approach to reconstruct a bounded and discontinuous two-dimensional current image from a magnetic field map, *J. Comput. Phys., 122,* 150–164, 1995.

Tillmann, A., A. Verweerd, A. Kemna, E. Zimmermann, and H. Vereecken, Non-invasive 3D conductivity measurements with MERIT, *Proc. Symp. Application of Geophysics to Engineering and Environmental Problems,* 516–522, Environ. Eng. Geophys. Soc., 2003.

Tillmann, A., R. Kasteel, A. Verweerd, E. Zimmermann, A. Kemna, and H. Vereecken, Non-invasive 3D conductivity measurements during flow experiments in columns with MERIT, *Proc. Symp. Application of Geophysics to Engineering and Environmental Problems,* 618–624, Environ. Eng. Geophys. Soc., 2004.

Tumanski, S., *Thin Film Magnetoresistive Sensors,* Institute of Physics Publishing, Ltd., 2001.

Valla, P., and A. Legchenko, One-dimensional modelling for proton magnetic resonance sounding measurements over an electrically conductive medium, *Journal of Applied Geophysics, 50,* 217–229, 2002.

Vouillamoz, J.-M., M. Descloitres, J. Bernard, P. Fourcassier, and L. Romagny, Application of integrated magnetic resonance sounding and resistivity methods for borehole implementation—A case study in Cambodia, *Journal of Applied Geophysics, 50,* 67–81, 2002.

Warsa, W., O. Mohnke, and U. Yaramanci, 3-D modelling of Surface-NMR amplitudes and decay times, In: *Water Resources and Environment Research,* ICWRER 2002, Eigenverlag des Forums für Abfallwirtschaft und Altlasten e.V., Dresden, 209–212, 2002.

Weichman, P.B., D.R. Lun, M.H. Ritzwoller, and E.M. Lavely, Study of surface nuclear magnetic resonance inverse problems, *Journal of Applied Geophysics, 50,* 129–147, 2002.

Yaramanci, U., G. Lange, and K. Knödel, Effects of regularisation in the inversion of Surface NMR measurements, *Proceedings of 60th Conference of European Association of Geoscientists & Engineers,* 10–18, 1998.

Yaramanci, U., G. Lange, and M. Hertrich, Aquifer characterisation using Surface NMR jointly with other geophysical techniques at the Nauen/Berlin test site, *Journal of Applied Geophysics, 50,* 47–65, 2002.

Yaramanci, U., G. Lange, and K. Knödel, Surface NMR within a geophysical study of an aquifer at Haldensleben (Germany), *Geophysical Prospecting, 47,* 923–943, 1999.

Zimmermann, E., G. Brandenburg, W. Glaas, and A. Kemna., Untersuchung hochempfindlicher magnetoresistiver Sensoren für geophysikalische Anwendungen, *Internal Report, KFA-ZEL-IB-500101,* Forschungszentrum Jülich GmbH, 2001.

Zimmermann, E., A. Verweerd, W. Glaas, A. Tillmann, and A. Kemna, MERIT—A new magneto-electrical resistivity imaging technique: (2) Magnetic sensors and instrumentation, *Proc. 3rd World Congress on Industrial Tomography,* 262–267, Univ. of Calgary/VCIPT/UMIST/Univ. of Leeds, 2003.

17 STOCHASTIC FORWARD AND INVERSE MODELING: THE "HYDROGEOPHYSICAL" CHALLENGE

YORAM RUBIN[1] and SUSAN HUBBARD[1,2]

[1]*Department of Civil and Environmental Engineering, University of California at Berkeley, Berkeley, CA 94720, rubin@ce.berkeley.edu*
[2]*Lawrence Berkeley National Laboratory, Earth Sciences Division, Berkeley, CA 94720 USA.*

17.1 Introduction

Successful integration of geophysical and hydrogeological datasets represents a recent and major breakthrough in hydrogeological site characterization. As discussed in Chapter 1 of this volume, the value of integrating these datasets for characterization lies in the extensive spatial coverage offered by geophysical techniques and in their ability to sample the subsurface in a minimally invasive manner. However, this breakthrough is associated with a few difficulties. One difficulty resides in the non-unique relationships that sometimes exist between hydrogeological and geophysical attributes; integration of hydrogeological and geophysical data under non-unique conditions has been investigated by Rubin et al. (1992), Copty et al. (1993), Hubbard et al. (1997) and Hubbard and Rubin (2000). This non-uniqueness can exist even under idealized conditions of error-free measurements in natural systems comprised of multiple hydrogeologically significant units (i.e., Prasad, 2003), and it is only exacerbated by measurement errors. In applications, the situation becomes even more difficult because the rock type at the location associated with the geophysical attribute is almost always unknown, and thus the applicable petrophysical model is also almost always unknown. Another difficulty stems from the disparity between the spatial resolution of the geophysical attributes and the scale that characterizes the hydrogeological attributes, collected for example, through boreholes (c.f., Ezzedine et al., 1999). This scale disparity hinders efforts to develop unique and accurate relations between the two types of measurements, and introduces another source of uncertainty.

It is obvious that the characterization techniques developed under such conditions must recognize these difficulties, and furthermore, that they must quantify their effects on estimation uncertainty. This leads us to formulate our data integration techniques under the very wide umbrella of stochastic techniques (Rubin, 2003). In this chapter, we shall present different concepts and ideas for dealing with these issues. The commonality among these concepts is in making uncertainty and spatial variability an integral part of the solution, in consistently attempting to address the challenge of estimation uncertainty, and in minimizing arbitrariness and subjectivity.

Y. Rubin and S.S. Hubbard (eds.), Hydrogeophysics, 487–511.

This chapter is organized as follows: We start by setting a general framework for forward modeling under conditions of uncertainty. This is done by casting the hydrogeological prediction problem in a stochastic framework, and thus opening the door for estimating variables such as pressure or concentration in terms of their expected values and estimation variances, at a minimum, or through their probability distribution functions, for a more complete characterization of uncertainty.

In hydrogeological applications, we can distinguish between dependent variables, such as concentration or soil moisture, and independent ones, such as hydraulic conductivity. The dependent variables are related, often through complex and nonlinear relationships, to the independent ones. Through these relationships, the uncertainty in estimating the independent variables can be propagated to the dependent variables. This error propagation is the essence of forward modeling: the goal is to express the uncertainty in the dependent variables as a function of the uncertainty in all the variables that affect them.

Our discussion will continue by focusing on methods for estimating the uncertainty in the independent variables. The goal here is to model the effects of spatial variability and data scarcity, while utilizing data obtained from conventional hydrogeological testing as well as data from geophysical surveys. The goal is also to utilize data of different quality, starting with precise measurements and ending with information of a more qualitative nature, commonly referred to as “soft” data. We refer to this as hydrogeophysical inverse modeling. We shall discuss alternative approaches for inverse modeling, such as the maximum likelihood and maximum *a posteriori* methods. By emphasizing their origin in Bayes’ theorem, we will be able to present entropy-based concepts intended to improve their accuracy and generality.

17.2 Forward Modeling with Parameter Error

Let us consider a random variable Z. This variable can be an independent variable, such as permeability or porosity, or a dependent variable, such as pressure head, concentration, or soil moisture. To model its uncertainty, we opt to define Z as a random variable and strive to model it through a probability distribution function (pdf), or through its statistical moments. The variable Z is characterized through its pdf $f_Z(z|\underline{\theta})$, where $\underline{\theta}$ is a vector of the parameters. This vector may include the parameters of spatial correlation models (e.g., a semivariogram model) employed for conditioning the pdf on measurements, as well as parameters of petrophysical, correlation or numerical models employed for relating between Z and other attributes or with primitive variables. Examples for such models are provided in Chapter 3 of this volume and in subsequent discussions given in this chapter.

Defined as a random variable, Z can be characterized through its moments. The *N-th* moment of Z is given by:

$$\left\langle Z^N \mid \underline{\theta} \right\rangle = \int_{-\infty}^{\infty} z^N f_Z(z \mid \underline{\theta}) dz \,. \tag{17.1}$$

The moments of Z conditional to measurements can be obtained by replacing $f_Z(z|\underline{\theta})$ with the conditional pdf $f_Z(z|\underline{\theta}, \{\underline{M}\})$, where $\{\underline{M}\}$ is the vector of measurements. For a given vector $\underline{\theta}$, $f_Z(z|\underline{\theta})$ or $f_Z(z|\underline{\theta}, \{\underline{M}\}$ can be obtained, for example, through Monte Carlo simulations (Rubin, 2003).

To account for parameter error, the parameter vector $\underline{\theta}$ is viewed as a realization of a random process characterized by the p-variate pdf $f_{\theta}(\theta_1,..,\theta_p)$, where p is the number of parameters in $\underline{\theta}$. The expected value in (17.1) must be taken over all possible combinations of parameters, weighted by their probabilities:

$$\left\langle Z^N \right\rangle = \int_{\theta} \int_{-\infty}^{\infty} z^N f_{Z,\underline{\theta}}(z,\underline{\theta}) dz d^p \underline{\theta} = \int_{\theta} \int_{-\infty}^{\infty} z^N f_Z(z \mid \underline{\theta}) f_{\underline{\theta}}(\theta_1,..,\theta_p) dz d^p \underline{\theta} \,, \tag{17.2}$$

where $\int_{\theta}$ implies integration over the entire vector space of $\underline{\theta}$, and $f_{Z,\underline{\theta}}$ is the joint pdf of Z and $\underline{\theta}$. The second equality in Equation (17.2) is based on Bayes' theorem. Integrating over the parameter space yields the marginal pdf of Z in the joint attribute-parameter space. This marginal is expected to be defined over a broader range than the one defined by the pdf of Z conditional on a given set of parameters, and it will lead to a larger variance, which reflects the effects of the parameter error. It is imperative to formulate inverse modeling strategies that can quantify the estimation errors of the parameters $\underline{\theta}$; an extensive discussion of various error types is provided in Rubin (2003, Chapter 13).

The integration in Equation (17.2) calls for a systematic screening of the multidimensional parameter space, which can be quite tedious. Some simplification is possible using the Monte Carlo integration. Applications of (17.2) are discussed in Rubin and Dagan (1992), Maxwell et al. (1999) and Rubin (2003).

17.3 Inverse Modeling

This section discusses different strategies for determining the statistical distributions of hydrogeological parameters and their estimation errors. It is common to employ direct measurements of the attribute for this purpose, as well as measurements that can be related to the attribute using forward models, and hence the name "inverse modeling." For example, inverse modeling of the spatial distribution of hydraulic conductivity may involve hydraulic conductivity and head measurements, and in that case will rely on the flow equation to relate between the two. Or it may employ a geophysical signal of some sort, and in this case will employ an equation for modeling the propagation of that signal and possibly petrophysical models. As our understanding of inverse modeling has evolved over time, it has become obvious that its success depends on the ability to incorporate information of all types—either "hard" information, such as direct measurements, or "soft" information, such as expert opinions given possibly in the form of pdfs, geophysical measurements, or information borrowed from geologically similar formations, used as preliminary (or prior) information.

Stochastic inverse modeling is a formulation of inverse modeling in a probabilistic framework. Employing a stochastic paradigm eliminates the problem of non-uniqueness (Carrera and Neuman, 1986a,c) and introduces a rational approach for modeling uncertainty (Rubin, 2003). Stochastic inverse models are commonly based on Bayes' theorem. Bayesian-related methods commonly include a statement of a prior pdf of the parameters, and a likelihood function that relates the attribute to the output of the forward model. Using the prior and the likelihood function, Bayes' theorem is then employed to define a posterior pdf (or *a-posteriori* pdf). Bayes' theorem serves to update the plausibility of a proposition as the state of information changes because of the availability of new data. The Bayesian framework is broad enough to allow for a remarkably large number of different interpretations.

Since prior information is an important component of Bayes' theorem, we shall start our discussion by considering methods for defining minimally subjective prior pdfs. This will help us in understanding the significance and implications of choosing a model for the prior pdf. Armed with this knowledge, we shall proceed to discuss the maximum likelihood (ML) and maximum *a posteriori* (MAP) methods, which offer alternative interpretation of Bayes' theorem.

A complementary classification of the various methods that we shall present is as follows: (1) The sequential approach, where the focus is on defining a pdf structure and identifying its parameters, and subsequently applying it for point estimation; (2) Methods that assume a pdf structure and focus on identifying its parameters, sometimes simultaneously with spatial (point) estimation.

17.3.1 BAYESIAN ESTIMATORS

A general statement of Bayes' theorem must include the prior and posterior pdfs. The starting point is of course the prior pdf. In this section we present several different approaches for estimating the prior pdf.

Let the pdf $g_{\underline{\Theta}}(\underline{\theta} \mid I)$ denote the prior pdf of the vector of parameters $\underline{\theta}$, which summarizes, in a probabilistic manner, our understanding of the values $\underline{\theta}$, and can assume direct or indirect measurements of these parameters based on other sources of information. Such information may include expert opinions or information borrowed from geologically similar sites (c.f., Chapters 2, 3 and 13 of Rubin, 2003) As soon as measurements at the site become available, they can be used for updating the prior pdf. This can be done by viewing the measurements as realizations of a space random function (SRF) X. The pdf of $\underline{\theta}$ conditioned on the vector of N measurements, $\boldsymbol{X}$, where $X_1=x_1,\ldots,X_N=x_N$, denoted by $f_{\underline{\Theta}|X_1,\ldots,X_N}(\underline{\theta}|x_1,\ldots,x_N,I)$, is called the posterior distribution of $\underline{\Theta}$, and it is given by:

$$f_{\underline{\Theta}|X_1,\ldots,X_N,I}(\underline{\theta}|x_1,\ldots,x_N,I)=\frac{f_{X_1,\ldots,X_N|\underline{\Theta},I}(x_1,\ldots,x_N|\underline{\theta},I)g_{\underline{\Theta}|I}(\underline{\theta}\,|\,I)}{f_{X_1,\ldots,X_N|I}(x_1,\ldots,x_N\,|\,I)}, \tag{17.3}$$

where $f_{X_1,\ldots,X_N}(x_1,\ldots,x_N\,|\,I)$ is the marginal of the joint pdf of $\boldsymbol{X}$ and $\underline{\Theta}$. In cases where several types of data are available, the vector $\boldsymbol{X}$ should be redefined accordingly, for example X_i, $i=1,\ldots,M$, for the M measurements of one attribute, and X_i, $i=1,\ldots,M+1,\ldots,N$, for the measurements of the other type.

The posterior distribution incorporates all information available, including measurements and prior knowledge. The function $f_{X_1,\ldots,X_N|\underline{\Theta},I}(x_1,\ldots,x_N|\underline{\theta},I)$ is the probability of making the observations for a given $\underline{\Theta}=\underline{\theta}$, and it is commonly referred to as the likelihood function. If we want to estimate $\underline{\theta}$, we need to adopt a decision rule, for example, we may take for estimator that $\underline{\theta}$ that maximizes the posterior distribution, or we could just estimate $\underline{\theta}$ with the mean or median of the posterior distribution. None of these choices reflects the width of the distribution. Rather, the pdf of the parameters should be considered in its entirety, and not just through a subset.

17.3.2 A SYSTEMATIC APPROACH TO DEFINING A PRIOR PROBABILITY

How do we estimate the prior pdf of unknown model parameters, $g_{\underline{\Theta}|I}(\underline{\theta}\,|\,I)$? The determination of the statistical properties of the unknown model parameters, and at a more fundamental level, the statistical model itself, are the major impediments to the effective use of probabilistic models. It is common to assume a model for the prior, and then to update it as additional information becomes available. But one can argue that the choice of the prior model is arbitrary. How significant is the choice of a prior? Jaynes (1968) stated, "Since the time of Laplace, applications of probability theory have been hampered by the difficulties in the treatment of prior information," and "Therefore, the logical foundations of decision theory cannot be put in fully satisfactory form until the old problem of arbitrariness (sometimes called "subjectiveness") in assigning prior probabilities is resolved."

Instead of postulating a prior pdf, we can infer it from measurements based on assumptions or axioms. One such approach is based on the principle of minimum relative entropy (MRE), the application of which to hydrogeology was pioneered by A. Woodbury and coworkers. The MRE approach (Woodbury and Rubin, 2000; Woodbury and Ulrych, 2000; Rubin, 2003) couples between Bayes' theorem and entropy measures of information. It represents the general approach of inferring a probability distribution from constraints that incompletely characterize that distribution. These constraints can be (but are not limited to be) in the form of averages or higher-order moments, following Equation (17.1). In the context of groundwater, such constraints can be, for example, the average and variance of the hydraulic conductivity obtained from a limited number of tests, from empirical determination, or from an informed guess.

Entropy is a measure of the amount of uncertainty in a pdf. The principle of MRE states that of all the probabilities that satisfy the given constraints, choose the one that has the highest entropy with respect to a known prior, which is the one that is the most uncommitted, or the least prejudiced, with respect to unknown information. The model identified based on MRE can be interpreted as the one which fits all the constraints and the one which has the greatest "spreadout" that can be realized in the greatest number of ways. The pdf thus identified can be used as priors for any of the inverse methods described below.

17.3.3 THE MAXIMUM LIKELIHOOD AND MAXIMUM *A-POSTERIORI* CONCEPTS IN INVERSE MODELING

In this section, we examine and formulate the inverse problem based on maximum likelihood (ML) and maximum *a posteriori* (MAP) concepts. Both methods are based on Bayes' theorem, but with different interpretations assigned to the prior and to the inferred parameter values.

17.3.3.1 *The Maximum Likelihood (ML) Method*

The ML formulation discussed below employs only information available from measurements and ignores prior information. In the absence of any prior information *I*, the likelihood function defined in Equation (17.3) becomes $f_{X_1,..,X_N|\underline{\Theta}}(x_1,..,x_N|\underline{\theta})$. Let us denote it by $L(\underline{\theta}\,|\,x_1,..,x_N)$. *L* gives the likelihood (or in fact, the pdf) that the random variables $X_1,..,X_N$, assume the particular values $x_1,..,x_n$. X_i can denote, for example, the logconductivity (with $X_i=X(\boldsymbol{b}_i)$ being the log-conductivity at location $\boldsymbol{b}_i$) and $\underline{\Theta}$, which will include the parameters of the *X* SRF (such as its mean, variance and the parameters of its correlation function). For a given vector $\underline{\theta}$, the particular values of the random variables most likely to occur are the values $x_1,..,x_N$ that maximize $f_{X_1,..,X_N|\underline{\Theta}}(x_1,..,x_N\,|\,\underline{\theta})$, or in fact, those values that define its mode.

Let us now consider a particular vector of observations $x_1,..,x_N$. We wish to find the particular combination of parameters of $\underline{\Theta}$, denoted by $\hat{\underline{\theta}}$, which maximizes the likelihood function $L(\underline{\theta}\,|\,x_1,..,x_N)$ or, in other words, maximizes the probability of observing $X_1,..,X_N$, equal to $x_1,..,x_N$, respectively. This vector, $\hat{\underline{\theta}}$, which is a function of the measurement vector $x_1, \ldots, x_N$, is called the maximum likelihood estimator of $\underline{\Theta}$. The maximum likelihood estimator is the solution for the set of equations:

$$\frac{\partial L(\underline{\theta})}{\partial \theta_j} = 0, j = 1,..p, \tag{17.4}$$

where p is the number of parameters in $\underline{\Theta}$. Both the logarithm of L and L have their maxima at the same value of $\underline{\Theta}$, and it is sometimes easier to find the maximum of the logarithm of the likelihood (or the minimum of the negative log-likelihood).

The pioneers in bringing the concept of maximum likelihood to hydrogeology are Peter Kitanidis and his coworkers. The foundations for their work were laid in Kitanidis and Vomvoris (1983), followed by Hoeksema and Kitanidis (1984, 1985), where they considered the problem of identifying the spatial distribution of the log conductivity.

One application of these ideas in hydrogeophysics is provided by Hubbard et al. (1997). It demonstrates the power of using Bayesian concepts in the case where non-unique petrophysical relationships exist. In that study, numerical experiments were performed to investigate the utility of ground-penetrating radar (GPR) data for saturation and permeability estimation under a range of hydrogeological conditions. As described in Chapter 7 of this volume, GPR is a noninvasive, high-resolution geophysical method. As described in Chapter 15 of this volume and by Huisman et al. (2003), estimates of dielectric constants can be extracted from both surface and crosshole GPR data. Petrophysical relations can then be used to transfer these dielectric constants into saturation or permeability estimates.

Petrophysical curves that illustrate a plausible relationship of saturation and dielectric constant as a function of facies (and corresponding permeability) for a synthetic sand-clay system are shown in Figure 17.1 (modified from Hubbard et al. 1997). These curves illustrate the non-uniqueness that sometimes exists when trying to map a geophysical attribute (such as dielectric constant obtained from radar data) into a hydrological estimate (such as saturation or facies permeability). For example, in the case study shown in Figure 17.1, for each dielectric constant value obtained from the radar data in the bimodal system under consideration, two values of saturation (S_1 and S_2) are possible, depending on whether the facies at that subsurface location is a clay or a sand, both with associated distinct permeability values. With two possible values for the saturation, the pdf of the saturation S can be expressed as follows:

$$f_S(s|S_1, S_2) = p\delta(s - S_1) + (1 - p)\delta(s - S_2) \tag{17.5}$$

where p denotes the probability to observe s=S_1, and δ is the Dirac delta function. The probability is related to the expected value of S, $\langle S \rangle$, through the relationship (Hubbard et al., 1997, Equation 9):

$$p = \frac{\langle S \rangle - S_2}{S_1 - S_2}. \tag{17.6}$$

Hence, once $\langle S \rangle$ is defined, the pdf of s is also defined. The saturation S that maximizes the probability to observe the measured dielectric constant is equal to S_1 if p is larger than 0.5, and to S_2 otherwise. The S thus chosen is the ML estimator, because we choose for $\underline{\hat{\theta}}$ the value of S that maximizes the probability to observe the measured dielectric constant. Hubbard et al. (1997) suggested a couple of alternatives for computing $\langle S \rangle$: the first was based on numerical infiltration simulations, and the second was based on kriging borehole saturation measurements.

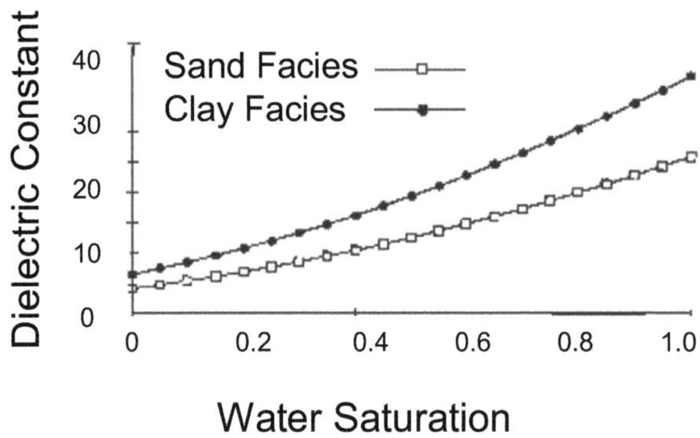

Figure 17.1. Example of a non-unique petrophysical relationship between dielectric constant and water content in a sand-clay system (modified from Hubbard et al., 1997, and reduced from Knoll et al., 1995)

The vector $\underline{\hat{\theta}}$ that minimizes *-lnL* is the ML estimator. Once it is determined, we can estimate the variances and covariances of the parameters. We can define the matrix $\underline{\underline{B}}$ of order p, where

$$B_{ij} = \frac{\partial}{\partial \theta_i}\left(\frac{\partial}{\partial \theta_j} ln\, L\right), \tag{17.7}$$

which is known as the Fisher information matrix. The matrix $\underline{\underline{\Sigma}} = \underline{\underline{B}}^{-1}$ is referred to as the Cramer-Rao lower bound of the estimation error covariance (Schweppe, 1973), and its (i,j) term provides an estimate for the estimation error's variance-covariance matrix for the vector $\underline{\theta}$. Once the parameters of the X SRF are identified, the spatial distribution of X, conditioned on the measurements, can be determined using geostatistical methods (see Chapter 3 of this volume) such as kriging (when only X measurements are available) and cokriging (when other types of measurements are also available).

17.3.3.2 *The Maximum* A Posteriori *(MAP) Approach*

MAP extends ML to situations where prior information does exist. This extension provides for a wide range of alternative interpretations and approaches. The MAP concept is illustrated in Figure 17.2. Figure 17.2a schematically shows an example of a prior logconductivity pdf at a single subsurface location. Figure 17.2b illustrates the joint distribution as well as the likelihood function. The geophysical data are given in the form of a precise measurement (solid line) and, alternatively, in the form of a distribution, representing imprecise data (dashed line). Uncertainty in the geophysical attribute can stem from many factors, such as the choice of tomographic data inversion approach and parameters, acquisition and data reduction errors, and variable coverage/sensitivities (c.f., Peterson, 2001; Alumbaugh et al., 2002; Binley et al., 2002). In the first case, the likelihood function is defined over a much narrower range of conductivity values, which will limit the range of physically-plausible conductivity values.

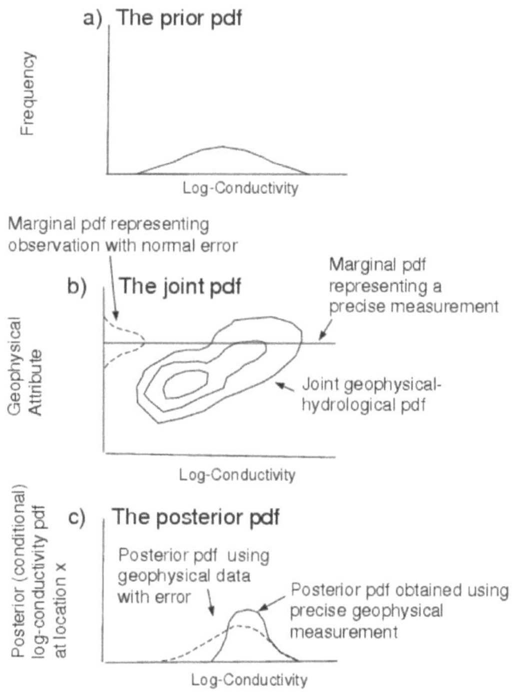

Figure 17.2. Example of the MAP approach to hydrogeophysical data fusion: (a) example of a prior logconductivity pdf at a single subsurface location; (b) example of the joint geophysical-hydrological pdf, as well as marginal pdf representing a geophysical observation with normal error (dashed line) and marginal pdf representing a precise measurement (solid line); (c) posterior pdfs at a single subsurface location obtained by fusing the prior logconductivity pdf, joint pdf, and marginal geophysical pdf within the MAP formalism. The solid posterior pdf was obtained using the precise geophysical measurement, while the dashed posterior pdf was obtained using the geophysical observation with normal error (shown in 17.2b).

Using the likelihood function for the precise information, the prior pdf can be updated using Bayes' theorem (17.3), to obtain the posterior pdf, as illustrated by the solid line in Figure 17.2c. We note that the high quality of the geophysical data renders this posterior pdf considerably sharper than the prior pdf (compare solid-line log-conductivity distributions in Figure 17.2a and 17.2c), because Bayes' theorem indicates that reconciliation between the prior pdf and the geophysical attribute is possible by assigning a larger probability to the higher log conductivities. In the case of imprecise geophysical data, as illustrated by the geophysical attribute distribution in Figure 17.2b, the improvement in the posterior pdf (shown by the dashed line in Figure 17.2c) is not as significant as when precise geophysical data are used. The posterior pdf is not as sharp because the improvement of the posterior compared to the prior depends on the amount of information provided by the geophysical data. This example illustrates the essence of MAP as a "reconciliation" between prior information, usually in the form of direct measurements of the attribute of interest, and additional information of varying quality. The information used for updating the prior pdf can be either empirically correlated or functionally related to the attribute of interest.

For discussion of the mathematical aspects of MAP, let us consider the case of identifying the spatial distribution for hydraulic conductivity based on measurements of conductivity as well as auxiliary information denoted by h. Such information can obviously include geophysical information collected through various types of geophysical surveys. Let us denote a finite sized vector of conductivities (or log conductivities, which is more commonly employed) and h measurements distributed over the aquifer by $\boldsymbol{z}=(\boldsymbol{z}_h,\boldsymbol{z}_a)$, with $\boldsymbol{z}_a$ denoting the vector of conductivity measurements and $\boldsymbol{z}_h$ denoting the h measurements. Let us now denote, by the vector of order p, the conductivities that we wish to estimate, for example, at the nodes of a numerical grid. Let us define the functional relationship between $\boldsymbol{z}_a$ and $\boldsymbol{a}$ in the form $\boldsymbol{z}_a=\boldsymbol{F}_a(\boldsymbol{a})+\boldsymbol{v}_a$, where $z_{a,i}$ and $v_{a,i}$ denote the i-th conductivity measurement and the measurement error, respectively. $\boldsymbol{F}_a$ represents the large-scale trend of the conductivity, given for example in the form of a spatial distribution of the conductivities over a grid or by a set of polynomials. Similarly, let us define the measurements vector $\boldsymbol{z}_h=\boldsymbol{F}_h(\boldsymbol{a})+\boldsymbol{v}_h$, where $\boldsymbol{F}_h$ represents the large-scale trend of the h, with $v_{h,i}$ denoting the measurement error at $\boldsymbol{x}$, owing to instrumentation and recording error, data reduction procedures, and the small-scale fluctuations of the actual point values from the corresponding large-scale trends. Note that $\boldsymbol{F}_a$ and $\boldsymbol{F}_h$ may include parameters whose values are unknown and need to be identified. The formulation below can be used to identify $\boldsymbol{a}$ at unknown locations, as well as the unknown parameter values, although in our discussion below we shall concentrate on $\boldsymbol{a}$. It is common to assume that the measurement errors are zero mean, and that $\boldsymbol{v}$ and $\boldsymbol{a}$ are uncorrelated. These two relationships can be summarized by the following measurement equation (cf., McLaughlin and Townley, 1996):

$$z = \begin{bmatrix} z_h \\ z_a \end{bmatrix} = \begin{bmatrix} F_h(a) \\ F_a(a) \end{bmatrix} + \begin{bmatrix} v_h \\ v_a \end{bmatrix} = F(a) + v. \tag{17.8}$$

The MAP estimator of $\boldsymbol{a}$, $\hat{\boldsymbol{a}}$, is the one most likely given the measurements (unlike ML, which seeks $\boldsymbol{a}$ that maximizes the probability of observing the measurements). It is defined based on Bayes' theorem by the posterior pdf:

$$f_{a|z}(a \mid z) = \frac{f_{z|a}(z|a) f_a(a)}{f_z(z)} = \frac{f_v(z - F(a)) f_a(a)}{f_z(z)}, \tag{17.9}$$

where f_a and $f_{a|z}$ are the prior and posterior pdfs of $\boldsymbol{a}$, $f_{z|a}$ is the conditional pdf of $\boldsymbol{a}$ given z, and f_z and f_v are the measurement and measurement-error pdfs, respectively. The second equality in Equation (17.9) follows from Equation (17.8) and from the independence of $\boldsymbol{a}$ and $\boldsymbol{v}$. The MAP estimate $\hat{\boldsymbol{a}}$ corresponds to the mode of $f_{a|z}$.

If f_a and f_v are multinormal with covariance matrices $\underline{\underline{c_a}}$ and $\underline{\underline{c_v}}$, the posterior pdf becomes:

$$f_{a|z}(a \mid z) = c(z) \boldsymbol{exp}\left\{-\frac{1}{2}[z - F(a)]^T \underline{\underline{c_v^{-1}}}[z - F(a)]\right\} \boldsymbol{exp}\left\{-\frac{1}{2}[a - \bar{a}]^T \underline{\underline{c_a^{-1}}}[a - \bar{a}]\right\} \tag{17.10}$$

where c(z) is a normalization factor not depending on $\boldsymbol{a}$, and $\bar{\boldsymbol{a}}$ is the prior mean of $\boldsymbol{a}$.

Assuming $\boldsymbol{v}$ to be normal is quite common, because it is the simplest model, requiring to state only mean and variance. However, it turns out that it is also the safest assumption, if all that is known about $\boldsymbol{v}$ is its mean and variance, since as can be shown using MRE considerations (see our earlier discussion in Section 17.3.2), normal distribution in that case is the least informative, and hence is minimally subjective. If the variance is not known precisely, (17.10) can be written as a pdf conditional on a given variance σ_v: $f_{a|z,\sigma_v}(a|z,\sigma_v)$, and the dependence on σ_v can be eliminated by taking the expected value over the range of σ_v (see Woodbury and Ulrych, 2000; Woodbury and Rubin, 2000).

It is convenient to find the MAP estimate by minimizing -2*ln*[$f_{a|z}$]. The MAP estimate $\hat{\boldsymbol{a}}$ is then the $\boldsymbol{a}$ that minimizes the weighted-least-squares criterion (Schweppe, 1973; Carrera and Neuman, 1986a,b; Tarantola, 1987):

$$[z - F(a)]^T \underline{\underline{c_v^{-1}}}[z - F(a)] + [a - \bar{a}]^T \underline{\underline{c_a^{-1}}}[a - \bar{a}], \tag{17.11}$$

which yields a compromise between the prior estimate (represented here by the term on the right) and the weighted-least-squares fit (which is summarized by the first term).

Numerical solutions for the above minimization problems are generally found with iterative search algorithms such as the Gauss-Newton method. Note that the number of unknown parameters may be large, on the order of the number of gridblocks. This number may be reduced by zonation, which calls for identifying areas of uniform conductivity (thus potentially reducing the number of unknowns), but this procedure is not without its problems, as we shall discuss below.

The concept underlying Equation (17.11) attaches a probabilistic interpretation to $\boldsymbol{v}$, which is viewed as a measurement error vector. If instead, we view the z as SRFs, with $\boldsymbol{F}$ representing their expected values, then z-$\boldsymbol{F}(\boldsymbol{a})$ contains the local fluctuations of the SRFs from their expected values, and it represents the effects of natural heterogeneity rather than measurement error. Furthermore, the matrices in (17.11) describe now the covariances between SRFs, which can be developed from physical principles (Rubin, 2003, Chapters 4 and 6). This is the concept underlying the inverse method of Kitanidis and Vomvoris (1983), and Rubin and Dagan (1987a,b; 1988).

Equation (17.11) can be used to identify other formulations of the inverse problem. One formulation is the least squares method. Least squares estimates minimize:

$$[z-\boldsymbol{F}(\boldsymbol{a})]^T \underline{\underline{c_v}}^{-1} [z-\boldsymbol{F}(\boldsymbol{a})], \tag{17.12}$$

and ignore the prior information term. The least squares estimator obtained by minimizing Equation (17.12) is equivalent to a Gaussian ML estimator with known statistical properties. We note from Equation (17.9) that the likelihood function is:

$$f_{z|a}(z \mid \boldsymbol{a}) = f_v[z-\boldsymbol{F}(\boldsymbol{a})], \tag{17.13}$$

and that the ML estimator is obtained by minimizing $-2ln f_{z|a}$. For a Gaussian f_v and a known covariance function, the ML estimator will be obtained by minimizing (17.13).

McLaughlin and Townley (1996) note that ML estimators are equivalent to MAP estimators when the prior covariance matrix $\underline{\underline{c_a}}$ (17.11) is arbitrarily large, implying that the prior information is uninformative. They note further that an uninformative prior is pessimistic, since we can usually put some bounds on the range of values that parameters such as conductivity can assume. Most regression and ML algorithms provide for such constraints in their search algorithm. Bounded least squares or ML estimators, which assume that f_v is Gaussian and that the prior probability is uniform between the specified bounds, are in fact equivalent to a MAP estimator, where the performance criteria to be minimized is given by:

$$[z-\boldsymbol{F}(\boldsymbol{a})]^T \underline{\underline{c_v}}^{-1} [z-\boldsymbol{F}(\boldsymbol{a})] - \boldsymbol{ln}\, f_a(\boldsymbol{a}), \tag{17.14}$$

and where the prior pdf f_a is a positive constant when $\boldsymbol{a}$ lies within the bounds, and 0 otherwise. Since this performance criterion imposes an infinite penalty on estimates lying outside the bounds, it will enforce the constraints. Hence, MAP offers flexibility by being able to handle prior information. A MAP generalization of the Fisher information matrix $\underline{\underline{B}}$ (see Equation 17.7) is provided in McLaughlin and Townley (1996) and Rubin (2003).

The error matrices $\underline{\underline{c}}_a$ and $\underline{\underline{c}}_v$ of Equation (17.11) pose a difficult challenge because they are generally not known, at least not *a priori*. MRE offers ideas on how to address this issue (see our discussion following Equation 17.10). Another way to approach this issue is by defining $\underline{\underline{c}}_a$ and $\underline{\underline{c}}_v$ as diagonal matrices, in the form $\underline{\underline{c}}_a = \sigma_a^2 \underline{\underline{I}}$ and $\underline{\underline{c}}_v = \sigma_v^2 \underline{\underline{I}}$, where σ_a^2 and σ_v^2 are the error variances and $\underline{\underline{I}}$ is the identity matrix. With these definitions, (17.11) becomes proportional to:

$$[z - \boldsymbol{F}(\boldsymbol{a})]^T [z - \boldsymbol{F}(\boldsymbol{a})] + \beta [\boldsymbol{a} - \bar{\boldsymbol{a}}]^T [\boldsymbol{a} - \bar{\boldsymbol{a}}] \tag{17.15}$$

where β is the ratio σ_v^2 / σ_a^2, and it can be used to change the relative weights of the two terms in Equation (17.15). A small σ_a^2 indicates a greater confidence in the prior, and hence a larger weighting in the form of a larger β. The parameter β can be determined by numerical experimentation. For example, we can use a training set of data to identify an optimal β. Equation (17.15) can be viewed as the weighted least squares representation of MAP, with the term on the right representing the prior, acting to regularize the solution, and the term on the left representing updating through the acquisition of additional measurements.

Hydrogeophysical inverse modeling based directly or indirectly on Equation (17.9) was included in the work of Ezzedine et al. (1999), who reported on a Bayesian method for using borehole and surface resistivity data for lithology estimation, and in Day-Lewis et al. (2000), who reported an approach for conditioning geostatistical simulation of high-permeability fracture zones in multiple dimensions to hydraulic-connection data inferred from wellbore tests, and in Chen et al. (2001) and Chen and Rubin (2003).

17.4 Point Estimation with Known Parameters

The formulations of Section 17.3 are commonly applied in two different modes. The first mode calls for identification of nodal values (e.g., conductivity values at nodes over a grid). The second mode calls for defining the parameters of interest as SRFs as a first step and their identification using ML or MAP, which is then followed by mapping the spatial distributions of the parameters using methods such as kriging or co-kriging (see Chapter 3 of this volume or Chapter 3 of Rubin, 2003). The second mode originated in the works of Kitanidis and co-workers and was subsequently pursued by Rubin and Dagan (1987a, b; 1988). One immediate advantage of the second mode is the reduction in the number of target parameters at the inversion stage. In other words, high resolution on the grid does not translate into a large number of target parameters, with its associated difficulties, including non-uniqueness, instability, and nonidentifiability (see Carrera and Neuman, 1986a–c).

For example, if we are interested in mapping the spatial distributions of the conductivity, then the second mode will include the following two steps: (1)

Identifying the parameters of the conductivity (or logconductivity) SRF, such as its mean and spatial covariance parameters, using all available data; and (2) Mapping the spatial distribution of the conductivity, conditional to all available data, using a statistical interpolation technique. Such mapping also includes characterization of uncertainty, which can be accomplished to varying degrees of completeness and accuracy.

As discussed in the previous section, MAP and ML formulations applied in the first mode produce MAP or ML estimates and estimation variances of the target parameters. As we shall show below, separating the identification problem from point estimation, as done in the second-mode applications, allows us to obtain more complete characterization of estimation uncertainty and better flexibility in conditioning the estimates on the various types of information available. Below, we discuss the second mode, and while its first stage was discussed in Section 17.3, we focus here on the stage of deriving spatial distributions over the computational grid, with the SRF parameters already known.

17.4.1 BAYESIAN POINT ESTIMATORS AND CONDITIONING ON “HARD” AND “SOFT” INFORMATION

The most straightforward approach to mapping spatial distributions is via kriging or co-kriging, which are discussed in Chapter 3 of this volume. As was documented in Chapter 3 (see also Chapter 3 in Rubin, 2003), kriging provides an estimate and a kriging variance. We should note the following: (1) The kriging variance can be viewed as a measure of uncertainty, but cannot be used to compute confidence intervals unless the estimation error is normally distributed. While this is often assumed, it is not part of the kriging algorithm. Hence, since kriging estimates are not characterized by any underlying statistical distributions, their characterization of uncertainty is lacking. (2) Kriging methods rely on two-point correlations, in the form of covariances or semivariograms, to project measurements onto unsampled locations. Such forms of correlation are efficient in capturing linear correlations, but fail in the presence of nonlinearity. Measures to amend for this limitation can be found in nonlinear transformations of the target variable, such as log-transforms or indicator transforms (Chapter 3 in Rubin, 2003), but such transforms are not always efficient in removing nonlinearity. This is particularly true for the case of inherent non-uniqueness in the correlation structure (i.e., stemming from physical principles, not from measurement errors). For example, Rubin et al. (1992) and Copty et al (1993) considered the ramifications of incorporating information from a non-unique relationship between permeability and seismic velocity in sandy clays for hydrogeological parameter estimation, where a given value of seismic velocity can be interpreted as an indication of high or low permeability, depending on the porosity.

To address these challenges, and to augment the discussion of Chapter 3, we choose to concentrate here on Bayesian conditioning. Let us consider an estimate of Z at $\boldsymbol{x}$, conditional to a measurement of Z at $\boldsymbol{x}'$, $Z(\boldsymbol{x}')$. The conditional pdf of $Z(\boldsymbol{x})$ is obtained from Bayes' theorem as follows:

$$f_{Z(x),Z(x')}(z \mid z') = \frac{f_{Z(x),Z(x')}(z, z')}{f_{Z(x')}(z')}. \tag{17.16}$$

Thus, $f_{Z(x),Z(x')}(z \mid z')dz$ is the probability to observe $Z(\boldsymbol{x})$ in the vicinity dz of z, given that $Z(\boldsymbol{x}')=z'$. When the bivariate distribution of $Z(\boldsymbol{x})$ and $Z(\boldsymbol{x}')$ is normal, the conditional pdf is also normal. Furthermore, in that case, the mean of the conditional distribution is equal to the kriging estimate, and the variance is equal to the kriging variance (with one important difference: since now we do have an underlying statistical distribution, the variance can be employed to determine confidence intervals, See Chapter 2 in Rubin, 2003). All this is also true when conditioning on a larger number of measurements. Finally, since Equation (17.16) models the joint distribution of Z at $\boldsymbol{x}$ and $\boldsymbol{x}'$ through a bivariate pdf and not through the two-point correlation, such as used by kriging estimators, it opens the door for dealing more effectively with nonlinear correlations (cf., Rubin et al., 1992; Copty et al.,1993; Ezzedine et al., 1999; Hubbard et al., 1997).

Heretofore, we have assumed that $Z(\boldsymbol{x}')$ is measured without error, and hence we refer to it as "hard" data. Equation (17.16) provides the means for conditioning on hard data. In applications, however, it is very likely that we would want to condition on a range of values for $Z(\boldsymbol{x}')$. Such a range of values may represent a distribution of values resulting from measurement errors, or a situation where $Z(\boldsymbol{x}')$ was not measured, but rather estimated (for example, by expert opinion, from geophysical data, or by analogy to a set of measurements collected under similar conditions). Many case studies illustrating different methods of extracting hydrogeological property information using geophysical data were given in Chapters 13–16 of this volume. We refer to such imprecise information as soft data, and identify the need for conditioning on soft information. Rubin (2003, Chapter 3) proposed to condition on the soft information by modeling it as a probabilistic distribution of $\overline{\boldsymbol{Z}}$, the estimator of $Z(\boldsymbol{x}')$, in the form of a pdf $f_{\overline{Z}}(\overline{z})$, defined over a range of values, A. The distribution of Z conditional to $\overline{Z}$ is thus (Chapter 3 in Rubin, 2003, Chapter 3 of this volume):

$$f_{Z(x)|Z(x')}(z \mid z' \in A) = \int_A f_{Z(x)|Z(x')}(z \mid z') f_{\overline{Z}}(z')d', \tag{17.17}$$

which implies taking the expected value of the conditional estimate over the range A. In the limiting case of precise measurements at $\boldsymbol{x}'$, the soft data pdf $f_{\overline{Z}}(\overline{z})$ is equal to the Dirac delta, $f_{\overline{Z}}(\overline{z}) = \delta(\overline{z} - z')$, and Equation (17.17) simplifies to (17.16).

The estimator in (17.17) is exact in the sense that as $\boldsymbol{x}$ approaches $\boldsymbol{x}'$, the conditional pdf approaches the pdf of the soft data, $f_{\overline{Z}}(\overline{z})$. In the limiting case of a Dirac,

$f_{\bar{Z}}(\bar{z}) = \delta(\bar{z} - z')$, $Z(\boldsymbol{x}')$ will approach z', which is the better-documented case of the exactitude property.

An example based on the Normal Linear Methods. As discussed in Chapter 1 of this volume, one of the challenges in using geophysical data for estimating hydrogeological parameters is understanding or developing a meaningful relationship between the two different types of measurements. Relationships between hydrogeological and geophysical parameters in sediments typical of near-surface environments have been a topic of research in recent years (e.g., Mavko et al., 1998; Marion et al., 1992; Knoll et al., 1995; Slater and Lesmes, 2002; Prasad 2003). Chen et al. (2001) presented a method to formulate a site-specific petrophysical model in a probabilistic fashion using likelihood functions. Their formulation was based on co-located geophysical and hydrogeological data collected at a site, and they used normal linear regression models to relate the disparate parameters. A review of their approach is given here.

Following Chen et al. (2001), let Y denote the SRF of the log conductivity, and let V_g, α, and V_s denote GPR velocity, GPR attenuation, and seismic velocity, respectively. The posterior pdf of $Y(\boldsymbol{x})$ is obtained using Bayes' theorem as follows:

$$\underbrace{f_Y'\left[y(\boldsymbol{x}) \mid v_g(\boldsymbol{x}), \alpha(\boldsymbol{x}), v_s(\boldsymbol{x})\right]}_{posterior} = C\underbrace{L\left[y(\boldsymbol{x}) \mid v_g(\boldsymbol{x}), \alpha(\boldsymbol{x}), v_s(\boldsymbol{x})\right]}_{Likelihood\ function}\underbrace{f_Y\left[y(\boldsymbol{x})\right]}_{prior} \tag{17.18}$$

Rubin et al. (1992) considered a simplified version of Equation (17.18), whereby the only data available for updating was the seismic velocity. Similar formulations can be written for different types of data combinations.

Note that the likelihood function in (17.18) is a trivariate pdf. It can be expressed as a product of three univariate conditional pdfs:

$$\begin{aligned} &L\left[y(\boldsymbol{x}) \mid v_g(\boldsymbol{x}), \alpha(\boldsymbol{x}), v_s(\boldsymbol{x})\right] = \\ &f_{v_g}\left[v_g(\boldsymbol{x}) \mid y(\boldsymbol{x})\right] f_\alpha\left[\alpha(\boldsymbol{x}) \mid y(\boldsymbol{x}), v_g(\boldsymbol{x})\right] \\ &f_{v_s}\left[v_s(\boldsymbol{x}) \mid y(\boldsymbol{x}), v_g(\boldsymbol{x}), \alpha(\boldsymbol{x})\right] \end{aligned} \tag{17.19}$$

With this representation, the problem of identifying a trivariate pdf is replaced with that of identifying three univariate pdfs.

The normal linear regression model reported in Chen et al. (2001) provides a systematic approach to the inference of the conditional pdfs shown in (17.19). In this approach, each of the univariate pdfs is modeled as a normal, nonstationary pdf. Consider, for example, one of the univariate pdfs forming the likelihood function:

$$f_{v_s}\left[v_s(\boldsymbol{x}) \mid y(\boldsymbol{x}), v_g(\boldsymbol{x}), \alpha(\boldsymbol{x})\right]. \tag{17.20}$$

It is assumed to be normal with a nonstationary mean $\mu(\boldsymbol{x})$, which is further assumed to be a member of the linear function space G, whose basis functions consist of m monomials, g_i, i=1,..,m, formed from combination of powers and products of $y(\boldsymbol{x})$, $v_g(\boldsymbol{x})$ and $\alpha(\boldsymbol{x})$, such as: 1, $y(\boldsymbol{x})$, $v_g(\boldsymbol{x})$, $\alpha(\boldsymbol{x})$, $y^2(\boldsymbol{x})$, $v_g^{\,2}(\boldsymbol{x})$, $\alpha^2(\boldsymbol{x})$, $y(\boldsymbol{x})v_g(\boldsymbol{x})$, $v_g(\boldsymbol{x})\alpha(\boldsymbol{x})$, etc. It is modeled as follows:

$$\mu(\boldsymbol{x})=\sum_{i=1}^{m}\beta_i g_i(\boldsymbol{x}). \tag{17.21}$$

The variance of the distribution is taken to be stationary. The final set of functions used to model the mean is determined by a model selection procedure, which will be described below.

The coefficients β_i are estimated by minimizing the residuals sum of squares:

$$\sum_{j=1}^{n}\left[v_s(\boldsymbol{x}_j)-\mu(\boldsymbol{x}_j)\right]^2. \tag{17.22}$$

Let $\underline{\beta}=(\beta_1,\beta_2,...,\beta_m)^T$ and $\underline{Z}=\left[v_s(\boldsymbol{x}_1),v_s(\boldsymbol{x}_2),...,v_s(\boldsymbol{x}_n)\right]^T$, where the exponent T denotes the transpose operator. The estimate $\hat{\underline{\beta}}$ of $\underline{\beta}$ that minimizes Equation (17.22) is given by:

$$\hat{\underline{\beta}}=\left(\underline{\underline{D}}^T\underline{\underline{D}}\right)^{-1}\underline{\underline{D}}^T\underline{Z}, \tag{17.23}$$

where $\underline{\underline{D}}$ is a design matrix, given by:

$$\begin{pmatrix} g_1(\boldsymbol{x}_1),g_2(\boldsymbol{x}_1),g_3(\boldsymbol{x}_1),...,g_m(\boldsymbol{x}_1) \\ g_1(\boldsymbol{x}_2),g_2(\boldsymbol{x}_2),g_3(\boldsymbol{x}_2),...,g_m(\boldsymbol{x}_2) \\ . \\ . \\ g_1(\boldsymbol{x}_n),g_2(\boldsymbol{x}_n),g_3(\boldsymbol{x}_n),...,g_m(\boldsymbol{x}_n) \end{pmatrix}. \tag{17.24}$$

Once $\hat{\underline{\beta}}$ is obtained, the mean and variance of the pdf $f_{v_s}\left[v_s(\boldsymbol{x})\,|\,y(\boldsymbol{x}),v_g(\boldsymbol{x}),\alpha(\boldsymbol{x})\right]$ are defined by:

$$\hat{\mu}(\boldsymbol{x})=\sum_{i=1}^{m}\hat{\beta}_i g_i(\boldsymbol{x}) \tag{17.25}$$

and:

$$\hat{\sigma}^2=\frac{1}{n-m}\sum_{j=1}^{n}\left[v_s(\boldsymbol{x}_j)-\mu(\boldsymbol{x}_j)\right]^2, \tag{17.26}$$

where n is the number of measurements and m the number of basis functions g_i.

Selecting and eliminating basis functions is the key to the normal linear regression model. Chen et al. (2001) includes all possible distinct monomials of powers up to 4 in the initial set of basis functions. By deleting some of the initial basis functions based on t-test and p-value selection criteria (Stone, 1995) and in an iterative manner, Chen et al. (2001) obtained the final set of basis functions.

In principle, an empirical approach such as we present here is at a disadvantage compared to methods that use physically based forward models to relate between the various attributes. However, in many cases, accurate and reliable mathematical models are difficult to construct, or are subject to uncertain boundary conditions.

An application of the Bayesian approach is discussed in great detail in Hubbard et al. (2001), and is briefly summarized here. Prior estimates of hydraulic conductivity were obtained from kriging flowmeter measurements available from wellbore at a U.S. Department of Energy Bacterial Transport Site in Virginia. Updating of the prior estimates was performed using 100 MHz radar velocity data available from crosshole radar tomograms collected between the wellbores (refer to Chapter 7 of this volume) as well as a likelihood relationship developed using co-located measurements near the wellbore. Figure 13.7, which is shown in Chapter 13 of this volume (modified from Hubbard et al., 2001), illustrates the estimates of the mean hydraulic conductivity posterior pdfs along two vertical cross sections at the study site. In Figure 13.7, hydraulic conductivity measurements from an electromagnetic wellbore flowmeter log are superimposed on top of the mean estimates obtained using the radar data within the Bayesian framework, showing good agreement. Because the sediments within the saturated section at this site are fairly homogeneous (consisting predominantly of sands with different levels of sorting), the range of hydraulic conductivity is only about an order of magnitude. Even under the relatively homogeneous conditions and with extensive wellbore flowmeter control available at this site, it was difficult to capture the variability of hydraulic conductivity using wellbore data alone.

Figure 17.3a illustrates the mean values of hydraulic conductivity at each point in space obtained through kriging measurements obtained from the five wellbores whose locations are shown on the figure, or the prior pdfs at each point in space. Figure 17.3b illustrates the contoured mean posterior pdf values at each point in space obtained by incorporating the radar tomography information into the Bayesian estimation procedure. This illustration highlights how incorporation of geophysical data can lead to dramatically different estimates of hydraulic conductivity than if only wellbore data were used in the estimation procedure, especially at distances of a few meters from direct wellbore control (such as the area between wellbores T2 and M3). Recent studies have attested to the value of the geophysically obtained posterior estimates (Figure 17.3b) for understanding subsurface bacterial transport (Mailloux et al., 2004) and for improving the prediction of chemical transport (Scheibe and Chien, 2003) at this site.

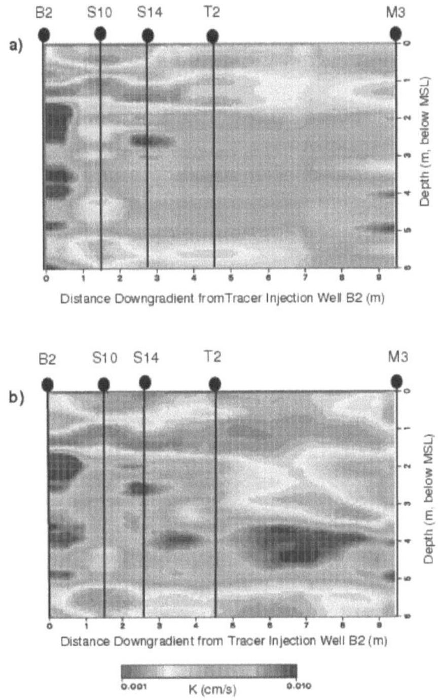

Figure 17.3(a). Contoured mean values of prior pdfs obtained from kriging dense wellbore flowmeter measurements; (b) Contoured mean values of posterior pdfs obtained using radar tomographic data conditioned to flowmeter data within a Bayesian estimation procedure (modified from Hubbard et al., 2001). Comparison of these figures shows that inclusion of the geophysical data significantly changes the estimates of hydraulic conductivity, especially at distances over ~1 m from direct wellbore control.

17.5 Model Selection

In our earlier discussion, reference was made to models (e.g., $\boldsymbol{F}$ in Equation 17.8), and we expand upon our model discussion in this section. Model development is a subjective process (see Chapter 1 in Rubin, 2003), often subject to large errors. Such errors, caused by data scarcity and errors in judgement, are unavoidable, but they should be minimized. Model selection is an attempt to address model error in a systematic manner. It requires formulation and comparison of alternative models, and elimination of all but the model that maximizes some optimality criteria. The idea (and hope) is that performing model selection will minimize modeling error. The most commonly used model selection methods include hypothesis testing (Mood et al., 1963; Luis and McLaughlin, 1992), cross-validation (Deutsch and Journel, 1998; Christensen and Cooley, 1999), Akaike's Information Criterion (AIC, see Akaike, 1974; Carrera and Neuman, 1986b) and Bayesian Information Criterion (BIC).

Cross-validation is the most intuitive approach, and it consists of dividing the data into two nonoverlapping subsets. The first subset, called the training subset, is used for model development and calibration, and a second, complementary, testing subset is used to evaluate the predictive accuracy of the calibrated model. This is done by comparing the model-predicted values at the locations where the testing data are, with the actual, measured values. Metrics are then developed based on model predictions and actual measurements, and are used to compare between models.

To demonstrate this idea, let us consider $z(\boldsymbol{x}_i)$, the measurement of the SRF Z at $\boldsymbol{x}_i$. A geostatistical model, identified using the training subset, is employed to obtain a kriging estimate at $\boldsymbol{x}_i$, z_i^*, and a kriging variance, σ_i^2 (see Chapter 3 of this volume). This is done repeatedly for all $\boldsymbol{x}_i$, i=1,..,N, where N denotes the number of data in the testing subset. At the end of this exercise, the following series of analysis can be performed:

- A correlation analysis between $z(\boldsymbol{x}_i)$ and z_i^*, i=1,..,N. This analysis can identify a systematic bias in the model. The absence of correlation is a positive indication.
- A correlation analysis between σ_i and $\left|z_i^* - z(\boldsymbol{x}_i)\right|$ for i=1,..,N. This analysis aims at evaluating the model's ability to predict correctly the estimation error. A high degree of correlation is a positive indication.
- A correlation analysis between σ_i and $z(\boldsymbol{x}_i)$ for i=1,..,N. This analysis aims to reveal heteroscedasticity. Any significant correlation here is an indication of a systematic bias (i.e., an unmodeled deterministic model component).
- Correlation analysis between the residuals $\left|z_i^* - z(\boldsymbol{x}_i)\right|$ and $\left|z_j^* - z(\boldsymbol{x}_j)\right|$ for i,j=1,..,N. In an ideal situation, there will not be any correlation.

Additional analyses are suggested in Kitanidis (1997). These analyses provide metrics that can be used to compare between models. None of them has the power to indicate optimality, but they can definitely indicate poor performance, and upon comparison between models, they can reveal which performs better.

Can we use the ML criterion (Section 17.3.3.1) to compare between models? ML leads to optimal parameters, in a maximum likelihood sense, for a given model structure without regard to how well the model represents the real system (Carrera and Neuman, 1986a-c). The estimated parameters may be meaningless if the model fails to capture the important characteristics of the aquifer. To compensate for poor model performance, the modeler may be tempted to increase the number of model parameters (e.g., number of gridblocks used in the numerical model of the aquifer), and hence the model complexity. Obviously, a larger number of parameters will produce a better fit, but not necessarily a better model, since an increase in the number of parameters leads to larger parameter variance and to instability (in the sense that the slightest changes in data will lead to significant changes in parameter values. Hence, the ML criterion cannot be used for model selection, since it accounts for goodness of fit but not for model simplicity.

The common model selection methods attempt to balance between simplicity, usually associated with the number of adjustable parameters, and goodness of fit (Forster, 2000). The main difference between these methods is in the relative weights they give to simplicity versus goodness of fit. The AIC intends to maximize entropy (See Section 17.3.2) and simplicity, and it selects the model that has the smallest AIC:

$$AIC\left(\hat{\underline{\theta}}\right) = -2\,ln\,L\left(\hat{\underline{\theta}}\right) + 2p\,, \tag{17.27}$$

where $\hat{\underline{\theta}}$ is the maximum likelihood estimate of the parameter vector, $L\left(\hat{\underline{\theta}}\right)$ is the corresponding maximum value of the likelihood criterion, and p is the number of parameters in $\hat{\underline{\theta}}$. The first term on the right-hand side of Equation (17.27) is the negative logarithm of the maximum value for the likelihood criterion, while the second term penalizes complexity, expressed as the number of parameters p.

The BIC method selects the *a posteriori* most probable model among the proposed alternatives. That model is the one that minimizes the *BIC* (Forster, 2000; Gaganis and Smith, 2001):

$$BIC\left(\hat{\underline{\theta}}\right) = n\!-\!2l\ \ L\left(\hat{\underline{\theta}}\right)\!+\!n\,p\,l\ \ (n), \tag{17.28}$$

where n is the number of measurements. Compared to *AIC*, *BIC* gives a greater weight, by a factor of $ln(n)/2$, to simplicity. Despite the difference, both *AIC* and *BIC* support the principle of parsimony in that, all other things being equal, the model with smallest p is chosen. The relative performance of *BIC* versus *AIC* depends on the nature of the problem, the size of the parameter vector, and the amount of available data on the dependent variables.

17.6 Recent Stochastic Hydrogeophysical Studies

In recent years, several studies have developed stochastic frameworks for using geophysical data in hydrogeological site characterization. Geostatistical approaches for fusing hydrogeological and geophysical data have been illustrated by many investigators (e.g., Poeter et al., 1997; Yeh et al., 2002). Bayesian approaches for estimating conductivities using head and conductivity measurements (as well as seismic data) have been reported by Rubin et al. (1992), Copty et al. (1993), Copty and Rubin (1995), Chen et al. (2001) and Chen and Rubin (2003). Hubbard et al. (1997) and Hubbard (1998) employed a Bayesian technique to incorporate GPR into mapping the spatial distribution of saturation and permeability. Knight et al. (1997) and Hubbard et al. (1999) combined geophysical and hydrogeological data for estimating the spatial correlation structure of the log-permeability. An application of these concepts to a field experiment on chemical and bacterial transport was described in Section 17.4.1 and in Hubbard et al. (2001). Purvance and Andricevic (2000) estimated the spatial distribution of the hydraulic conductivity and its spectral density using electrical and hydraulic conductivities. Day-Lewis et al. (2000) reported an application in fractured rock. Hyndman et al., (1994) and Hyndman and Harris (1996) used stochastic techniques and geophysical data to delineate hydrogeological units, and Hyndman and Gorelick (1996) used a stochastic approach to jointly invert geophysical and tracer test data. More recently, other investigators have experimented with the use of the sampling-based Monte Carlo Markov Chain (MCMC) method to integrate often complex and nonlinear geophysical and hydrogeological data sets. For example, the MCMC approach was used by Chen et al. (2004) for estimating sediment geochemical parameters, such as Fe(II) and Fe(III), using GPR attenuation data, through the dependence of both radar attenuation and sediment geochemistry on lithology.

These recent studies highlight the power and potential of stochastic methods for providing the systematic framework needed to handle complexities that sometimes arise in fusing disparate hydrogeological-geophysical data sets, such as those associated with spatial variability, measurement error, model discrimination, and conceptual model uncertainty.

17.7 Concluding Remarks

The stochastic paradigm offers a rational framework for analyzing spatial variability and the consequences of data scarcity, and hence for subsurface characterization. In this context, Bayes' theorem offers a flexible platform, allowing for alternative formulations, and for incorporation of complementary data of various types. There are several elements of Bayes' theorem that are needed for a successful application (which we reviewed here), including a systematic approach for modeling prior information and the likelihood function, and a rigorous model selection. For a systematic modeling of prior information, we presented the MRE concept. The idea here is to select for the prior pdf the one with the largest entropy among all pdfs (relative to a benchmark prior) that satisfy a given set of data constraints, and in doing that, to minimize the infusion of information that is not data-based and to eliminate potential bias. This is a critical element in integration between hydrological and geophysical data. MRE is a relatively new concept in hydrogeological applications. For model selection, we reviewed a few classical, well-documented methods such as the Akaike's Information Criterion and the Bayesian Information Criterion.

Our primary goal in this chapter was to show the application of stochastic concepts for fusion of hydrological and geophysical data. Although investigators are just starting to explore this subject, there is already a substantial body of evidence that illustrates the potential of such endeavors. The existence of such potential is intuitively appealing: geophysics offers a view of the subsurface from a perspective other than hydrogeology, and thus, when reconciled with the hydrogeological perspective, can help obtain a more coherent image. What are the keys for unleashing this potential? First and foremost, in our opinion, is the availability of a systematic approach for estimation and data integration. Second is a substantial improvement in our understanding of the petrophysical models that relate between geophysical and geological attributes. An improved understanding will allow better usage of the geophysical data and will permit us to benefit from various geophysical surveys, exploring each for its proven strength. A third key element is the availability of true joint hydrogeological-geophysical inversion approaches, which are distinguished from the sequential approaches used now, whereby geophysical surveys are inverted, and the outcome is then viewed as a starting point for hydrogeological inversion. True joint inversion approaches will introduce geological constraints on the geophysical inversion, which will lead to better accuracy. Additional discussion is provided in Hubbard and Rubin (2002) and in Chapter 1 of this volume.

Acknowledgments

The authors would like to acknowledge the support of NSF EAR-0087802 and DOE DE-AC03-76SF00098.

References

Akaike, H., A new look at the statistical model identification, *IEEE Trans. Autom. Control*, AC-19, 716–723, 1974.

Alumbaugh, D., P. Change, L. Paprocki, J. Brainard, R.J. Glass, and C.A. Rautman, Estimating moisture contents in the vadose zone using cross-borehole ground penetrating radar: a study of accuracy and repeatability, *Water Resour. Res., 38*, doi: 2001WRR001028, 2002.

Binley, A., P. Winship, L.J. West, M. Pokar and R. Middleton, Seasonal variation of moisture content in unsaturated sandstone inferred from borehole radar and resistivity profiles, *Journal of Hydrology,* 267(3–4), 160–172, 2002.

Carrera, J., and S.P. Neuman, Estimation of aquifer parameters under transient and steady state conditions, 1, Maximum likelihood method incorporating prior information, *Water Resour. Res., 22*(2), 199–210, 1986a.

Carrera, J., and S.P. Neuman, Estimation of aquifer parameters under transient and steady state conditions, 2, Uniqueness, stability and solution algorithms, *Water Resour. Res., 22*(2), 211–227, 1986b.

Carrera, J., and S.P. Neuman, Estimation of aquifer parameters under transient and steady state conditions, 3, Application to synthetic and field data, *Water Resour. Res., 22*(2), 228–242, 1986c.

Chen, J., S. Hubbard, and Y. Rubin, Estimating the hydraulic conductivity at the South Oyster site based on the normal linear regression model, *Water Resour. Res., 37*(6), 1603–1613, 2001.

Chen, J., and Y. Rubin, An effective Bayesian model for lithofacies estimation using geophysical data, *Water Resour. Res., 39*(5), 2003.

Chen, J., S. Hubbard, Y. Rubin, C. Murray, E. Roden, and E. Majer, Geochemical characterization using geophysical data: a case study at the South Oyster bacterial transport site in Virginia, *Water Resour. Res* , 2004 (submitted).

Christensen, S., and R.L. Cooley, Evaluation of prediction intervals for expressing uncertainties in groundwater flow model predictions, *Water Resour. Res. 35*(9), 2627–2639, 1999.

Copty, N., Y. Rubin, and G. Mavko, Geophysical-hydrological identification of field permeabilities through Bayesian updating, *Water Resour. Res., 29*(8), 2813–1725, 1993.

Copty, N., and Y. Rubin, A stochastic approach to the characterization of lithofacies from surface seismic and well data, *Water Resour. Res., 31*(7), 1673–1686, 1995.

Day-Lewis, F., P.A. Hsieh, and S.M. Gorelick, Identifying fracture-zone geometry using simulated annealing and hydraulic connection data, *Water Resour. Res. 36*(7), 1707–1721, 2000.

Deutsch, C. and A.G. Journel, *GSLIB: Geostatistical Software Library and Users Guide*, Oxford Univ. Press, New York, 1998.

Ezzedine, S., Y. Rubin, and J. Chen, Bayesian method for hydrogeological site characterization using borehole and geophysical survey data: Theory and application to the Lawrence Livermore National Laboratory Superfund site, *Water Resour. Res., 35*(9), 2671–2683, 1999.

Forster, M.R., Key concepts in model selection: Performance and generalizability, *J. Math Psychol., 44*, 205–231, 2000.

Gaganis, P., and L. Smith, A Bayesian approach to the quantification of the effect of model error on the predictions of groundwater models, *Water Resour. Res., 37*(9), 2309-2322, 2001

Hoeksema, R.J., and P.K. Kitanidis, An application of the geostatistical approach to the inverse problem in two-dimensional groundwater modeling, *Water Resour. Res., 20*(7), 1003–1020, 1984.

Hoeksema, R.J., and P.K. Kitanidis, Analysis of the spatial structure of properties of selected aquifers, *Water Resour. Res., 21*(4), 563–572, 1985.

Hubbard, S., Y. Rubin, and E. Majer, Ground penetrating radar-assisted saturation and permeability estimation in bimodal systems, *Water Resour. Res., 33*(5), 1997.

Hubbard, S., Stochastic characterization of hydrogeological properties using geophysical data, Ph.D. Dissertation, University of California at Berkeley, 1998.

Hubbard, S., Y. Rubin, and E. Majer, Spatial correlation structure estimation using geophysical and hydrological data, *Water Resour. Res., 35*(6), 1709–1725, 1999.

Hubbard, S., and Y. Rubin, Integrated hydrogeological-geophysical site characterization techniques, *J. Contam. Hydrology, 45*, 3–34, 2000.

Hubbard, S. and Y. Rubin, Hydrogeophysics: state-of-the-discipline, *EOS, 83*(51), 602–606, 2002.

Huisman, J.A., S.S. Hubbard, J.D. Redman, and A.P. Annan, Measuring soil water content with ground penetrating radar: A review, *Vadose Zone Journal, 4*(2), 476–491, 2003.

Hubbard, S., J. Chen, J. Peterson, E.L. Majer, K.H. Williams, D.J. Swift, B. Mailloux, and Y. Rubin, Hydrogeological characterization of the South Oyster Bacterial Transport Site using geophysical data, *Water Resour. Res., 37*(10), 2431–2456, 2001.

Hyndman, D.W., and J.M. Harris, Traveltime inversion for the geometry of aquifer lithologies, *Geophysics, 61*(6), 1996.

Hyndman, D.W., J.M. Harris, and S.M. Gorelick, Coupled seismic and tracer test inversion for aquifer property characterization, *Water Resour. Res., 30*(7), 1965–1977, 1994.

Hyndman, D.W., and S.M. Gorelick, Estimating lithologic and transport properties in three dimensions using seismic and tracer data: The Kesterson aquifer, *Water Resour. Res., 32*(9), 2659–2670, 1996.

Kitanidis, P.K., *Introduction to Geostatistics*, Cambridge University Press, 1997

Kitanidis, P.K., and E.G. Vomvoris, A geostatistical approach to the inverse problem in groundwater modeling, *Water Resour. Res., 19*, 677—690, 1983.

Knight, R., P. Tercier, and H. Jol, The role of ground penetrating radar and geostatistics in reservoir description, *Leading Edge, 16*(11), 1576–1583, 1997.

Knoll, M., R. Knight, and E. Brown, Can accurate estimates of permeability be obtained from measurements of dielectric properties? SAGEEP Annual Meeting *Extended Abstracts*, Environ. and Eng. Geophys. Soc., Englewood, CO., 1995.

Luis, S.J., and D. McLaughlin, A stochastic approach to model validation, *Adv. Water Resour., 15*, 15–32, 1992.

Mailloux, B.J., M.E. Fuller, T.C. Onstott, J. Hall, H. Dong, M.F. DeFlaun, S.H. Streger, R.K. Rothmel, M. Green, D.J.P. Swift, and J. Radke, The role of physical, chemical and microbial heterogeneity on the field-scale transport and attachment of bacteria, Accepted for Publication in *Water Resour. Res.*, 2003.

Marion, D., A. Nur, H. Yin, and D. Han, Compressional velocity and porosity in sand-clay mixtures, *Geophysics, 57*, 554–563, 1992.

Mavko, G., T. Mukerji, and J. Dvorkin, *The Rock Physics Handbook: Tools for Seismic Analysis in Porous Media*, Cambridge University Press, New York, 1998.

Maxwell, R., W. Kastenberg, and Y. Rubin, Hydrogeological site characterization and its implication on human risk assessment, *Water Resour. Res., 35*(9), 2841–2855, 1999.

McLaughlin, D., L.R. Townley, A reassessment of the groundwater inverse problem, *Water Resour. Res., 32*(5), 1131–1161, 1996.

Mood, A.M., F.A. Graybill, and D.C. Boes, *Introduction to the Theory of Statistics, 3rd Edition*, McGraw-Hill, New York., 1963.

Peterson, J.E., Jr., Pre-inversion corrections and analysis of radar tomographic data, *Journal of Env. and Eng. Geophysics, 6*, 1–17, 2001.

Poeter, EP, S.A. McKenna, and W.L. Wingle, Improving groundwater project analysis with geophysical data, *The Leading Edge*, 1675–1681, November 1997.

Prasad, M., Velocity-permeability relations within hydraulic units, *Geophysics, 68*(1),108–117, 2003.

Purvance, D.T., and R. Andricevic, Geoelectric characterization of the hydraulic conductivity field and its spatial structure at variable scales, *Water Resour. Res., 36*(10), 2915–1924, 2000.

Rubin, Y., and G. Dagan, Stochastic identification of transmissivity and effective recharge in steady groundwater flow, 1, Theory, *Water Resour. Res., 23*(7), 1175–1192, 1987a.

Rubin, Y., and G. Dagan, Stochastic identification of transmissivity and effective recharge in steady groundwater flow, 2, Case study, *Water Resour. Res., 23*(7), 1193–1200, 1987b.

Rubin, Y., G. Mavko, and J. Harris, Mapping permeability in heterogeneous aquifers using hydrological and seismic data, *Water Resour. Res., 28*(7), 1192–1700, 1992.
Rubin, Y., *Applied Stochastic Hydrogeology*, Oxford University Press, 2003.

Rubin, Y., and G. Dagan, Conditional estimation of solute travel time in heterogeneous formations: Impact of the transmissivity measurements, *Water Resour. Res., 28*(4), 1033–1040, 1992.
Rubin, Y., and G. Dagan, Stochastic analysis of the effects of boundaries on spatial variability in groundwater flows: 1. Constant head boundary, *Water Resour. Res., 24*(10), 1689–1697, 1988.
Scheibe, T.D., and Y. Chien, An evaluation of conditioning data for solute transport prediction, Accepted for Publication in *Ground Water*, 2003.
Schweppe, F.C., *Uncertain Dynamic Systems*, Prentice-Hall, Englewood Cliffs, N.J., 1973.
Slater, L., and D.P. Lesmes, Electrical-hydraulic relationships observed for unconsoli-dated sediments, *Water Resour. Res., 38*(10), 1213, doi: 10.1029/2001WR001075, 2002.
Stone, C., *A Course in Probability and Statistics*, Duxbury, Boston, Mass., 1995.
Tarantola, A., *Inverse Problem Theory*, Elsevier, New York, 1987.
Woodbury, W., and Y. Rubin, A full-Bayesian approach to parameter inference from tracer travel time moments and investigation of scale-effects at the Cape Cod experimental site, *Water Resour. Res., 36*(1), 159–171, 2000.
Woodbury, A.D., and T.J. Ulrych, A full-Bayesian approach to the groundwater inverse problem for steady state flow, *Water Resour. Res., 36*(8), 2081–2093, 2000.
Yeh, T.-C., J., S. Liu, R.J. Glass, K. Baker, J. R. Brainard, D. L. Alumbaugh, and D. LaBrecque, A geostatistically based inverse model for electrical resistivity surveys and its application to vadose zone hydrology, *Water Resour. Res., 38*(12), 1278, December 2002.

INDEX

NATO Advanced Study Institute
Hydrogeophysics
Třešť, Czech Republic
July 2002

JIM BEAM

Water Science and Technology Library

1. A.S. Eikum and R.W. Seabloom (eds.): *Alternative Wastewater Treatment.* Low-Cost Small Systems, Research and Development. Proceedings of the Conference held in Oslo, Norway (7–10 September 1981). 1982 ISBN 90-277-1430-4
2. W. Brutsaert and G.H. Jirka (eds.): *Gas Transfer at Water Surfaces.* 1984 ISBN 90-277-1697-8
3. D.A. Kraijenhoff and J.R. Moll (eds.): *River Flow Modelling and Forecasting.* 1986 ISBN 90-277-2082-7
4. World Meteorological Organization (ed.): *Microprocessors in Operational Hydrology.* Proceedings of a Conference held in Geneva (4–5 September 1984). 1986 ISBN 90-277-2156-4
5. J. Němec: *Hydrological Forecasting.* Design and Operation of Hydrological Forecasting Systems. 1986 ISBN 90-277-2259-5
6. V.K. Gupta, I. Rodríguez-Iturbe and E.F. Wood (eds.): *Scale Problems in Hydrology.* Runoff Generation and Basin Response. 1986 ISBN 90-277-2258-7
7. D.C. Major and H.E. Schwarz: *Large-Scale Regional Water Resources Planning.* The North Atlantic Regional Study. 1990 ISBN 0-7923-0711-9
8. W.H. Hager: *Energy Dissipators and Hydraulic Jump.* 1992 ISBN 0-7923-1508-1
9. V.P. Singh and M. Fiorentino (eds.): *Entropy and Energy Dissipation in Water Resources.* 1992 ISBN 0-7923-1696-7
10. K.W. Hipel (ed.): *Stochastic and Statistical Methods in Hydrology and Environmental Engineering.* A Four Volume Work Resulting from the International Conference in Honour of Professor T. E. Unny (21–23 June 1993). 1994
    10/1: Extreme values: floods and droughts ISBN 0-7923-2756-X
    10/2: Stochastic and statistical modelling with groundwater and surface water applications ISBN 0-7923-2757-8
    10/3: Time series analysis in hydrology and environmental engineering ISBN 0-7923-2758-6
    10/4: Effective environmental management for sustainable development ISBN 0-7923-2759-4
    Set 10/1–10/4: ISBN 0-7923-2760-8
11. S.N. Rodionov: *Global and Regional Climate Interaction: The Caspian Sea Experience.* 1994 ISBN 0-7923-2784-5
12. A. Peters, G. Wittum, B. Herrling, U. Meissner, C.A. Brebbia, W.G. Gray and G.F. Pinder (eds.): *Computational Methods in Water Resources X.* 1994 Set 12/1–12/2: ISBN 0-7923-2937-6
13. C.B. Vreugdenhil: *Numerical Methods for Shallow-Water Flow.* 1994 ISBN 0-7923-3164-8
14. E. Cabrera and A.F. Vela (eds.): *Improving Efficiency and Reliability in Water Distribution Systems.* 1995 ISBN 0-7923-3536-8
15. V.P. Singh (ed.): *Environmental Hydrology.* 1995 ISBN 0-7923-3549-X
16. V.P. Singh and B. Kumar (eds.): *Proceedings of the International Conference on Hydrology and Water Resources* (New Delhi, 1993). 1996
    16/1: Surface-water hydrology ISBN 0-7923-3650-X
    16/2: Subsurface-water hydrology ISBN 0-7923-3651-8

16/3: Water-quality hydrology ISBN 0-7923-3652-6
16/4: Water resources planning and management ISBN 0-7923-3653-4
Set 16/1–16/4 ISBN 0-7923-3654-2

17. V.P. Singh: *Dam Breach Modeling Technology*. 1996 ISBN 0-7923-3925-8
18. Z. Kaczmarek, K.M. Strzepek, L. Somlyódy and V. Priazhinskaya (eds.): *Water Resources Management in the Face of Climatic/Hydrologic Uncertainties*. 1996 ISBN 0-7923-3927-4
19. V.P. Singh and W.H. Hager (eds.): *Environmental Hydraulics*. 1996 ISBN 0-7923-3983-5
20. G.B. Engelen and F.H. Kloosterman: *Hydrological Systems Analysis*. Methods and Applications. 1996 ISBN 0-7923-3986-X
21. A.S. Issar and S.D. Resnick (eds.): *Runoff, Infiltration and Subsurface Flow of Water in Arid and Semi-Arid Regions*. 1996 ISBN 0-7923-4034-5
22. M.B. Abbott and J.C. Refsgaard (eds.): *Distributed Hydrological Modelling*. 1996 ISBN 0-7923-4042-6
23. J. Gottlieb and P. DuChateau (eds.): *Parameter Identification and Inverse Problems in Hydrology, Geology and Ecology*. 1996 ISBN 0-7923-4089-2
24. V.P. Singh (ed.): *Hydrology of Disasters*. 1996 ISBN 0-7923-4092-2
25. A. Gianguzza, E. Pelizzetti and S. Sammartano (eds.): *Marine Chemistry*. An Environmental Analytical Chemistry Approach. 1997 ISBN 0-7923-4622-X
26. V.P. Singh and M. Fiorentino (eds.): *Geographical Information Systems in Hydrology*. 1996 ISBN 0-7923-4226-7
27. N.B. Harmancioglu, V.P. Singh and M.N. Alpaslan (eds.): *Environmental Data Management*. 1998 ISBN 0-7923-4857-5
28. G. Gambolati (ed.): *CENAS. Coastline Evolution of the Upper Adriatic Sea Due to Sea Level Rise and Natural and Anthropogenic Land Subsidence*. 1998 ISBN 0-7923-5119-3
29. D. Stephenson: *Water Supply Management*. 1998 ISBN 0-7923-5136-3
30. V.P. Singh: *Entropy-Based Parameter Estimation in Hydrology*. 1998 ISBN 0-7923-5224-6
31. A.S. Issar and N. Brown (eds.): *Water, Environment and Society in Times of Climatic Change*. 1998 ISBN 0-7923-5282-3
32. E. Cabrera and J. García-Serra (eds.): *Drought Management Planning in Water Supply Systems*. 1999 ISBN 0-7923-5294-7
33. N.B. Harmancioglu, O. Fistikoglu, S.D. Ozkul, V.P. Singh and M.N. Alpaslan: *Water Quality Monitoring Network Design*. 1999 ISBN 0-7923-5506-7
34. I. Stober and K. Bucher (eds): *Hydrogeology of Crystalline Rocks*. 2000 ISBN 0-7923-6082-6
35. J.S. Whitmore: *Drought Management on Farmland*. 2000 ISBN 0-7923-5998-4
36. R.S. Govindaraju and A. Ramachandra Rao (eds.): *Artificial Neural Networks in Hydrology*. 2000 ISBN 0-7923-6226-8
37. P. Singh and V.P. Singh: *Snow and Glacier Hydrology*. 2001 ISBN 0-7923-6767-7
38. B.E. Vieux: *Distributed Hydrologic Modeling Using GIS*. 2001 ISBN 0-7923-7002-3

39. I.V. Nagy, K. Asante-Duah and I. Zsuffa: *Hydrological Dimensioning and Operation of Reservoirs*. Practical Design Concepts and Principles. 2002 ISBN 1-4020-0438-9
40. I. Stober and K. Bucher (eds.): *Water-Rock Interaction*. 2002 ISBN 1-4020-0497-4
41. M. Shahin: *Hydrology and Water Resources of Africa*. 2002 ISBN 1-4020-0866-X
42. S.K. Mishra and V.P. Singh: *Soil Conservation Service Curve Number (SCS-CN) Methodology*. 2003 ISBN 1-4020-1132-6
43. C. Ray, G. Melin and R.B. Linsky (eds.): *Riverbank Filtration*. Improving Source-Water Quality. 2003 ISBN 1-4020-1133-4
44. G. Rossi, A. Cancelliere, L.S. Pereira, T. Oweis, M. Shatanawi and A. Zairi (eds.): *Tools for Drought Mitigation in Mediterranean Regions*. 2003 ISBN 1-4020-1140-7
45. A. Ramachandra Rao, K.H. Hamed and H.-L. Chen: *Nonstationarities in Hydrologic and Environmental Time Series*. 2003 ISBN 1-4020-1297-7
46. D.E. Agthe, R.B. Billings and N. Buras (eds.): *Managing Urban Water Supply*. 2003 ISBN 1-4020-1720-0
47. V.P. Singh, N. Sharma and C.S.P. Ojha (eds.): *The Brahmaputra Basin Water Resources*. 2004 ISBN 1-4020-1737-5
48. B.E. Vieux: *Distributed Hydrologic Modeling Using GIS*. Second Edition. 2004 ISBN 1-4020-2459-2
49. M. Monirul Qader Mirza (ed.): *The Ganges Water Diversion: Environmental Effects and Implications*. 2004 ISBN 1-4020-2479-7
50. Y. Rubin and S.S. Hubbard (eds.): *Hydrogeophysics*. 2005 ISBN 1-4020-3101-7
51. K.H. Johannesson (ed.): *Rare Earth Elements in Groundwater Flow Systems*. 2005 ISBN 1-4020-3233-1

MIX
Papier aus verantwortungsvollen Quellen
Paper from responsible sources
FSC® C105338

If you have any concerns about our products,
you can contact us on
ProductSafety@springernature.com

In case Publisher is established outside the EU,
the EU authorized representative is:
Springer Nature Customer Service Center GmbH
Europaplatz 3, 69115 Heidelberg, Germany

Printed by Libri Plureos GmbH
in Hamburg, Germany